ANSYS Mechanical APDL for Finite Element Analysis

ANSYS Mechanical APDL for Finite Element Analysis

Mary Kathryn Thompson, PhD

John M. Thompson, PhD, PE

Butterworth-Heinemann
An imprint of Elsevier

Butterworth-Heinemann is an imprint of Elsevier
The Boulevard, Langford Lane, Kidlington, Oxford OX5 1GB, United Kingdom
50 Hampshire Street, 5th Floor, Cambridge, MA 02139, United States

British Library Cataloguing-in-Publication Data
A catalogue record for this book is available from the British Library

Library of Congress Cataloging-in-Publication Data
A catalog record for this book is available from the Library of Congress

ISBN: 978-0-12-812981-4

For Information on all Butterworth-Heinemann publications
visit our website at https://www.elsevier.com/books-and-journals

Publisher: Joe Hayton
Acquisition Editor: Brian Guerin
Editorial Project Manager: Katie Chan
Production Project Manager: Kiruthika Govindaraju
Cover Designer: Mary Kathryn Thompson, John M. Thompson, Mark Rogers

Typeset by MPS Limited, Chennai, India

To Veronica

Mother, Wife, Inspiration

Contents

Preface

Commercial finite element programs provide a powerful and extensive collection of tools for the design, analysis, and optimization of complex engineering systems. Unfortunately, learning to use finite element software can be a difficult and time-consuming process. This book was written to reduce the learning curve associated with ANSYS Mechanical APDL and to prepare you to use the program and its documentation as quickly and painlessly as possible.

This book was written using the Academic Research version of ANSYS Mechanical APDL 17.2. Since ANSYS Mechanical APDL was designed to be backward compatible, you should be able to use the book with future versions of the program as well.

We have marked information in the book that we believe is likely to become out of date quickly, such as specific product names and the number of reference manuals for a given product, with a superscript asterisk (*) for your reference. An appendix has also been included to help you to identify specific information in the program documentation, even if the documentation numbering changes. However, the authors of this book are not involved with changes that may be made in the future to the ANSYS Mechanical APDL software or the related documentation. Therefore the asterisks and appendix are for guidance only and may not identify all future changes in such software or documentation.

Each chapter in the book includes suggested readings from the ANSYS Mechanical APDL user manuals. Most chapters are also accompanied by one or more hands-on exercises. This book does not duplicate or replace the information in the documentation. Instead, it is intended to help you to understand and use the documentation. Thus the reading assignments in the ANSYS user manuals are strongly recommended. To help you develop comfort and confidence as an independent analyst, each exercise introduces new skills and increases the complexity of the models while gradually decreasing the amount of guidance given. We recommend completing all of the exercises, even if they are not immediately relevant to your work.

This book can be used for self-study or as part of a formal course in finite element applications. It is suitable for professional engineers and engineering students at all levels. When you have finished the book, you should be able to use the present ANSYS documentation to build, solve, and postprocess sophisticated finite element models on your own.

The world is full of interesting and important problems. This book gives you some of the tools that you need to help solve them. Good luck!

Mary Kathryn Thompson and John M. Thompson
Pittsburgh, PA, USA

Acknowledgments

This book was originally developed as the course text for a graduate level IAP course at the Massachusetts Institute of Technology (MIT). The goal of that course was to prepare students to perform research quality finite element analyses with ANSYS in 5 days or less. We began writing in 2002 and began teaching with the book in 2004. In 2008 the book and the course moved to Korea. In 2012 the book moved to Denmark. And in 2016 it moved home again to the United States. Over the years, we have received a tremendous amount of support from wonderful people all over the world. We would like to acknowledge some of them here.

The software licenses for the MIT IAP course and for the development of this book were provided through an ANSYS Academic Partnership. We are grateful for the generous support that we have received from ANSYS, Inc. over the years. We would especially like to acknowledge the late Mr. Jerry Bittner, former Director of ANSYS Global Technical Services, who started this journey with us, and Dr. Paul Lethbridge, Senior Manager for Academic and Start Up Programs at ANSYS, who saw it through to the end. We would also like to thank Sheryl Ackerman, Cordell Blackshere, Vishal Ganore, Helen Renshaw, and the rest of the incredible team, past and present, at ANSYS Worldwide Headquarters in Canonsburg, PA, USA.

We would like to thank Prof. Alexander H. Slocum, Prof. Rohan Abeyaratne, Prof. David M. Parks, Prof. Lallit Anand, Prof. Thomas Peacock, and the MIT Department of Mechanical Engineering; Prof. Heekyung Park, Prof. Chung-Bang Yun, and the KAIST Department of Civil and Environmental Engineering; and Prof. Hans Nørgaard Hansen, Prof. Leonardo De Chiffre, and the DTU Department of Mechanical Engineering for their encouragement and support of this project as it has traveled around the world.

We would like to acknowledge Dr. Veronica V. Thompson for her tremendous assistance in editing this book. She has helped us to become the writers we are.

We would like to thank Anne M. Thompson and Will C. Lauer for all of their logistical and moral support over the years. We could not have done it without them.

We would like to thank Courtney S. Bermack, Jeffrey Chambers, Chad Foster, Marissa Jacovich, Christina Laskowski, Michael Mischkot, A. Zachary Trimble, and Antonio Vicente for their help in testing and improving various versions of this book. Their help and feedback is greatly appreciated.

We would like to thank our wonderful publication team at Elsevier, especially Brian Guerin, Katie K. Chan, and Kiruthika Govindaraju who made this process a pleasure.

Finally, we would like to thank Prof. Nam P. Suh, Prof. Donald A. Norman, Prof. Sami Kara, and Prof. Guan Heng Yeoh. They have all provided crucial advice and opportunities at critical times. We are glad to have traveled with them on this journey.

Chapter

1

Introduction to ANSYS and Finite Element Modeling

Suggested Reading Assignments:
None

CHAPTER OUTLINE

This chapter provides an introduction to finite element analysis and the ANSYS Mechanical APDL family of software. It begins with an overview of the finite element method, its benefits, and its limitations. It summarizes the current ANSYS Mechanical APDL products and program capabilities. Finally, it describes the program's evolution and how that influences the use of ANSYS, Inc. products.

1.1. What Is the Finite Element Method?

The finite element method (FEM) is a mathematical technique for setting up and solving systems of partial differential (or integral) equations. In engineering, the finite element method is used to divide a system whose behavior cannot be predicted using closed form equations into small pieces, or elements, whose solution is known or can be approximated. The finite element method requires the system geometry to be defined by a number of points in space called nodes. Each node has a set of degrees of freedom (temperature, displacements, etc.) that can vary based on the inputs to the system. These nodes are connected by elements that define the mathematical interactions of the degrees of freedom (DOFs). For some elements, such as beams, the closed form solution is known. For other elements, such as continuum elements, the interaction among

ANSYS Mechanical APDL for Finite Element Analysis.
DOI: http://dx.doi.org/10.1016/B978-0-12-812981-4.00001-0

the degrees of freedom is estimated by a numerical integration over the element. All individual elements in the model are combined to create a set of equations that represent the system to be analyzed. Finally, these equations are solved to reveal useful information about the behavior of the system.

Just as a regular polygon approaches a perfect circle as the number of sides approaches infinity, a finite element model approaches a perfect representation of the system as the number of elements becomes infinite. Since it is impossible to divide the system into an infinite number of elements, the finite element method produces the exact solution to an approximation of the problem that you want to solve. When the number of elements becomes sufficiently large, the approximation becomes good enough to use for engineering analysis. However, this may increase the number of equations to be solved beyond the point where it is practical or desirable to solve them by hand. For this reason, the finite element method is associated with computer programs that set up, solve, and visualize the solutions of these large sets of equations for you.

1.2. Why Use the Finite Element Method?

The cost, in terms of the manpower and computer resources, required to set up and solve a finite element model for a simple problem like a cantilever beam is very high compared to the benefit. Simple problems can—and should—be solved with simple methods (or obtained from engineering handbooks). But not all problems are simple. For example, if a bridge is built using a simple truss supported by two piers, the deflections and stresses in the bridge can be found using information taught in an introductory statics and strength of materials class. But as the complexity of the truss increases, solving this problem using the engineering fundamentals becomes more difficult, leaving the analyst with long hours of error-prone calculations. As system complexity continues to increase, closed-form analysis rapidly becomes impossible. The real benefit of finite element analysis lies in the ability to solve arbitrarily complex problems for which analytical solutions are not available or which would be prohibitively time consuming and expensive to solve by hand.

1.3. Basic Procedure for Finite Element Analysis

There are 10 basic steps in any finite element analysis. First, the solid model geometry is created, the element type(s) and material properties are defined, and the solid model geometry is meshed to create the finite element model. In ANSYS, these steps are performed in the Preprocessor (PREP7). Next, loads and constraints are applied, solution options are defined, and the problem is solved. These steps are performed in the Solution processor (SOL). After the solution is ready, the results are plotted, viewed, and exported in one of the postprocessors (POST1 or POST26). Finally, the results are compared to first-order estimates, closed-form solutions, mathematical models, or experimental results to ensure that the output of the program is reasonable and as expected. (Processors will be addressed in more detail in chapter 2.)

/PREP7

1. Define the Solid Model Geometry
2. Select the Element Types
3. Define the Material Properties
4. Mesh

/SOLUTION

5. Define the Boundary Conditions
6. Define the Loads
7. Set the Solution Options
8. Solve

/POST1 or /POST26

9. Plot, View, and Export the Results
10. Compare and Verify the Results

It is sometimes possible to omit one or more steps. For example, the default solution options are often sufficient for a simple analysis. It is possible to perform some steps out of order. For example, the element types and material properties can be defined in either order. Similarly, the loads and boundary conditions can be defined in either order. It is occasionally necessary to perform these steps out of order. For example, solid model geometry is not required for a finite element analysis. When the nodes and elements are generated directly, the element type(s) must be specified before the geometry can be created. Finally, complicated analyses may involve multiple trips through one or more processors.

For simplicity, this 10-step procedure will be used in this book whenever possible.

1.4. Engineering Software—Not an Engineer

As with all computer programs, the quality of your results will depend on the quality of your model. This includes the accuracy of the material properties, the appropriateness of the material models, how closely the simulated geometry and loads match the actual geometry and loads, and the validity of the simplifications and assumptions made. Simply put, Garbage In = Garbage Out. Finite element software programs can be thought of as very sophisticated calculators that help you to analyze engineering systems that could not otherwise be evaluated. They integrate the section properties of the system with the material properties to generate the equations to be solved. They convert the applied loads to the appropriate forms and apply them to the specified DOFs. They solve the generated system of equations. And, they help you to visualize and understand the results. But a finite element program will not comment on the validity of any assumptions made in setting up the model as long as the laws of physics are not violated. It also will not ensure that you are using the correct laws of physics for a given problem. Any errors that the program reports will be associated with the use of the program, and not with the physical or analytical system. In addition, it will not provide any commentary on the quality or implications of the results. Finite element software is only a tool. In the end, you, and you alone, are responsible for determining whether or not the results of your finite element model can be used to make or justify engineering decisions.

1.5. A Brief History of ANSYS and Finite Element Analysis

The finite element method was first proposed in the early 1940s as a numerical technique for solving partial differential equations. At that time, a mesh of elements could be defined and the interaction of the elements could be used to create the system of equations to be solved. However, the system of equations still had to be solved by hand. This limitation rendered the finite element method an academic curiosity until the early 1960s when computers that could solve large systems of simultaneous equations started to become available. This made it possible to apply the finite element method to general problems. As a result, interest in using the finite element method in engineering practice began to grow.

Early finite element programs were specialty codes that were developed to solve a specific type of problem. They generally contained a single element type (e.g., beams, axisymmetric shells, or plane stress solids) and included a single type of physics (structural, thermal, etc.). This limited the type of problem that each program could solve. It also meant that there were no standard analysis tools. It was common for different groups in the same organization to use different computer programs. In many cases, each group of engineers developed and used its own finite element code. This led to concerns about the compatibility of results from different programs, the overall quality of those results, and whether the engineers' time was being used efficiently.

1.5.1. The Development of NASTRAN

In 1965, the United States National Aeronautics and Space Administration (NASA) issued a request for proposals to create a computer program that could be used by all of its engineering organizations to solve a variety of structural problems related to the development of lunar exploration technology. The resulting program was known as NASTRAN. In 1969, NASA began to develop coupled thermal-structural capabilities in order to predict the optical performance of a large space telescope system that was exposed to changing orbital thermal conditions. By 1971, NASTRAN® was available for commercial use. It is still the default finite element program in the aerospace industry today.

1.5.2. The Development of ANSYS

While NASA was focused on lunar exploration, Westinghouse Electric Corporation was developing nuclear reactors for space propulsion and nonconventional energy production. Like their aerospace counterparts, the Westinghouse mechanical and nuclear engineers needed to predict transient stresses and displacements in reactor systems due to thermal and pressure loads. Dr. John Swanson, then an employee at the Westinghouse Astronuclear Labs in Pittsburgh, believed that an integrated, general-purpose finite element program would save both time and money when doing these types of calculations. He began developing such a program, called STASYS, for Westinghouse in 1969.

In 1970, John Swanson left Westinghouse and founded Swanson Analysis Systems, Inc. (SASI) where he continued to develop a commercial general-purpose finite element program that he called ANSYS®. The original version of ANSYS contained 40 elements of various types (springs, dampers, beams, bricks, etc.) including several elements with thermal degrees of freedom. Westinghouse became ANSYS's first customer by the end of the year. The program was rapidly adopted by other companies and became the default finite element program for much of the power industry. Today, ANSYS products are used in all major engineering fields including the aerospace, automotive, chemical processing, construction, consumer goods, electronics, energy, health care, offshore, marine, and materials industries.

1.5.3. The Evolution of ANSYS

With every new release since 1970, new features and functionality have been added to ANSYS. Many additions were specifically developed for the program. For example, the first elements with thermoelectric (1975) and electromagnetic (1983) DOFs were developed by ANSYS engineers. Some capabilities have been added by interfacing ANSYS with other programs. For example, computational fluid dynamics (CFD) capabilities were first added in 1989 by building an interface between SASI's ANSYS and Compuflo's FLOTRAN. Similarly, explicit dynamics capabilities were added in 1996 by developing an interface between ANSYS and Livermore Software Technology Corporation's LS-DYNA™. Finally, some capabilities have been added by incorporating other programs into ANSYS. For example, SASI purchased Compuflo in 1992 and FLOTRAN was fully integrated into the program by 1994 (revision 5.1).

In 1994, SASI was sold to TA Associates and the company was renamed ANSYS, Inc. This introduced a need to distinguish the company from its flagship product. It also marked a major shift in the strategic development of the company's software.

In the late 1990s, ANSYS, Inc. began to move from a single product to a portfolio of simulation products by acquiring other businesses. For example, they acquired ICEM CFD in 2000, CFX in 2003, and Fluent in 2006 to strengthen their computational fluid dynamics offerings. Similarly, ANSYS, Inc. purchased Century Dynamics in 2005 to add AUTODYN® to its suite of explicit dynamics capabilities. The company has also continued to invest in the development of new products and technologies. For example, in 2009 (revision 12.0), ANSYS, Inc. introduced a new explicit dynamics product named ANSYS Explicit STR™.

In the 1990s, development also began on a new user-friendly platform that would become the ANSYS Workbench environment. Workbench was intended to combine the strengths of existing ANSYS, Inc. technology with new capabilities including improved solid modeling and more robust CAD importation. From this point on, new technology that was purchased from other companies was no longer integrated into the original ANSYS environment. Instead, new products and new capabilities were to be integrated into the Workbench environment and all ANSYS, Inc. products would interface with each other using Workbench.

1.6. ANSYS Today*

Today, ANSYS, Inc. offers a wide variety of computer-aided engineering products. Some are for general use while others offer capabilities that are specifically designed for certain applications like electronics, turbo machinery, and offshore structures. Some can be used to perform all steps of a finite element analysis while others offer support for a specific stage of analysis like solid modeling or meshing. Some products use the same underlying technology but access it through different (or multiple) user interfaces. Finally, some products are built on the same platform but have different capabilities depending on the licensing options.

This book provides an introduction to the family of products offered by ANSYS, Inc. that evolved from the original ANSYS finite element software program. Throughout this book, the term "ANSYS" is used to refer to any (or all) of the ANSYS, Inc. commercial and academic products that provide access to the general-purpose structural, thermal, and/or multiphysics finite element simulation capabilities via the original (non-Workbench) user interface. Collectively, this group is known as the ANSYS Mechanical APDL family and is referred to as such in the program documentation.

Today, the ANSYS Mechanical APDL family includes ANSYS Mechanical Enterprise, ANSYS Mechanical Premium, and ANSYS Mechanical Pro. It also includes versions of the program intended for university use, such as ANSYS Student, ANSYS Academic Teaching, ANSYS Academic Research, and ANSYS Academic Associate. The ANSYS Mechanical APDL product portfolio and the names of those products are constantly evolving. Therefore, this part of the book will always be out of date. For up-to-date product information, please refer to the company website (http://www.ansys.com).

The exercises in this book are limited to structural and thermal analyses. However, the information in the book is equally valid for analyses that use other physics (fluid, electric, magnetic, and low-frequency electromagnetic) and multiphysics (thermal-fluid, piezoelectric, acoustic-structural, electromagnetic-thermal-structural, etc.) capabilities. As a result, information related to these capabilities will be referred to in the text.

Because explicit dynamics capabilities can be accessed through the ANSYS Mechanical APDL family of products (with an appropriate license), we will make occasional references to ANSYS LS-DYNA® in this book. It will be referred to as "LS-DYNA" for simplicity.

Because much of the underlying technology is the same, a considerable amount of the information presented in this book is still relevant to products like ANSYS DesignSpace® that exclusively use the ANSYS Workbench environment. However, this book does not address the Workbench environment in any detail.

1.7. ANSYS Licensing

As noted above, access to ANSYS, Inc. products is based on product licenses. You can download and install all of the software that ANSYS, Inc. offers from the ANSYS Customer Portal. (See chapter 2 for more details.) However, you cannot run any ANSYS, Inc. products without a valid license. Each ANSYS license specifies which ANSYS, Inc. products or capabilities can be used, the maximum number of elements that may be included in a model, and how many people can use the software at the same time. It may also limit the physical distance that you may be from the license server while using the software or impose other restrictions. The software in the ANSYS Mechanical APDL family is basically the same for all products. Only the licenses are different.

1.8. Functionality and Features of the ANSYS Mechanical APDL Family

The features of the products in the ANSYS Mechanical APDL family are constantly evolving and new capabilities are added regularly. This section outlines some of the functionality and features that you can expect in the full multiphysics version of the program.

1.8.1. Can ANSYS...?

New users often want to make sure that the program can be used for their intended application(s) so the very first question that they usually ask is "Can ANSYS do...?". As long as you want the program to do engineering analysis (and not the dishes), the answer is usually "Yes." It may not be quick or easy. You may need licenses for additional ANSYS, Inc. products. You might need a more powerful computer. You might need to write some code. You might even need to recompile or relink the program. But there is very little engineering analysis that ANSYS can't do given enough time, resources, and creativity.

1.8.2. Steady-State and Time-Dependent Analyses

ANSYS supports a variety of steady-state and time-dependent analyses. These include static analyses where inertia effects are not included and dynamic analyses where inertia effects are important. It permits two types of static analyses: single step analyses where all loads are applied at the same time and multistep static analyses where different loads can be applied or removed with each load step. Multistep analyses allow multiple combinations of loads to be solved in a single run. They also allow loads to be applied gradually in nonlinear analyses, such as creep analysis, where time is important but mass effects are not.

ANSYS permits nonlinear transient dynamic analysis where the response of the system is time dependent due to changing loads and other system nonlinearities, mode-frequency analyses where the outputs are the natural frequencies (eigenvalues) and mode shapes (eigenvectors) of the system, spectrum analyses to model phenomena such as earthquakes where the loads applied to the system are frequency dependent, harmonic analyses where the excitation loads are harmonic (sinusoidal), and analyses with random vibrations.

1.8.3. Physics Capabilities*

The full multiphysics version of ANSYS offers structural, thermal, fluid, electric, magnetic, and electromagnetic physics capabilities.

Structural analyses may include linear and nonlinear buckling; fracture; composites; fatigue; and contact with and without friction, gaskets, joints, pretension, and spot welds. They can involve geometric nonlinearities such as large strain and large deflection, and may use linear and nonlinear material models including rate-dependent and rate-independent plasticity, hyperelasticity, viscoelasticity, and creep. Structural analyses where time is important but time steps are very short, such as high-speed impacts and explosions, should be performed using one of the ANSYS explicit dynamics products such as ANSYS LS-DYNA® or ANSYS AUTODYN®.

Pure thermal analyses may include conduction, convection, radiation, phase change, or some combination of the four.

Fluid analyses may include laminar and turbulent compressible and incompressible flow, multiphase flow, free surfaces, porous media, fans or pumps, smooth or rough walls, cavitation, multiple species transport, particle tracing, and swirl. Other types of fluid analyses should be performed using one of the ANSYS CFD products, such as ANSYS CFD®, ANSYS Fluent®, ANSYS CFX®, ANSYS CFD-Flo™, or ANSYS Polyflow®.

Electromagnetic analyses may include electric fields, magnetic fields, alternating current (AC), direct current (DC), far fields, and electric circuits. Other types of electromagnetic analyses, especially those involving high-frequency analysis, should be performed using an ANSYS Electronics product such as ANSYS HFSS™ or ANSYS Maxwell®.

Finally, analyses using multiple (coupled) physics are possible. The ANSYS Mechanical APDL family supports acoustic, acoustic-structural, electromagnetic, electromagnetic-fluid, electromagnetic-thermal (for induction heating), electromagnetic-thermal-structural (for MEMS), electrostatic-structural, fluid-structural, magnetic-structural, piezoelectric (for ultrasonic transducers), piezoresistive, thermal-electric (for resistive heating), thermal-fluid, thermal-structural, and thermal-electric-structural analyses. Coupled analyses using thermal-electric-fluids and electromagnetic-thermal-fluids can only be performed using ANSYS Fluent at this time.

Access to these features depends on your license and new features may be added at any time. For up-to-date information, contact your local ANSYS representative or see the company website for more details.

1.8.4. Special Features

Members of the ANSYS Mechanical APDL family offer some or all of the following special features: APDL, probabilistic design, optimization, submodeling, substructuring, user materials, and user programmable features.

The ANSYS Parametric Design Language (APDL) is one of the most powerful features of ANSYS. It allows you to define some or all parts of your model (geometry, material properties, loads, etc.) as parameters. Creating and solving a new variation of a parameterized model is as simple as changing a few parameter values and rerunning the model. This makes ANSYS a powerful tool for engineering analysis, optimization, root cause analysis, and for the design of new systems and technologies. APDL also allows you to build and execute macros, run macros as ANSYS commands, operate on parameter arrays, and do simple logic (if, then, else, do, repeat, etc.).

Probabilistic design allows you to randomly vary certain input parameters in order to model indeterminate features such as surface finish.

Optimization allows you to automate the process of varying model parameters and rerunning the solution in order to identify the best design for a given situation. Optimization in Mechanical APDL requires a parameterized input file.

Submodeling is a two-step process in which the results from a large model with a coarse mesh are used as the boundary conditions for a small model with a fine mesh. This is useful for problems (like those involving stress concentrations) where the inclusion of a sufficiently refined mesh in the large model would make the solution prohibitively expensive. For submodeling, two solutions are required. The first solution is needed to obtain the results for the coarse model. These results are then used as boundary conditions in the second model to obtain results in the area of interest. The creation of the boundary conditions for a submodel, which requires the interpolation of the results from the coarse model, is done by ANSYS and does not require manual calculations by the user.

Substructuring is a technique in which a stiffness matrix is used to replace a large section of a finite element model in order to reduce the overall cost of the analysis. Substructures were originally called superelements and should be thought of as such. To perform an analysis using substructuring, first a finite element model is created for the substructure. The master degrees of freedom are identified and the model is solved to create the substructure matrix. Next, the substructure is defined and used as an element in the larger model. Finally, the detailed response of the substructure can be expanded using the analysis files from step 1 and the results file from step 2 if desired. Substructures remain linear for the entire analysis but can be used with other nonlinear elements in nonlinear analyses.

ANSYS also includes a number of user programmable features that allow you to create a customized version of ANSYS. User programmable features include customized elements, loading conditions, material models, and commands. For example, ANSYS users are not restricted to the material constitutive models contained in the ANSYS material library. You can create your own material models from handbook values, experimental values, or tabulated data. All ANSYS user programmable features require you to relink the program. This means that you will need the Fortran, C, and C++ compilers used to create the ANSYS release version of the program that you are using. Details may be found in the Guide to ANSYS User-Programmable Features, which has been incorporated into the Mechanical APDL Programmer's Reference.

The combination of APDL, user materials, and user programmable features allows ANSYS users to develop custom applications and macros. Applications and macros are collections of ANSYS commands that can be used to solve a particular class of problems or perform a certain set of commonly used functions. Both applications and macros accept parameterized arguments as part of the input. This makes them very flexible and powerful tools.

1.9. ANSYS: Backward Compatibility and Legacy Code

Applications, macros, and customized versions of ANSYS can be and are created by all types of users. However, they are most commonly associated with large corporations that have the resources to invest in their creation and enough product variations to justify the investment. Some industrial applications contain a large number of ANSYS commands, are well verified, and have been used for years (or decades) by the analyst or the company.

If the ANSYS development team were to delete an element or a feature that is used in an application (or an input file), either the model would no longer run or it would produce meaningless or erroneous results. The time and cost to update and repair these applications (macros, input files, etc.) with every revision of ANSYS would quickly become prohibitive. As a result, ANSYS, Inc. goes to great lengths to ensure that the program is fully backward compatible (i.e., that a newer version of the program can accept input generated by older versions of itself). In particular, ANSYS, Inc. ensures that:

- Binary database files can be read by the next release of the program. (More than one release is not guaranteed.)
- ASCII (plain text) input files from earlier versions of the program will run in later versions and will give the same results.
- When elements or capabilities are to be removed, an announcement in the program is made two releases prior to removal and the elements or capabilities are undocumented one release prior to removal.
- Elements, commands, and other features are rarely deleted from the program. 'Removing' elements, commands, and features usually involves removing them from the graphical user interface and the documentation. The functionality continues to be available via the command prompt and input files.

The implications of this policy are important, if not immediately obvious. First, the program that you run today still contains code that was written in 1970. This means that ANSYS does not look, or act, like most other software programs. This can be a source of annoyance to new users, but it is also a benefit. Any model that you build today will run tomorrow, next week, next month, next year, and next decade—as long as you save your input as an input file instead of as a binary database file. (See chapter 2 for more details about ANSYS file formats.)

Since the program is timeless, so is this book. New features that are added to the program after the publication date of this book will not be included, but all information pertaining to existing features will be valid for the foreseeable future. In addition, menu paths and program graphics may change with future revisions of the program and some options will inevitably be hidden. But since nothing is ever truly removed from the program, all of the exercises in this book will continue to run—as long as you run them using the ANSYS commands provided at the end of the exercises instead of following the step-by-step instructions using the graphical user interface.

Finally, in order to harness the power of ANSYS and to truly understand how the program functions, you will need to learn to use ANSYS commands and to read and write input files. Commands and input files are addressed in detail in chapter 2, and will be referred to throughout the book. Writing, editing, and debugging input files are discussed in chapter 10. The other advanced features listed in section 1.8.4 are beyond the scope of this book.

Chapter

2

Interacting with ANSYS

Suggested Reading Assignments:
Mechanical APDL Operations Guide: Chapters 2–4
Mechanical APDL Command Reference: Chapters 1–3

CHAPTER OUTLINE

This chapter introduces the organization and behavior of ANSYS. It presents multiple ways to interact with the program and summarizes the types of feedback that ANSYS provides. It also answers common questions including "How do I save files?," "Where is the Undo button?," and "Where can I find help?." Most importantly, it explains why ANSYS may be different from other software programs that you have used and how you can benefit from these differences.

Although this chapter was designed for readers who are new to ANSYS, it can be a little overwhelming. Do not be discouraged. Much of what you do not understand now will be clearer after you have some experience with the program. We suggest reading chapter 2 once to gain an appreciation for the program's structure and capabilities, and then moving on to the exercise that follows. You can always read chapter 2 again at a later time. Finally, we strongly recommend re-reading chapter 2 before starting chapter 10. Much of the information below is needed to work with input and batch files and is not repeated at the end of the book.

ANSYS Mechanical APDL for Finite Element Analysis.
DOI: http://dx.doi.org/10.1016/B978-0-12-812981-4.00002-2

2.1. ANSYS Simulation Environments

When ANSYS was first released in 1970, it was a command-driven batch-processing program that ran in a data center environment. The user supplied input to the program in the form of a batch file (originally a box of punch cards) and the program printed the results on paper. There was no interaction between the user and the program once the input was submitted. A primitive graphical user interface (GUI) was added to the Preprocessor at Release 4.0 (1984). Interactive processing, where immediate graphical feedback was available for each command issued, became the default mode of operation at Release 5.0 (1993). But the internal behavior of the program remained the same. Each button (or series of buttons) within the GUI issued a command with complete syntax to the program.

ANSYS remains a command-driven program today. And, it can still be run in both batch and interactive mode. In batch mode, you supply input to the program in the form of a batch file (now a plain text file beginning with a **/BATCH** command) and the program writes the results to the results file. Batch mode runs ANSYS in the background of your computer. It does not permit you to open the GUI, so it provides no real time feedback. In interactive mode, you supply input to the program through the GUI menus, the GUI command prompt, and/or input files. ANSYS runs in the foreground of the computer and allows you to see both text and graphical feedback as your commands are executed. Running simulations in batch mode is faster than using interactive mode and is preferable for large problems that have long running times, but creating and postprocessing models is easier in interactive mode.

The ANSYS Product Launcher (Figure 2.1) is the recommended way to choose your simulation environment and start the program. The "ANSYS" option in the drop down box at the top of the Product Launcher opens the program in interactive mode while the "ANSYS Batch" option runs the program in batch mode (Figure 2.2). Interactive mode also can be launched directly from the Windows Start Menu or from a shortcut on your computer desktop. Batch mode also can be launched via your operating system command prompt.

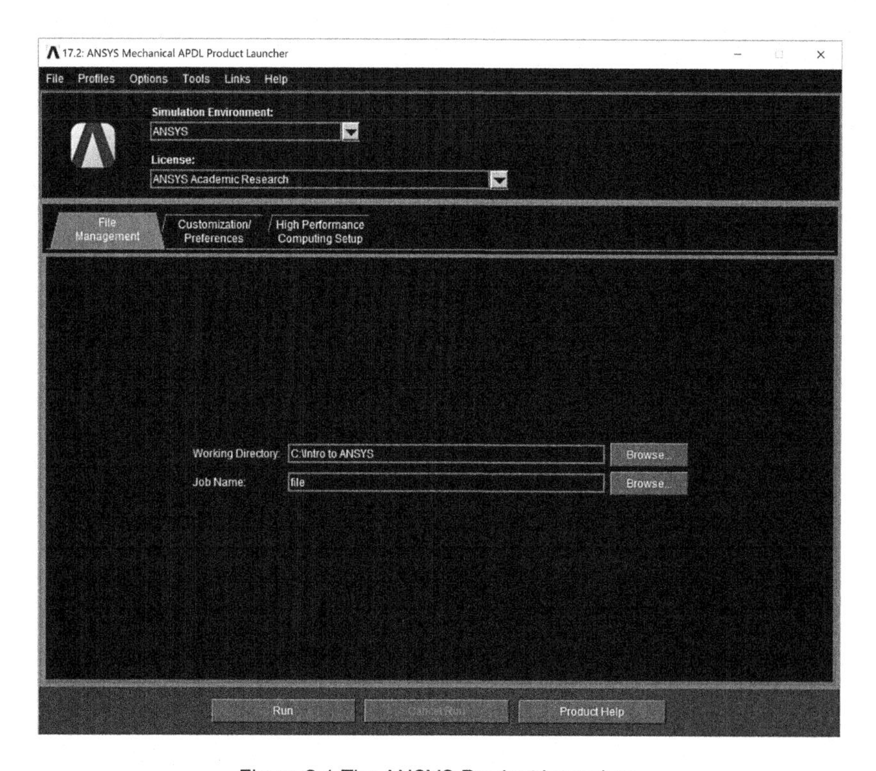

Figure 2.1 The ANSYS Product Launcher.

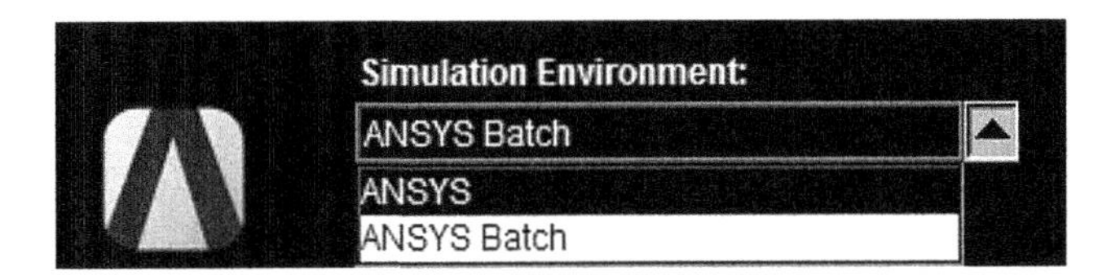

Figure 2.2 Choosing an ANSYS Simulation Environment using the Product Launcher.

2.2. Communicating with ANSYS

There are three ways to communicate with (i.e., issue commands to) ANSYS in the original environment: the graphical user interface (officially referred to as the Mechanical APDL user interface or the M-APDL UI), the command prompt located within the M-APDL UI, and input and batch files. You can also communicate with ANSYS using the ANSYS Workbench environment but that discussion is beyond the scope of this book.

2.2.1. ANSYS Commands

ANSYS commands are alphanumeric strings that provide instructions or information to the program. All ANSYS commands begin with an ANSYS verb, which also serves as the command name. Most commands also require or accept arguments to specify and control their behavior.

Commands can contain up to 640 characters in 20 fields. The verb and each argument occupy a field of the command. Commas are used to separate the fields. For example, `MP,EX,1,30e6` defines a Young's modulus (EX) value of 30e6 for material #1. **MP** is the command verb for the specification of linear material properties. "EX," "1," and "30e6" are the arguments for the command.

Command names are between one and ten characters long. However, you only need to supply enough characters to identify a unique command. For example, the **RECTNG** command creates a rectangle by two sets of coordinates. `RECTNG,0,1,0,1` creates a 1x1 square with the lower left hand corner at the origin (0,0). `RECT,0,1,0,1` is an equivalent syntax and performs the same operation. In most cases, the first four or five characters of the command name are sufficient to identify the command. If you do not provide enough characters to identify the command to use, ANSYS will issue a warning informing you that the command was not recognized and will ignore the instruction.

ANSYS commands are not case sensitive. They can be written in all uppercase letters, all lower case letters, or a mix of upper and lower case letters. Empty spaces are ignored. For example, the command `MP,EX,1,30e6` is equivalent to `mp, ex, 1, 30e6`. Anything to the right of an exclamation point (!) is assumed to be a comment and is also ignored.

The data supplied to a field can be numerical (real or integer), alphabetic, or alphanumeric (containing both numbers and letters). Numerical data may be input in exponential form. The sign after the E character is used to define the sign of the exponent (i.e., 1e2 = 100.0, 1e-2 = 0.01). If numerical data is entered into a field without a decimal point, a decimal point is assumed to exist after the right-most digit of the number (i.e., 1e2 = 1.0e2 = 100.0).

If a numerical expression is input into a field that requires numerical data, ANSYS will evaluate the expression and use the result. For example, `F,ALL,FX,(100/2)` (apply a load of 100/2 units in the x direction to all nodes) is equivalent to the command `F,ALL,FX,50`. Parentheses are not required, but they are strongly recommended and often necessary to ensure the correct order of operations.

If alphabetic or alphanumeric data is input into a field that expects or requires numerical data, ANSYS will check the input against the list of valid labels. If a valid label is not found, it will check the input against the list of defined parameters. If a parameter with a valid numerical value is found, the parameter value will be used for the field. If a parameter with a valid numerical value is not found, ANSYS will issue a warning and set the value of the field to a value that is approximately zero. (Parameterization in general is beyond the scope of this book.)

If nothing is supplied in a field, ANSYS will use the default argument if one exists. For example, `DLIST` is equivalent to `DLIST,ALL` (list the constraints for all nodes in the currently active set) because the default label for the second field in the **DLIST** command is "ALL." The default value for numerical fields is often zero. For example, `K` (define a keypoint) is equivalent to `K,,0,0,0` (define a keypoint at (0,0,0)). However, the default for numerical fields is not always zero. For example, in solid modeling commands like the **K** command, the second field usually defaults to the next available entity number. You should always check the documentation to determine the default value before leaving an argument blank. If no default exists and a value is not required, the field will be ignored. If an argument is required and no default exists, ANSYS will issue a warning and the command will be ignored.

Some commands have undocumented fields (i.e., no information about the field is provided in the documentation). These fields are either associated with features that have been undocumented or have been included for use by ANSYS developers. In general, you should never enter data into any field that is undocumented.

Finally, some commands are undocumented (i.e., no information about the command is provided in the documentation). These commands are either associated with features that have been undocumented or have been included for use by ANSYS developers. In general, you should never use an undocumented command.

2.2.2. The Graphical User Interface

The GUI is a collection of windows, menus, buttons, and toolbars that allow you to interact with the program using a keyboard and a mouse. Although many features of the GUI can be customized to suit your needs (see Sections 4.3 and 6.3 of the Mechanical APDL Operations Guide), the default layout is used in this book.

2.2.2.1. The Output Window

When you start ANSYS in interactive mode, two windows open. The Output Window (Figure 2.3) is a small window with a black background and white text. It displays all text output from ANSYS and tracks the program's progress as it attempts to calculate a solution. Closing this window will terminate the program abnormally and will leave a lock file behind, so it is best to keep it in the background and refer to it when necessary.

Mechanical APDL 17.2 Output Window

```
ANSYS Academic Research
Release 17.2

Point Releases and Patches installed:

ANSYS, Inc. Products Release 17.2
ANSYS Mechanical Products Release 17.2
ANSYS Customization Files for User Programmable Features Release 17.2
ANSYS, Inc. License Manager Release 17.2


        *****  ANSYS COMMAND LINE ARGUMENTS  *****
 INITIAL JOBNAME                 = file


 DESIGNXPLORER REQUESTED
 START-UP FILE MODE              = READ
 STOP FILE MODE                  = READ
 GRAPHICS DEVICE REQUESTED       = win32
 GRAPHICAL ENTRY                 = YES
 LANGUAGE                        = en-us
 INITIAL DIRECTORY = C:\Intro to ANSYS

RELEASE= Release 17.2           BUILD= 17.2       UP20160718    VERSION=WINDOWS x64
```

Figure 2.3 The Output Window.

2.2.2.2. GUI Toolbars and Menus

The GUI appears a few moments after the Output Window opens. The GUI consists of several toolbars and menus, a command prompt, a text display area, and a graphics display window (Figure 2.4).

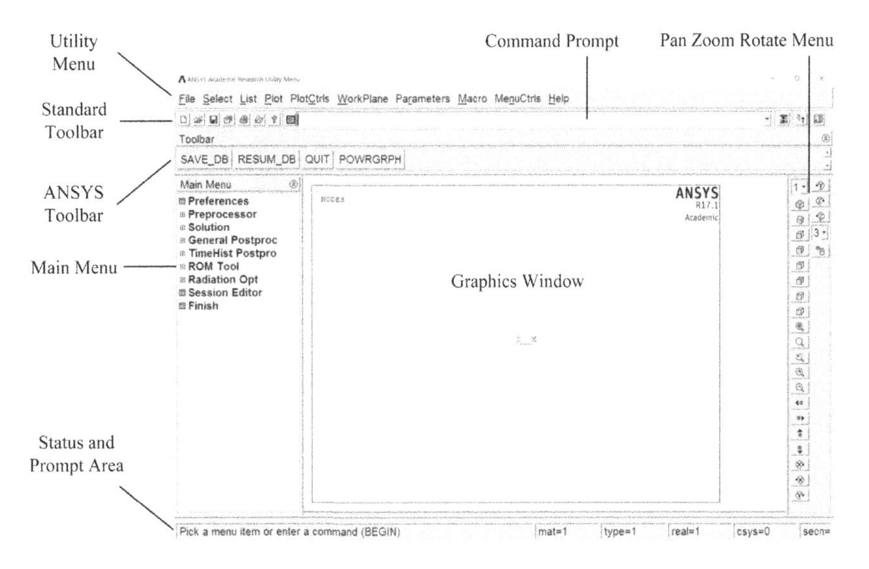

Figure 2.4 The ANSYS Mechanical APDL User Interface (i.e., the GUI).

The Utility Menu is located at the top of your screen. This menu allows you to manipulate files (save the database, clear the database, read input files, etc.) and to manipulate the model (select model entities, list the selected entities, list parameters, modify the view and orientation of the model on the screen, etc.).

The Standard Toolbar is located below the Utility Menu. This toolbar contains a series of shortcut buttons for commonly used functions from the Utility Menu.

The command prompt is located below the Utility Menu and to the right of the Standard Toolbar. The command prompt allows you to issue commands directly to the program.

The ANSYS Toolbar is located below the Standard Toolbar. This toolbar contains a second set of shortcut buttons. By default, the toolbar includes buttons for saving a database, resuming a database, quitting the program, and toggling PowerGraphics. You can add other buttons by using the ***ABBR** command.

The Main Menu is located below the ANSYS Toolbar. The Main Menu contains the commands needed to create, modify, solve, and postprocess the model.

The Graphics Window is located to the right of the Main Menu. It is the largest feature in the GUI. This window is where all graphics for the program are displayed. It also allows graphical picking (i.e., selecting solid model and finite element model components with the mouse). The default background of the Graphics Window is black. For clarity, this book shows the Graphics Window with a white background. You can toggle the Graphics Window background color by using the GUI path: **Utility Menu > PlotCtrls > Style > Colors > Reverse Video**.

The Status and Prompt Area is located below the Main Menu and the Graphics Window. This area displays information, prompts, and instructions based on the current status of the program. For some dialog boxes, it also displays the name and a brief description of the command to be executed. This information is often valuable and should not be ignored.

Finally, the Pan Zoom Rotate Menu is located to the right of the Graphics Window. This toolbar contains a set of shortcuts for the commands normally accessed via the GUI path: **Utility Menu > PlotCtrls > Pan Zoom Rotate**.... It controls the view and orientation of objects in the Graphics Window.

2.2.2.3. Advantages and Disadvantages of using the GUI

For new users, the GUI is the easiest and most intuitive way to interact with ANSYS. The GUI only presents options that are appropriate for the model entities, attributes, and results that have been defined or are available. The other options are hidden to make the program easier to navigate. Once an operation has been selected, the GUI guides you through the process of supplying the information required to define the arguments and execute the relevant command(s). The GUI provides immediate feedback and visualization when features (solid model geometry, nodes and elements, boundary conditions, etc.) are added to the model. And, the GUI allows you to view the model to determine its status and to understand its results.

However, there are also some disadvantages associated with the GUI. Issuing a command through the GUI is usually slower than issuing the command directly. For example, defining element type #1 to be a PLANE182 element via the GUI requires a total of 6 mouse clicks to navigate through the program's menus and dialog boxes. The same task using commands requires 8 characters to be entered into the command prompt: `ET,1,182`. In addition, the GUI provides less insight into what the program is actually doing. For example, if multiple commands can be used to accomplish the same task, it is often unclear which command the GUI will use a priori. Similarly, when specifying an option in a dialog box, the GUI often issues a series of commands to set all possible options instead of setting only the one that was modified. Finally, some program capabilities are not accessible via the GUI.

2.2.3. The GUI Command Prompt

The command prompt allows you to issue commands directly to ANSYS from within the GUI. Simply type the command in the command prompt and press enter. Although issuing commands using the command prompt is often faster than using the GUI menu paths, it requires more planning and knowledge of the command being issued. For example, the full syntax for a command must be supplied or ANSYS will issue an error and ignore the command. Since this can be difficult to remember, ANSYS displays a syntax reminder above the GUI command line when you begin to type a new command (Figure 2.5).

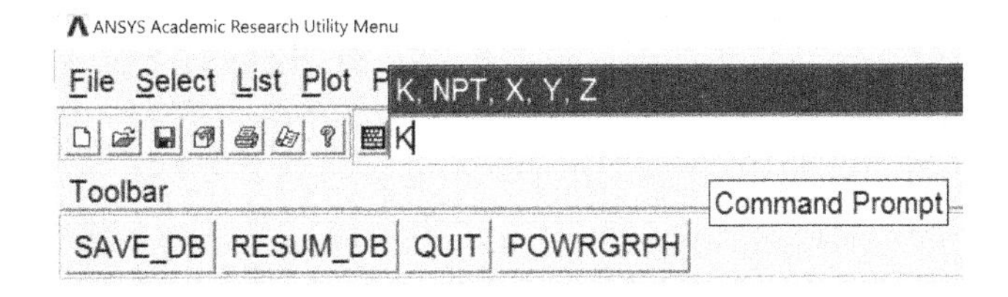

Figure 2.5 The GUI Command Prompt.

The command prompt keeps a record of all commands issued during the current session (Figure 2.6). You can click on the down arrow located at the right end of the command line to display these commands. Clicking on one of the entries in this list places the selected command into the command prompt. Once the selected command has been placed in the command prompt, it can be modified if desired. Finally, the selected or modified command can be issued by pressing enter.

Figure 2.6 GUI Command Prompt History.

As you become more comfortable with ANSYS commands, you may find it more convenient to type (or retrieve) frequently used commands in the command prompt rather than clicking through the GUI menus.

2.2.4. Input Files and Batch Files

Input files and batch files are collections of commands (one command per line) written in a plain text file with a .inp or a .txt extension. Input files are intended to be executed in interactive mode while batch files are intended to be executed in batch mode. To enable batch execution, batch files must begin with a **/BATCH** command on the first line of the file. Input and batch files are otherwise identical from the user's point of view.

Input and batch files can be run in interactive mode using the GUI path: **Utility Menu > File > Read Input from...** or by issuing the **/INPUT** command. (The **/BATCH** command is ignored in interactive mode.) In addition, the full or partial contents of an input or batch file can be copied and pasted into the GUI command prompt. Pressing enter will execute all pasted commands. When input and batch files are executed in interactive mode, the program sends text (command responses, notes, warnings, errors, and other messages) to the Output Window and plots solid model entities, finite element model entities, and boundary conditions in the Graphics Window. Watching an input or batch file run in interactive mode is like watching a movie on fast-forward. You can see what is happening, but you are unable to do anything (except suspend processing in the case of an error) until the program has reached the end of the file.

Input and batch files can be run in batch mode from the ANSYS Product Launcher. (The Product Launcher issues the **/BATCH** command for you.) Batch files can also be run from your operating system command prompt. (See Section 3.4.1 of the Mechanical APDL Operations

Guide for details.) When batch files are executed in batch mode, the program sends text (command responses, notes, warnings, errors, and other messages) to the output file. No real time feedback is provided.

Since its presence is never a problem and its absence can be, we recommend including a **/BATCH** command at the beginning of every file. ANSYS automatically issues a **/BATCH** command at the beginning of every session log file for the same reason.

2.3. How ANSYS Communicates with You

Although this book emphasizes how you can communicate with ANSYS, it is important to recognize that ANSYS also attempts to communicate with you. In interactive mode, ANSYS provides information in five ways: it plots entities in the Graphics Window, it displays information in the Status and Prompt Area, it creates pop up dialog boxes (Figure 2.7), it displays text output (messages, command responses, etc.) in the Output Window (Figure 2.8), and it writes warning and error information to the error file (file.err). In batch mode, ANSYS writes text output to the output file (file.out) and writes warning and error information to the error file (file.err). (See section 2.5 for more information on the error and output files.)

ANSYS provides five levels of feedback depending on the importance of the information. These levels are: INFOrmation, NOTEs, WARNINGs, ERRORs and FATAL errors. Limited documentation of these messages can be found with the ***MSG** command in the Mechanical APDL Commands Reference.

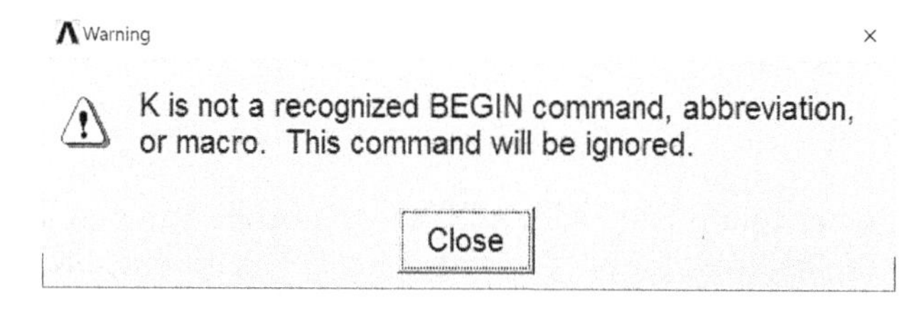

Figure 2.7 Example Warning Pop Up Dialog Box.

```
*** WARNING ***                         CP =      21.949   TIME= 12:27:19
K is not a recognized BEGIN command, abbreviation, or macro.  This
command will be ignored.
```

Figure 2.8 Example Warning in the Output Window.

2.3.1. INFO Level Feedback

INFO level feedback is information that may or may not be of interest to the user. INFO messages appear without headers in the Output Window in interactive mode and are written to the output file in batch mode. INFO messages do not appear in pop up dialog boxes and they are not written to the error file.

2.3.2. NOTE Level Feedback

NOTE level feedback is non-critical information that may be needed to use the program more effectively. For example, NOTEs are issued when you do not reset the program defaults. This occurs in static analyses when you fail to specify the solution end time. Since time is required for all nonlinear analyses, this results in a note that says "Note: Present time 0 is less than or equal to the previous time. Time will default to 1." NOTEs also provide information about solver defaults and alternatives to user-selected program options. NOTEs do not appear in pop up dialog boxes, but they are displayed with a NOTE header in the Output Window in interactive mode and are written to the output file in batch mode. NOTEs are not written to the error file.

2.3.3. WARNING Level Feedback

WARNING level feedback is information concerning a mistake in the use of the program. For example, WARNINGs are issued if a unique command cannot be identified, if a command name is misspelled, if a command is issued in the wrong processor (see section 2.4), if a command is missing an argument that has no default value, or if a command cannot be followed for any other reason. WARNINGs are issued in pop up dialog boxes. Each new WARNING pop up dialog box will replace the previous one. When commands are read from an input file or pasted as a group in the command prompt, this can happen so quickly that some WARNING messages are missed. However, the information is not lost. WARNING level feedback is also displayed in the Output Window with a header or is written to the output file. Finally, WARNINGs are written to the error file.

WARNINGs are usually considered to be non-critical and the program will continue to run after a WARNING is issued. However, WARNINGs should not be ignored as they can affect the results of the analysis. WARNINGs are promoted to ERRORs in batch mode.

2.3.4. ERROR Level Feedback

ERROR level feedback is information concerning a mistake in the use of the program that is more critical than a WARNING. Examples of ERRORs include meshing failures in preprocessing or missing required data (like material properties) at solution time. ERROR information is considered to be critical to the analysis and the program cannot proceed until the ERROR is corrected. ERRORs are presented in pop up dialog boxes that must be dismissed by the user. They are displayed in the Output Window with a header or are written to the output file. ERRORs are also written to the error file.

ERRORs are promoted to FATAL errors in batch mode and will cause a batch run to terminate at the earliest “clean exit” point.

2.3.5. FATAL Level Feedback

FATAL level feedback informs the user that the program has encountered a critical error. This means that the program cannot proceed and will terminate the run immediately. For example, running out of disk space during solution will cause a FATAL error. FATAL level information is displayed in the Output Window with a header or written to the output file. It is also written to the error file. In interactive mode, a FATAL pop up dialog box may or may not appear before termination. It may be necessary for you to review the error file to identify the problem that caused termination.

2.4. ANSYS Program Structure

2.4.1. Levels and Processors

The locations of the various items and menus in the ANSYS GUI reflect the internal structure and behavior of the program. ANSYS has two basic levels: the Begin Level and the Processor Level (Figure 2.9). The ANSYS documentation refers to the Begin Level as the gateway into and out of ANSYS. A better analogy would be to think of the Begin Level as the hallway in a house and the processors as different rooms. When you enter the program, you start at the Begin Level. From here, you can move to any one of the processors. However, the various processors, or rooms, are not connected. To change processors, you must first go back to the hallway or the Begin Level.

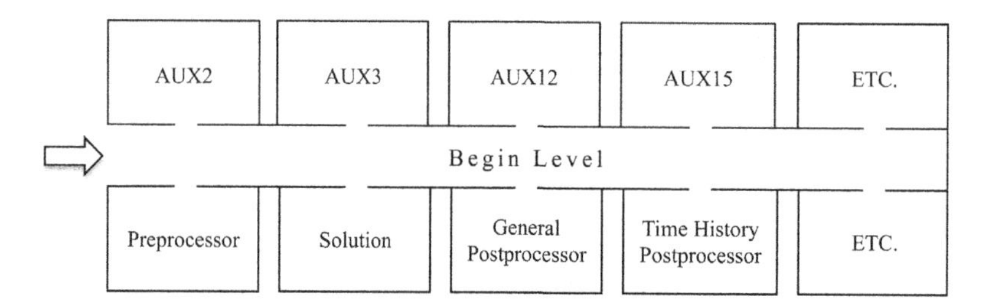

Figure 2.9 ANSYS Program Structure: Begin Level and Processors.

Most ANSYS commands are limited to a specific level or processor. Just as you can only cook in the kitchen and wash clothes in the laundry room, you can only build model geometry from the Preprocessor and solve the model from the Solution processor. However, some commands are processor independent. They can be issued at any time and from any location in the program. These are referred to as global commands.

In general, the ANSYS GUI handles the navigation between the processors and the Begin Level for you. For example, if you use the mouse to click on the Preprocessor option in the Main Menu, the Preprocessor menu tree will open. This indicates that the program has issued the **/PREP7** command to enter the Preprocessor. All other processor menu trees within the Main Menu will remain closed. If you want to access a feature in the Solution processor, you must click on the Solution processor option in the Main Menu. When you do this, the GUI issues a **FINISH** command to exit the Preprocessor and closes its menu tree. Then, it issues the **/SOLU** command to enter the Solution processor. This causes the Solution processor menu tree to open. Since you cannot be in two processors at the same time, the GUI will never permit you to open or view more than one processor menu tree at a time.

Because these navigation commands have been embedded in the GUI, the task of moving between processors is almost invisible to new users. However, when you start to work with the GUI command prompt and input or batch files, you will need to issue the navigation commands (**/PREP7**, **FINISH**, etc.) instead of relying on the GUI to issue them for you. You will also be responsible for issuing commands from the appropriate level and processor. If you try to issue an ANSYS command from the wrong processor, from the wrong level, or at an inappropriate time, the program will issue a warning and ignore the command.

2.4.2. The ANSYS Database

The development of interactive mode at Release 5.0 made it necessary to be able to add, list, display, modify, and delete model information quickly and easily at any level and in any processor. The ANSYS database was developed to meet this need. The ANSYS database contains all input data (geometry, material properties, boundary conditions, etc.) for the current analysis. The database can contain one set of results (displacements, stresses, etc.) at a time. By default, it contains the last set of results that was generated. Otherwise, it will contain the last set of results that was loaded from the results file. The database also contains any postprocessing data (parameters, element tables, etc.) that was created for the analysis. Almost all ANSYS commands access or modify the database in some way.

The ANSYS database has four characteristics that are important to new users. First, the database is not automatically saved to a file. If you close the program without saving the database, the memory will be cleared and the information contained within the database will be lost. Second, only one model can exist in the database at a given time. If you want to create a new model without restarting the program, you must clear the database and start a new one. This can be done using the GUI path: **Utility Menu > File > Clear & Start New**... or by issuing the **/CLEAR** command. Third, the database only represents the current state of the model. It does not keep any records of the model's previous state(s). To keep a record of the database's history, you must save the database or archive the model for each desired state. (See section 2.6 for details.) Finally, saving the database will not necessarily save all of your results. To protect your results, you must always keep a copy of the results file.

2.4.3. Types of Commands and Their Locations

The Mechanical APDL Command Reference lists 15* different categories of ANSYS commands: 4 Begin Level groupings, 8 Processor Level groupings, and 3 sets of specialty commands.

2.4.3.1. Begin Level Commands

The 4 types of Begin Level commands are session commands, database commands, graphics commands, and APDL commands. These commands are accessible through the Utility Menu in the GUI. Session commands control the characteristics of each ANSYS session. They allow you to change the jobname, copy and delete files, enter a processor, etc. Database commands interact with and clear the database. They allow you to select entities, create components, use and modify the working plane, change the coordinate system, etc. Graphics commands control the graphical output from the program. They allow you to modify the size and scale of the display; add or remove labels, notes, and symbols to the GUI display; specify contour plot values; etc. APDL commands allow you to define parameters and arrays, create and use macro commands, incorporate do-loops and if-then-else statements, perform matrix operations, and more.

Many Begin Level commands begin with a slash (/) or a star (*). Slash commands supply general control instructions to ANSYS. Most slash commands can be used in any level of the program but some can only be used at the Begin Level. Commands that begin with a star (*) are commands in the ANSYS Parametric Design Language. Star commands can be used at any level of the program.

2.4.3.2. Processor Level Commands

The four most commonly used processors are the Preprocessor, the Solution processor, the General Postprocessor, and the Time History Postprocessor. Commands for creating the model are in the Preprocessor (PREP7). These include commands for creating a solid model; defining material properties, elements, and real constants; and meshing the model. Commands for setting the solution options and for solving the model are found in the Solution processor (SOL). Commands associated with applying boundary conditions are found in both the Preprocessor and Solution GUI tree menus and can be issued from either processor. This is a legacy feature that stems from the time before ANSYS Release 5.0 when the Preprocessor was used to create the batch file for solution. Officially, boundary conditions are considered to be solution options. In this book, boundary conditions will only be applied in the Solution processor.

Commands for viewing, evaluating, manipulating, and exporting the model results are located in the postprocessing modules. The General Postprocessor (POST1) allows you to plot and list results (displacements, stresses, temperatures, etc.) for the entire model or for a section of the model based on a single loading condition or at a single point in time. Viewing POST1 results is like looking at a photograph. You can view the results from other loading conditions or at other points in time but you must look at each one separately. You can animate results from within the General Postprocessor if you want to see them in sequence. In contrast, the Time History Postprocessor (POST26) allows you to plot or list the variation of a single result (displacement, stress, temperature, etc.) at a single node or element over time. Viewing results in POST26 is like looking at a line graph.

Figure 2.10 shows examples of the output from the General Postprocessor and the Time History Postprocessor for a transient thermal analysis of a combustion chamber. Figure 2.10 (left) shows the temperature distribution in the combustion chamber at the end of the analysis when the combustion chamber has reached steady state. You can see that the maximum temperature is located at the inner edge of the nozzle. This plot was generated in the General Postprocessor (POST1). Figure 2.10 (right) shows the temperature at four different locations in the chamber, including the location of maximum temperature, as a function of time. This plot was generated in the Time History Postprocessor (POST26).

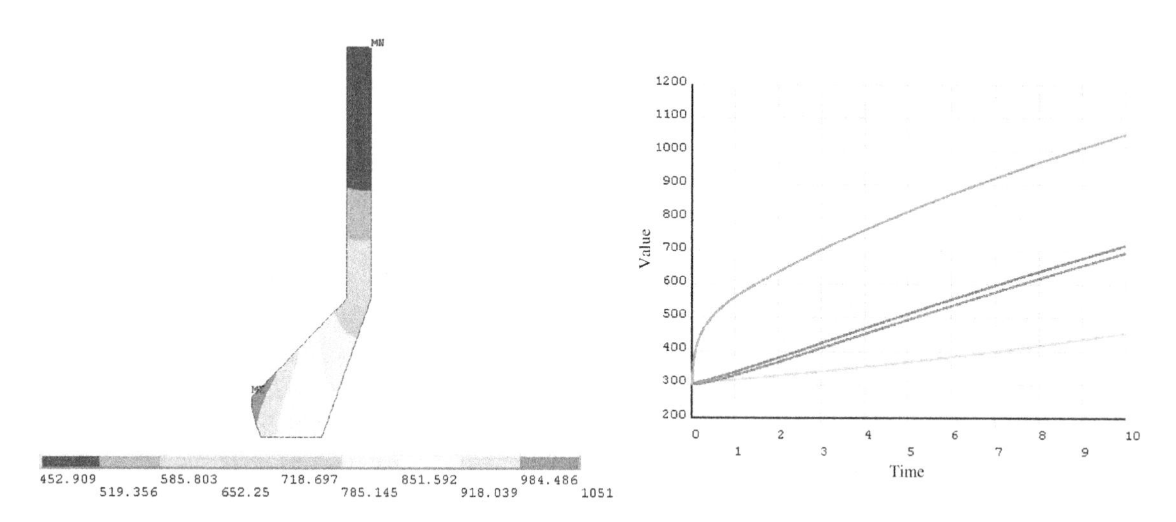

Figure 2.10 Temperature as a Function of Space (POST1—left) and as a Function of Time (POST26—right) for an Axisymmetric Combustion Chamber.

The remaining processors are specialty processors and have specific functionality. AUX2 examines binary files. AUX15 imports IGES files. Both are accessed through the File option in the Utility Menu. The other processors are accessed through the Main Menu.

Some processor level commands, including those associated with listing and plotting, can be accessed from any part of the program. These commands are located in the Utility Menu along with the other global commands.

2.5. ANSYS File Structure

ANSYS creates and uses many different files during an analysis. Some files are temporary and others are permanent (i.e., you must choose to delete them). Some files are for your use and others are exclusively for the use of the program. Some files, like the output and error files, are written in ASCII format and can be opened in an ordinary text editor. Others, like the database file and the results file, are written in binary form and are intended to be opened from within ANSYS.

The program places all of the files that it creates in the working directory. You can change the working directory from the ANSYS Product Launcher or by using the GUI path: **Utility Menu > File > Change Directory**. . ..

All files created by ANSYS are named jobname.ext. The jobname defaults to "file" but can be changed from the Product Launcher or the Utility Menu. The extension ranges from 2 to 4 characters and indicates the type of file. For example, file.db is the default name for the database file while file.log is the default name for the log file. We recommend that you specify a new working directory and/or jobname at the beginning of every new analysis. This will prevent existing files from being accidentally overwritten by a subsequent analysis.

The file types that you will encounter most frequently as a new user are the database file, the log file, the lock file, the error file, the output file, and the results file. You can learn about the other ANSYS file types and file management in ANSYS in Chapter 19 of the Mechanical APDL Basic Analysis Guide and in Section 2.4 of the Mechanical APDL Operations Guide.

2.5.1. The Database File

The ANSYS database exists in the computer's memory. The database file (file.db) is a copy of some or all of the information in the ANSYS database. You can save a copy of the database at any time except when the model is actively solving. By default, the **SAVE** command saves all model data, solution data, and postprocessing data. When you quit ANSYS in interactive mode, a pop up dialog box will appear asking if you want to save the database. In this dialog box, the default is to save only the model data ("Save Geom + Loads"). In general, we recommend that you save before quitting and "Save Everything" on exit.

2.5.2. The Log File

All commands issued during an interactive session are written to the log file (file.log). The log file is created (for a new analysis) or opened (for an analysis with an existing jobname in the working directory) when you start the program. It is closed when you exit the program. At the beginning of each session, ANSYS adds a time stamp to the log file and continues writing new commands to the old log file. This ensures a complete record of your work starting from the last time the log file was deleted. Log files are never overwritten or deleted by ANSYS. You cannot delete the log file while the program is open.

The log file is written in ASCII format. You can open it from the working directory using any text editor. You can also list the session log file by following the GUI path: **Utility Menu > File > List > Log File. . ..**

The commands issued during the current session are also stored in the ANSYS database (i.e., in memory). This information can be written to the database command log file (file.lgw) using the GUI path: **Utility Menu > File > Write DB log file** . . . or by issuing the **LGWRITE** command. In addition to allowing you to choose the name and location of the new database log file, the Write Database Log dialog box has a drop down menu at the bottom of the window with two options: "Write essential commands only" and "Write non-essential cmds as comments." The first option removes all non-essential ANSYS commands that were issued during the session (listing and plotting commands, commands issued through the Pan Zoom Rotate menu, etc.). The second option leaves those commands in the file, but comments them out by placing an exclamation point (!) in front of them.

2.5.3. The Lock File

The lock file (file.lock) is a temporary file that prevents ANSYS from running more than one job with the same jobname from the same directory. It is created when the program is opened and deleted when the program is closed. If the program crashes or if you quit improperly, the lock file will not be deleted as expected. When you start the program with a lock file in the working directory, you will receive a message in the Output Window asking if you want to delete the lock file. If the lock file is associated with an analysis that is currently in progress and you want to start a new analysis, say 'no', change the working directory and/or the jobname, and launch a new version of ANSYS for the this analysis. Otherwise, you can say 'yes' to delete the file and proceed as usual.

2.5.4. The Error File

The error file (file.err) records all warning and error messages that are issued by ANSYS. The error file is created (for a new analysis) or opened (for an analysis with an existing jobname in the working directory) when you start the program. It is closed when you exit the program. The error file is written in ASCII format and can be read by any text editor. You should always review the error file following a batch run. It can also be helpful to review the error file if you close a dialog box prematurely, when debugging a model, or in the event of a program crash.

2.5.5. The Output File

The output file (file.out) records the text output from ANSYS in batch mode. You can direct the text from the Output Window to the output file in interactive mode using the GUI path: **Utility Menu > File > Switch Output to > File** . . . or by using the **/OUTPUT** command. You can also print results information to the output file in batch mode using the **OUTPR** command. The output file is written in ASCII format and can be read by any text editor.

2.5.6. The Results File

The results file records the results of an analysis. The extension of the results file reflects the degrees of freedom in the analysis: file.rst is created for structural and coupled field analyses, file.rth is created for thermal and diffusion analyses, and file.rmg is created for magnetic field analyses. The different file extensions allow ANSYS to use the results from one analysis as the loads for another in coupled-field analyses.

All of the data required to postprocess a model (including the finite element geometry and the element attribute data) is saved to the results file. This means that a results file can be postprocessed without a copy of the database file. However, the solid model geometry and solid model loads are not stored on the results file. Thus, all solid model information will be lost unless you save a copy of the database or have an input file to recreate it.

You can read any results file and load the selected information into the database by following the GUI path: **Main Menu > General Postproc > Data & File Opts**. You can also access the results file for the current jobname via the GUI path: **Main Menu > General Postproc > Read Results**.

2.6. Saving Files and Results in ANSYS

There are four ways to save your work in ANSYS: (1) reading and writing database files, (2) reading and writing archive files, (3) saving and rerunning the log file, and (4) creating input and batch files.

2.6.1. Saving Database Files

You can write the information in the database to a binary database file using the GUI path: **Utility Menu > File > Save as Jobname.db** (Figure 2.11 left) or by clicking the SAVE_DB button in the ANSYS Toolbar. If you use the "save as" option (**Utility Menu > File > Save as**. . .), you may specify a different file name in place of the default jobname. You can also save some or all of the database using the **SAVE** command. The database file can be read back into the program using the GUI paths: **Utility Menu > File > Resume from**. . . or **Utility Menu > File > Resume Jobname.db**. . .. You can also resume a database using the **RESUME** command.

Database files are very convenient to use. However, they can be very large and difficult to move between computers, and they can become corrupted over time. In addition, binary files, including database files, are not backward compatible and cannot be read using versions of ANSYS that are older than the one used to create them. Database files also have limited forward compatibility. They are only guaranteed to be readable by the next release of ANSYS. We recommend saving a copy of the database in addition to (but not as a replacement for) archive, log, input, or batch files.

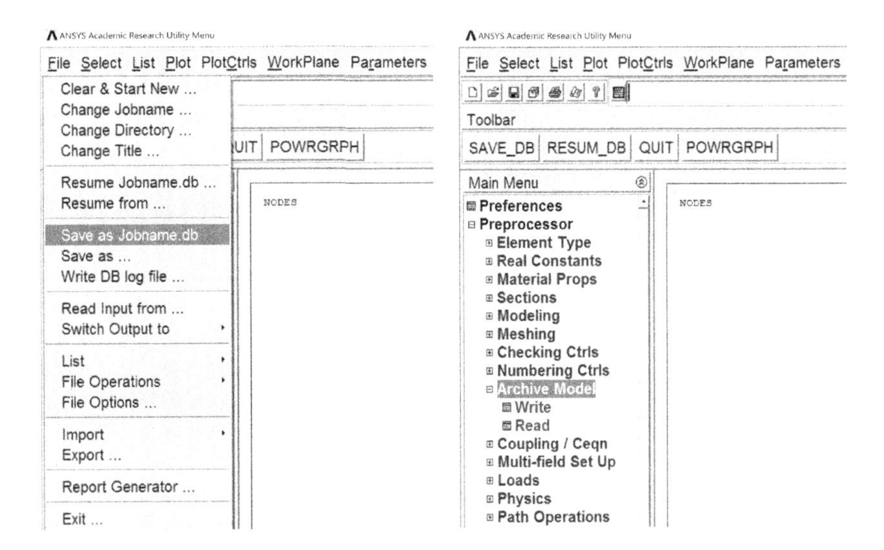

Figure 2.11 Saving (left) and Archiving (right) Database Files in the GUI.

2.6.2. Archiving Models

You can archive a model's geometry, material properties, loads, and components in a plain text file using the GUI path: **Main Menu > Preprocessor > Archive Model > Write** (Figure 2.11 right) or by issuing the **CDWRITE** command. Since archive files are forward compatible, this is the traditional method for transferring an analysis from one computer, operating system, or program release version to another. Archive files can be read by using the GUI path: **Main Menu > Preprocessor > Archive Model > Read** or the **CDREAD** command.

2.6.3. Rerunning Log Files

The log file that is generated during each ANSYS session can be used as an input file to recreate a database. This method results in a relatively compact file that can be emailed and modified with any text editor. However, using an unedited log file as an input file will re-run all previously issued commands—not just the useful ones.

2.6.4. Creating Input and Batch Files

Finally, input and batch files can be written from scratch or can be created by editing a log file. This is the best way to save and share your models. Input and batch files are small, easily emailed, and rarely corrupted. They can always be used with newer versions of the program. In addition, input and batch files show how a given model was created and outline all of the assumptions that went into the model. (Database files can only give another user the current state of the model.) As a result, it is also easier to debug a model that was created using an input or batch file than one created using the GUI. For more information about writing, debugging, documenting, and running input and batch files, see chapter 10.

2.7. Where is the Undo Button?

It is common for new ANSYS users to express frustration at their inability to find the "undo" button. And this makes perfect sense because there is no "undo" button.

When ANSYS commands are executed, the database is changed in ways that sometimes cannot be undone. For example, the **AADD** (area add) command creates a new single area from existing areas. It creates new keypoints and lines if they are needed. It also deletes the original areas and any unneeded keypoints and lines unless you specify otherwise. Deleting these features removes them from the database. Once this process is complete, the original entities—and the information required to re-create them—are lost. The change cannot be undone in the traditional sense. For this reason, having an "undo" button is impractical.

Since nothing is ever deleted from the log file, there is a record of all the work that you did up to (and including) the command that you want to undo. You can take advantage of this record by using the **UNDO** command. Typing `UNDO,NEW` in the command prompt brings up a text editor with all commands issued since the last **SAVE** to the database file. You may delete the unwanted commands and save the remaining desirable input. When you click "OK," the saved database is resumed and the remaining desirable input is read into the program. This brings you back to the program state before the unwanted commands were issued.

Some commands can be reversed by using opposing commands. For example, "delete" will undo "create" commands and "clear mesh" will undo "mesh" commands. But in the early stages of modeling, you may find it easier to clear the database and start again.

Once you begin to work with input files, the lack of an undo button ceases to be an issue. You will learn to write, edit, and use input files in chapter 10. Until then, we recommend saving your database early, often, and especially before issuing 'risky' commands.

2.8. How Do You Specify Units?

It is also common for new ANSYS users to ask how to specify the units for their analyses. ANSYS does not have an internal or required system of units. Geometry, material properties, loads, and boundary conditions are stored in the program as unit-less scalar quantities. It is your responsibility to supply all numerical values to the program in a consistent system of units. You may choose any system of units. However, we recommend using SI units to minimize the chances of errors in unit conversion. In the ANSYS documentation, the SI system is referred to as the MKS system (Meters, Kilograms, Seconds).

There are a few exceptions to this recommendation. First, the default value for the Stefan-Boltzman constant is in US customary units. You must supply a value for the Stefan-Boltzman constant if you wish to perform analyses with radiation using a different system of units. In addition, some systems of units will cause some calculations in the program to exceed the maximum or minimum allowable value. The best example of this is with MEMS and micro scale simulation. To counteract this problem, ANSYS has developed the microMKSV system (microMeters, Kilograms, Seconds, Volts) and the microMSVfA system (microMeters, grams, Seconds, Volts, fAmps). These conversions scale all of the values back into the acceptable range. Finally, for magnetic field problems, the free-space permittivity will need to be set for your chosen system of units.

There is a **/UNITS** command. It can be used to remind yourself which systems of units you have chosen, but it will not impact the values that you enter into the program. For more information on units, see Section 1.2 of the Mechanical APDL Coupled-Field Analysis Guide and the **EMUNIT** entry in the Command Dictionary at the end of the Mechanical APDL Commands Reference.

2.9. Where to Find Help: The ANSYS Documentation

ANSYS, Inc. offers an extensive set of documentation for its software. It is installed in HTML format with hypertext links when you install the software. You can access it from the Windows Start Menu (or equivalent), by clicking the Product Help button at the bottom of the ANSYS Product Launcher, or by using the Help drop down menu in the Mechanical APDL GUI Utility Menu. You can download PDF copies of the documentation from the ANSYS Customer Portal (see section 2.10 for more information). In addition, the documentation for commands and elements can be accessed using the **HELP** command.

No matter how averse you might be to the idea of using the documentation, how well you know the program, or how long you have been using the program, there is no way to avoid using the ANSYS documentation. There are too many features, options, assumptions, and commands in the program. By the end of this book, you should be familiar with the documentation and be comfortable using it to find information related to your analysis.

The ANSYS Mechanical APDL documentation is a collection of 31 books* divided into roughly five categories: reference manuals, programmer's guides, examples manuals, analysis guides, and training manuals. It is indexed and searchable, giving you access to thousands of pages of information with the added convenience of turning to a page in a different manual at the push of a button.

The Help Viewer (Figure 2.12) is divided into two sections. The left side has a table of contents that allows you to navigate through the documentation and choose a page to view. The right side of the window acts like a web browser, displaying the chosen information on a series of tabs.

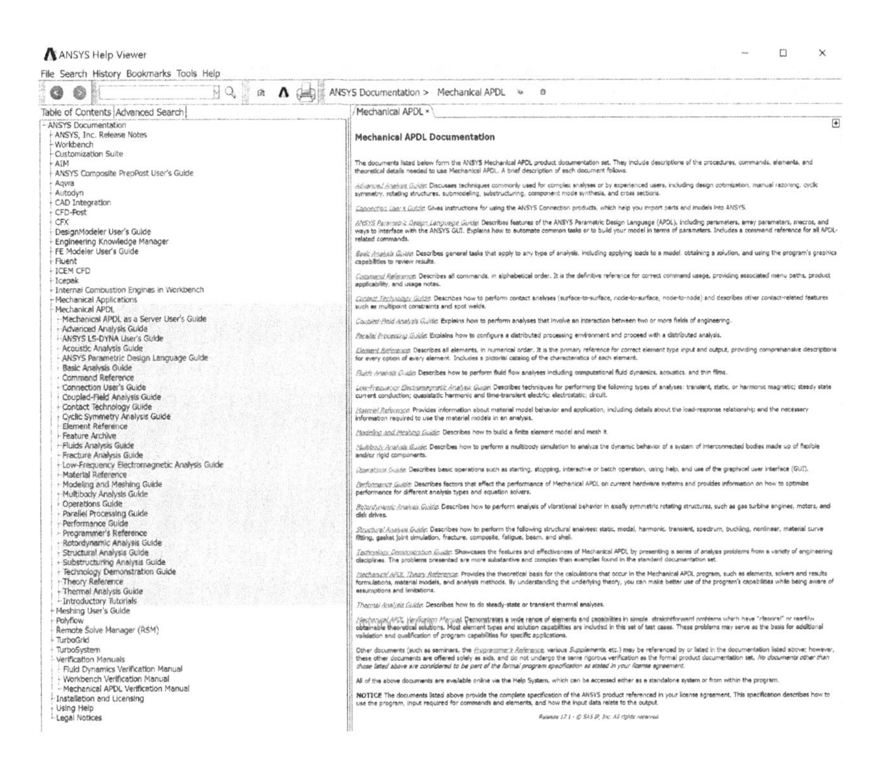

Figure 2.12 The ANSYS Online Documentation Window.

2.9.1. Reference Manuals

The ANSYS Mechanical APDL documentation includes four reference manuals: the Mechanical APDL Command Reference, the Mechanical APDL Element Reference, the Mechanical APDL Material Reference, and the Mechanical APDL Theory Reference. The Command Reference contains information on ANSYS commands usage, syntax, and command groupings. Most of the book is a dictionary of ANSYS commands listed alphabetically with full syntax, menu paths, and usage notes. The Element Reference contains information on choosing, defining, and using elements in an analysis. The second half of the book is an element library with specific information on each element in ANSYS including a description, picture, input information, output information, assumptions, restrictions, and notes. The Material Reference provides information about modeling various material behaviors and using advanced material models in an analysis. The Theory Reference documents the implementation of finite element theory in ANSYS and outlines the theoretical basis for calculations in the program. This includes element formulations, results formulations, and details about the solvers, material models, and analysis methods.

The Command, Element, and Material References are very helpful. You may find yourself using them often. The Theory Reference is used primarily by expert analysts who need to understand the assumptions made in the program in order to push the limits of the software.

2.9.2. Programmers Manuals

ANSYS allows advanced users to customize the program by adding specialized elements, material models, and commands. The programmer's manuals provide information about the programming interfaces and languages available. The two programmer's manuals are the ANSYS Parametric Design Language Guide and the Mechanical APDL Programmer's Reference. The APDL Guide includes information on defining and using parameters, and using APDL to write macro commands. It also has an APDL command reference with information on all APDL related commands. The Programmer's Reference is divided into two parts: The Guide to Interfacing with ANSYS and the Guide to ANSYS User-Programmable Features. It provides information about directly interacting with the ANSYS database at the Fortran level and modifying or extending the program's capabilities.

2.9.3. Examples Manuals

The ANSYS examples manuals are a good resource for additional practice and example problems. There are three examples manuals: the ANSYS Mechanical APDL Introductory Tutorials, the Mechanical APDL Verification Manual, and the Mechanical APDL Technology Demonstration Guide. The Introductory Tutorials and the Technology Demonstration Guide are grouped with the rest of the Mechanical APDL Documentation. The Mechanical APDL Verification Manual is grouped with the Fluid Dynamics and Workbench Verification Manuals. The Verification Manuals can be found at the bottom of the left column in the ANSYS online documentation window.

The Tutorials manual contains a set of examples that introduce many of the physics capabilities within ANSYS. The tutorials have step-by-step instructions with screen shots of both the Graphics Window and the dialog boxes used during the tutorial. The tutorials require between one half hour and two hours to complete. The estimated time to completion is listed at the beginning of each tutorial. This book does not use any of the ANSYS tutorials for example problems to maximize the number of resources available to you.

The Verification Manual is a collection of 293* problem descriptions and input files (and counting). ANSYS is being improved constantly. As changes are made to the software, the program is run against a large set of problems to ensure that all features of the program work correctly after the changes are implemented. The problems in the verification manual are a small subset of those verification problems and are provided to demonstrate the program's capabilities.

The Verification Manual can be a powerful tool for beginning users. We encourage you to read through these problems and use them as guides for similar analyses. However, the problems in the verification manual are not as user-friendly as those in the ANSYS Tutorials. There are fewer comments in the verification input files and you may be required to spend some time with the Command Reference to understand and reproduce the problem using the GUI. Also, almost all verification problems use direct generation of nodes and elements instead of building and meshing solid model geometry. (See chapter 3 for more information on direct generation.)

The Technology Demonstration Guide showcases a variety of sophisticated problems that have been solved with ANSYS. The input files for these problems are available for download from the ANSYS Customer Portal.

2.9.4. Analysis Guides

The analysis guides contain information about using ANSYS. The Mechanical APDL Operations Guide, the Mechanical APDL Modeling and Meshing Guide, and the Mechanical APDL Basic Analysis Guide are physics independent. The information in them is valid for all types of analyses. Reading all three guides cover-to-cover may be beneficial (although time consuming) for the new user. The other guides focus on a single physics environment or a specific type of application and should be read or used as necessary.

Each analysis guide can function either as a part of the ANSYS documentation set or as a separate entity. For this reason, some information will be repeated in each of the physics related analysis guides and some information may be distributed across several guides instead of being concentrated in one place. This is one of the reasons why HTML-based documentation is so important for ANSYS.

2.9.5. The Feature Archive

As noted in chapter 1, ANSYS is fully backward compatible. Features are rarely removed from the code. Nevertheless, the program continues to improve and expand. Old commands, features, and elements are routinely replaced with newer, more efficient, and more accurate versions. The old versions are classified as legacy features or undocumented features. Access to these features in the GUI is usually removed and active development ceases. Legacy features continue to receive limited support. Documentation for legacy features can be found in the Mechanical APDL Feature Archive. Undocumented features receive no support and all information related to them is removed from the documentation.

There are good reasons to use legacy and undocumented features. But unless you know what they are, you should limit yourself to the "current technology" features in the program.

2.9.6. Additional Documentation*

The ANSYS Mechanical APDL documentation also includes the Mechanical APDL as a Server User's Guide, the Mechanical APDL Connection User's Guide, the Mechanical APDL Parallel Processing Guide, and the Mechanical APDL Performance Guide. The Connection User's Guide provides information about importing solid model geometry from other software programs like CATIA® and SolidWorks® directly into ANSYS. The Server User's Guide, the Performance Guide, and the Parallel Processing Guide address issues related to ANSYS hardware.

2.10. Where to Get Extra Help: ANSYS Technical Support*

Another source of help is the ANSYS Customer Support service. Commercial users usually purchase a full support contract. This gives them access to support engineers and other resources by telephone and online via the ANSYS Customer Portal (http://support.ansys.com).

University users only have access to the Customer Portal using their customer number. If you are using ANSYS on a university license, you can determine your customer number by issuing the command `/STATUS,TITLE` in the GUI command prompt.

You should only use technical support as a last resort. You should first try to solve your problem by reading the documentation, debugging your code, or asking friends and coworkers for help. Questions sent to technical support should be well defined and include a description of the system to be modeled, assumptions made in the modeling, a well commented input file, and a summary of what you have done to solve the problem on your own.

Exercise 2-1

Static Axial Loading of a Notched Plate in Tension

Overview

In this exercise, you will perform a steady-state structural analysis of a notched rectangular plate that is loaded axially in tension. The plate is made of 6061-T6 aluminum with a Young's modulus of 73.1 GPa and a Poisson's ratio of 0.33. It is 15 cm wide and 10 cm tall with 1 cm diameter notches on each side. The notches are centered horizontally on the plate midplane and centered vertically on the upper and lower plate edges (Figure 2-1-1). The plate is thin so this can be treated as a two-dimensional (2D) problem. The geometry will be created using top down solid modeling techniques including Boolean operations.

A load of 1 MPa will be placed on the right edge of the plate. The left edge of the plate will be constrained in x. This constraint will create a reaction force that will act as the second 1 MPa load on the left. The plate will deform under the applied load, growing longer and thinner. This will force the notches to become ovals and a stress concentration will appear near the center of each notch.

The ratio of the height of the plate (10 cm) to the diameter of the notches (1 cm) is 10 to 1. Thus, the expected elastic stress concentration $K_t = \sigma_{max}/\sigma_{nom}$ for this problem is 2.735 and the expected gross stress concentration $K_g = \sigma_{max}/\sigma_{gross}$ for this problem is 3.03.

Figure 2-1-1 Schematic of the Notched Plate with Applied Loads.

ANSYS Mechanical APDL for Finite Element Analysis.
DOI: http://dx.doi.org/10.1016/B978-0-12-812981-4.00014-9

Model Attributes

Material Properties for 6061-T6 Aluminum

- Young's modulus—7.310e10 Pa
- Poisson's ratio—0.33

Loads

- 1 MPa tensile load on the right edge of the plate

Constraints

- No displacement of the left edge of the plate in the horizontal (x) direction
- No displacement of the lower left corner of the plate in the vertical (y) direction

File Management

This analysis begins with the establishment of a good file management system. You will create a directory called "Intro-to-ANSYS" on your computer's desktop. You will then create one subdirectory for each exercise where ANSYS can save all its files. You will also specify a different jobname for each exercise in this book. These steps greatly reduce the possibility of accidentally overwriting important files and help to organize your work.

Create a New Folder on Your Desktop Named "Intro-to-ANSYS"

- Right click on the desktop and follow the path: New > Folder
- Change the name of the new folder from "New Folder" to "Intro-to-ANSYS"

Create a New Folder in the "Intro-to-ANSYS" Folder Named "Exercise2-1"

- Right click in the "Intro-to-ANSYS" folder and follow the path: New > Folder
- Change the name of the new folder from "New Folder" to "Exercise2-1"

Open a New Session of ANSYS Using the Mechanical APDL Product Launcher*

- Windows Start Menu > All Programs > ANSYS 17.2 > Mechanical APDL Product Launcher

Or

- Windows Start Menu > All Apps > ANSYS 17.2 > Mechanical APDL Product Launcher

Ensure That the Simulation Environment Is Set to "ANSYS" (Figure 2-1-2)

Change the Working Directory to the New "Exercise2-1" Folder

- Ensure that the "File Management" tab is selected in the Product Launcher
- Click the "Browse" button to the right of the Working Directory path
- Choose the "Exercise2-1" folder
- Click OK

This sets the "Exercise2-1" folder as your ANSYS working directory for all future sessions. All output files, including the log file and the results files, will be saved in this directory.

Change the Jobname to "Exercise2-1"

Click Run to Start ANSYS

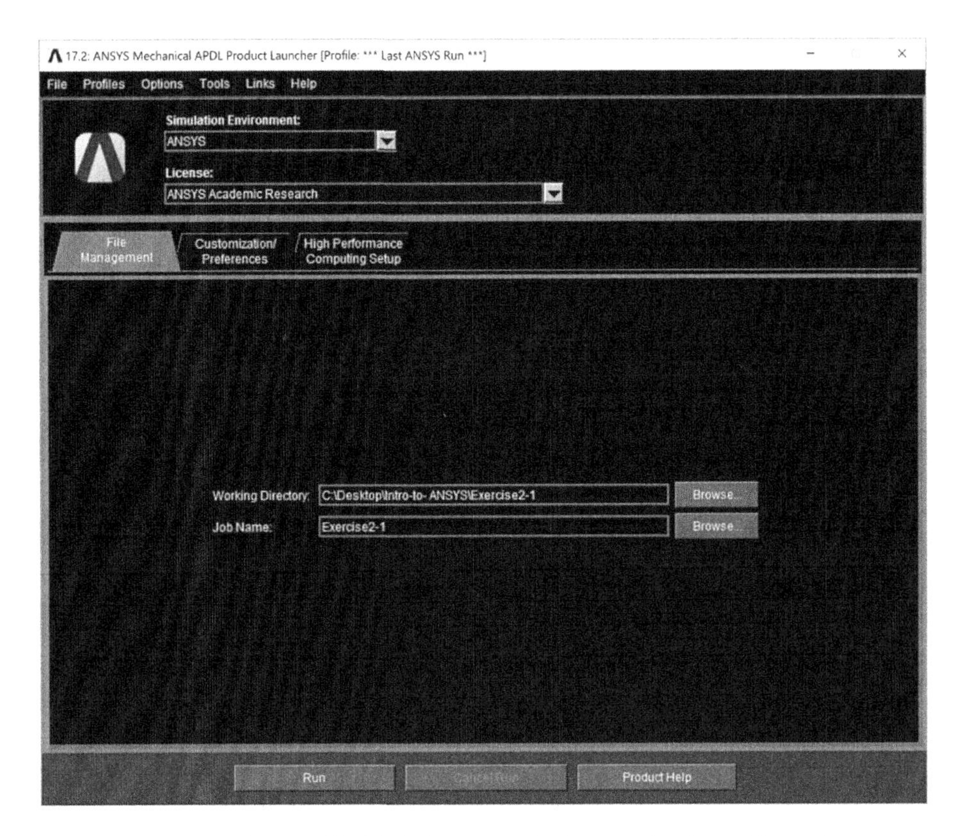

Figure 2-1-2 Starting ANSYS in Interactive Mode using the Product Launcher.

Before You Begin

If you make a mistake that you do not know how to correct, you can start over by following the path:

- Utility Menu > File > Clear and Start New... (Figure 2-1-3)
- Click OK to clear the database and start a new analysis
- Click Yes to close the pop up dialog box

You can also press the return key twice to accept the default values and close the dialog boxes.

If you have saved your work at least once, you can restart the analysis from the save point by following the path:

- Utility Menu > File > Resume from...
- Choose "Exercise2-1.db"

If you have saved more than once, you will also see a file named Exercise2-1.dbb. This is the database backup file for this exercise. When you issue a ***SAVE*** *command, the old database file extension is changed to .dbb and the new database is saved as a new .db file. Only one backup file is kept. Older files are overwritten.*

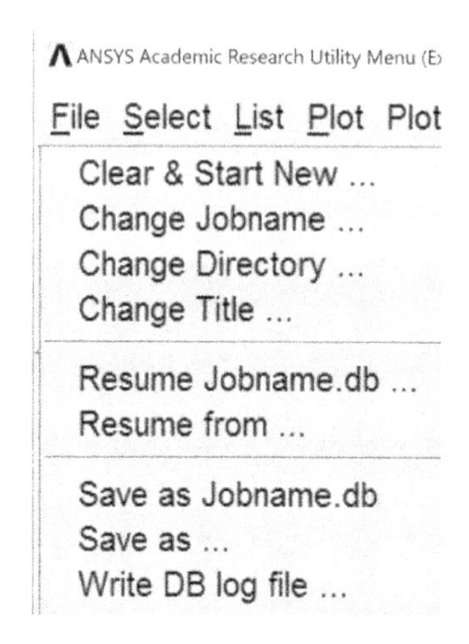

Figure 2-1-3 Saving and Resuming from the Utility Menu.

Step 1: Define Geometry

1-1. Create a rectangle to represent the plate

- In the Main Menu, follow the path:
- Preprocessor > Modeling > Create > Areas > Rectangle > By 2 Corners
- Enter 0 for WP X (Figure 2-1-4, left)
- Enter 0 for WP Y

WP is an abbreviation for working plane. The working plane will be discussed in chapter 3.

- Enter 0.15 for Width
- Enter 0.10 for Height
- Click OK

The lower left corner of the rectangle is specified by its working plane coordinates (0,0). ANSYS calculates the coordinates for the upper right corner of the rectangle using the width and height provided. Since the MKS system (Meters, Kilograms, Seconds) is used for all SI models in this book, the geometry must be defined in meters instead of centimeters.

1-2. Create a circle to represent the bottom notch

- Preprocessor > Modeling > Create > Areas > Circle > Solid Circle
- Enter 0.15/2 for WP X (Figure 2-1-4, center)

If you enter a simple mathematical expression into a dialog box, ANSYS will calculate and use the resulting value.

- Enter 0 for WP Y
- Enter 0.005 for Radius

ANSYS does not require the first zero before a decimal point or trailing zeros after the last significant figure to be entered. In this book, leading and trailing zeros are included for clarity.

- Click Apply

Clicking "Apply" instead of "OK" performs the operation but keeps the dialog box open

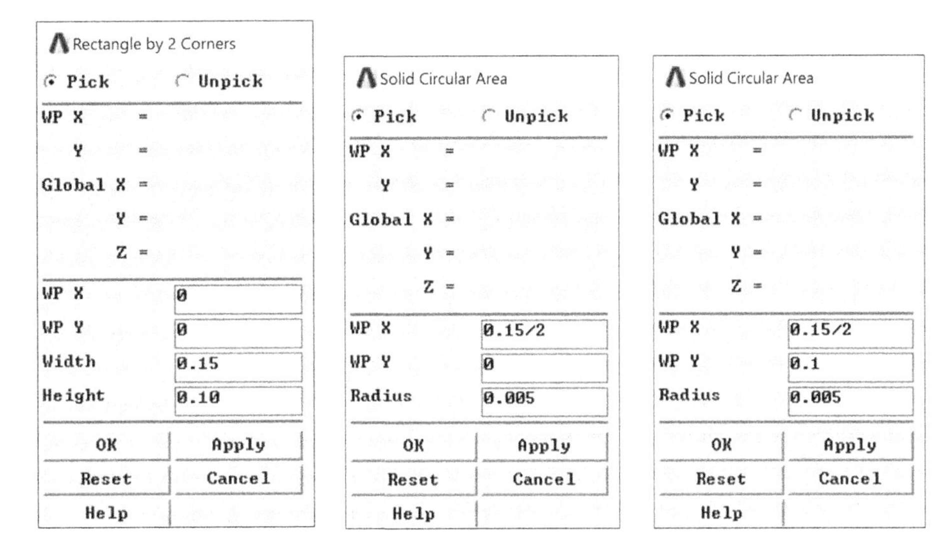

Figure 2-1-4 Defining Solid Model Entities: Rectangle (left), Bottom Circle (center), Upper Circle (right).

1-3. Create a circle to represent the upper notch

- Enter 0.15/2 for WP X (Figure 2-1-4, right)
- Enter 0.1 for WP Y
- Enter 0.005 for Radius
- Click OK

1-4. Save your progress

- ANSYS Toolbar > SAVE_DB

ANSYS will issue a Note in the Output Window after each ***SAVE*** *command (Figure 2-1-5).*

```
ALL CURRENT ANSYS DATA WRITTEN TO FILE NAME= Exercise2-1.db
 FOR POSSIBLE RESUME FROM THIS POINT
```

Figure 2-1-5 Output Window Save Confirmation.

1-5. Subtract Areas 2 and 3 (the circles) from Area 1 (the rectangle)

- Preprocessor > Modeling > Operate > Booleans > Subtract > Areas

Use the mouse for graphical picking:

- Click on Area 1 (the plate) with the mouse near the centroid

The area will change color when you click on it.

- Click OK
- Click on Area 2 (the lower circle) with the mouse near the centroid
- Click on Area 3 (the upper circle) with the mouse near the centroid
- Click OK

Or, you may choose the areas directly by entering their entity numbers:

- Enter 1 in the text box
- Click OK
- Enter 2,3 in the text box
- Click OK

ANSYS assigns numbers to solid model entities (points, lines, areas, and volumes) in the order in which they were created using the lowest entity number available. Since the plate was created first, it is Area 1. The two circles are Areas 2 and 3. The syntax of the subtract command is ***ASBA****, NA1, NA2, SEPO, KEEP1, KEEP2. NA1 is the base area for subtraction. NA2 is the area to be subtracted. When using the GUI, the first window selects the base area from which to subtract. Clicking OK resets the graphical picking operation and allows you to select the areas to be subtracted.*

The completed solid model is shown in Figure 2-1-6.

Figure 2-1-6 Completed Solid Model.

1-6. Save the solid model geometry

- ANSYS Toolbar > SAVE_DB

Step 2: Define Element Types

2-1. Define the elements type to use for this model

- Preprocessor > Element Type > Add/Edit/Delete
- Click the "Add" button to bring up the Library of Element Types
- In the left-hand column under "Structural Mass," choose "Solid" (Figure 2-1-7)
- In the right-hand column, choose "8 Node 183"
- Click OK
- Click the "Close" button in the Add/Edit/Delete dialog box

These steps define element type #1 as a PLANE183 element. This is a 2D structural solid that can represent plane stress and plane strain. It has eight nodes with two degrees of freedom (displacement in x and y) at each node. It is well suited for irregular meshes. Only one element type is needed for this analysis.

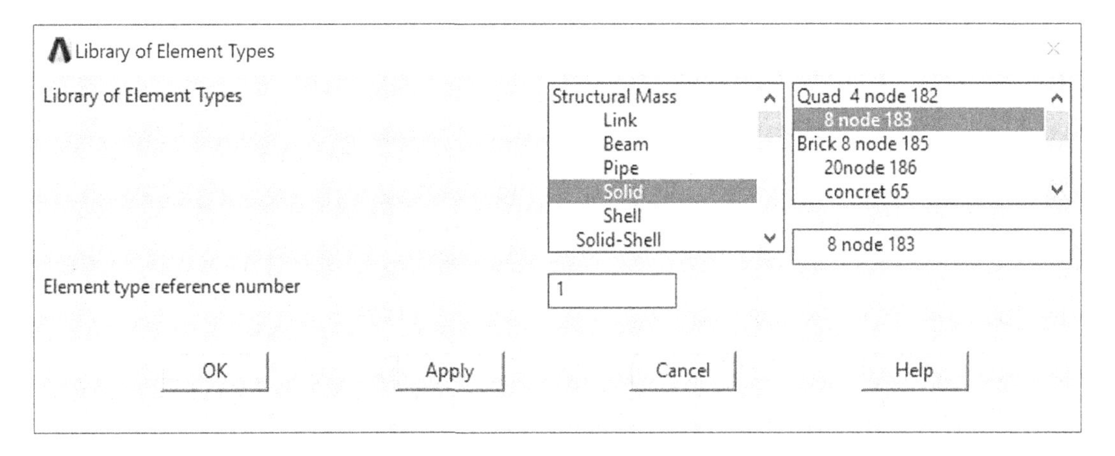

Figure 2-1-7 Defining the Element Type.

Step 3: Define Material Properties

3-1. Create a linear elastic material model for 6061-T6 aluminum

- Preprocessor > Material Props > Material Models
- In the right column, follow the path: Structural > Linear > Elastic > Isotropic (Figure 2-1-8)

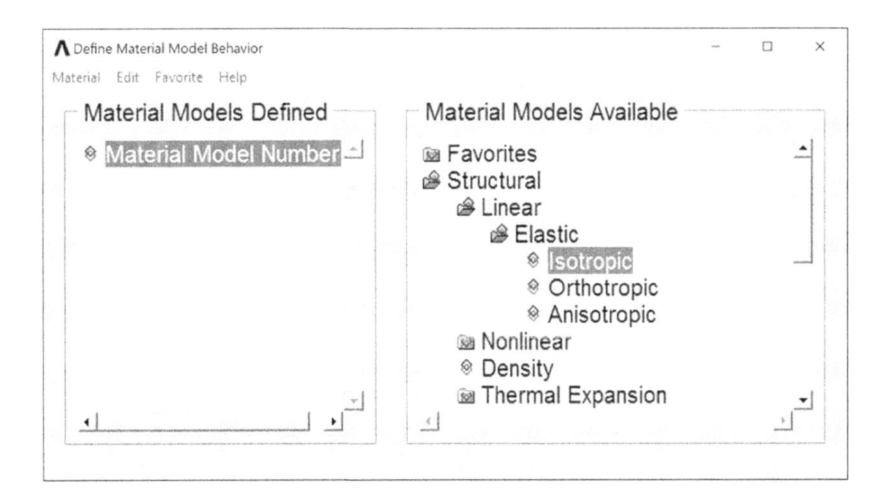

Figure 2-1-8 Choosing a Material Model.

- Click in the EX box and type 7.310e10 (Figure 2-1-9)
- Click in the PRXY box and type 0.33
- Click OK
- Close the Material Models dialog box by clicking the "X" button in the top right corner

EX stands for Young's modulus (E) in the x direction. PRXY stands for the major Poisson's ratio in the xy plane. Since aluminum is an isotropic material (i.e., material properties are independent of orientation), only one value for each material property is required. See chapter 5 *for more details.*

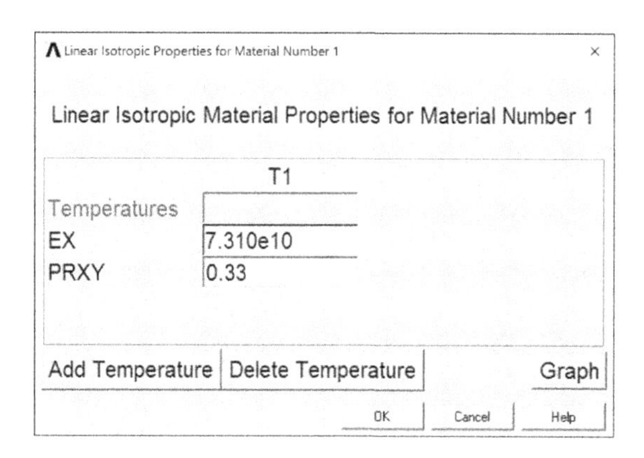

Figure 2-1-9 Specifying Material Properties.

3-2. Save your progress

- ANSYS Toolbar > SAVE_DB

Step 4: Mesh

4-1. Create the mesh for the finite element model

- Preprocessor > Meshing > Mesh Tool
- Check the "Smart Size" box in the Mesh Tool dialog box (Figure 2-1-10, left)
- Adjust the Smart Size Slider to 1
- Click the "Mesh" button
- Click the "Pick All" button in the Mesh Areas dialog box (Figure 2-1-10, center)

The initial mesh is shown in Figure 2-1-10 (top right).

If you want to experiment with Smart Sizing, you can clear the mesh using the following steps:

- Preprocessor > Meshing > Mesh Tool
- Click the "Clear" button
- Click the "Pick All" button in the Clear Areas dialog box

*The Mesh Tool automatically issues an **EPLOT** (element plot) command so you can see the elements that were created during meshing. ANSYS will continue to display the element plot until a new plot command is issued. After you clear the mesh, there will be no elements to plot, so you will see a blank screen with the word "Elements" in the upper left corner. To redisplay the area plot, you can follow the GUI path: **Utility Menu > Plot > Areas** or type* `APLOT` *in the command prompt.*

After you are done experimenting, you can re-mesh the model using the procedure described in step 4-1.

4-2. Refine the mesh for the finite element model

- Bring the Mesh Tool back to the foreground or reopen it if closed
- Confirm that the "Refine at" option in the Mesh Tool dialog box is set to "Elements" (Figure 2-1-10, left)
- Click the "Refine" button in the Mesh Tool dialog box
- Click the "Pick All" button
- Choose a "LEVEL Level of refinement" of 3 (Figure 2-1-11)
- Click OK

The final mesh is shown in Figure 2-1-10 *(bottom right).*

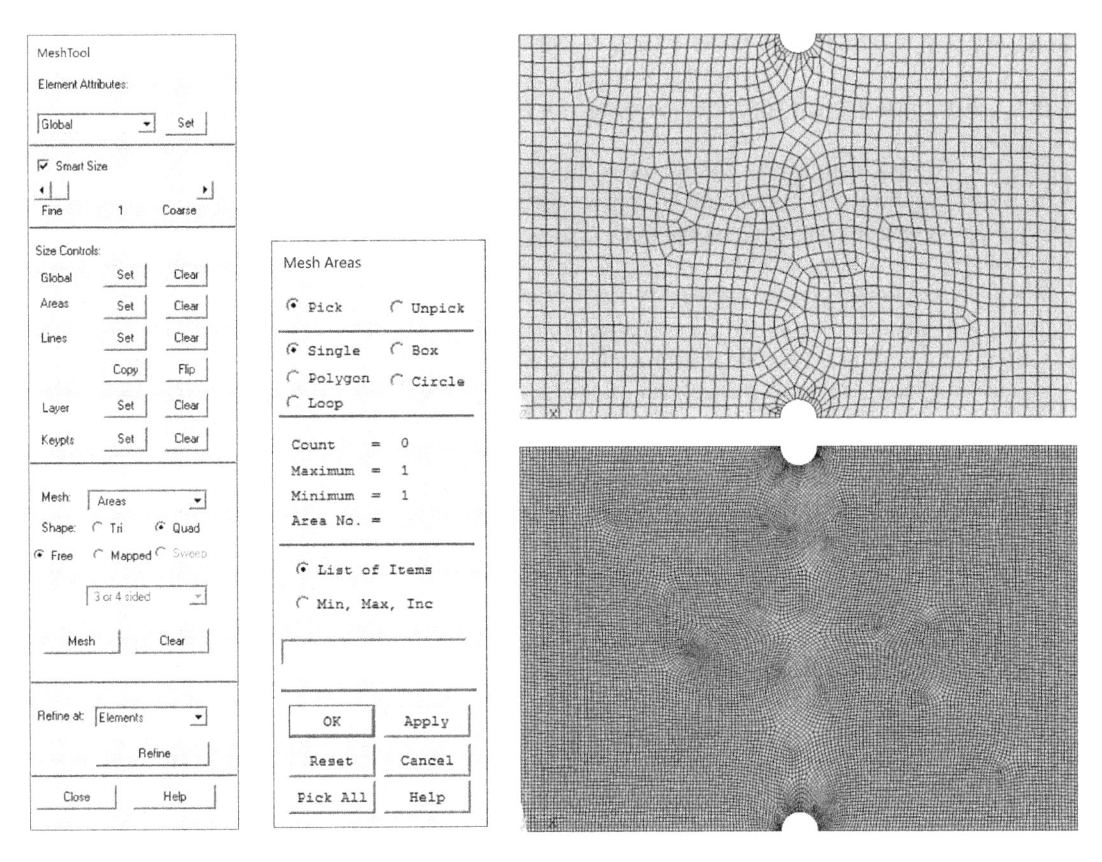

Figure 2-1-10 Mesh Tool (left), Mesh Areas Dialog Box (center), and Resulting Meshes (right).

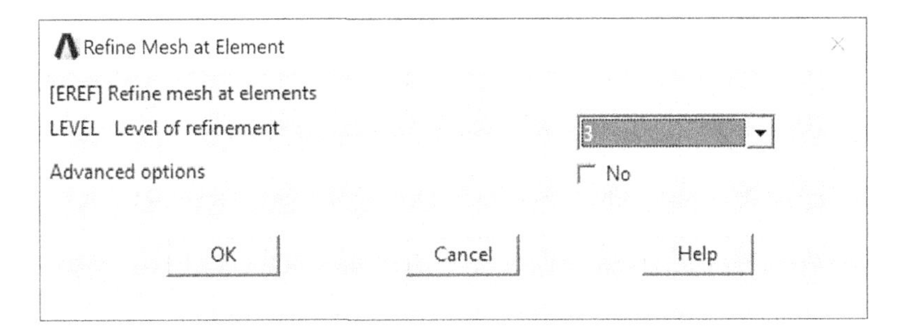

Figure 2-1-11 Refine Mesh Dialog Box.

Because you used Smart Sizing to create the mesh, it is possible that your mesh will be slightly different from the one in this book. Since the results are calculated using the finite element mesh, your results or the color contours used to display your results may also be slightly different from the ones in this book. As long as the differences are well below 1%, there is no cause for concern.

4-3. Save your finite element mesh

- ANSYS Toolbar > SAVE_DB

Step 5: Apply Constraint Boundary Conditions

5-1. Apply degree of freedom constraints to the model

- In the Main Menu, follow the path:
- Solution > Define Loads > Apply > Structural > Displacement > On Lines
- Select the left edge on the plate (the vertical line at $x=0$)

The line should now be highlighted in the Graphics Window. The "Count" field in the picking dialog box (also known as the Picker) should have incremented to "1" and the "Line No." field should also contain a number (Figure 2-1-12, left). (This is Line 4 at revision 17.2.) This allows you to verify that you have successfully chosen a line and to confirm that it was the correct line.

- Click OK
- In the "Lab2 DOFs to be constrained" field select UX (Figure 2-1-12, right)
- For the "VALUE Displacement value" enter 0
- Click OK

Three blue triangles pointing to the right should appear along the left edge of the plate. These indicate the presence and direction of a structural constraint and confirm that your boundary condition was applied successfully.

If you issue a new plot command (or if the program issues one for you), these symbols will disappear. You can reconfirm that the displacement has been successfully applied by following the path: ***Utility Menu > List > Loads > DOF Constraints > On All Lines*** *or by typing* `DLLIST, ALL` *in the command prompt.*

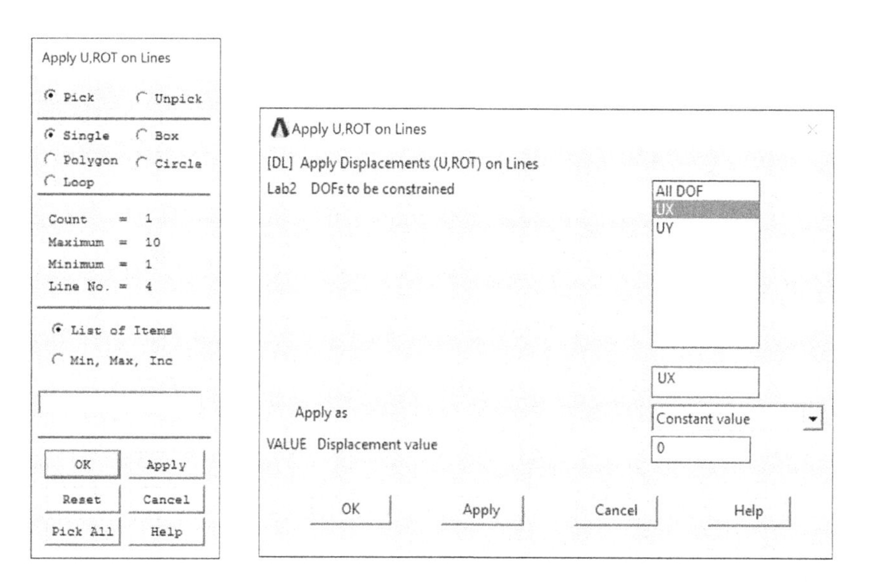

Figure 2-1-12 Applying Line Constraints in X.

Objects in the real world have six degrees of freedom. They can move in each of the coordinate directions (x, y, and z) and they can rotate about each of the coordinate axes. 3D structural finite element models also have six physical degrees of freedom, each of which must be constrained at one or more locations before the equation solver can generate a solution. This problem is assumed to be 2D and thus only has three physical degrees of freedom: translation in the x and y directions and rotation about the z axis.

The boundary condition that you just applied prevents the left edge of the plate from moving in the horizontal direction and thus constrains the model in x. By applying the constraint to a line instead of a point, you defined multiple horizontal constraints (one at each node on the line). This prevents rotation about the z axis and thus constrains the second degree of freedom. The left edge is still allowed to expand and contract in y as the plate deforms. This prevents the model from being overconstrained. Overconstraining this model would distort the stress field and produce incorrect results. See chapter 8 *for more details.*

5-2. Apply additional degree of freedom constraints to the model

- Solution > Define Loads > Apply > Structural > Displacement > On Keypoints
- Pick the keypoint at the bottom left corner of the model
- Click OK (Figure 2-1-13, left)
- In the "Lab2 DOFs to be constrained" field choose UY (Figure 2-1-13, right)
- For the "VALUE Displacement value" enter 0
- Click OK

A blue triangle pointing up should appear under the constrained keypoint to indicate that the structural boundary condition was successfully applied. This boundary condition prevents that keypoint (and that corner of the plate) from moving in the vertical direction without restricting any other type of movement or deformation in the plate. This provides the third constraint for the model.

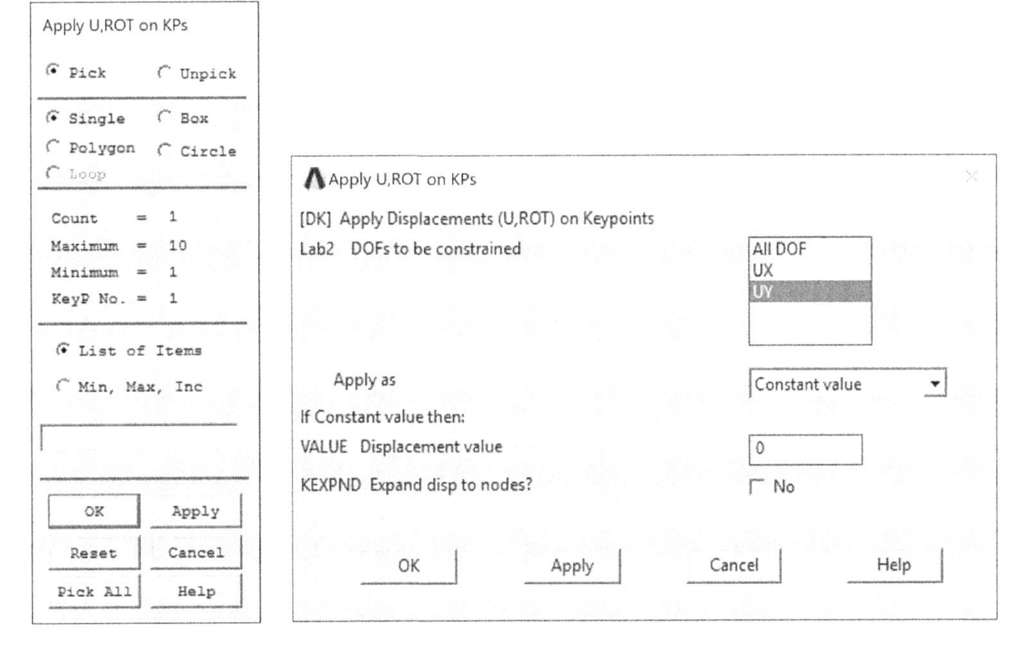

Figure 2-1-13 Applying Keypoint Constraints in Y.

5-3. Save your constraints

- ANSYS Toolbar > SAVE_DB

Step 6: Apply Load Boundary Conditions

6-1. Apply the load to the right edge of the plate

- Solution > Define Loads > Apply > Structural > Pressure > On Lines
- Use the mouse to click on the right edge of the plate (the vertical line at $x = 0.15$)
- Click OK (Figure 2-1-14, left)
- In the first text box labeled "If Constant value then: VALUE Load PRES value" enter −1e6 (Figure 2-1-14, right)
- Click OK

ANSYS applies pressures to lines as if the system were 1 unit thick in the third dimension. Thus, the value for the applied pressure does not need to be adjusted for this analysis.

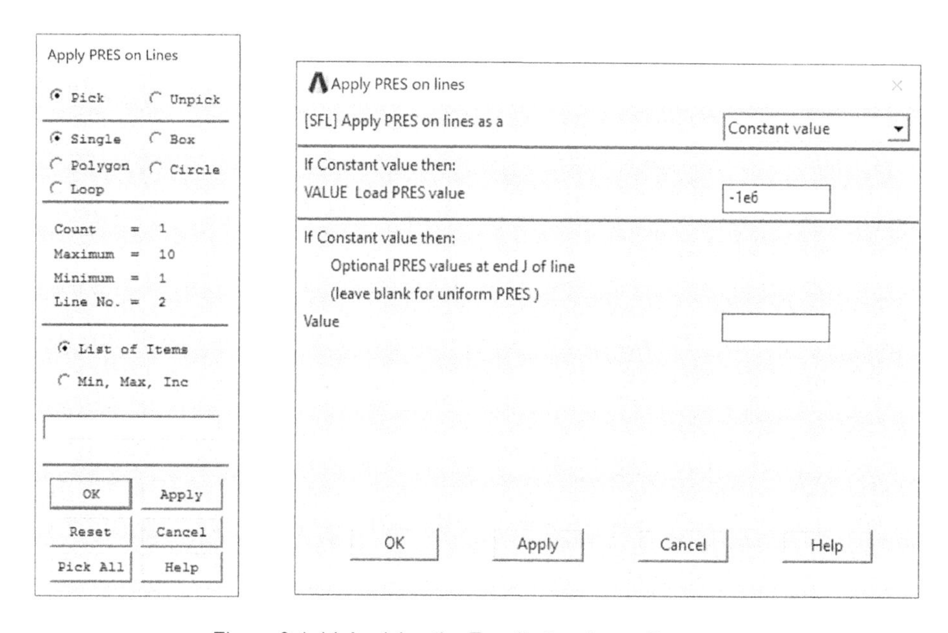

Figure 2-1-14 Applying the Tensile Load as a Pressure.

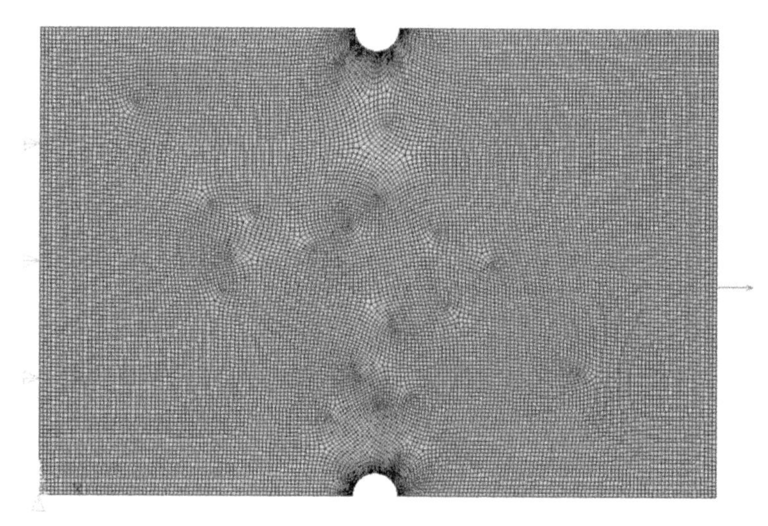

Figure 2-1-15 Element Plot with Applied Boundary Conditions.

A red arrow pointing to the right should appear on the right side of the plate to confirm the presence and direction of the applied load (Figure 2-1-15). *ANSYS assumes that any pressure applied to a model is compressive unless otherwise specified. Thus, a negative value for pressure is required to apply a load in tension.*

Applying equal and opposite forces to a body should keep it from moving. However, the computers that solve finite element models have round off error. While the round off error is typically small (~1e–20 units), it is sufficiently large to cause problems for the solver. As a result, it is common for models that involve equal and opposite forces to be constrained on one side and loaded on the other side. That approach is used here.

6-2. Save your loads

- ANSYS Toolbar > SAVE_DB

Step 7: Set the Solution Options

The default solution options can be used for this analysis.

Step 8: Solve

8-1. Select everything in your model

- Utility Menu > Select > Everything

You can confirm that this command was successfully issued in the Output Window. It should say: "Select all entities of type = all and below."

8-2. Solve

- Solution > Solve > Current LS
- Click OK

8-3. Close the /Status Command Window

8-4. Close the note informing you that the solution is done

8-5. Save your results

- ANSYS Toolbar > SAVE_DB

Step 9: Postprocess the Results

9-1. Plot the von Mises stress distribution in the plate

- In the Main Menu, follow the path:
- General Postproc > Plot Results > Contour Plot > Nodal Solu
- Choose "Stress" (Figure 2-1-16)
- Scroll down and choose "von Mises stress"
- For the Scale Factor choose "True scale"
- Click OK

By default, ANSYS plots the results using an automatically calculated scaling factor to make the physical behavior of the system easier to understand. In this exercise, we plotted the results at the true scale (i.e., using a scaling factor of 1) to show the true deformation of the plate.

Contour Nodal Solution Data

Item to be contoured

Favorites
Nodal Solution
DOF Solution
Stress
X-Component of stress
Y-Component of stress
Z-Component of stress
XY Shear stress
YZ Shear stress
XZ Shear stress
1st Principal stress
2nd Principal stress
3rd Principal stress
Stress intensity
von Mises stress

Undisplaced shape key
Undisplaced shape key: Deformed shape only
Scale Factor: True Scale 1

Additional Options

OK Apply Cancel Help

Figure 2-1-16 Contour Plot Menu.

Depending on your graphics settings, you should see one of the plots shown in Figure 2-1-17. *The plot on the left was generated using a Win32 graphics driver and uses 9 color contours to indicate the stress levels in the model. The plot on the right was generated using a 3D graphics driver and uses 128 colors with the same 9 color legend. The graphics driver can be changed via the Customization/Preferences tab in the Mechanical APDL Product Launcher.*

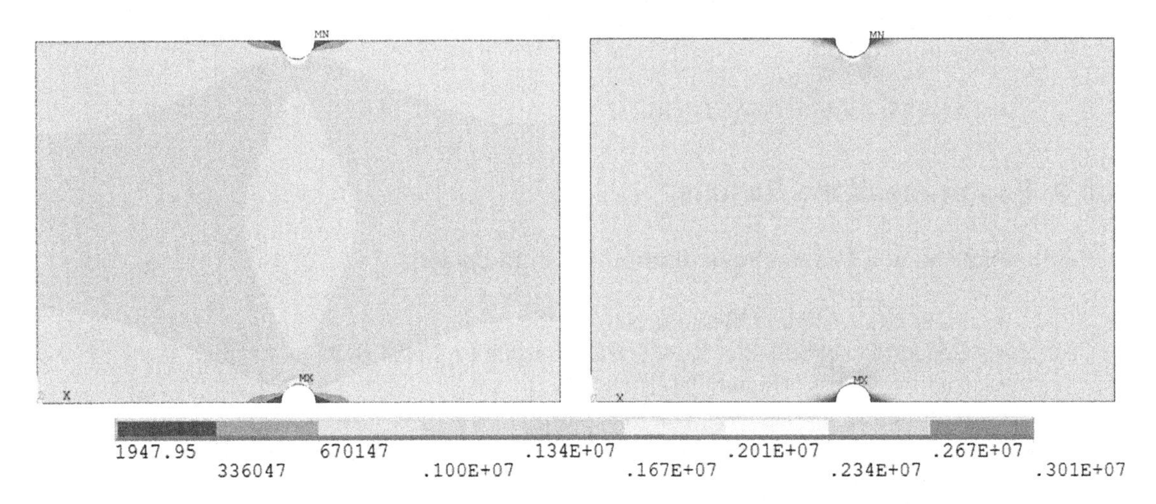

Figure 2-1-17 von Mises Stress Plot: Win32 Graphics (left) and 3D Graphics (right).

The model geometry has perfect quarter symmetry. However, we applied nonsymmetric boundary conditions. Therefore, the model results are also slightly asymmetric. As long as the deviations from a perfectly symmetrical solution are small, they can be neglected.

Step 10: Compare and Verify the Results

The von Mises stress plots show that the maximum stress in the plate is 3.01e6 Pa and is located near the base of the notches. The analytical solution predicts that the maximum stress should be 2.735 times the nominal stress (1e6*0.1 m N/m/0.09 m = 1.11e6 Pa) or 3.0388e6 Pa. The analytical solution also predicts that the maximum stress should be 3.03 times the gross stress (1e6 N/m) or 3.03e6 N/m. Comparing these values shows that the numerical solution is within 0.6% of the analytical solution. Therefore, the model is in excellent agreement with the theory and can be used for engineering design and analysis.

Close the Program

- Utility Menu > File > Exit...
- Choose "Quit - No save!"
- Click OK

In this exercise, the postprocessing operations were limited to plotting. Plotting does not affect the database, therefore there is no need to save again at this point.

Sample Input File

*The following is an input file that lists the commands for exercise 2-1. Exclamation points are used to indicate comments in the file. ANSYS will not process anything to the right of an exclamation point. Commas are used to separate the arguments of a command. Spaces and empty lines are ignored by the program. The **OUTPR** command has been included for use in* chapter 9. *Input files will be discussed in more detail in* chapter 10.

```
/BATCH                  ! Enable the file to run in batch mode

/PREP7                  ! Enter the Preprocessor
BLC4,0,0,0.15,0.1       ! Create a rectangle 0.15 m long and 0.10 m high
                        ! with bottom left corner at (0,0)
CYL4,0.075,0,0.005      ! Create circle w/radius 0.005 at (0.075,0)
CYL4,0.075,0.1,0.005    ! Create circle w/radius 0.005 at (0.075,0.1)
ASBA,1,ALL              ! Subtract Areas 2 and 3 from Area 1
ET,1,PLANE183           ! Use PLANE183 elements
MP,EX,1,7.31e10         ! Define Young's modulus for material #1
MP,PRXY,1,0.33          ! Define Poisson's ratio for material #1
SMRT,1                  ! Element size determined by Smart Size of 1
AMESH,ALL               ! Mesh all areas
EREF,ALL,,,3            ! Level 3 refinement for all elements in model
FINISH                  ! Finish and Exit the Preprocessor

/SOLU                   ! Enter the Solution processor
LSEL,S,LOC,X,0          ! Select line at x = 0
DL,ALL,,UX,0            ! Constrain line from movement in x
KSEL,S,LOC,X,0          ! Select all keypoints at x = 0
KSEL,R,LOC,Y,0          ! Reselect all keypoints at y = 0
                        ! Results in a single selected KP at (0,0)
DK,ALL,UY,0             ! Constrain keypoint movement in y
LSEL,S,LOC,X,0.15       ! Select line at x = 0.15
SFL,ALL,PRES,-1e6,      ! Apply pressure (tension) to line
ALLSEL                  ! Select everything
OUTPR,ALL,ALL           ! Output all solution items for all substeps
SOLVE                   ! Solve the model
FINISH                  ! Finish and Exit Solution

/POST1                  ! Enter the General Postprocessor
/DSCALE,ALL,1           ! Use true scale (scale displacements by 1.0)
PLNSOL,S,EQV,0,1.0      ! Plot equivalent (von Mises) stress
FINISH                  ! Finish and Exit the Postprocessor

SAVE                    ! Save the database
!/EXIT                  ! Exit ANSYS
```

Chapter

3

Creating and Importing Geometry

Suggested Reading Assignments:
Mechanical APDL Modeling and Meshing Guide: Chapters 1, 2, 3, 4, 5, 6, and 9

CHAPTER OUTLINE

In this chapter, you will learn how to create solid model geometry using the ANSYS native solid modeler, how to import geometry from other solid modeling software packages, and how to skip the solid modeling process altogether and directly generate the finite element mesh. You will also learn how to view, manipulate, and interact with solid models in ANSYS.

3.1. Considerations for Model Geometry

Before you can begin to create your model geometry, there are a number of decisions that must be made, including whether to create a solid model or generate the nodes and elements directly, whether to create the solid model in ANSYS or import it from another program, how many dimensions to include in the model, and which geometric details to include in the model.

3.1.1. Choosing Direct Generation or Solid Modeling

Usually, the first step in a finite element analysis is to create a solid model that represents the system geometry. The solid model is then meshed to create the finite element entities (nodes and elements) necessary to create and solve the partial differential equations in the model. But a solid model is not required. Nodes and elements can be defined directly in ANSYS. This process is called direct generation.

ANSYS Mechanical APDL for Finite Element Analysis.
DOI: http://dx.doi.org/10.1016/B978-0-12-812981-4.00003-4

In the early days of finite element analysis, direct generation was the only option for creating the model geometry. Today, direct generation is used to create small models and models that use specialty elements. These models often run faster than models with solid model geometry and take up less memory since the solid modeling and meshing operations are not performed. In addition, direct generation gives you complete control over the numbering and locations of the individual nodes in the model. However, direct generation is usually impractical for large and complicated models. It is also not recommended for novice FE analysts since it requires you to assume responsibility for the nodal spacing, which determines the mesh quality and thus the accuracy of the solution.

3.1.2. Choosing Whether to Create or Import Solid Model Geometry

If you choose to build a solid model, you must decide if you will create it using the ANSYS native solid modeler or if you will build it in another program and import it into ANSYS. The ANSYS native solid modeler was developed specifically for ANSYS. It gives you maximum control over the geometry and allows complete parameterization of the model. This permits a more robust and convenient selection of geometric entities inside the model, and can improve your ability to mesh, apply boundary conditions, and postprocess the model. When combined with APDL, the ANSYS native solid modeler can also allow you to do things that would be impossible in most commercial solid modeling packages. We build most of our Mechanical APDL models using the ANSYS native solid modeler because of these advantages. However, the ANSYS native solid modeler was developed decades before the modeling packages that are available today. It is now old technology and no significant improvement is expected in the near future. In addition, it may not be the best choice if you already have a solid model that you want to use or if you are very attached to your current CAD/CAE package.

3.1.3. Choosing the Dimensionality of the Model

The next decision to make when planning the geometry for your analysis is whether to build the model in three dimensions (3D) or if a one-dimensional (1D) or two-dimensional (2D) model will suffice. Lower order models have simpler solid model geometry and fewer degrees of freedom after meshing, so they generally require less time to solve. However, lower order models often require more complex boundary conditions and element definitions, which can increase the cost of building and postprocessing the model. For example, nodal forces need to be calculated and applied per unit depth for analyses using plane stress or plane strain, and per radian for axisymmetric analyses. The expected behavior of the system (plane stress, plane strain, plane stress with thickness, generalized plane strain, etc.) may need to be specified using element key options. And, postprocessing may require the creation and use of element tables. Finally, lower order modeling is only possible if an appropriate element is available for use.

Lower order models were very common in the early days of finite element analysis when computing power was limited. As a result, many lower order and lumped elements were created to facilitate their use. Although computing and finite element technologies have come a long way since the 1970s, model size is still a consideration. Your ANSYS license may limit the maximum number of nodes and elements that you can include in a model. Even if you can create a very large model, your ability to solve it is limited by the amount of memory available on your computer. And, if you can solve it, your computer's processing power will limit how quickly you can find the solution. Unless you have access to a high-performance computing cluster or can take advantage of expandable cloud-based computation resources with ANSYS® Enterprise Cloud™, you should consider using lower order elements when possible.

3.1.3.1. Characteristics of 1D, 2D, and 3D Models

One-dimensional models are meshed with simple lumped elements where the required behavior is included via the assumptions of the problem or has been incorporated into the element itself.

The underlying solid model geometry is composed of keypoints and lines. Examples include structures composed of simple beams, spring and damper systems, piping networks, and electric circuits.

Two-dimensional models can be used when the physical situation involves long, thin, planar structures or symmetric structures. The underlying solid model geometry is composed of key-points, lines, and areas. Examples include plates (Figure 3.1), shells, and pressure vessels. 2D models yield excellent results for plane stress, plane strain, and 1D or 2D heat conduction.

Three-dimensional models are used when the geometry does not have a constant cross sectional area or when the boundary conditions and loads are three-dimensional. The underlying solid model geometry is composed of keypoints, lines, areas, and volumes.

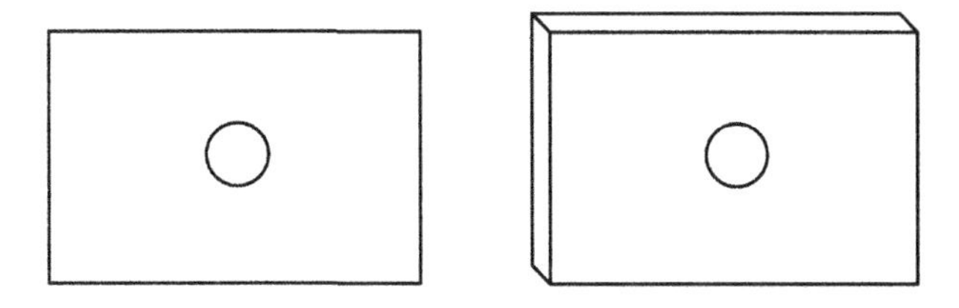

Figure 3.1 Plate with a Central Hole: Lower Order 2D Geometry (left) and Full 3D Geometry (right).

3.1.3.2. Reducing Model Size Using Symmetry

In many cases, the size and/or dimension of your geometry can also be reduced by taking advantage of symmetry. Many objects and systems display some kind of geometric symmetry. Reflective symmetry is achieved by mirroring an object, feature, or motif across a line or plane. Rotational symmetry is achieved by rotating the same object, feature, or motif at a constant angle around a central axis. Flowers, turbines, and end mills all display rotational symmetry. Axisymmetry is a special type of rotational symmetry where a one or two-dimensional shape is rotated 360 degrees about a central axis. Light bulbs, candlesticks, soda bottles, and pressure vessels frequently exhibit axisymmetry. Translational symmetry is achieved by repeating the same object, feature, or motif at a constant interval. Architectural pillars, finned heat exchangers, comb drives, and saw blades all exhibit translational symmetry.

If an object is symmetric in geometry, boundary conditions (loads and constraints), and material properties, then only the base shape needs to be modeled. The influence of the remainder of the object can be included in the model via key options or symmetry boundary conditions. This can substantially reduce the size of your model and the time and resources required to solve it. Thus, symmetric models can result in enormous cost savings without sacrificing accuracy.

Figure 3.2 shows two examples where symmetry can be used to create reduced models. The first example shows a three-dimensional thick walled cylinder. Since this is an axisymmetric object, it can be modeled as a 2D rectangle with a suitable offset from the axis of symmetry (the model y axis). Before this model could be solved, an element key option to indicate axisymmetric behavior would need to be set. The second example shows the plate with a central hole from Figure 3.1. Since this object exhibits reflective symmetry in two dimensions, only one-quarter of the 2D geometry needs to be included in the model. Symmetry boundary conditions along the axes of symmetry (the top and right most lines in the model) are needed to account for the remaining three-quarters of the object. The center of the hole does not need to be located at the origin.

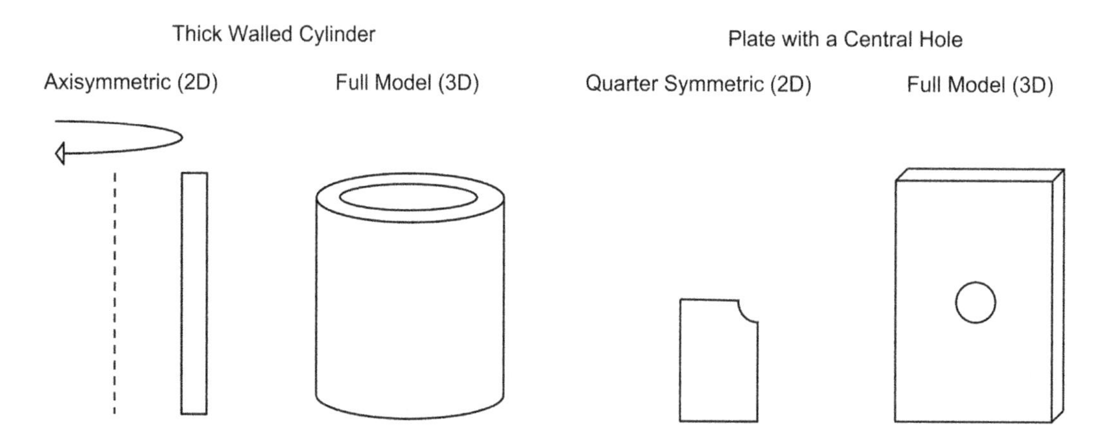

Figure 3.2 Using Symmetry to Create Lower Order Models: Axisymmetry (left) and Quarter Symmetry (right).

3.1.4. Choosing How Much Detail to Include

Finally, you must choose how much detail to include in your model geometry. Solid models often include fine details, like fillets and chamfers, that are important for visualizing or manufacturing the final product but that have little or no impact on the performance or behavior of the system. When fine details add substantial cost to the finite element model but do not add any value to the analysis, they can and should be excluded. There are no rules of thumb about when you can simplify the geometry for a finite element analysis. It will depend on the system, on the physics involved, and on what you hope to learn from the model. Ultimately, this is an engineering decision and it needs to be made by an engineer. But the decision should be made before you start to build or import your solid model.

3.2. Creating Model Geometry

There are three ways to define geometry in the ANSYS Mechanical APDL environment: direct generation of nodes and elements, bottom-up solid modeling using the ANSYS native solid modeler, and top-down solid modeling using the ANSYS native solid modeler. These options are not mutually exclusive. Different solid modeling techniques can be used or combined to create different parts of your solid model within ANSYS.

You can also create solid model geometry using ANSYS DesignModeler™ or ANSYS SpaceClaim® in the ANSYS Workbench environment but that discussion is beyond the scope of this book.

3.2.1. Direct Generation of Nodes and Elements

In finite element geometry, nodes are the most basic entity and are used to create the elements in the model. Point (0D) elements are defined by one node. Line (1D) elements are defined by two nodes. Area (2D) elements are defined by three or more nodes. And, volume (3D) elements are defined by four or more nodes.

Nodes and elements can be directly generated in ANSYS using the GUI path: **Main Menu > Preprocessor > Modeling > Create**. The GUI offers options to create nodes in the active coordinate system by typing in the coordinates; on the working plane by typing in the coordinates or by graphical picking; at curvature centers by typing in the coordinates or by graphical picking; and on keypoints by typing in the keypoint number or by graphical picking. Nodes can also be created using the **N** command.

Once nodes have been created, they can be used to generate more nodes. This can be accomplished by filling in the space between two nodes with more nodes using the GUI path: **Main Menu > Preprocessor > Modeling > Create > Nodes > Fill between Nds** or by using the **FILL** command. New nodes can also be created by copying and then offsetting existing nodes using the GUI path: **Main Menu > Preprocessor > Modeling > Copy > Nodes > Copy** or the **NGEN** command.

Nodes can be used to create elements by following the GUI path: **Main Menu > Preprocessor > Modeling > Create > Elements > Auto Numbered > Thru Nodes** or by using the **E** command. Once elements have been created, they can be extruded to create more elements by following the GUI path: **Main Menu > Preprocessor > Modeling > Copy > Elements** or by using the **EGEN** command.

Example 3.1 lists the commands for the direct generation of a simple line of nodes and elements. The first command enters the Preprocessor. The second command defines PLANE182 as element type #1. The third and fourth commands create two nodes: node #1 at the origin (0,0,0) and node #11 at (20,0,0). The **FILL** command creates the nodes in between (Figure 3.3). The **NGEN** command creates a second line of nodes by offsetting the first line of nodes 1 unit in the y direction (Figure 3.4). The **E** command creates a single element connecting nodes 1, 2, 22, and 21 (Figure 3.5). Finally, the **EGEN** command creates elements between the remaining nodes in the model (Figure 3.6). No solid model geometry is used in this example.

Example 3.1

```
/PREP7                      ! Enter the Preprocessor
ET,1,182                    ! Use PLANE182 elements
N,1,0,0,0                   ! Create node 1 at the origin (0,0,0)
N,11,20,0,0                 ! Create node 11 at (20,0,0)
FILL                        ! Fill in the nodes between the last 2 nodes
NGEN,2,20,1,11,1,0,1        ! Generate nodes once (2-1 times). Increment
                            ! each node number by 20 starting with node 1 and
                            ! ending with node 11. Offset each node location
                            ! by 1 unit in y.
E,1,2,22,21                 ! Define the element connecting nodes 1,2,22,21
EGEN,10,1,-1                ! Generate elements nine (10-1) times from the
                            ! previously defined nodes
```

1 x 2 3 4 5 6 7 8 9 10 11

Figure 3.3 Nodes Created using the **N** and **FILL** Commands.

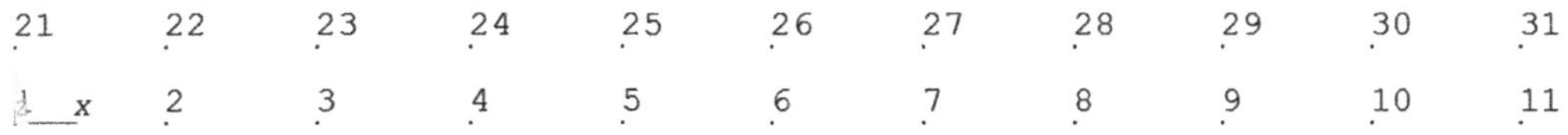

21 22 23 24 25 26 27 28 29 30 31

1 x 2 3 4 5 6 7 8 9 10 11

Figure 3.4 Additional Nodes Created using the **NGEN** Command.

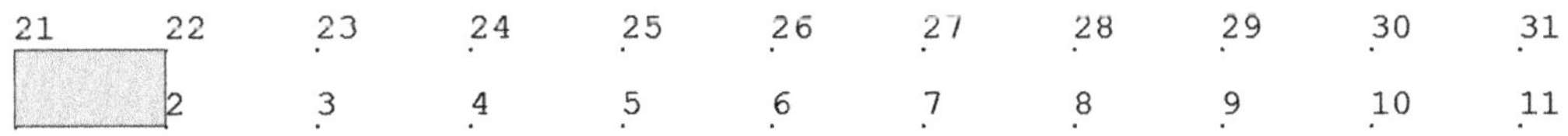

21 22 23 24 25 26 27 28 29 30 31

2 3 4 5 6 7 8 9 10 11

Figure 3.5 A Single Element Created using the **E** Command.

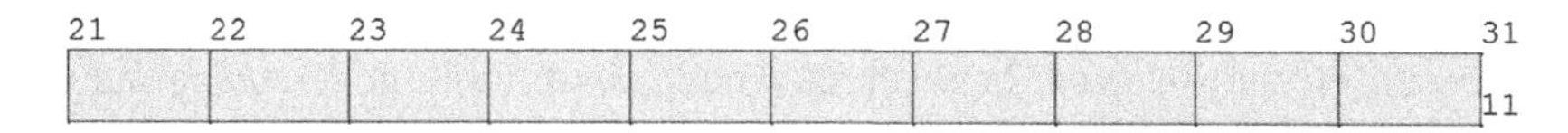

Figure 3.6 A Row of Elements Created using the **EGEN** Command.

You can enter each of these commands exactly as they are written into the GUI command prompt to create these nodes and elements. The text after the exclamation points contains comments to remind you or to inform others about the purpose and function of the commands. The comments do not need to be entered into the program and will be ignored if they are provided. If you have an electronic version of this book, you can also copy and paste the entire block of code into the command line. As long as the text copies and pastes with the original line breaks, the program will interpret it correctly.

3.2.2. Creating Model Geometry from the Bottom-Up

Bottom-up solid modeling is similar to the direct generation of nodes and elements except on a larger scale. In solid model geometry, a keypoint (a point in space) is the most basic unit. Lines are defined by two or more keypoints. Areas are defined by three or more lines (three or more keypoints). And, volumes are defined by two or more areas. Once created, keypoints, lines, and areas can be extruded to create lines, areas, and volumes. Keypoints can also be created by filling in the space between two keypoints. Solid modeling geometry can be created using the GUI path: **Main Menu > Preprocessor > Modeling > Create**.

ANSYS will automatically generate intermediate geometry when creating higher order entities. For example, if a rectangular area is defined by four keypoints, the lines connecting those keypoints will be generated in addition to the new area. If a cubic volume is defined by eight keypoints, the lines connecting those keypoints and the areas connecting those lines will be generated in addition to the new volume.

Bottom-up solid modeling is our preferred method for building solid models in ANSYS. It is very robust and gives you complete control over the model geometry. It can be used at any time, but it is especially useful for the modification of existing models and for parameterization.

Example 3.2 lists the commands for the creation of the same finite element model from Example 3.1 using bottom-up solid modeling. The first command enters the Preprocessor. The four **K** commands create keypoints to represent the four corners of the area (Figure 3.7). The **A** command creates the area bounded by the keypoints (Figure 3.8). The **ET** command defines PLANE182 as element type #1. The **ESIZE** command defines the default length of each element edge to be 2 units, resulting in a total of 10 elements. Finally, the **AMESH** command meshes the area to create the nodes and elements from Example 3.1 (Figure 3.9). You can see the node numbers by turning node numbering on using the **/PNUM** command and replotting the elements using the **EPLOT** command (Figure 3.10).

Example 3.2

```
/PREP7                  ! Enter the Preprocessor
K,1,0,0,0               ! Create Keypoint 1 at the origin (0,0,0)
K,2,20,0,0              ! Create Keypoint 2 at (20,0,0)
K,3,20,1,0              ! Create Keypoint 3 at (20,1,0)
K,4,0,1,0               ! Create Keypoint 4 at (0,1,0)
A,1,2,3,4               ! Create an area bounded by Keypoints 1,2,3,4
ET,1,182                ! Use PLANE182 elements
ESIZE,2                 ! Set default element edge length to 2 units
AMESH,ALL               ! Mesh all areas
/PNUM,NODE,1            ! Turn node numbering on
EPLOT                   ! Plot the elements in the model
```

Figure 3.7 Keypoints Created using the **K** Command.

Figure 3.8 A Single Area Created using the **A** Command.

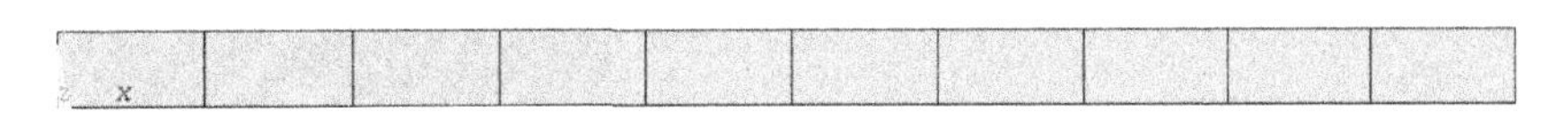

Figure 3.9 A Row of Elements Created using the **MESH** Command.

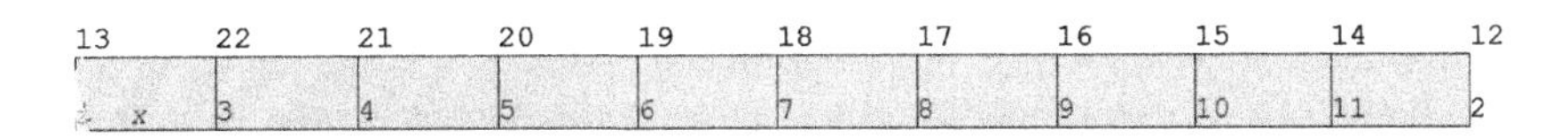

Figure 3.10 Element Plot with Node Numbering Activated.

Although the geometries and finite element meshes are the same for Example 3.1 and Example 3.2, the node numbering is not the same. In Example 3.1, we defined the upper node numbers to be equal to 20 plus the bottom node numbers. Thus, the maximum node number in the model is $(11+20)$ or 31. In Example 3.2, the program chose the order in which to create the nodes and assigned the node numbers sequentially. This resulted in a maximum node number of 22. If you examine Figure 3.10 carefully, you will also see that the program's placement of the node numbers is a little unintuitive.

Example 3.3 is a more elegant version of the previous example using the **KGEN** command.

Example 3.3

```
/PREP7                     ! Enter the Preprocessor
K,1,0,0,0                  ! Keypoint 1 is located at the origin (0,0,0)
K,2,20,0,0                 ! Keypoint 2 is located at (20,0,0)
KGEN,2,1,2,1,,1,,2         ! Generate a second line of keypoints using
                           ! the already defined key points
A,1,2,4,3                  ! Create an area bounded by Keypoints 1,2,3,4
ET,1,182                   ! Use PLANE182 elements
ESIZE,2                    ! Set default element edge length to 2 units
AMESH,ALL                  ! Mesh all areas
/PNUM,NODE,1               ! Turn node numbering on
EPLOT                      ! Plot the elements in the model
```

3.2.3. Creating Model Geometry from the Top Down

Top down solid modeling is based on the creation and modification of two and three-dimensional primitives. Primitives are basic shapes (rectangles, circles, blocks, spheres, cones, etc.) that are defined by a set of parameters (height, width, thickness, radius, etc.). All lower order entities (keypoints, lines, etc.) are automatically created within the primitive.

Example 3.4 uses the **RECT**(a)**NG**(le) command to create a 20x1 area from the previous examples.

Example 3.4

```
/PREP7                     ! Enter the Preprocessor
RECTNG,0,20,0,1            ! Define a rectangle w/ corners at (0,0) and (20,1)
ET,1,182                   ! Use PLANE182 elements
ESIZE,2                    ! Set default element edge length to 2 units
AMESH,ALL                  ! Mesh all areas
/PNUM,NODE,1               ! Turn node numbering on
EPLOT                      ! Plot the elements in the model
```

3.3. Boolean Operations

During top-down solid modeling, more complicated shapes are created from primitives through Boolean operations. Boolean operations available in the ANSYS native solid modeler include:

Intersect—Creates new entities where all of the selected entities overlap.

Add—Creates a new entity that contains all parts of the originals (union, join, or summation).

Subtract—If the entities are of the same order (e.g., ASBA—area subtract base area), then a new entity (area) is created. If the entities are of a different order (e.g., ASBL—area subtract base line), then the larger order entity (area) is divided.

Divide—Divides a higher order entity at its intersection with a lower order entity.

Glue—Causes overlapping entities to share boundaries and removes redundant entities.

Overlap—An add operation that does not delete the original overlapping entities.

Partition—An overlap operation that does not delete nonoverlapping entities.

These definitions may seem confusing, and even redundant, at first. Section 5.4 in the Mechanical APDL Modeling and Meshing Guide has excellent illustrations and examples of all the Boolean operations allowed in ANSYS.

Example 3.5 uses top-down solid modeling and Boolean operations to create a 2D plate with a central hole. First, the **RECTNG** command is used to create a 2x3 area with the lower left corner at the origin. Next, the **CYL**(inder)**4** command creates a circle centered at (1,1.5) with a radius of 0.5. Finally, the **ASBA** (area subtract base area) command is used to subtract Area 2 (the circle) from Area 1 (the rectangle) to create a new area (Area 3). These operations are illustrated in Figure 3.11.

Example 3.5

```
/PREP7                    ! Enter the Preprocessor
/PNUM,AREA,1              ! Turn area numbering on
RECTNG,0,2,0,3            ! Define a rectangle w/ corners at (0,0) and (2,3)
CYL4,1,1.5,0.5            ! Define a circle w/ center at (1,1.5), radius 0.5
ASBA,1,2                  ! Subtract Area 2 from Area 1
APLOT                     ! Plot the areas in the model
```

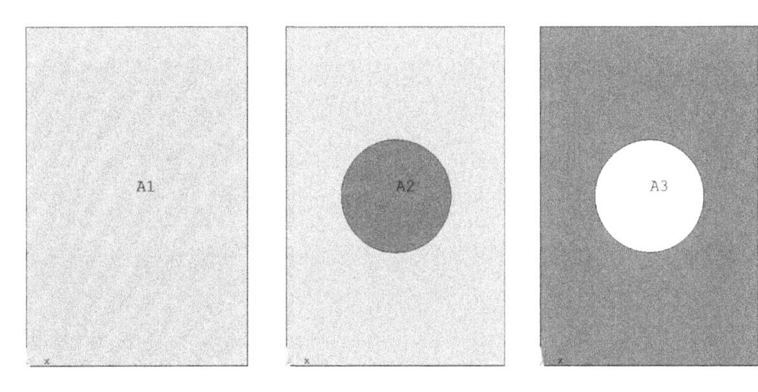

Figure 3.11 Area Plots with Area Numbering Activated: A Rectangle (Area 1) (left); The Rectangle (Area 1) with a Circle (Area 2) (center); and A Rectangle with a Central Hole (Area 3) (right).

Although this operation involves two entities of the same order (area and area), Boolean operations are not limited to entities of the same order. For example, lines may intersect lines (**LINL**), areas (**LINA**), and volumes (**LINV**). Areas may intersection areas (**AINA**) and volumes (**AINV**). And, volumes may intersect volumes (**VINV**). These operations may have different effects when used on the same and on different order entities.

Example 3.6 uses top-down solid modeling to create two areas from a line and an area. The **RECTNG** command is used to create the original area (Area 1). Bottom-up solid modeling is used to create two keypoints and a line that connects them (Line 5). Finally, a Boolean operation (**ASBL**—area subtract base line) is used to create two areas (Areas 2 and 3) from the area and the line. These operations are illustrated in Figure 3.12.

3.3.1. Boolean Options

Boolean operations require you to decide whether or not to keep the original entities in the model. By default, ANSYS deletes the original entities after a Boolean operation is performed. In Figure 3.12, you can see that Area 1 and Lines 1, 3, and 5 were deleted after the **ASBL** operation in Example 3.6.

Deletion of the original entities can be undesirable in some cases. For example, you might want to assign a load to a line at the intersection of two perpendicular areas. If the line does not exist, it can be created using the **AINA** (area intersect area) command. However, both original areas will be deleted in the process. The deletion or preservation of the original entities can be controlled using Boolean options. `BOPTN,KEEP,YES` preserves the original entities while `BOPTN,KEEP,NO` will delete the original entities after the new entity has been created. You can access the Boolean option settings through the GUI path: **Main Menu > Preprocessor > Modeling > Operate > Booleans > Settings**.

Example 3.6

```
/PREP7              ! Enter the Preprocessor
/PNUM,AREA,1        ! Turn area numbering on
RECTNG,0,2,0,4      ! Define a rectangle w/ corners at (0,0) and (2,4)
K,5,1,-1            ! Keypoint 5 is located at (1,-1)
K,6,1,5             ! Keypoint 6 is located at (1,5)
L,5,6               ! Define a line between Keypoints 5 and 6
/PNUM,LINE,1        ! Turn line numbering on
LPLOT               ! Plot the lines in the model
ASBL,1,5            ! Intersect Area 1 with Line 5
/PNUM,LINE,0        ! Turn line numbering off
APLOT               ! Plot the areas in the model
```

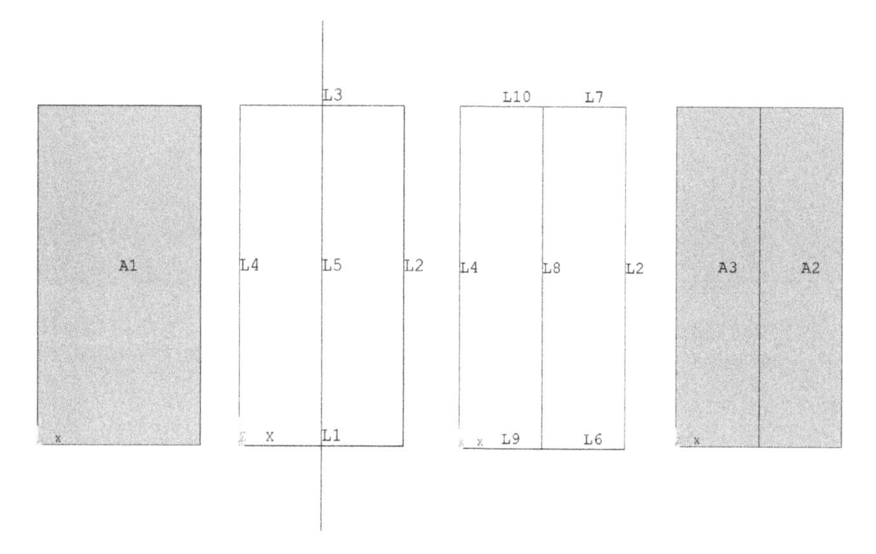

Figure 3.12 Plots from Example 3.6 with Numbering Activated: Plot of a Rectangular Area (left), Line Plot for the Area and an Additional Line (left center); Line Plot After ASBL Operation (right center); and Area Plot After ASBL Operation (right).

3.3.2. Number Merging

ANSYS does not assume that two identical entities (keypoints, lines, nodes, elements, etc.) that occupy the same space at the same time are the same entity. 'Duplicate' entities that are created during solid modeling and/or retained using Boolean options remain separate until the program is told to merge them. This procedure is referred to as number merging.

Example 3.7 lists the commands to create a bimetallic cantilever beam built from two rectangular areas.

Example 3.7

```
/PREP7                  ! Enter the Preprocessor
RECTNG,0,10,0,1         ! Define a rectangle w/ corners at (0,0) and (10,1)
RECTNG,0,10,1,2         ! Define a rectangle w/ corners at (0,1) and (10,2)
```

The first **RECTNG** command creates four keypoints numbered 1 through 4 and four lines numbered 1 through 4. The second **RECTNG** creates four keypoints numbered 5 through 8 and four lines numbered 5 through 8. As a result, there will be two keypoints at location (0,1) and two keypoints at location (10,1). There will also be two lines (Lines 3 and 5) at $y=1$ connecting each set of keypoints (Figure 3.13).

If these two areas are meshed, each line will be meshed separately. Since there will be no shared nodes between the two areas, there will be no loads transferred across this boundary. If the beam is constrained along the left edge and a download force is applied to the lower right corner of the bottom beam, only the bottom beam will deform (Figure 3.14).

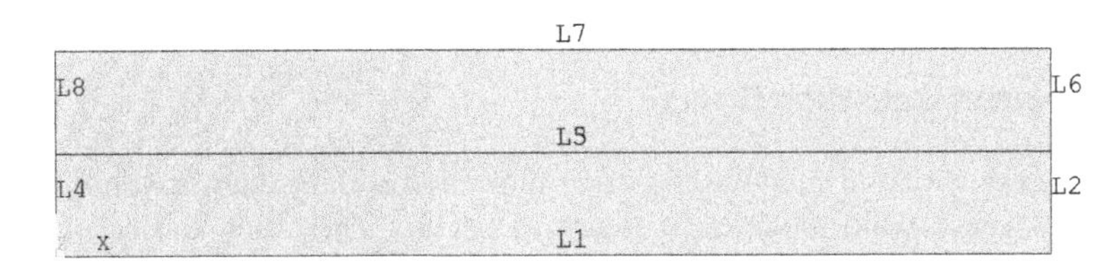

Figure 3.13 Bimetallic Cantilever Beam Without Number Merging.

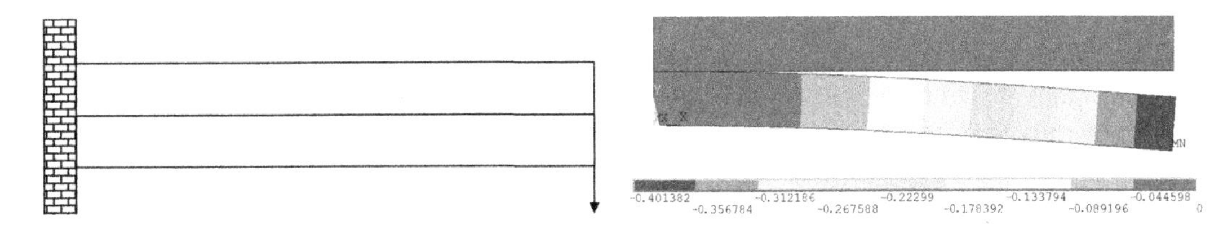

Figure 3.14 Vertical Displacement of a Simply Supported Bimetallic Cantilever Beam without Number Merging.

When situations like this exist, the model entities must be merged to form a continuous model using GUI path: **Main Menu > Preprocessor > Numbering Ctrls > Merge Items** or by using the **NUMMRG** command.

Entities in ANSYS must be merged in a specific order. If a finite element model exists, the nodes must be merged first using `NUMMRG,NODE` or the GUI equivalent. The elements must be merged next using `NUMMRG,ELEM`. Then, the solid model keypoints can be merged using `NUMMRG,KP`. Merging the keypoints automatically merges the higher level solid model entities (lines, areas, and volumes) without the need for additional operations. Failure to follow this merging order will likely result in a corrupted database.

The command `NUMMRG,ALL` merges all entities beginning with the nodes and thus avoids the problems associated with merging order. However, `NUMMRG,ALL` also merges the element attributes like element types, material properties, and real constants. This is usually both unexpected and undesirable so `NUMMRG,ALL` should be used after meshing with caution.

Because of the required merging order and the potential for accidental element attribute merging, we recommend number merging before meshing whenever possible.

3.3.3. Numbering in Boolean Operations

Top-down solid modeling is best for creating large, simple models quickly and easily. One command can create a 3D solid. Using bottom-up modeling, it might take ten commands (creating key points, lines, and areas) to create the same 3D solid. However, since each Boolean operation changes the numbering of the solid model entities by deleting the original entities and creating new ones, the use of large numbers of Boolean operations can make it difficult to keep track of the current state of the geometry. Boolean operations also offer less control over which entities are retained or deleted, and which boundaries and keypoints those entities share. Finally, Boolean operations do not permit you to specify the identifying numbers for the entities created. This can be a problem when selecting by entity number because the entity numbering may be different in different revisions of ANSYS. See chapter 7 for more details.

3.3.4. Boolean Operations: Model First, Mesh Second

Boolean operations are not permitted on meshed components. If you have already meshed your model, you will have to clear the mesh, perform the Boolean operation, and then re-mesh the model.

3.3.5. Boolean Operation Errors

Although top-down solid modeling is often more convenient than bottom-up solid modeling in the ANSYS native solid modeler, it is not as robust. There are known errors in the ANSYS native solid modeler associated with some Boolean operations, especially when small tolerances are involved. For this reason, we consider Boolean operations to be 'risky' procedures in ANSYS and recommend that you always save your database before performing one.

Since the ANSYS native solid modeler is no longer under active development, these errors may not be fixed. If you contact ANSYS Customer Support about these errors, they will usually suggest seeking an alternative way to create your solid model. Most Boolean 'errors' are user errors and can be debugged with a little bit of effort. But if you encounter a true Boolean error, we recommend looking for a (bottom-up) work-around and moving on with your analysis.

3.4. Deleting Solid Model Geometry

Solid model geometry can be deleted in ANSYS in much the same way that it was created. The GUI path **Main Menu > Preprocessor > Modeling > Delete** opens a submenu with options to delete keypoints, hard points, lines only, lines and below, areas only, areas and below, volumes only, volumes and below, nodes, elements, pretension elements, and concatenations. Deleting an entity "only" will remove the entity itself, but leave all of the lower order entities that were contained within that entity. For example, deleting an "area only" would delete the area but leave all of the lines and keypoints contained within the area. Deleting an entity "and below" will remove the entity and all of the lower order entities that were contained within it.

3.5. Importing Solid Model Geometry*

It is possible to import solid model geometry that was created in another CAD/CAE package into ANSYS rather than creating it within the program. However, the native file formats of commercial solid modeling packages are not intrinsically compatible with ANSYS. To import a solid model into ANSYS, you must export your solid model geometry as an IGES file, use ANSYS connection functionality, or import your model using the ANSYS Workbench environment.

3.5.1. Importing Solid Models Using IGES Files

IGES (Initial Graphics Exchange Specification) files are saved in a vendor-neutral format that is used to transfer models between various CAD/CAE systems. Most solid modeling packages can export IGES files and ANSYS can read them without additional software or licenses. This was the traditional default import option for most users in the classic ANSYS Mechanical APDL environment.

To import an IGES file into ANSYS, you must first clear the database (**Utility Menu > File > Clear & Start New**...) and then import the file (**Utility Menu > File > Import > IGES**...). If you do not clear the database before importing the geometry, you may corrupt your database. Once solid model geometry has been imported into ANSYS, it can only be sent back to the original CAD/CAE package as an IGES file.

This part of the program is under active development and is being improved regularly. For up-to-date information, see Chapter 6 of the Mechanical APDL Modeling and Meshing Guide or the company website (http://www.ansys.com).

3.5.2. Importing Solid Models Using Connection Products

You can import geometry from certain CAD/CAE programs using one of the ANSYS connection products by following the path **Utility Menu > File > Import >** and then choosing the appropriate file type. This will usually result in a better import than the IGES option. For more information, see the Mechanical APDL Connection User's Guide or the company website.

3.5.3. Importing CAD Using ANSYS Workbench and DesignModeler

The final way to import complex CAD/CAE models into ANSYS is to use ANSYS DesignModeler. Depending on the program, you may be able to directly import geometry, to exchange files back and forth, or to make changes to the same model from within both the CAD package and ANSYS Workbench using plug-ins. This is a good option if you are planning to regularly import complex solid models into ANSYS.

3.6. Coordinate Systems

In ANSYS, all geometric entities (both created and imported) exist and are displayed within a coordinate system (CS). All operations on geometric entities are performed within a coordinate system. All loads are applied in a coordinate system. And, all results are calculated, stored, and postprocessed in a coordinate system.

ANSYS uses six types of coordinate systems to perform these activities: global coordinate systems, local coordinate systems, display coordinate systems, nodal coordinate systems, element coordinate systems, and results coordinate systems. The global and local coordinate systems locate nodes and keypoints in space. The display coordinate system is used for listing and plotting the model data. The nodal coordinate system orients the degrees of freedom and the nodal results data. The element coordinate system orients the material properties and the element results data. And, the results coordinate system is used for listing, plotting, and other postprocessing of the results data.

The most important coordinate systems for creating and importing geometry are the global, local, and display coordinate systems. The remaining coordinate systems will be addressed in chapter 9.

3.6.1. Global Coordinate Systems

Global coordinate systems are the ANSYS equivalent of an absolute reference frame. They are used to define the coordinate locations of nodes and keypoints in space. They can also be used to identify or select solid model and finite element model entities based on their location(s) in space. (For more information on selection, see chapter 7.) There are four predefined global coordinate systems in ANSYS:

CS 0—Global Cartesian (X, Y, Z)

CS 1—Global cylindrical (R, θ, Z)

CS 2—Global spherical (R, θ, φ)

CS 5—Global cylindrical (R, θ, Y)

The global coordinate systems are all right handed and share the same global origin (0,0,0).

All new entities are created and all existing entities are selected in the active coordinate system. Only one coordinate system can be active at a given time. By default, the active coordinate system is CS 0 (Global Cartesian). You can change the active coordinate system by using the **CSYS** command or the GUI path: **Utility Menu > WorkPlane > Change Active CS to**.

3.6.2. Local Coordinate Systems

If none of the global coordinate systems are suitable for a given task, you can define a local coordinate system to simplify the creation or verification of your model. This usually involves a translation and/or rotation of one of the global coordinate systems. However, toroidal local coordinate systems can also be created. Local coordinate systems are assigned a CS number by the user. The first 10 CS numbers are reserved for the program's use. Thus, local coordinate systems must be assigned a number greater than or equal to 11. When a new coordinate system is defined, it becomes the new active coordinate system. To create a local coordinate system, use the GUI path: **Utility Menu > WorkPlane > Local Coordinate Systems > Create Local CS** or use the **LOCAL** command.

3.6.3. The Display Coordinate System

The display coordinate system is the coordinate system used by the program to display plots and to list the coordinates of the nodes and keypoints. As a result, it affects the display of geometric entities and changes the appearance of keypoint, line, area, and volume plots. By default, the display coordinate system is CS 0 (Global Cartesian). The display coordinate system is *not affected* by changing the active coordinate system associated with geometry creation.

There are times when changing the display coordinate system is useful. For example, if you created a set of new nodes or keypoints using a different coordinate system, you may wish to align the display coordinate system with the active coordinate system and then the list the nodes or keypoints to verify that they were created correctly. However, we live and think in a Cartesian world. Therefore, we recommend changing the display coordinate system rarely and with care. We also recommend changing the display coordinate system back to Global Cartesian immediately after completing the required operations in the alternate display coordinate system.

To change the display coordinate system, use the **DSYS** command or the GUI path: **Utility Menu > WorkPlane > Change Display CS to**.

3.7. The Working Plane

Although ANSYS can create and solve 3D models, some aspects of the program can only operate in two dimensions at any given time. This two dimensional space is called the working plane. Graphical picking (i.e., the ability to choose a model feature with the mouse) can only be done on the working plane. Graphical picking chooses the model feature whose centroid is on or closest to the working plane. All area primitives are created on the working plane. All volume primitives are created with the base on the working plane. For example, Figure 3.15 shows a conical primitive with its base on the working plane. Some divide Boolean operations use the working plane. You can also use the working plane to define a new active, display, or local coordinate system.

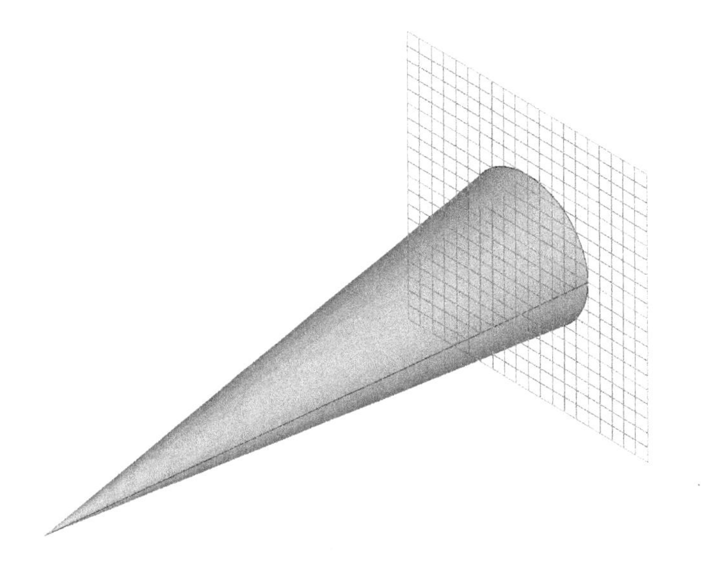

Figure 3.15 A Conical Primitive with its Base on the Working Plane.

By default, the working plane is the *xy* plane in the Global Cartesian coordinate system. The working plane can be moved by offsetting the working plane, by aligning the working plane with a previously defined plane, or by defining the working plane by coordinates or keypoints. The working plane can be defined in Cartesian or polar coordinates.

To change the working plane, use the GUI path: **Utility Menu > WorkPlane > Offset WP to** or **Utility Menu > WorkPlane > Align WP with**.

Example 3.8 uses a local coordinate system and a working plane offset to create the geometry from Example 3.5. In Example 3.5, the rectangle and the circle were created in the default coordinate system (CS 0) and on the default working plane (aligned with CS 0). In Example 3.8, the rectangle is created in CS 0. Next, a local coordinate system is created at the center of the rectangle (CS 11). Then, the working plane is aligned with the new active coordinate system, and the circle is created at the center of the new working plane. Finally, the circle is subtracted from the rectangle as usual.

Example 3.8

```
/PREP7                    ! Enter the Preprocessor
RECTNG,0,2,0,4            ! Define a rectangle w/ corners at (0,0) and (2,4)
LOCAL,11,0,1,2,0          ! Create a local Cartesian coordinate system (#11)
                          ! with origin at (1,2). This becomes the active CS.
WPCSYS                    ! Align the working plane with the active CS
CYL4,0,0,0.5              ! Define a circle w/ center at (0,0), radius 0.5
ASBA,1,2                  ! Subtract Area 2 from Area 1
```

3.8. Solid Model Viewing

ANSYS offers a number of tools for viewing and manipulating your model. These are all located in the GUI Utility Menu under the List, Plot, and PlotCtrls menus.

3.8.1. List

Each entity in ANSYS is created with certain properties, including a number (assigned by default based on the order in which it was created), geometrical properties (locations for keypoints, lengths for lines, areas for areas, etc.), physical properties (material numbers, etc.), and the lower order entities that are located within it. The List menu (**Utility Menu > List**) can be used to list these values for each type of entity in ANSYS: keypoints, lines, areas, volumes, nodes, elements, components, properties, loads, and results.

Listing is a convenient way to confirm whether or not an entity, boundary condition, or load has been created or deleted. For example, you could list the keypoints created in Example 3.5 by using the menu path **Utility Menu > List > Keypoint > Coordinates Only**. This would display the *x, y, z* coordinates and the *xy, yz,* and *zx* rotation angles shown in Figure 3.16.

KLIST Command

File

LIST ALL SELECTED KEYPOINTS. DSYS= 0

NO.	X,Y,Z LOCATION			THXY,THYZ,THZX ANGLES		
1	0.000000	0.000000	0.000000	0.0000	0.0000	0.0000
2	2.000000	0.000000	0.000000	0.0000	0.0000	0.0000
3	2.000000	3.000000	0.000000	0.0000	0.0000	0.0000
4	0.000000	3.000000	0.000000	0.0000	0.0000	0.0000
5	1.500000	1.500000	0.000000	0.0000	0.0000	0.0000
6	1.000000	2.000000	0.000000	0.0000	0.0000	0.0000
7	0.5000000	1.500000	0.000000	0.0000	0.0000	0.0000
8	1.000000	1.000000	0.000000	0.0000	0.0000	0.0000

Figure 3.16 List of Keypoints Created in Example 3.5.

Similarly, the menu path **Utility Menu > List > Areas** lists the information for all of the areas in the model. In Example 3.5, the first area that was generated was the rectangle. By default, this was assigned to be Area 1. The second area that was generated was the circle. This was assigned to be Area 2. Then we used a Boolean operation to subtract Area 2 from Area 1. This created a new area (Area 3) and deleted the original areas (Areas 1 and 2). Thus, Area 3 is the only one listed by the **ALIST** command (Figure 3.17). The ALIST Command dialog box shows that the area has two loops (one for the outer rectangle and the other for the inner circle) and is composed of 8 lines (4 for the outer rectangle and 4 for the inner circle).

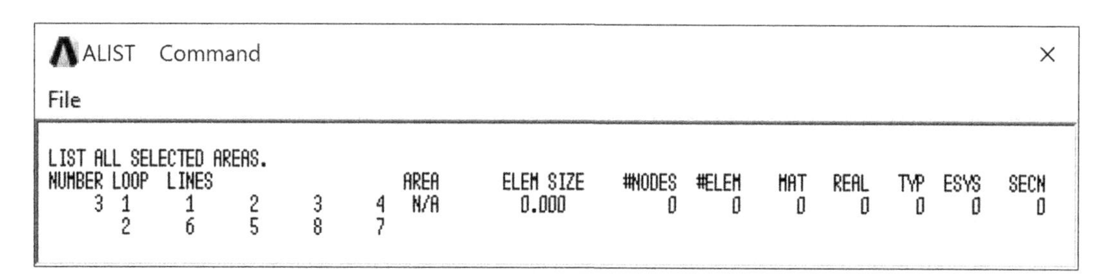
ALIST Command

File

LIST ALL SELECTED AREAS.

NUMBER	LOOP	LINES				AREA	ELEM SIZE	#NODES	#ELEM	MAT	REAL	TYP	ESYS	SECN
3	1	1	2	3	4	N/A	0.000	0	0	0	0	0	0	0
	2	6	5	8	7									

Figure 3.17 List of Areas Created in Example 3.5.

There is one listing command for each entity type in ANSYS:

- The **KLIST** command lists the specified keypoints or all keypoints in the active set.
- The **LLIST** command lists the specified lines or all lines in the active set.
- The **ALIST** command lists the specified areas or all areas in the active set.
- The **VLIST** command lists the specified volumes or all volumes in the active set.
- The **NLIST** command lists the specified nodes or all nodes in the active set.
- The **ELIST** command lists the specified elements or all elements in the active set.

3.8.2. Plot

Plotting is another convenient way to confirm whether or not an entity has been created or deleted. There is one plotting command for each entity type in ANSYS:

- The **KPLOT** command plots the specified keypoints or all keypoints in the active set.
- The **LPLOT** command plots the specified lines or all lines in the active set.
- The **APLOT** command plots the specified areas or all areas in the active set.
- The **VPLOT** command plots the specified volumes or all volumes in the active set.
- The **NPLOT** command plots the specified nodes or all nodes in the active set.
- The **EPLOT** command plots the specified elements or all elements in the active set.

These can be accessed via the GUI path: **Utility Menu > Plot**. By default, lines are plotted as colored solid (if unmeshed) or dashed (if meshed) lines. Areas, volumes, and elements are plotted as turquoise shapes outlined in black (on a white background) or in white (on a black background).

3.8.3. PlotCtrls

The PlotCtrls menu allows you to customize the appearance of the plots generated by ANSYS. The two PlotCtrls submenus that are most important for solid modeling are the Plot Numbering Controls menu and the Pan Zoom Rotate menu.

3.8.3.1. Plot Numbering Controls

Plot Numbering Controls (Figure 3.18) allow you to choose whether or not to display numbering-related information for each entity type in a plot. Turning numbering ON for an entity type causes each entity of that type to appear in a different color and/or with its entity number next to it. For example, Figure 3.11 shows three area plots with area numbering turned on (`/PNUM,AREA,1`) and with the numbering shown with both colors and numbers (`/NUMBER,0`). You can display entities using colors only (`/NUMBER,1`), using numbers only (`/NUMBER,2`) (Figure 3.12), or without numbering (`/NUMBER,-1`). Numbering labels appear as close to the entity centroid as possible. This occasionally results in labels being placed on top of each other (Figure 3.13). Keypoints and nodes are always shown in white (on a black screen) or in black (on a white screen) regardless of the coloring option specified. Plot Numbering Controls can be accessed using the GUI path: **Utility Menu > PlotCtrls > Numbering**....

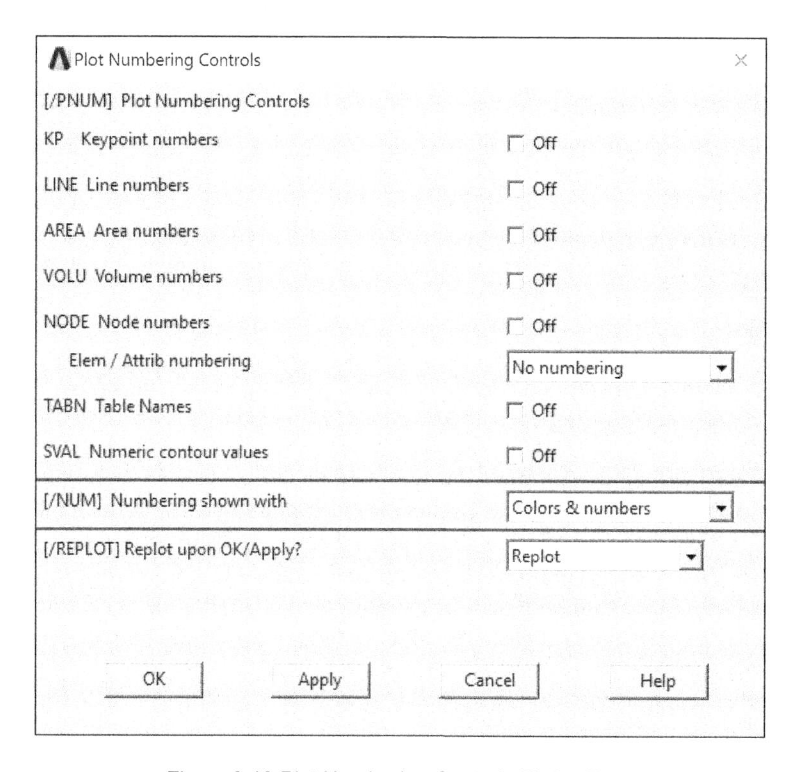

Figure 3.18 Plot Numbering Controls Dialog Box.

3.8.3.2. Pan Zoom Rotate Menu

The Pan Zoom Rotate menu (Figure 3.19, left) allows you to pan (move), zoom, and rotate the image in the Graphics Window. The Pan Zoom Rotate menu can be found using the GUI path: **Utility Menu > PlotCtrls > Pan Zoom Rotate**.... Many of these commands also have shortcut buttons on the right side of the GUI (Figure 3.19, right).

By default, there is only one view open in the ANSYS Graphics Window at any given time. However, the Window drop down menu in the Pan Zoom Rotate dialog box allows up to five views to be open at once. All of these views will be contained within the Graphics Window but each can be controlled independently. ANSYS refers to these views as "windows" in the Pan Zoom Rotate dialog box and in the documentation.

The nine buttons immediately below the Window drop down menu are named views that you can choose from. These include top, bottom, front, back, left, right, isometric, and oblique views, and the view from the current working plane.

There are three ways for you to zoom in and out of the view. The first method is to use the four buttons just above the arrow keys in the center of the menu. These buttons offer Zoom, Box Zoom, Win Zoom, and Back Up options. The second method is to use the buttons between the arrow keys. The button with the small circle zooms out and the button with the large circle zooms in. The third method is to use Dynamic Mode, which is controlled by your mouse wheel or equivalent scrolling device.

Dynamic Mode is activated by checking the box in the Pan Zoom Rotate menu or the corresponding shortcut (shaped like a mouse) in the Graphics Window. Once it is active, you can pan the image on the screen by clicking and dragging the left mouse button. You can rotate the image in all three dimensions by clicking and dragging the right mouse button.

The arrow keys are pan buttons. They move the image on the screen up, down, left, or right without changing the scaling of the image.

Below the arrow keys are six rotate buttons. These rotate the image on the screen about the *x*, *y*, and *z* axes by a positive or negative angle. These can be difficult to use with great accuracy. For better accuracy, use Dynamic Mode.

The Rate (of Change for Model Manipulation) slider determines how much and how fast the model moves in Dynamic Mode. You can adjust the Rate (in increments of 1 from 1 to 100) using the slider in the Pan Zoom Rotate Menu. You can also select from Rates of 1, 15, 30, 45, 60, 75, 90, and 100 from the drop down shortcut in the Graphics Window.

Finally, if the results of your efforts are not as expected, you can "Reset" the image or scale the image to "Fit" in the Graphics Window using the buttons at the bottom of the Pan Zoom Rotate menu.

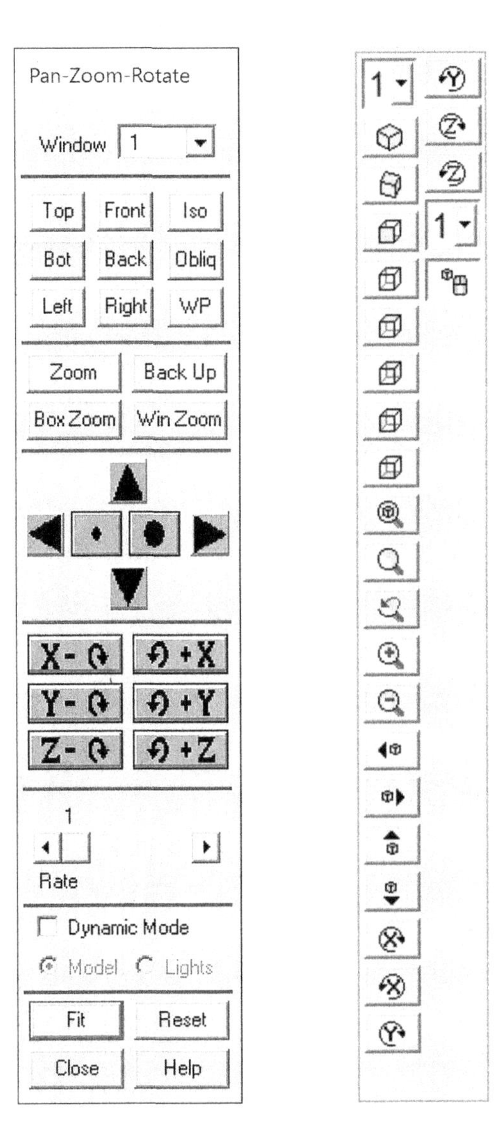

Figure 3.19 Pan Zoom Rotate Menu (left) and Pan Zoom Rotate Shortcut Buttons in the Graphics Window (right).

Exercise 3-1

Bottom-Up Solid Modeling of a Plate With a Central Hole Using Quarter Symmetry

Overview

This exercise involves the steady-state structural analysis of a rectangular plate with a central hole loaded axially in tension. This is another common problem used to demonstrate mechanical stress concentrations. Again, the plate is made of 6061-T6 aluminum with a Young's modulus of 73.1 GPa and a Poisson's ratio of 0.33. It is 15 cm wide and 10 cm tall. The hole is 2 cm in diameter. The plate is thin so this can be treated as a 2D problem. In the previous exercise, you used top-down solid modeling techniques to create a full 2D model. In this exercise, you will take advantage of the two axes of symmetry to create a quarter model using bottom-up solid modeling (Figure 3-1-1).

A load of 1 MPa will be placed on the left edge of the plate. A symmetry boundary condition will be applied to the right edge of the plate. This will create a reaction force that acts as the second 1 MPa load on the right. Again, the plate will deform under the applied load, growing longer and thinner. The central hole will become an oval and a stress concentration will appear on both sides. The expected elastic stress concentration K_t for this problem is 3.14.

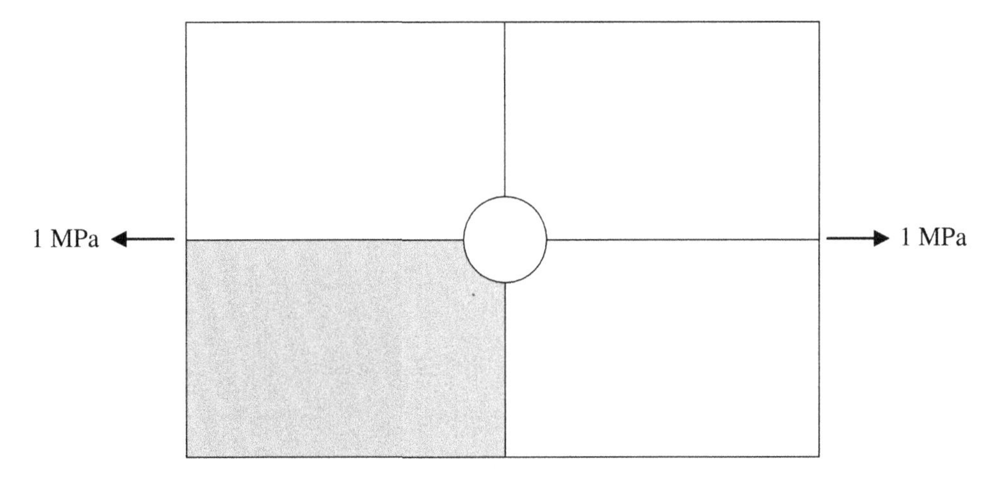

Figure 3-1-1 Schematic of the Quarter Plate with Applied Loads.

ANSYS Mechanical APDL for Finite Element Analysis.
DOI: http://dx.doi.org/10.1016/B978-0-12-812981-4.00015-0

Model Attributes

Material Properties for 6061-T6 Aluminum

- Young's modulus—7.310e10 Pa
- Poisson's ratio—0.33

Loads

- 1 MPa tensile load on the left edge of the plate

Constraints

- Symmetry constraints on the two axes of symmetry

File Management

Create a new folder in your "Intro-to-ANSYS" folder named "Exercise3-1"

Open a new session of ANSYS using the Mechanical APDL Product Launcher

Change the Working Directory to the new "Exercise3-1" folder

Change the Jobname to "Exercise3-1"

Click Run to start ANSYS

Step 1: Define Geometry

1-1. Create keypoints to define the lower left corner of the plate

- Preprocessor > Modeling > Create > Keypoints > In Active CS
- Leave the "NPT Keypoint number" box empty to automatically assign KP numbers (Figure 3-1-2)
- For the "X,Y,Z Location in active CS" enter (0,0,0)
- Click Apply to create the first keypoint

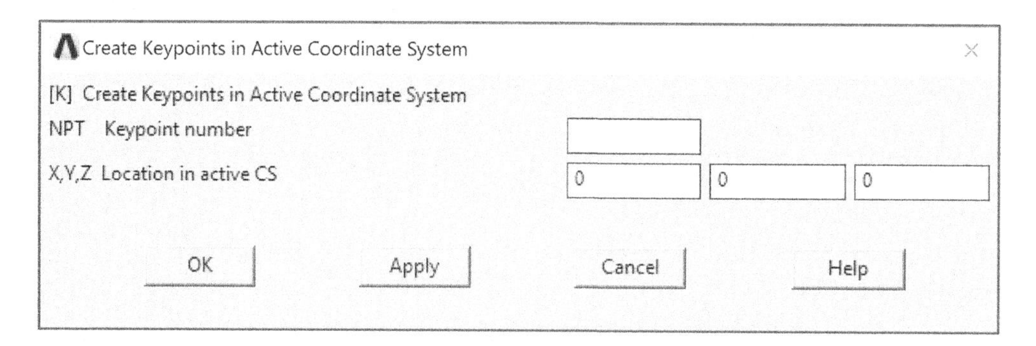

Figure 3-1-2 Keypoint Coordinate Input Dialog Box.

The default value for a keypoint coordinate is zero. Thus for the first keypoint, you can also just click "Apply" without providing any values.

A dialog box that is left open in ANSYS by clicking "Apply" instead of "OK" can be hidden behind another window. The "Raise Hidden" button to the right of the Command Prompt will bring the dialog box back to the front of your screen (Figure 3-1-3).

Figure 3-1-3 Raise Hidden Icon.

- Enter (0,0.05,0) as the coordinates for the 2nd KP
- Click Apply
- Enter (0.075,0,0) as the coordinates for the 3rd KP
- Click Apply
- Enter (0.075,0.05,0) as the coordinates for the 4th KP
- Click Apply

1-2. Create keypoints to define the intersection of the hole with the plate

- Enter (0.065,0.05,0) as the coordinates for the 5th KP
- Click Apply
- Enter (0.075,0.04,0) as the coordinates for the 6th KP
- Click OK

1-3. Turn on keypoint and line numbering

- Utility Menu > PlotCtrls > Numbering...
- Click the check box labeled "KP Keypoint numbers"
- Click the check box labeled "LINE Line numbers"

The labels on the check boxes should change from "Off" to "On."

This operation ensures that entity numbers will be displayed near the centroid of each keypoint and line in the model. This provides more detailed feedback during solid modeling.

- Click OK

1-4. Create lines to represent the edges of the plate

- Preprocessor > Modeling > Create > Lines > Lines > Straight Line
- Click KP 1 then click KP 2 in the Graphics Window to create Line 1

Keypoint 1 is located under the global XYZ triad at the origin.

- Click KP 1 then click KP 3 in the Graphics Window to create Line 2
- Click KP 3 then click KP 6 in the Graphics Window to create Line 3
- Click KP 2 then click KP 5 in the Graphics Window to create Line 4
- Click OK

1-5. Create a cylindrical local coordinate system

- Utility Menu > WorkPlane > Local Coordinate Systems > Create Local CS > By 3 Keypoints +
- Click on KP 4

Keypoint 4 may be hidden by the ANSYS license information in the upper right corner of the Graphics Window. You can click the "Zoom Out" button in the Pan Zoom Rotate shortcut menu to reveal KP 4.

- Click on KP 5
- Click on KP 6
- Choose "Cylindrical 1" from the "KCS Type of coordinate system" drop down menu
- Click OK

By default, lines are created "straight" in the currently active coordinate system. By defining a cylindrical local coordinate system, "straight" lines created in this system will be arcs.

1-6. Create an arc to represent the hole in the plate

- Preprocessor > Modeling > Create > Lines > Lines > In Active Coord
- Click on KP 5
- Click on KP 6
- Click OK

Or

- Type "L,5,6" in the command prompt above the Graphics Window

*You could have also used the **LARC** command (**Preprocessor > Modeling > Create > Lines > Arcs > By End KPs & Rad**) to create the arc between Keypoints 5 and 6. The result would be the same.*

1-7. Return the active coordinate system to Global Cartesian

- Utility Menu > WorkPlane > Change Active CS to > Global Cartesian

1-8. Create an area to represent the quarter plate with hole

- Preprocessor > Modeling > Create > Areas > Arbitrary > By Lines
- Click on all five lines to define the perimeter of the area
- Click OK

The completed solid model is shown in Figure 3-1-4.

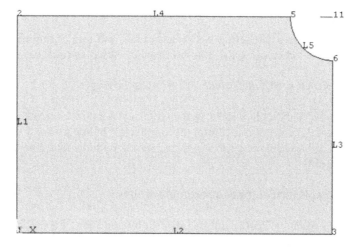

Figure 3-1-4 Completed Solid Model.

1-9. Save the model geometry

- ANSYS Toolbar > SAVE_DB

Step 2: Define Element Types

2-1. Define the element type to use for this model

- Preprocessor > Element Type > Add/Edit/Delete
- Click the "Add..." button to bring up the Library of Element Types
- In the left-hand column, choose "Solid"
- In the right-hand column, choose "8 Node 183"
- Click OK
- Close the Element Types dialog box

Step 3: Define Material Properties

3-1. Create a linear elastic material model for 6061-T6 aluminum

- Preprocessor > Material Props > Material Models
- Click on Structural
- Click on Linear
- Click on Elastic
- Click on Isotropic
- Click in the EX box and type 7.310e10
- Click in the PRXY box and type 0.33
- Click OK
- Close the material model dialog box

3-2. Save your progress

- ANSYS Toolbar > SAVE_DB

Step 4: Mesh

4-1. Create the mesh for the finite element model

- Preprocessor > Meshing > MeshTool
- Check the "Smart Size" button in the Mesh Tool dialog box
- Adjust the Smart Size Slider to 3
- Click the "Mesh" button
- Click the "Pick All" button in the Mesh Areas dialog box

The resulting mesh is shown in Figure 3-1-5. *This mesh is adequate for most analyses. However, the goal of this analysis is to investigate the stress concentration at the hole so the mesh should be refined in that region. You will use the "Refine" option in the Mesh Tool to improve the mesh around the hole.*

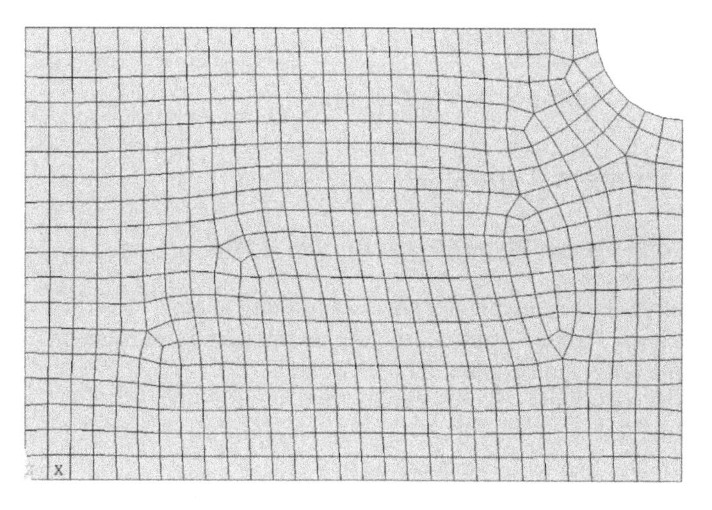

Figure 3-1-5 Finite Element Mesh.

4-2. Refine the mesh along Line 5

- Bring the Mesh Tool back to the foreground or reopen it if closed
- In the fifth section of the Mesh Tool, change “Refine at: Elements” to “Refine at: Lines”
- Click the “Refine” button
- Click on Line 5 (the arc)
- Click OK
- Change “LEVEL Level of refinement” to 2
- Click OK
- Close the Mesh Tool

The final mesh is shown in Figure 3-1-6.

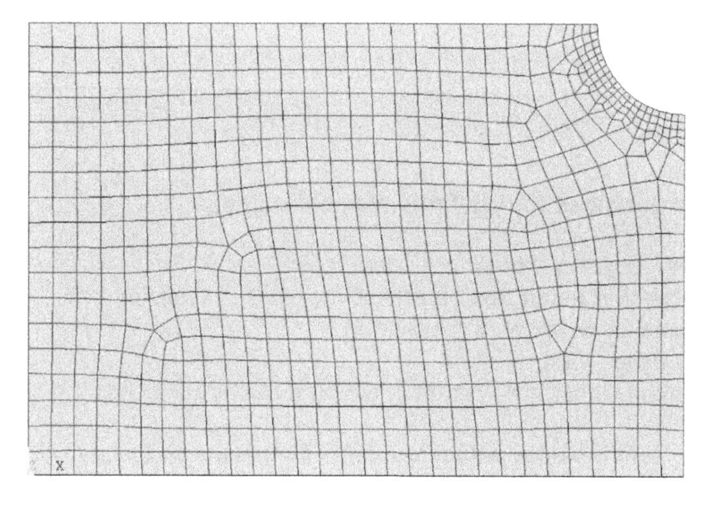

Figure 3-1-6 Refined Finite Element Mesh.

4-3. Save your finite element mesh

- ANSYS Toolbar > SAVE_DB

Step 5: Apply Constraint Boundary Conditions

5-1. Apply symmetry constraints to the model

- Solution > Define Loads > Apply > Structural > Displacement > Symmetry B.C. > On Lines
- Click on Line 4 (top) which represents the axis of symmetry in the vertical direction
- Click on Line 3 (right) which represents the horizontal axis of symmetry
- Click OK

Three S's should appear on each of the selected lines to indicate that symmetry boundary conditions have been applied. In structural analysis, a symmetry boundary condition means no translation across the plane of symmetry and no rotation about the plane of symmetry. You could have defined these constraints directly rather than applying symmetry boundary conditions. For this analysis, the planes of symmetry are the top edge and the right edge of the model. Thus, you could have set translation of the top edge in the y direction and translation of the right edge in the x direction to zero. These boundary conditions would also prohibit rotation about the z axis and provide all of the necessary constraints for the model.

5-2. Save your constraints

- ANSYS Toolbar > SAVE_DB

Step 6: Apply Load Boundary Conditions

6-1. Apply the load to the left edge of the plate

- Solution > Define Loads > Apply > Structural > Pressure > On Lines
- Use the mouse to click on the left edge of the plate
- Click OK
- In the first text box labeled "If Constant value then: VALUE Load PRES value" enter −1e6
- Click OK

The finite element model with the applied boundary conditions is shown in Figure 3-1-7.

6-2. Save your loads

- ANSYS Toolbar > SAVE_DB

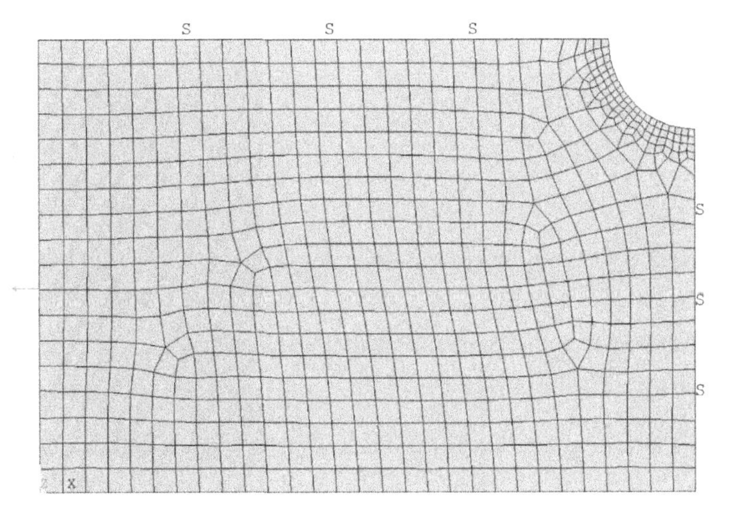

Figure 3-1-7 Element Plot with Applied Boundary Conditions.

Step 7: Set the Solution Options

The default solution options can be used for this analysis.

Step 8: Solve

8-1. Select everything in your model

- Utility Menu > Select > Everything

8-2. Solve

- Solution > Solve > Current LS

8-3. Save your results

- ANSYS Toolbar > SAVE_DB

Step 9: Postprocess the Results

9-1. Plot the von Mises stress distribution in the plate

- General Postproc > Plot Results > Contour Plot > Nodal Solu
- Choose "Stress"
- Scroll down and choose "von Mises stress"
- For the Scale Factor choose "True scale"
- Click OK

Step 10: Compare and Verify the Results

The von Mises stress plot (Figure 3-1-8) shows that the maximum stress in the plate is 0.317e7 (3.17 MPa) and is located along the central hole. The analytical solution predicts that the maximum stress should be 3.14 times the nominal stress (3.14x1.0 MPa) or 3.14 MPa. Comparing these values shows that the numerical solution is within 1% of the theory. Therefore, the model is in excellent agreement with the theory and can be used for engineering design and analysis.

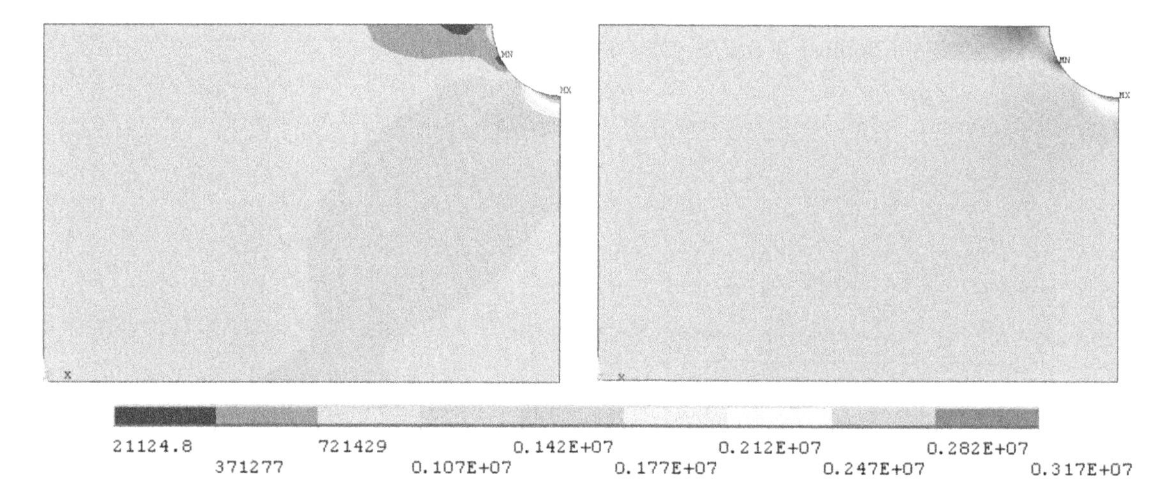

Figure 3-1-8 von Mises Stress Plot: Win32 Graphics (left) and 3D Graphics (right).

Close the Program

- Utility Menu > File > Exit…
- Choose "Quit - No save!"
- Click OK

Sample Input File

```
/PREP7                    ! Enter the Preprocessor
K,,0,0,0                  ! Create keypoint at (0,0,0)
K,,0,0.05,0               ! Create keypoint at (0,0.05,0)
K,,0.075,0,0              ! Create keypoint at (0.075,0,0)
K,,0.075,0.05,0           ! Create keypoint at (0.075,0.05,0)
K,,0.065,0.05,0           ! Create keypoint at (0.065,0.05,0)
K,,0.075,0.04,0           ! Create keypoint at (0.075,0.04,0)
/PNUM,KP,1                ! Turn on keypoint numbering
/PNUM,Line,1              ! Turn on line numbering
LSTR,1,2                  ! Create line from KP 1 and KP 2
LSTR,1,3                  ! Create line from KP 1 and KP 3
LSTR,3,6                  ! Create line from KP 3 and KP 6
LSTR,2,5                  ! Create line from KP 2 and KP 5
CSKP,11,1,4,5,6,1,1       ! Create local coordinate system (#11) from
                          ! Keypoints 4, 5 and 6
L,5,6                     ! Create line from KP 5 and KP 6
CSYS,0                    ! Active coordinate system: Global Cartesian
AL,ALL                    ! Create area from all lines
ET,1,PLANE183             ! Use PLANE183 elements
MP,EX,1,7.31e10           ! Define Young's modulus for material #1
MP,PRXY,1,0.33            ! Define Poisson's ratio for material #1
SMRT,3                    ! Element size determined by Smart Size of 3
AMESH,ALL                 ! Mesh all areas
LREFINE,5,,,2,1,1,1       ! Refine mesh (level 2) along Line 5
FINISH                    ! Finish and Exit Preprocessor
/SOLU                     ! Enter the Solution Processor
LSEL,S,LOC,X,0.075        ! Select line of symmetry at X = 0.075
LSEL,A,LOC,Y,0.050        ! Also select line of symmetry at Y = 0.05
DL,ALL,,SYMM              ! Apply symmetry boundary condition to lines
LSEL,S,LOC,X,0            ! Select line at X = 0
SFL,ALL,PRES,-1e6,        ! Apply pressure (tension) to line
ALLSEL                    ! Select everything
SOLVE                     ! Solve the model
FINISH                    ! Finish and Exit Solution
/POST1                    ! Enter the General Postprocessor
/DSCALE,ALL,1             ! Use true scale (scale displacements by 1.0)
PLNSOL,S,EQV,0,1.0        ! Plot equivalent (von Mises) stress
FINISH                    ! Finish and Exit Postprocessor
SAVE                      ! Save the database
!/EXIT                    ! Exit ANSYS
```

Exercise 3-2

Top-Down Solid Modeling of a Pipe Flange Using Symmetry

Overview

In this exercise, you will create a finite element model of a pipe flange using top-down solid modeling techniques. Although the flange itself is axisymmetric, the bolt circle is not. Instead, you will take advantage of the fact that the part is symmetric about two axes and model only 90 degrees of the pipe flange.

The pipe and the flange are made from high carbon steel with a thermal conductivity of 40 W/mK. The pipe is a hollow cylinder 25 cm long, with an inner radius of 8.75 cm and an outer radius of 10 cm. The flange is a hollow cylinder 2.5 cm thick, with an inner radius of 10 cm and an outer radius of 25 cm. The bolt holes are located 20 cm from the center of the flange. They are spaced at an interval of 30 degrees and have a radius of 1.25 cm (Figure 3-2-1).

The fluid inside the pipe is assumed to be at 95°C (368 K) with a convective heat transfer coefficient of 1000 W/m^{2}K. The pipe is cooled via natural convection with a heat transfer coefficient of 20 W/m^{2}K. The surrounding air is at 25°C (298 K). The goal is to find the steady-state temperature distribution and thermal flux in the pipe flange.

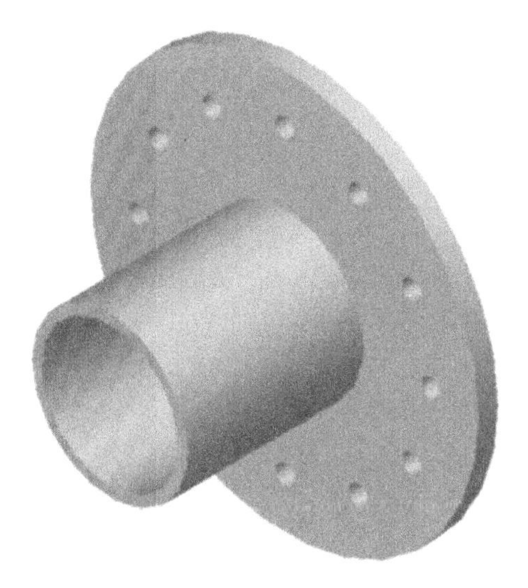

Figure 3-2-1 Pipe Flange.

ANSYS Mechanical APDL for Finite Element Analysis.
DOI: http://dx.doi.org/10.1016/B978-0-12-812981-4.00016-2

Model Attributes

Material Properties for High Carbon Steel

- Thermal conductivity—40 W/mK

Loads

- Thermal convection of 1000 W/m^{2}K with a bulk temperature of 95°C on the inside of the pipe
- Thermal convection coefficient of 20 W/m^{2}K with a bulk temperature of 25°C on the outside of the pipe/pipe flange

File Management

Create a new folder in your "Intro-to-ANSYS" folder named "Exercise3-2"

Open a new session of ANSYS using the Mechanical APDL Product Launcher

Change the Working Directory to the new "Exercise3-2" folder

Change the Jobname to "Exercise3-2"

Click Run to start ANSYS

Step 1: Define Geometry

1-1. Create the pipe cylinder

- Preprocessor > Modeling > Create > Volumes > Cylinder > By Dimensions
- Enter 0.10 for RAD1
- Enter 0.0875 for RAD2
- Enter 0 for Z1
- Enter 0.25 for Z2
- Enter 0 for THETA1
- Enter 90 for THETA2
- Click Apply

1-2. Adjust the view

- Click the "Isometric View" button at the top of the Pan Zoom Rotate shortcut menu (Figure 3-2-2, left)
- Click the "Fit View" button in the middle of the Pan Zoom Rotate shortcut menu (Figure 3-2-2, right)

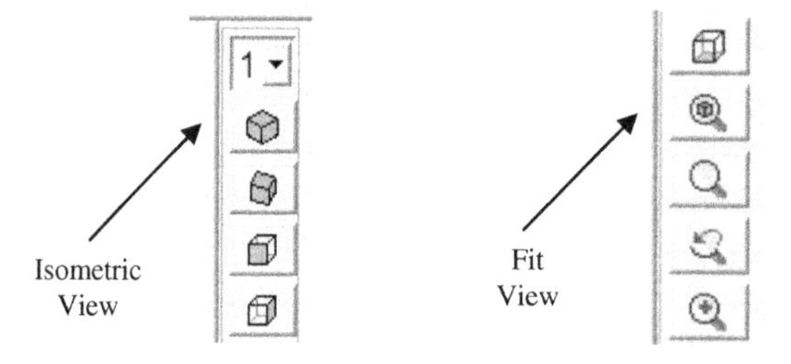

Figure 3-2-2 Isometric View (left) and Fit View (right) Buttons in the Graphic Window's Pan Zoom Rotate Shortcut Menu.

Or

- Utility menu > PlotCtrls > Pan Zoom Rotate...
- Click the "Iso" button in the second section
- Click the "Fit" button at the bottom of the menu
- Click "Close"

1-3. Bring the Create Cylinder by Dimensions dialog box back to the foreground

- Click the "Raise Hidden" button to the right of the command prompt or reopen the dialog box if needed

1-4. Create the flange cylinder

- Enter 0.25 for RAD1
- Enter 0.0875 for RAD2
- Enter 0 for Z1 and 0.025 for Z2
- Enter 0 for THETA1
- Enter 90 for THETA2
- Click OK

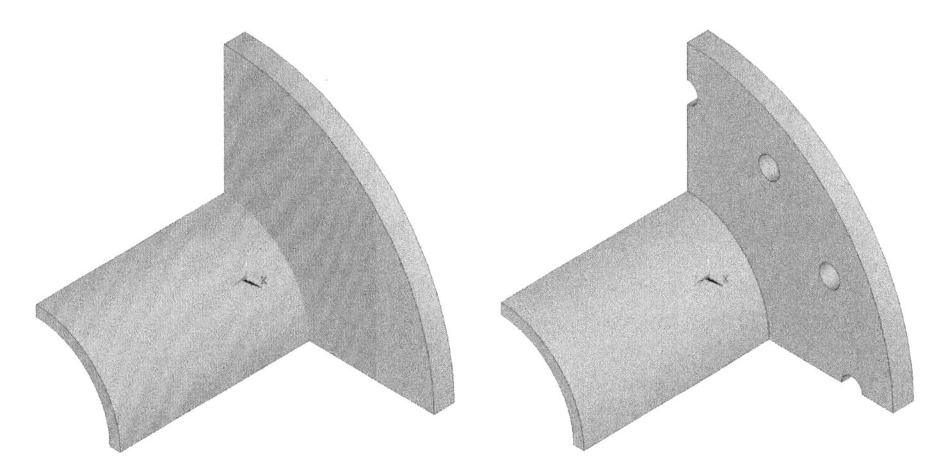

Figure 3-2-3 Solid Model for Two Quarter Cylinders (left) and Complete Solid Model (right).

This will create a volume that overlaps the existing pipe (Figure 3-2-3, left). Recall that in ANSYS a physical overlap does not mean that two bodies are connected. If they are meshed independently, they will have no nodes or elements in common and no loads will be transferred between them. These cylinders could be added to become a single part. However, this would make selecting the individual volumes difficult during postprocessing. Instead, you will perform a Boolean overlap operation. This will fully connect the two volumes (creating a third volume where they overlap) without losing the ability to distinguish one from the other.

1-5. Overlap the two cylinders to create the pipe flange

- Preprocessor > Modeling > Operate > Booleans > Overlap > Volumes
- Click Pick All

1-6. List the volumes in the model to confirm that the overlap operation was successful

- Utility menu > List > Volumes
- Close the VLIST Command window when you are finished

Three volumes should be listed: volumes 3, 4, and 5. Volumes 1 and 2 were deleted as part of the Boolean operation.

1-7. Save your progress

- ANSYS Toolbar > SAVE_DB

1-8. Create a solid cylinder to represent the first bolt hole

- Preprocessor > Modeling > Create > Volumes > Cylinder > Solid Cylinder
- Enter 0 for WP X
- Enter 0.2 for WP Y
- Enter 0.0125 for Radius
- Enter 0.05 for Depth
- Click OK

This cylinder will eventually be subtracted from the pipe flange to create the first bolt hole. Its length is not important as long as it passes completely through the pipe flange.

1-9. Change the coordinate system from Global Cartesian to Global Cylindrical

- Utility Menu > WorkPlane > Change Active CS to > Global Cylindrical

1-10. Copy the volume that will become the bolt hole to create the bolt circle

- Preprocessor > Modeling > Copy > Volumes
- Click the bolt hole volume
- Click OK
- For "ITIME Number of Copies" (including the original) enter 4
- For "DY Y-offset in active CS" enter −30

In cylindrical coordinates, y represents the angle θ in degrees.

- Click OK

1-11. Subtract the bolt hole volumes to create the bolt holes

- Preprocessor > Modeling > Operate > Booleans > Subtract > Volumes
- Click on the pipe flange

Clicking near the center of the areas will give you the greatest chance of picking the correct entity.

- Click OK
- Click on the four cylinders representing the bolt holes
- Click OK

1-12. Rotate the model

- Click the "Dynamic Model Mode" button at the bottom of the Pan Zoom Rotate menu located to the right of the Graphics Window (Figure 3-2-4, left)

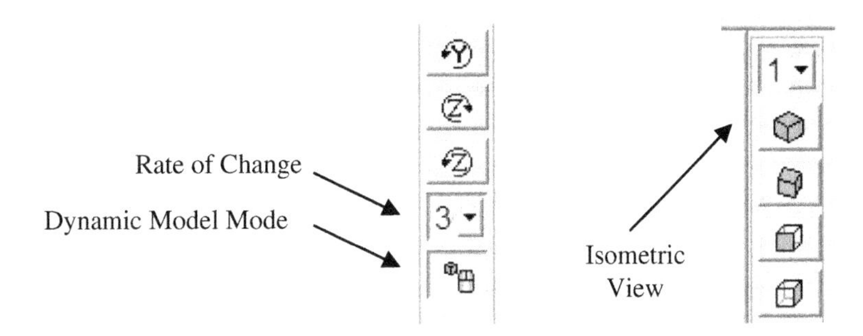

Figure 3-2-4 Rate of Change and Dynamic Model Mode Buttons (left) and Isometric View Button (right) in the Graphic Window's Pan Zoom Rotate Menu.

- Click on an object in the Graphics Window using the left mouse button
- While holding the left mouse button, move the mouse. The model should translate (pan) in the Graphics Window
- Click on an object in the Graphics Window using the right mouse button
- While holding the right mouse button, move the mouse. The model should rotate in the Graphics Window.
- Click the "Isometric View" button in the Pan Zoom Rotate menu to return to the isometric view (Figure 3-2-4, right)

Depending on your graphics and mouse settings, you may need to adjust the Rate of Change for Model Manipulation in order to effectively use Dynamic Mode. You can do this using the Rate slider in the Pan Zoom Rotate dialog box from the PlotCtrls menu or using the Rate of Change drop down box in the Pan Zoom Rotate menu in the Graphics Window (Figure 3-2-4, left).

The completed solid model is shown in Figure 3-2-3 (right)

1-13. Save the model geometry

- ANSYS Toolbar > SAVE_DB

Step 2: Define Element Types

2-1. Define the element type required for this model

- Preprocessor > Element Type > Add/Edit/Delete
- Click the "Add..." button to bring up the Library of Element Types
- In the left column, scroll down until you see "Thermal Mass" and then choose the "Solid" option below it
- At the bottom of the right-hand column, choose "Tet 10node 87"
- Click OK
- Close the Element Types dialog box

SOLID87 is a 3D thermal solid that can be used in thermal analyses. It has 10 nodes with one degree of freedom (temperature) at each node. Because of its tetrahedral shape, this element is well suited to irregular meshes.

Step 3: Define Material Properties

3-1. Create a thermal material model for high carbon steel

- Preprocessor > Material Props > Material Models
- Click on Thermal
- Click on Conductivity
- Click on Isotropic
- Click in the KXX box and enter 40
- Click OK
- Close the material model dialog box

KXX stands for thermal conductivity (k) in the x direction. Since steel is an isotropic material and this is a static analysis, KXX is the only material property required.

3-2. Save your progress

- ANSYS Toolbar > SAVE_DB

Step 4: Mesh

This problem requires a fine mesh to produce accurate results. Some ANSYS licenses (especially educational versions) may not support the number of elements needed for this problem. If you have a very low element limit, use a Smart Size of 3 in step 4-1 and skip steps 4-2-2 and 4-2-3.

4-1. Create the mesh for the finite element model

- Preprocessor > Meshing > MeshTool
- Check the "Smart Size" button in the Mesh Tool dialog box
- Adjust the Smart Size Slider to 1
- Click the "Mesh" button
- Click the "Pick All" button in the Mesh Areas dialog box

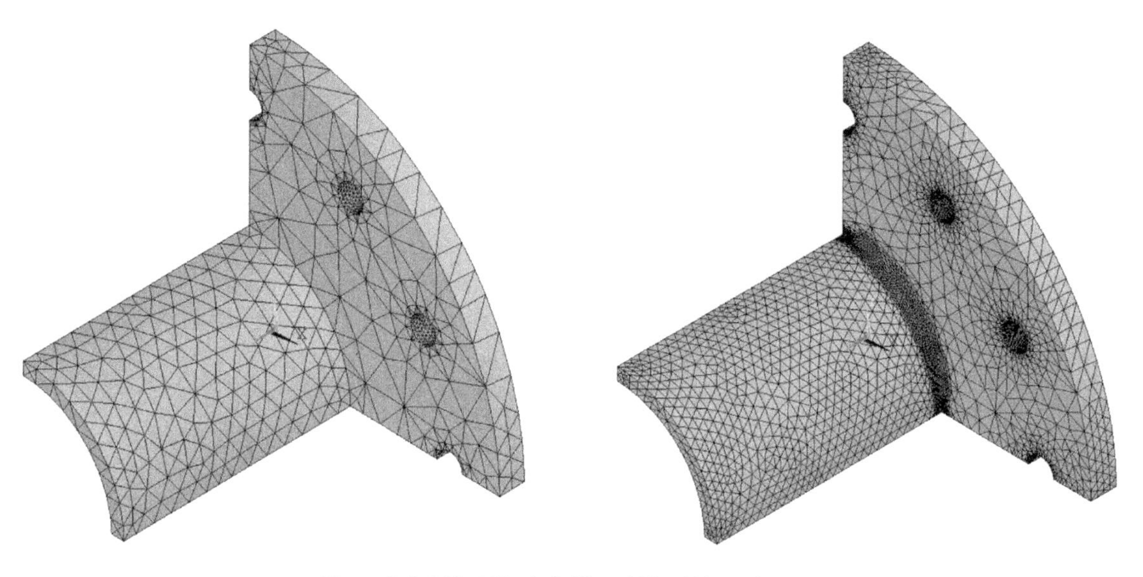

Figure 3-2-5 First Mesh (left) and Final Mesh (right).

This finite element mesh (Figure 3-2-5, left) is fairly coarse. In particular, there is only one element through the thickness of both the pipe and the flange. In order to adequately model the temperature gradient through the pipe wall, you will need to refine the mesh.

4-2. Refine the mesh

4-2-1. Refine the mesh in the entire model

- Bring the Mesh Tool back to the foreground or reopen it if you have closed it
- Ensure that the "Refine at:" option is set to "Elements"
- Click the "Refine" button
- Click the "Pick All" button in the Mesh Areas dialog box
- Ensure that the level of refinement is set to "1 (Minimal)"
- Click OK

The refined mesh has approximately 39,000 elements and is sufficient to capture all temperature gradients in the model. However, heat flux calculations are very sensitive to the finite element mesh. Thus, a very fine mesh is required to accurately estimate the maximum heat flux. Since a very high heat flux is expected where the outer edge of the pipe intersects the pipe flange, you must refine the mesh further in this area.

The next mesh refinement will result in approximately 84,000 elements in your model. If your ANSYS license has a relatively low element limit, then you may skip steps 4-2-2 and 4-2-3.

4-2-2. Identify the number of the line where the outer edge of the pipe intersects the flange

- Utility Menu > Plot > Areas
- Utility Menu > PlotCtrls > Numbering...
- Turn line numbering on by clicking the "LINE Line numbers" check box
- Click OK

4-2-3. Refine the mesh along the line where the pipe intersects the flange

- Bring the mesh tool back to the foreground or reopen it if you have closed it
- Change the "Refine at:" option to "Lines"
- Click the "Refine" button
- Click the line along the outer edge of the pipe where the pipe intersects the flange or type the line number (Line 31 in revision 17.2) in the "List of Items"
- Click OK

The line number where the pipe intersects the flange may be hidden until you select the line. It should appear when you click on the line from the Picker.

If you make an error during graphical picking, you can reset the graphical picking operation by clicking the "Reset" button in the Picker.

- Set the level of refinement to 3
- Click OK

The final finite element mesh is shown in Figure 3-2-5 (right).

4-3. Save your finite element mesh

- ANSYS Toolbar > SAVE_DB

Step 5: Apply Constraint Boundary Conditions

5-1. Apply symmetry constraints to the model

For symmetric thermal analyses, the temperature gradient on one side of the plane of symmetry must be equal to the gradient on the other side of the plane of symmetry. No heat flux is possible without a temperature gradient; therefore all symmetric surfaces are adiabatic. If you do not apply any thermal boundary conditions to a surface in ANSYS, that surface is treated as adiabatic by default. Thus, no special symmetry boundary conditions are needed in this model.

5-2. Apply temperature constraints to the model

There are no temperature constraints for this problem. The temperatures are defined as part of the thermal loads.

Step 6: Apply Load Boundary Conditions

6-1. Apply convection boundary conditions to the inside of the pipe

The areas associated with the inside of the pipe are too difficult to select via graphical picking. You will select them based on their locations instead.

6-1-1. Select the areas associated with the inside of the pipe ($r = 0.0875$)

- Utility Menu > Select > Entities...
- Choose "Areas" as the entity type to select from the first drop down menu
- Choose "By Location" as the selection method from the second drop down menu

Note that the dialog box options change to reflect the selection method.

- Ensure that the "X coordinates" radio button is selected
- Enter 0.0875 in the "Min,Max" text box
- Ensure that "From Full" is selected
- Click OK

Since ANSYS is still operating in cylindrical coordinates, the x coordinate represents the radial distance from the origin.

6-1-2. Plot the model areas to ensure that the correct areas were selected

- Utility Menu > Plot > Areas

6-1-3. Apply convection boundary conditions to the inside of the pipe

- Solution > Define Loads > Apply > Thermal > Convection > On Areas
- Click "Pick All"
- In the text box labeled "VALI Film coefficient" enter 1000
- In the text box labeled "VAL2I Bulk temperature" enter 95
- Click OK

A red grid should appear along the inside of the pipe to indicate that the load has been applied successfully.

6-2. Select all entities in the model

- Utility Menu > Select > Everything

6-3. Turn area numbering on

- Utility Menu > PlotCtrls > Numbering ...
- Turn line numbering off by unchecking the "LINE Line numbers" box
- Turn area numbering on by checking the "AREA Area numbers" box
- Click OK

6-4. Apply convection boundary conditions to the outside of the pipe and the flange

- Solution > Define Loads > Apply > Thermal > Convection > On Areas
- Click on the outer edge of the flange (Area 9 in revision 17.2)
- Click on the front face of the flange (Area 45 in revision 17.2)
- Click on the outer surface of the pipe (Area 18 in revision 17.2)

You can also select these areas by typing the area numbers separated by commas in the "List of Items" text box

- Click OK
- In the text box labeled "VALI Film coefficient" enter 20
- In the text box labeled "VAL2I Bulk temperature" enter 25
- Click OK

*Blue grids should appear along the three outer surfaces of the pipe flange. The colors of the contours indicate the convection heat coefficient values. If the lines are all the same color, replot the image by following the GUI path **Utility Menu > Plot > Replot** or by typing `REPLOT` in the command prompt.*

6-5. Turn area numbering off

- Utility Menu > PlotCtrls > Numbering...
- Turn area numbering off by clicking the "AREA Area numbers" check box
- Click OK

6-6. Save your loads

- ANSYS Toolbar > SAVE_DB

Step 7: Set the Solution Options

The default solution options can be used for this analysis.

Step 8: Solve

8-1. Select everything in your model

- Utility Menu > Select > Everything

8-2. Solve

- Solution > Solve > Current LS

8-3. Save your results

- ANSYS Toolbar > SAVE_DB

Step 9: Postprocess the Results

9-1. Plot the temperature distribution in the model

- General Postproc > Plot Results > Contour Plot > Nodal Solu
- In "Item to be contoured" choose "DOF Solution > Nodal Temperature"
- For the Scale Factor choose "True scale"
- Click OK

*If there are any graphical distortions in the resulting plot, replot the image by following the GUI path **Utility Menu > Plot > Replot** or by typing `REPLOT` in the command prompt. Clicking any button in the Pan Zoom Rotate menu will also prompt a replot.*

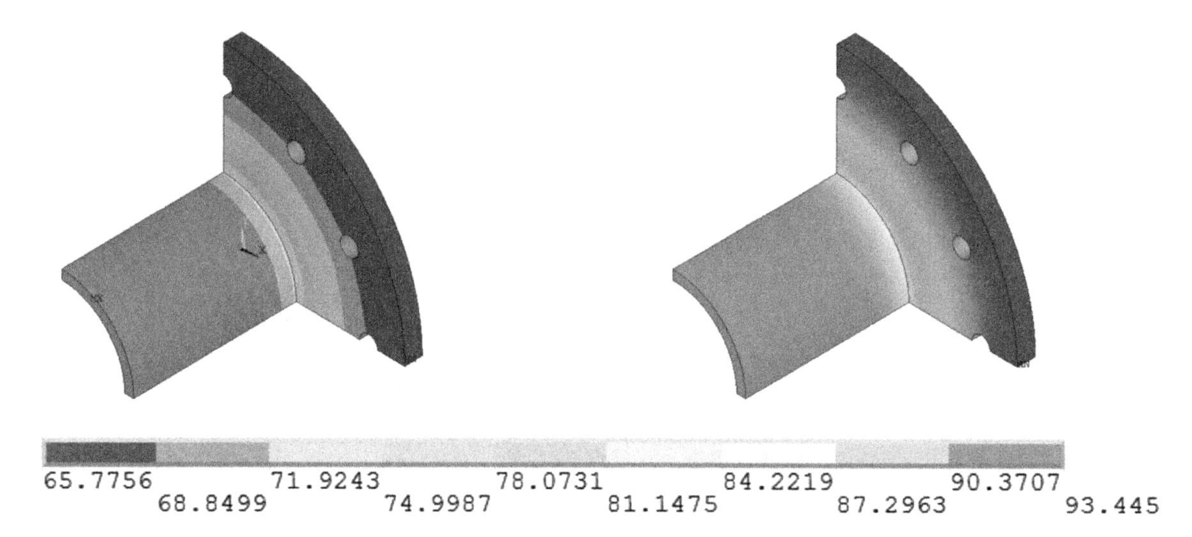

Figure 3-2-6 Temperature Contour Plot: Win32 Graphics (left) and 3D Graphics (right).

In the temperature contour plot (Figure 3-2-6), you can see that the pipe temperature is relatively uniform. The maximum temperature in the model is 93.445°C. This indicates a 1.5°C temperature drop in the fluid near the pipe wall. The minimum temperature in the flange is 65.77°C. Thus, the pipe flange should not be touched with bare skin.

9-2. Plot the thermal flux distribution in the plate

- General Postproc > Plot Results > Vector Plot > Predefined
- Ensure that "Thermal Flux TF" is selected in "Item Vector item to be plotted"
- Click OK

In the thermal flux vector plot (Figure 3-2-7), you can see that most of the heat transfer within the model is from the pipe section to the flange section. Thus, the flange is acting like a finned heat exchanger.

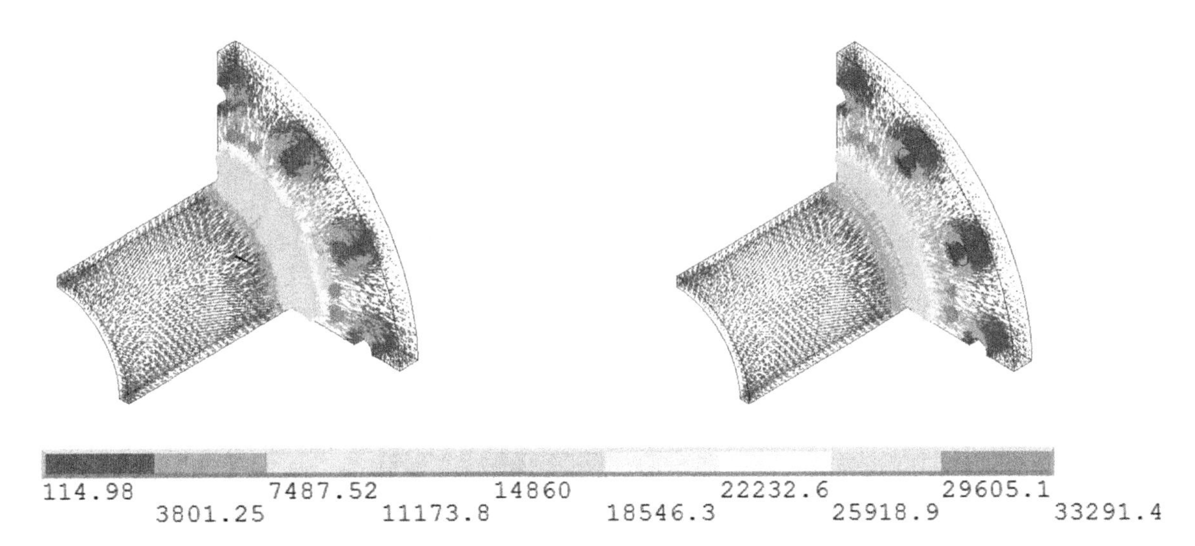

Figure 3-2-7 Thermal Flux Vector Plot: Win32 Graphics (left) and 3D Graphics (right).

9-3. Plot the temperature distribution in the flange

By default, results are plotted for the entire model. However, it is sometimes desirable to look at only a portion of the model. To do so, you must select the part of the model that you wish to view before plotting.

9-3-1. Select the flange volume

- Utility Menu > Select > Entities...
- Choose "Volumes" from the drop down menu
- Choose "By Num/Pick" as the selection method
- Ensure that "From Full" is selected
- Click Apply
- Click on the flange or type the volume number (Volume 8 at revision 17.2) in the "List of Items" text box
- Click OK

9-3-2. Select the elements attached to the selected volume

- Bring the Select Entities menu back to the foreground or reopen it if closed
- Change the entities to select from "Volumes" to "Elements"
- Change the selection method from "By Num/Pick" to "Attached to"
- Click the "Volumes" radio button
- Ensure that "From Full" is selected
- Click OK

9-3-3. Verify that the elements in the flange were selected

- Utility Menu > Plot > Elements

9-3-4. Plot the temperature distribution in the flange

- General Postproc > Plot Results > Contour Plot > Nodal Solu
- In "Item to be contoured" choose "DOF Solution > Nodal Temperature"
- Click OK

9-3-5. Change to the front view

- Click the "Front View" button in the Pan Zoom Rotate menu (Figure 3-2-8, left)
- Click on the "Fit" button in the Pan Zoom Rotate menu if necessary (Figure 3-2-8, right)

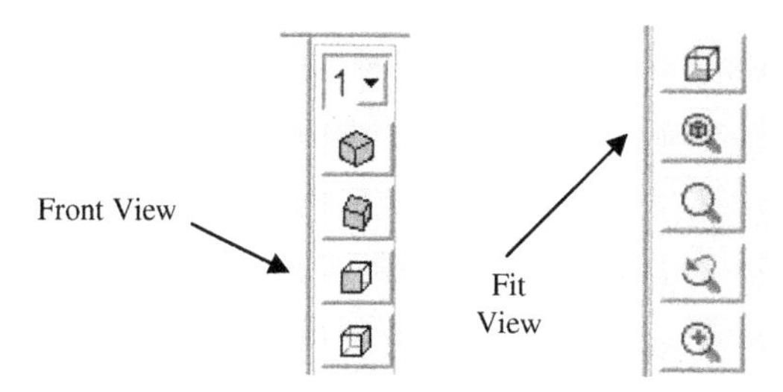

Figure 3-2-8 Front View (left) and Fit View (right) Buttons in the Graphic Window's Pan Zoom Rotate Menu.

The color contours in this plot (Figure 3-2-9) are closer together near the inner radius of the pipe flange and further apart near the outer radius. This implies that the temperature gradient through the radius of the pipe flange is greater near the pipe and smaller toward the outer edge. The temperature distribution through the flange thickness is relatively uniform near the outer edge.

Note that the maximum value of the temperature contour scale at the bottom of Figure 3-2-9 is different than the one in Figure 3-2-6. This is because the contour limits are based on the minimum and maximum values of the currently selected set of nodes and elements and not on the global results values in the model.

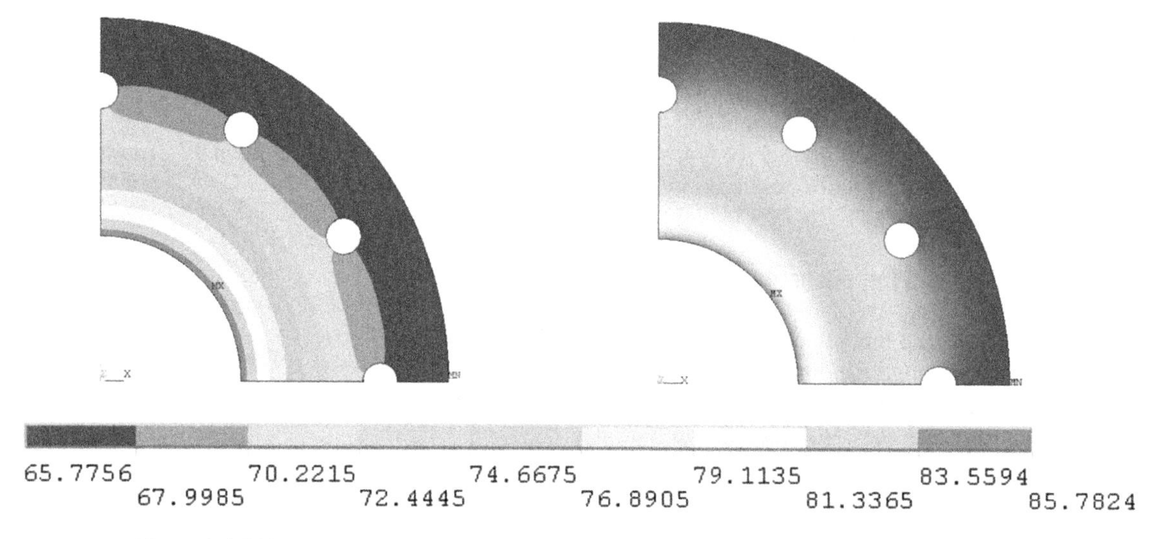

Figure 3-2-9 Temperature Plot of the Flange: Win32 Graphics (left) and 3D Graphics (right).

9-4. Plot the temperature distribution in the pipe

Although there are three volumes in this model, there are only two entities (the flange and the pipe). Thus, any elements that are not attached to the flange must be attached to the pipe. To plot the temperature distribution in the pipe, you can invert the selected element set and then select the nodes attached to those elements.

9-4-1. Invert the selected nodes and elements

- Utility Menu > Select > Entities...
- Ensure that "Elements" are the entity to select
- Click the "Invert" button

If the element set was successfully inverted, the text in the Output Window should be similar to that shown in Figure 3-2-10.

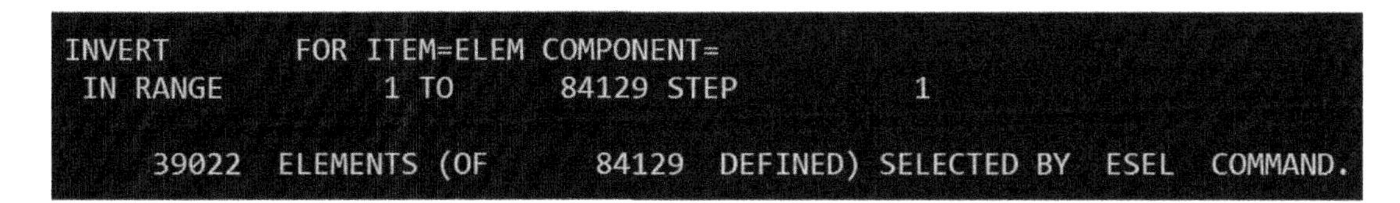

INVERT FOR ITEM=ELEM COMPONENT=
IN RANGE 1 TO 84129 STEP 1
39022 ELEMENTS (OF 84129 DEFINED) SELECTED BY ESEL COMMAND.

Figure 3-2-10 Output Window Confirmation of Element Inversion.

9-4-2. Verify that the elements in the pipe were selected

- Utility Menu > Plot > Elements
- Click the "Isometric View" button in the Pan Zoom Rotate menu to return to the isometric view

If you do not see the elements associated with the pipe or if the confirmation is not present in the Output Window, you may either click the "Invert" button again or select the elements manually:

- Utility Menu > Select > Everything
- Utility Menu > Plot > Volumes
- Utility Menu > Select > Entities...
- Choose "Volumes" as the entity to be selected
- Choose "By Num/Pick"
- Choose "From Full"
- Click Apply
- Click the "Dynamic Model Mode" button in the Pan Zoom Rotate menu to rotate the model until you can see the underside of the pipe
- Uncheck the "Dynamic Model Mode" button so you can proceed with graphical picking
- Click on the two volumes associated with the pipe or type "3,4" in the text box
- Click OK

This operation will select the two volumes associated with the pipe.

- Bring the Select Entities menu back to the foreground or reopen it if closed
- Choose "Elements" as the entity to be selected
- Choose "Attached To"
- Choose "Volumes"
- Choose "From Full"
- Click OK

This will select the elements attached to the volumes associated with the pipe.

- Utility Menu > Plot > Elements

This will ensure that your selecting operation was successful.

9-4-3. Plot the temperature distribution in the pipe

- General Postproc > Plot Results > Contour Plot > Nodal Solu
- In "Item to be contoured" choose "DOF Solution > Nodal Temperature"
- Click OK
- Click on the "Dynamic Model Mode" button at the bottom of the Pan Zoom Rotate menu
- Hold the right mouse button and rotate pipe to show the inside of the pipe

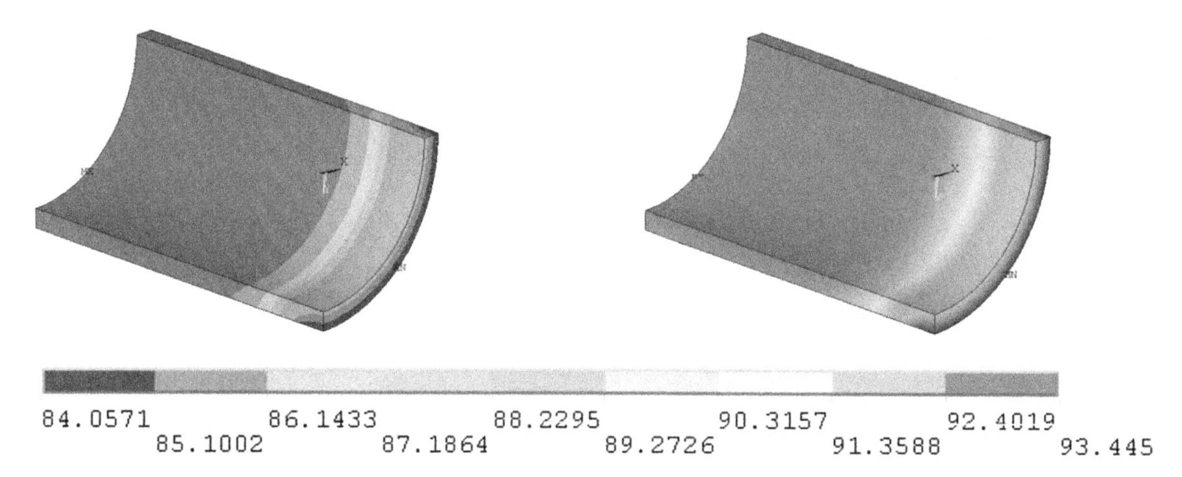

Figure 3-2-11 Temperature Plot of the Pipe: Win32 Graphics (left) and 3D Graphics (right).

Note that both the maximum and minimum values of the temperature contour scale at the bottom of Figure 3-2-11 are different than the ones in Figure 3-2-9. Also note that the temperature through the pipe is not uniform where it is in contact with the pipe flange due to the asymmetric heat transfer associated with conduction into the flange and convection into the surroundings.

9-5. Plot the through wall temperature distribution in the pipe far from the flange

As an analyst, you may also be interested in the temperature distribution through the wall of the pipe.

9-5-1. Change to the isometric view

- Click on the "Isometric View" button in the Pan Zoom Rotate menu

9-5-2. Select the area at the hot end of the pipe

- Utility Menu > Select > Entities...
- Change the type of entity to select from "Element" to "Areas"
- Change "Attached to" to "By Num/Pick"
- Ensure that "From Full" is selected
- Click Apply
- Click on the area at the hot end of the pipe (Area 2 at release 17.2)
- Click OK

9-5-3. Select the nodes attached to the area at the hot end of the pipe

- Bring the Select Entities menu back to the foreground or reopen it if closed
- Change "Areas" to "Nodes"
- Change "By Num/Pick" to "Attached to"
- Change radio button to "Areas, all"
- Click Apply

Picking "Areas, interior" will select all of the nodes attached to the area but exclude the nodes that are on the lines and keypoints. "Areas, all" will select all of the nodes attached to the area including those on the lines and at the keypoints.

9-5-4. Select the elements attached to the selected nodes

- Bring the Select Entities menu back to the foreground or reopen it if closed
- Change "Nodes" to "Elements"
- Ensure that "Attached to" and "Nodes" are selected
- Click OK

9-5-5. Change to the front view

- Click on the "Front View" button in the Pan Zoom Rotate menu
- Click on the "Fit" button in the Pan Zoom Rotate menu if necessary

9-5-6. Plot the temperature through the cross section of the pipe

- General Postproc > Plot Results > Contour Plot > Nodal Solu
- In "Item to be contoured" choose "DOF Solution > Nodal Temperature"
- Click OK

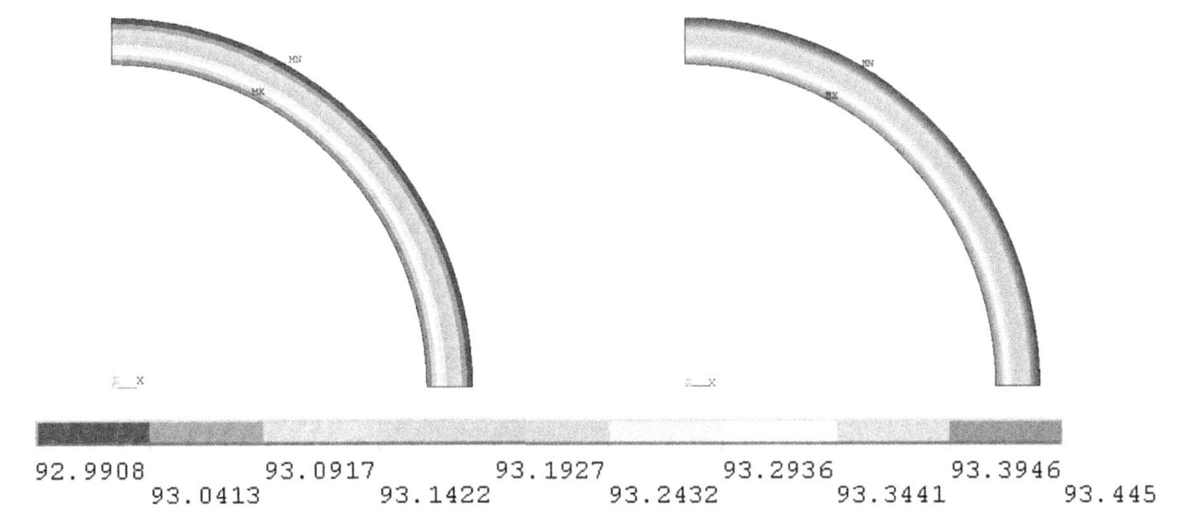

Figure 3-2-12 Temperature Plot Through the Pipe Thickness (Hot End): Win32 Graphics (left) and 3D Graphics (right).

Step 10: Compare and Verify the Results

Although calculating the closed form solution for this model is difficult, there are some simple checks that can be performed to ensure that the model is performing as expected. Because the incoming fluid is 95°C, the temperature at the inside edge of the pipe at the inlet in the model should be just below 95°C. This is confirmed by the fact that the maximum value in the model is located at the pipe inlet and has a value of 93.445°C (Figures 3-2-6, 3-2-11, and 3-2-12). Because the exterior of the pipe is being cooled, we expect the outer edge of the pipe to be cooler than the inner edge. This is confirmed in Figure 3-2-12. Because the pipe flange acts as a fin, we expect the temperature drop in the pipe to increase near the flange. This is confirmed in Figure 3-2-11. Finally, we expect the temperature drop in the flange to be much greater than in the pipe. This is confirmed in Figure 3-2-9. Collectively, these plots show that the model is behaving as expected. These checks do not guarantee that the values in the model are correct. (For example, the supplied convection coefficients could still have been input incorrectly.) However, they do indicate that there are no sign errors in the model and that all boundary conditions were correctly applied.

Close the Program

- Utility Menu > File > Exit...
- Choose "Quit - No save!"
- Click OK

Sample Input File

```
/PREP7                              ! Enter the Preprocessor
/VIEW,1,1,1,1                       ! Set view to isometric (1,1,1)
CYLIND,0.1,0.0875,0,0.25,0,90       ! Create the pipe volume
CYLIND,0.25,0.0875,0,0.025,0,90     ! Create the flange volume
VOVLAP,ALL                          ! Overlap the created volumes
CM,PIPEFLANGE,VOLU                  ! Create a component of the volumes
VSEL,NONE                           ! Unselect all volumes

CYL4,0,0.2,0.0125,,,,.05            ! Create a volume for the first bolt hole
CSYS,1                              ! Change to global cylindrical CS
VGEN,4,ALL,,,,-30,,,0               ! Generate volumes for the other bolt holes
VSEL,ALL                            ! Select all volumes
VSBV,PIPEFLANGE,ALL                 ! Subtract bolt hole cylinders from flange
ET,1,SOLID87                        ! Use SOLID87 elements
MP,KXX,1,40                         ! Define thermal conductivity for material #1
SMRT,1                              ! Use Element Smart Size of 1
VMESH,ALL                           ! Mesh all areas
EREF,ALL,,,1,0,1,1                  ! Refine mesh for all elements
LSEL,S,LOC,Z,0.025                  ! Select all lines at z=0.025
CSYS,1                              ! Change to cylindrical coordinate system
LSEL,R,LOC,X,0.1                    ! Reselect all lines at r=;0.1
LREFINE,ALL,,,3                     ! Refine mesh at selected line (level 3)
ALLSEL,ALL                          ! Select everything in the model

FINISH                              ! Finish and Exit Preprocessor
/SOLU                               ! Enter the Solution Processor
ASEL,S,LOC,X,0.0875                 ! Select all areas inside the pipe radius
```

```
SFA,ALL,1,CONV,1000,95          ! Apply heat convection to all selected areas
KSEL,S,LOC,X,0.1,1              ! Select all keypoints at the pipe OD and beyond
KSEL,R,LOC,Z,0.025,1            ! Reselect keypoints at flange top and beyond
KSEL,A,LOC,X,0.25               ! Add the keypoints at the flange outer diameter
LSLK,S,1                        ! Select all lines defined by the keypoint set
ASLL,S,1                        ! Select all areas defined by the line set
SFA,ALL,1,CONV,20,25            ! Apply heat convection to all selected areas
ALLSEL                          ! Select everything
SOLVE                           ! Solve the model

FINISH                          ! Finish and Exit Solution
/POST1                          ! Enter the General Postprocessor
PLNSOL,TEMP                     ! Plot temperature contours in the model
PLVECT,TF,,,,VECT,ELEM,ON,0     ! Plot thermal flux vectors
VSEL,S,LOC,X,0.1,0.25           ! Select volume associated with the flange
ESLV,S                          ! Select the elements attached to the volume
PLNSOL,TEMP                     ! Plot temperature contours in the flange
ESEL,INVE                       ! Invert the selected element set
PLNSOL,TEMP                     ! Plot temperature contours in the pipe
ASEL,S,LOC,Z,0.25               ! Select the area at the hot end of the pipe
NSLA,S                          ! Select the nodes attached to the area
ESLN,S                          ! Select the elements attached to the nodes
/VIEW,1,,,1                     ! Set view to front (0,0,1)
/AUTO,1                         ! Fit the view to the Graphics Window
PLNSOL,TEMP                     ! Plot temperature contours

FINISH                          ! Finish and Exit Postprocessor
SAVE                            ! Save the database
!/EXIT                          ! Exit ANSYS
```

Exercise 3-3

Structural Analysis of a Simple Warren Truss Using Direct Generation

Overview

In this exercise, you will create a finite element model of a simple Warren truss using direct generation of the nodes and elements. The model will be built using the US customary system (in, lbf s^2/in, s, °F). The truss is 20 in. wide and 10 in. tall. Each member has a 1x1 in. cross section (Figure 3-3-1). The truss is made of aluminum 6061-O with a Young's modulus of 1e7 psi and a Poisson's ratio of 0.33.

The truss will be modeled using spar (truss) elements. This allows uniaxial tension and compression within the members, but no bending of the members. All joints are pinned and can rotate freely. The pin in the lower left corner of the truss is fixed in space. The lower right corner of the truss is supported by rollers. A downward force of 1000 lbf is applied to the bottom center joint of the truss. Because there are no out-of-plane boundary conditions, the truss will be modeled in 2D.

Given these assumptions and boundary conditions, the truss is statically determinate. All displacements and forces can be calculated by hand. Selected results are calculated at the end of the exercise for verification.

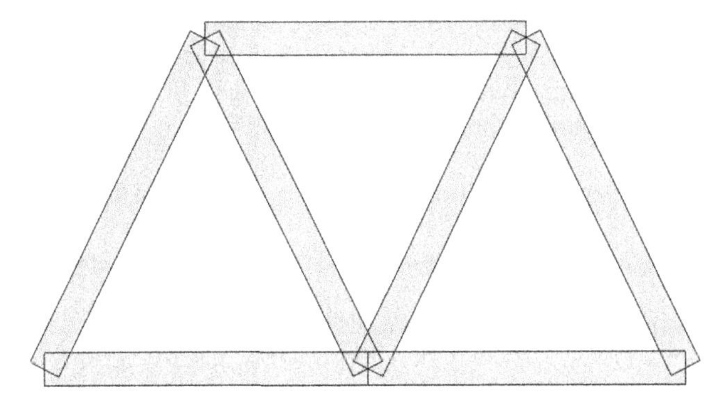

Figure 3-3-1 Schematic of the Warren Truss.

ANSYS Mechanical APDL for Finite Element Analysis.
DOI: http://dx.doi.org/10.1016/B978-0-12-812981-4.00017-4

Model Attributes

Material Properties for 6061-O aluminum

- Young's modulus—1e7 psi
- Poisson's ratio—0.33

Loads

- 1000 lbf downward load on the lower center joint

Constraints

- No displacement on the lower right corner of the truss
- No displacement on the lower left corner of the truss in the vertical direction
- No out-of-plane displacements

File Management

Create a new folder in your "Intro-to-ANSYS" folder named "Exercise3-3"

Open a new session of ANSYS using the Mechanical APDL Product Launcher

Change the Working Directory to the new "Exercise3-3" folder

Change the Jobname to "Exercise3-3"

Click Run to start ANSYS

Step 1: Geometry

This analysis directly generates the nodes and elements for the finite element model, therefore there is no need to create solid model geometry.

Step 2: Define Element Types

2-1. Define the element type required for this model

- Preprocessor > Element Type > Add/Edit/Delete
- Click the "Add..." button to bring up the Library of Element Types
- In the left column under "Structural Mass" choose "Link"
- In the right column, choose "3D finit stn 180"
- Click OK
- Close the Element Types dialog box

LINK180 is a 3D spar with two nodes and three structural DOFs (ux, uy, and uz) at each node.

2-2. Assign section properties to the link elements

- Preprocessor > Sections > Link > Add
- For the Link Section ID, enter 1
- Click OK

- For [SECDATA] Section Data Link area, enter 1
- Click OK

This specifies a cross-sectional area of 1 in^2 for the spars.

Step 3: Define Material Properties

3-1. Define the material properties for aluminum 6061-O

- Preprocessor > Material Props > Material Models
- Click on Structural > Linear > Elastic > Isotropic
- For EX, enter 1e7
- For PRXY, enter 0.33
- Click OK
- Close the material model dialog box

Poisson's ratio is not normally required for analyses involving beams or spars, however ANSYS will issue an error and refuse to solve the model unless a Poisson's ratio is provided.

3-2. Save your progress

Step 4: Mesh (Create Nodes and Elements)

There is no solid model geometry to mesh in this analysis. Instead, we will directly define the nodes and elements for the finite element model.

4-1. Create Node 1 at (0,0)

- Preprocessor > Modeling > Create > Nodes > In Active CS
- Enter 1 for NODE Node number
- Enter 0 in the first text box for X,Y,Z Location in active CS
- Enter 0 in the second text box for X,Y,Z Location in active CS
- Click Apply

4-2. Create Node 2 at (10,0)

- Reopen the Create Nodes in Active Coordinate System dialog box if closed
- Enter 2 for NODE Node number
- Enter 10 in the first text box for X,Y,Z Location in active CS
- Enter 0 in the second text box for X,Y,Z Location in active CS
- Click OK

4-3. Create Node 3 at (20,0)

- Enter the following command into the command prompt:

```
N,3,20,0
```

Press enter to execute the command.

4-4. Create the remaining nodes

- Enter the following commands into the command prompt:

```
N,4,5,10
N,5,15,10
```

4-5. Create an element between Nodes 1 and 2

- Preprocessor > Modeling > Create > Elements > Auto Numbered > Thru Nodes
- Click on Nodes 1 and 2 or enter "1,2" in the List of Items text box
- Click Apply

4-6. Create an element between Nodes 2 and 3

- Click on Nodes 2 and 3 or enter "2,3" in the List of Items text box
- Click OK

4-7. Create the remaining elements

- Enter the following commands into the command prompt:

```
E,1,4
E,4,2
E,2,5
E,5,3
E,4,5
```

The completed finite element model is shown in Figure 3-3-2.

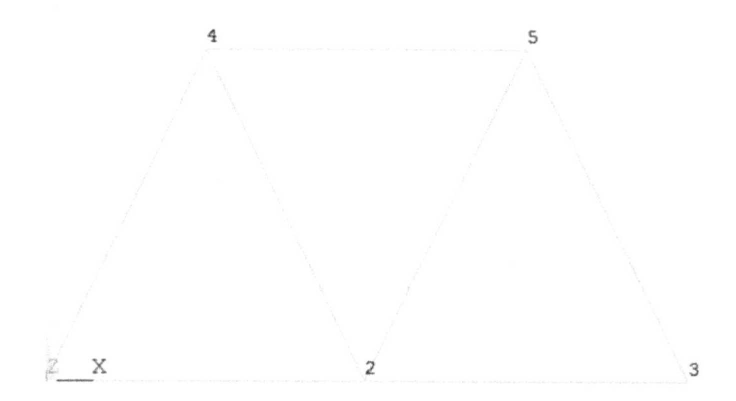

Figure 3-3-2 Finite Element Model Geometry with Node Numbering On.

Step 5: Apply Constraint Boundary Conditions

5-1. Constrain the displacement of the node in the lower left corner in *x* and *y*

- Solution > Define Loads > Apply > Structural > Displacement > On Nodes
- Click on the node at the origin (Node 1) or type "1" in the text box
- Click OK
- For "Lab2 DOFs to be constrained" choose "UX" and "UY"
- For "VALUE Displacement value" enter 0
- Click OK

5-2. Constrain the displacement of the node in the lower right corner (Node 3) in *y*

- Enter the following command into the command prompt:

```
D,3,UY,0
```

This specifies a displacement constraint (D) on Node 3 for the y degree of freedom (UY) with a value of zero (0).

5-3. Constrain the displacement of all nodes in the model in z

- Enter the following command into the command prompt:
```
D,ALL,UZ,0
```

This specifies a displacement constraint (D) on all nodes in the active set (ALL) for the z degree of freedom (UZ) with a value of zero (0).

Since we are using a 3D element, ANSYS will not solve the model until at least one constraint is applied in the z direction. Since we are performing a 2D analysis, we do not want or expect movement in the z direction. However, computer round off errors may generate displacements along the third axis even if no forces are applied. Therefore, we must constraint the model to prevent out-of-plane movement. Constraining the z displacements for all nodes in the model effectively removes this degree of freedom from the solution.

Step 6: Apply Load Boundary Conditions

6-1. Apply a downward force of 1000 lbf to the lower center node (Node 2)

- Solution > Define Loads > Apply > Structural > Force/Moment > On Nodes
- Click on the node at the center of the bottom line (Node 2) or type "2" in the text box
- Click OK
- For "Lab Direction of force/mom" choose "FY"
- For "VALUE Force/moment value" enter −1000
- Click OK

We usually recommend applying boundary conditions by location rather than entity number because it is more robust. However, we specified the node numbers when the nodes were created. Therefore, these entity numbers are known and stable, and can be used for selection and the application of loads and constraints.

6-2. Turn element shape display on

- Utility Menu > PlotCtrls > Style > Size and Shape...
- Turn "[/ESHAPE] Display of element shapes based on real constant descriptions" on
- Click OK

6-3. Plot the elements in the model

- Utility Menu > Plot > Elements

The completed finite element model with the applied boundary conditions is shown in Figure 3-3-3.

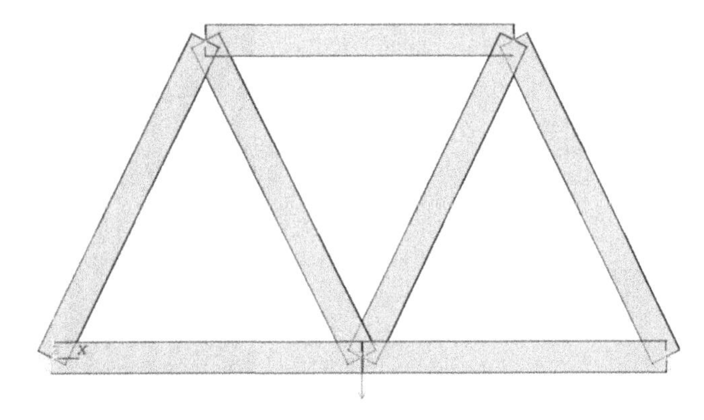

Figure 3-3-3 Finite Element Model Geometry with Boundary Conditions and Element Shapes On.

Step 7: Set the Solution Options

The default solution options can be used for this analysis.

Step 8: Solve

8-1. Select everything in your model

- Utility Menu > Select > Everything

8-2. Solve

- Solution > Solve > Current LS

8-3. Save your results

- ANSYS Toolbar > SAVE_DB

Step 9: Postprocess the Results

When evaluating trusses, we are generally interested in the maximum displacement and the axial forces and stresses in the members. In order to calculate these manually, we also need to know the reaction forces and moments in the system.

9-1. List the nodal reaction solutions

- General Postproc > List Results > Reaction Solu
- Leave "All items" selected
- Click OK

The list of nodal reaction forces (Figure 3-3-4) shows that there are reaction forces of 500 lbf at Nodes 1 and 3 to balance the downward applied load of 1000 lbf at Node 2. There is also an unexpected reaction force listed for Node 1 in the x direction. This was probably generated due to round off error in the program. Since the value of this reaction force is effectively zero, it can be neglected.

PRRSOL Command

File

```
PRINT REACTION SOLUTIONS PER NODE

 ***** POST1 TOTAL REACTION SOLUTION LISTING *****

 LOAD STEP=     1  SUBSTEP=     1
  TIME=    1.0000      LOAD CASE=   0

 THE FOLLOWING X,Y,Z SOLUTIONS ARE IN THE GLOBAL COORDINATE SYSTEM

   NODE       FX           FY          FZ
      1 -0.48317E-12   500.00       0.0000
      2                             0.0000
      3                500.00       0.0000
      4                             0.0000
      5                             0.0000

TOTAL VALUES
VALUE  -0.48317E-12  1000.0       0.0000
```

Figure 3-3-4 List of Nodal Reaction Forces.

9-2. Plot the total displacement in the truss

- General Postproc > Plot Results > Contour Plot > Nodal Solu
- In "Item to be contoured" choose "DOF Solution > Displacement vector sum"
- Click OK

While the displacement in the y direction is symmetric, the displacement in the x direction is not. Therefore, the plot of the displacement vector sum (Figure 3-3-5) is also asymmetric.

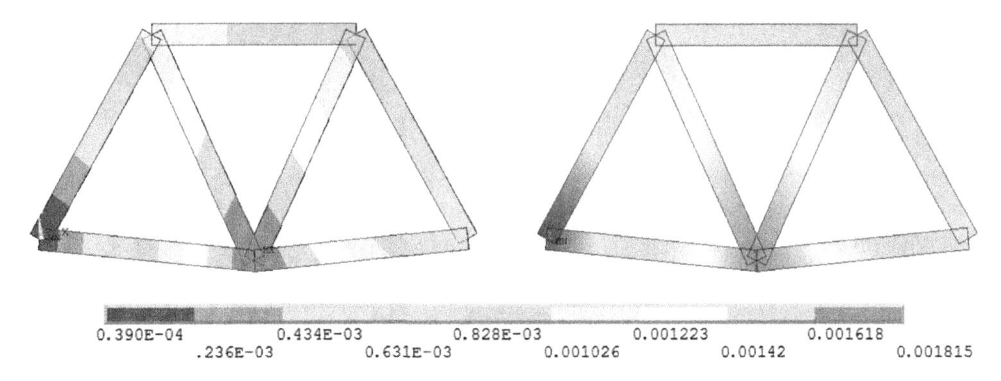

Figure 3-3-5 Plot of Displacement Vector Sum: Win32 Graphics (left) and 3D Graphics (right).

9-3. List the displacement of the joints (nodes) in the truss

- General Postproc > List Results > Nodal Solution
- In "Item to be listed" choose "DOF Solution > Displacement vector sum"
- Click OK

The list of nodal displacements (Figure 3-3-6) confirms that the boundary conditions were correctly applied and that the deformations are almost perfectly symmetric about the midplane.

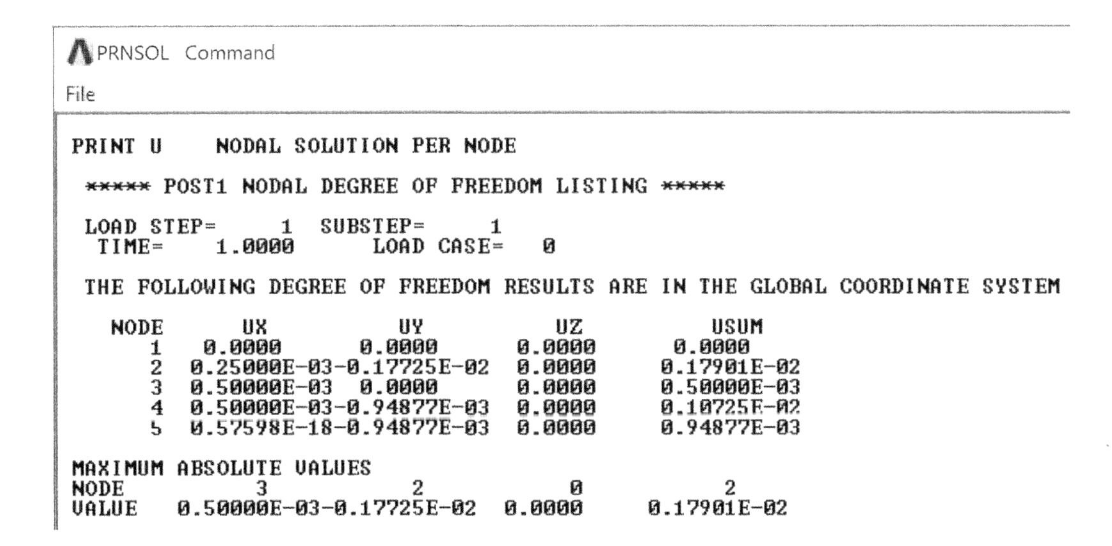

PRNSOL Command

File

```
PRINT U     NODAL SOLUTION PER NODE

 ***** POST1 NODAL DEGREE OF FREEDOM LISTING *****

 LOAD STEP=     1  SUBSTEP=     1
  TIME=    1.0000      LOAD CASE=   0

 THE FOLLOWING DEGREE OF FREEDOM RESULTS ARE IN THE GLOBAL COORDINATE SYSTEM

    NODE       UX          UY          UZ          USUM
       1   0.0000      0.0000       0.0000       0.0000
       2  0.25000E-03-0.17725E-02  0.0000      0.17901E-02
       3  0.50000E-03  0.0000      0.0000      0.50000E-03
       4  0.50000E-03-0.94877E-03  0.0000      0.10725E-02
       5  0.57598E-18-0.94877E-03  0.0000      0.94877E-03

MAXIMUM ABSOLUTE VALUES
NODE          3           2          0            2
VALUE   0.50000E-03-0.17725E-02  0.0000      0.17901E-02
```

Figure 3-3-6 List of the Nodal Displacements.

Note the slight discrepancy between the listed nodal values and the maximum values in the nodal plot. The degree of freedom solution for this problem was only calculated at the five nodes. Therefore the listed degree of freedom results for these five nodes are the "true" values for this model. The values shown in the nodal plots represent the averaged nodal solution, which was derived from the degree of freedom results using the element shape functions. This always involves some approximation and can introduce small discrepancies in the results.

9-4. List the component forces for the members of the truss

- General Postproc > List Results > Element Solution
- Scroll to the bottom of the window
- In "Item to be listed" choose "Element Solution > All Available force items"
- Click OK

The list of element coordinate forces (Figure 3-3-7) shows the forces applied to each element in the three Cartesian coordinates (x, y, and z). The values are reported for each node of each element in the model.

```
PRESOL Command
File

PRINT FORC ELEMENT SOLUTION PER ELEMENT

***** POST1 ELEMENT NODE TOTAL FORCE LISTING *****

 LOAD STEP=      1  SUBSTEP=      1
  TIME=    1.0000        LOAD CASE=   0

 THE FOLLOWING X,Y,Z FORCES ARE IN GLOBAL COORDINATES

 ELEM=       1  FX          FY          FZ
      1   250.00      0.0000      0.0000
      2  -250.00      0.0000      0.0000

 ELEM=       2  FX          FY          FZ
      2   250.00      0.0000      0.0000
      3  -250.00      0.0000      0.0000

 ELEM=       3  FX          FY          FZ
      1  -250.00     -500.00      0.0000
      4   250.00      500.00      0.0000

 ELEM=       4  FX          FY          FZ
      4   250.00     -500.00      0.0000
      2  -250.00      500.00      0.0000

 ELEM=       5  FX          FY          FZ
      2   250.00      500.00      0.0000
      5  -250.00     -500.00      0.0000

 ELEM=       6  FX          FY          FZ
      5  -250.00      500.00      0.0000
      3   250.00     -500.00      0.0000

 ELEM=       7  FX          FY          FZ
      4  -500.00      0.0000      0.0000
      5   500.00      0.0000      0.0000
```

Figure 3-3-7 List of the Element Coordinate Forces.

9-5. List the component forces for the members of the truss

- General Postproc > List Results > Element Solution
- Scroll to the bottom of the window
- In "Item to be listed" choose "Element Solution > Line Element Results > Element Results"
- Click OK

Figure 3-3-8 shows the complete element solution for each line element, including the coordinates for the center of each element (xc, yc, zc), the resulting (axial) force, the resulting (axial) stress, and the resulting elastic equivalent strain (EPEL). The forces and stresses have the same value since the spars have a unit cross section. There is no applied temperature and therefore there is no equivalent thermal strain (EPTH) in this model.

```
PRESOL Command
File

PRINT ELEM ELEMENT SOLUTION PER ELEMENT

***** POST1 ELEMENT SOLUTION LISTING *****

 LOAD STEP      1  SUBSTEP=      1
 TIME=    1.0000          LOAD CASE=   0

EL=       1 NODES=        1         2 MAT=      1  XC,YC,ZC=  5.000       0.000       0.000      AREA=  1.0000                       LINK180
 FORCE=  250.00      STRESS=  250.00      EPEL= 0.25000E-04
 TEMP=     0.00     0.00  EPTH=  0.0000

EL=       2 NODES=        2         3 MAT=      1  XC,YC,ZC=  15.00       0.000       0.000      AREA=  1.0000                       LINK180
 FORCE=  250.00      STRESS=  250.00      EPEL= 0.25000E-04
 TEMP=     0.00     0.00  EPTH=  0.0000

EL=       3 NODES=        1         4 MAT=      1  XC,YC,ZC=  2.500       5.000       0.000      AREA=  1.0000                       LINK180
 FORCE= -559.02      STRESS= -559.02      EPEL=-0.55902E-04
 TEMP=     0.00     0.00  EPTH=  0.0000

EL=       4 NODES=        4         2 MAT=      1  XC,YC,ZC=  7.500       5.000       0.000      AREA=  1.0000                       LINK180
 FORCE=  559.02      STRESS=  559.02      EPEL= 0.55902E-04
 TEMP=     0.00     0.00  EPTH=  0.0000

EL=       5 NODES=        2         5 MAT=      1  XC,YC,ZC=  12.50       5.000       0.000      AREA=  1.0000                       LINK180
 FORCE=  559.02      STRESS=  559.02      EPEL= 0.55902E-04
 TEMP=     0.00     0.00  EPTH=  0.0000

EL=       6 NODES=        5         3 MAT=      1  XC,YC,ZC=  17.50       5.000       0.000      AREA=  1.0000                       LINK180
 FORCE= -559.02      STRESS= -559.02      EPEL=-0.55902E-04
 TEMP=     0.00     0.00  EPTH=  0.0000

EL=       7 NODES=        4         5 MAT=      1  XC,YC,ZC=  10.00       10.00       0.000      AREA=  1.0000                       LINK180
 FORCE= -500.00      STRESS= -500.00      EPEL=-0.50000E-04
 TEMP=     0.00     0.00  EPTH=  0.0000
```

Figure 3-3-8 List of the Element Solutions for All Line Elements in the Model.

Step 10: Compare and Verify the Results

Since the model is deterministic, we can use the schematics in Figure 3-3-9 to calculate the reaction forces and the axial forces in all members of the truss.

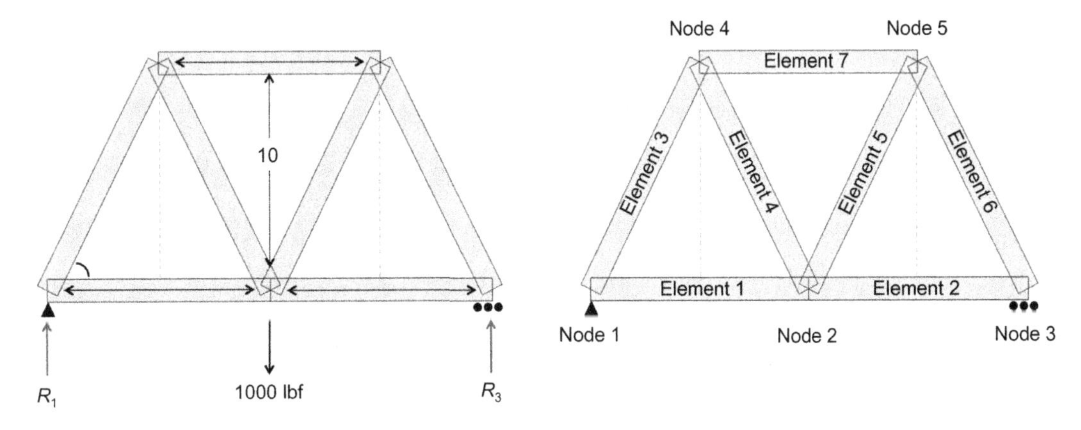

Figure 3-3-9 Schematics of the Truss for Calculating the Analytical Solution.

By inspection, the reaction forces at nodes 1 and 3 should be 500 lbf. This can be verified by setting the sum of the moments about node 1 equal to zero and then solving for the reaction forces at nodes 3 and 1. This is also consistent with the results shown in Figure 3-3-4.

$$\Sigma M_1 = 0 \qquad R_1 + R_3 = 1000$$
$$20R_3 = (10)(1000) \qquad R_1 = 1000 - 500$$
$$R_3 = 500 \qquad R_1 = 500$$

The axial forces in each of the members can be calculated by summing the forces at each of the nodes in the horizontal and vertical directions. This predicts an axial tension in Element 3 of −559.01 lbf and an axial tension in Element 1 of 250 lbf. This is consistent with the results shown in Figure 3-3-8.

$$\Sigma F_{1\text{vert}} = 0 \qquad \Sigma F_{1\text{horz}} = 0$$
$$R_1 = F_{\text{N1}-4} \sin\theta \qquad F_{\text{N1}-2} = F_{14} \cos\theta$$
$$F_{\text{N1}-4} = -500/\sin(\text{atan}(2)) \qquad F_{\text{N1}-2} = 559.01 \cos(\text{atan}(2))$$
$$F_{\text{N1}-4} = F_{\text{E3}} = -559.01 \qquad F_{\text{N1}-2} = F_{\text{E1}} = 250$$

Similar calculations can be made to verify the rest of the results from this model. Based on the agreement between the theoretical and numerical results, we can conclude that the model is in excellent agreement with the theory and can be used for engineering design and analysis.

Close the Program

- Utility Menu > File > Exit...
- Choose "Quit - No save!"
- Click OK

Sample Input File

```
/PREP7               ! Enter the Preprocessor
ET,1,LINK180         ! Use LINK180 elements
SECTYPE,1,LINK       ! Define section #1 for link elements
SECDATA,1            ! Use a cross section of 1 in2
N,1,0,0              ! Create node #1 at (0,0)
N,2,10,0             ! Create node #2 at (10,0)
N,3,20,0             ! Create node #3 at (20,0)
N,4,5,10             ! Create node #4 at (5,10)
N,5,15,10            ! Create node #5 at (15,10)
E,1,2                ! Create an element between Nodes 1 and 2
E,2,3                ! Create an element between Nodes 2 and 3
E,1,4                ! Create an element between Nodes 1 and 4
E,4,2                ! Create an element between Nodes 4 and 2
E,2,5                ! Create an element between Nodes 2 and 5
E,5,3                ! Create an element between Nodes 5 and 3
E,4,5                ! Create an element between Nodes 4 and 5
MP,EX,1,1e7          ! Define Young's modulus for material #1
MP,PRXY,1,0.33       ! Define Poisson's ratio for material #1

/SOLU                ! Enter the Solution processor
D,1,UX,0,,,,UY       ! Constrain Node 1 in x and y
```

```
D,3,UY,0              ! Constrain Node 3 in y
D,ALL,UZ,0            ! Constrain all nodes in the z direction
F,2,FY,-1000          ! Apply downward load of -1000 lbf to Node 2
ALLSEL                ! Select everything
SOLVE                 ! Solve the steady state model
FINISH                ! Finish and Exit Solution
/POST1                ! Enter the General Postprocessor
PLNSOL,U,SUM,0,1      ! Plot the nodal displacement vector sum
PRNSOL,U,COMP         ! Print the nodal structural displacements
PRESOL,FORC           ! Print all available element force items
PRESOL,ELEM           ! Print all available element results

SAVE                  ! Save the database
!/EXIT                ! Exit ANSYS
```

Chapter

4

Elements and Element Input

Suggested Reading Assignments:
Mechanical APDL Element Reference: Chapter 2, Sections 2.1–2.5
Mechanical APDL Element Reference: Chapter 3, Section 3.1
Mechanical APDL Basic Analysis Guide: Chapter 1, Sections 1.1.2 and 1.1.3
Mechanical APDL Modeling and Meshing Guide: Chapter 2, Section 2.2

CHAPTER OUTLINE

In the previous chapter, we discussed some of the assumptions that can be used to simplify and reduce the dimensionality of a finite element model. For example, if a beam has a constant cross section, it can be modeled as a one-dimensional body. If a beam is symmetric with a uniform cross section, it can be modeled as a two-dimensional body. Or, it can be modeled as three-dimensional body. If created correctly, all three types of models will produce the same results. However, they will have different solid model geometry and must be meshed using different element types (Figure 4.1).

In this chapter, you will learn about the element types available in ANSYS, their properties, and their element families. You will learn how to find and use the element documentation. You will also learn how to choose, define, and delete elements; how to define real constants; and how to define the section properties for your elements.

ANSYS Mechanical APDL for Finite Element Analysis.
DOI: http://dx.doi.org/10.1016/B978-0-12-812981-4.00004-6

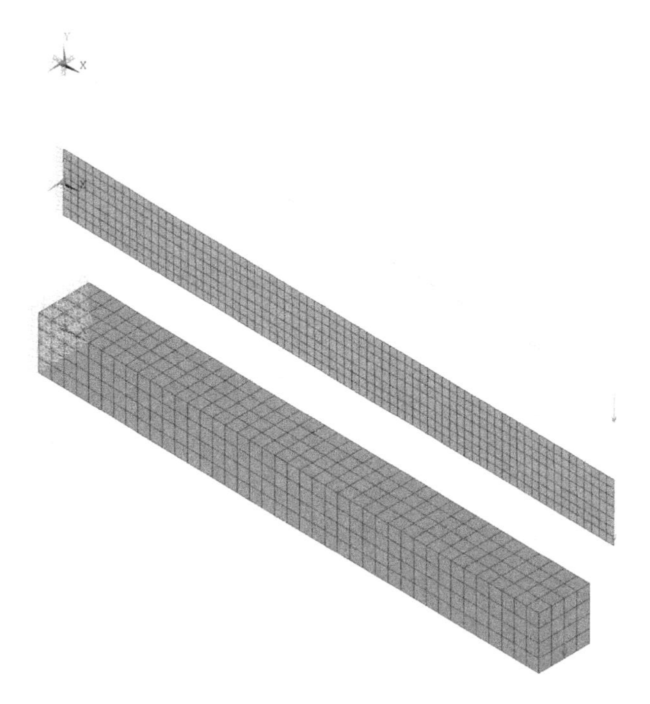

Figure 4.1 Element Plots of a Square Cantilever Beam Modeled using 1D Beam Elements (BEAM189, top), 2D Continuum Elements (PLANE183, center), and 3D Continuum Elements (SOLID185, bottom).

4.1. Element Classification in ANSYS

Elements formulations in ANSYS are classified based on whether they use current technology or legacy technology and whether they are intended for general use or for special-purpose applications. These characteristics determine the location of each element's documentation and whether the element can be accessed through the GUI.

4.1.1. Current-Technology Elements

Most of the element types offered by ANSYS are based on current state-of-the-art technology and are intended for general use. For this reason, they are referred to as current-technology elements. The documentation for current-technology elements can be found in the Mechanical APDL Element Reference. These elements can be accessed normally through the GUI, the command prompt, input files, and batch files.

4.1.2. GUI-Inaccessible Elements

A small number of current-technology elements (including REINF263–265 and SURF251–252) cannot be accessed through the GUI element library. These are referred to as GUI-inaccessible elements. GUI-inaccessible elements must be defined via the command prompt, input files, or batch files using the **ET** command.

4.1.3. Legacy Elements

Legacy elements are older elements whose underlying technology is from an earlier time. Each legacy element has a recommended current-technology replacement. For example, PLANE182 is the current-technology version of PLANE42. Similarly, SOLID185 is the current-technology version of SOLID45. Legacy elements are left in the program for the benefit of analysts who understand the elements' behavior and limitations and who have old input files and macro commands that still might be in use. New analysts, models, and input files should use current-technology elements.

To discourage their use, legacy elements are removed from the GUI element library and their documentation is transferred from the Mechanical APDL Element Reference to the Mechanical APDL Feature Archive. Legacy elements continue to receive limited support but may be undocumented in the future. Legacy elements cannot be accessed through the GUI.

4.1.4. Undocumented Elements

Undocumented elements are legacy elements that have been replaced by improved technology. Undocumented elements are not supported and are not guaranteed to function after being undocumented. To prevent their use, undocumented elements are removed from the GUI element library and all references to them are removed from the documentation. LINK1 is an example of an undocumented element.

4.1.5. Superelements

As noted in chapter 1, superelements are specialty elements that can be used to replace a large section of a finite element model in order to reduce the overall cost of the analysis. Currently, there is only one superelement: MATRIX50. Like other current-technology elements, superelements can be accessed normally through the GUI, the command prompt, input files, and batch files.

4.1.6. User Elements

Finally, user elements are specialty elements that allow you to define your own element formulation. The old user elements are USER100 through USER105. These are documented in the Mechanical APDL Programmer's Reference in Part II: Guide to User-Programmable Features. The newer USER300 is included in the Mechanical APDL Element Reference. User elements cannot be accessed through the GUI.

4.2. The ANSYS Element Library

The ANSYS Element Library in the Mechanical APDL Element Reference is the best source of information for every current-technology element in the program. You can access the program documentation from the Windows Start Menu, by clicking the Product Help button at the bottom of the Product Launcher, or by clicking the Help button in the GUI Utility Menu. Once the documentation is open, you can access the Element Library by following the documentation path: **Mechanical APDL > Element Reference > I. Element Library**. You can also reach the entry for an individual element from the command prompt by using the **HELP** command followed by a comma and the name of the element. For example, `HELP,PLANE55` or `HELP,55` will open the PLANE55 entry in the Element Library.

Figure 4.2 shows a typical entry in the ANSYS Element Library. The name of the element (PLANE55) and its physics (2D Thermal Solid) are listed at the top left side of the entry. This is followed by a description of the element's use and other elements that can be used in its place. At the top right are the product codes that identify which members of the ANSYS Mechanical APDL family can use the element. (See section 4.5 for more information on ANSYS product codes.) Each element is also accompanied by a schematic diagram that identifies the node locations, the face numbering for surface loadings, and any permitted degenerate shapes. The schematic for PLANE55 and its degenerate triangular shape (Figure 55.1 of the Mechanical APDL Element Reference) follows the element description.

PLANE55

2-D Thermal Solid

Compatible Products: – | Pro | Premium | Enterprise | Ent PP | Ent Solver | –
Product Restrictions

PLANE55 Element Description

PLANE55 can be used as a plane element or as an axisymmetric ring element with a 2D thermal conduction capability. The element has four nodes with a single DOF, temperature, at each node.

The element is applicable to a 2D, steady-state, or transient thermal analysis. The element can also compensate for mass transport heat flow from a constant velocity field. If the model containing the temperature element is also to be analyzed structurally, the element should be replaced by an equivalent structural element (such as PLANE182).

A similar element with mid-side node capability is PLANE77. A similar axisymmetric element which accepts nonaxisymmetric loading is PLANE75.

An option exists that allows the element to model nonlinear steady-state fluid flow through a porous medium. With this option the thermal parameters are interpreted as analogous fluid flow parameters. See PLANE55 in the *Mechanical APDL Theory Reference* for more details about this element.

Figure 55.1: PLANE55 Geometry

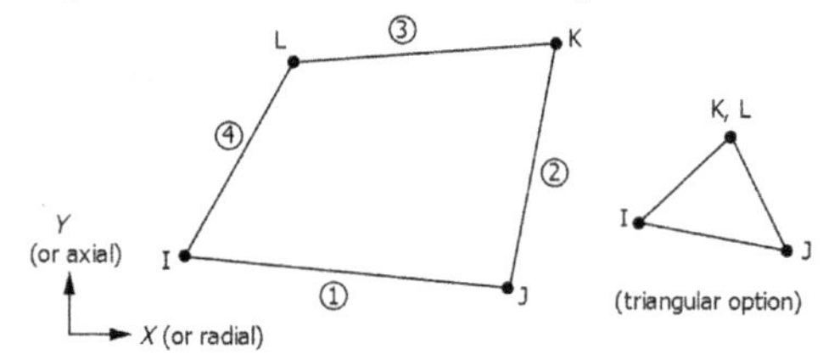

Figure 4.2 ANSYS Element Library Entry Excerpt for PLANE55: Element Description.

4.3. Element Properties

Each ANSYS element has a number of properties including its name, characteristic and degenerate shapes, number of nodes, degrees of freedom, real constants, key options, material properties, permitted loads, and special features. An overview of these features can be found in the input summary of each element in the ANSYS Element Library. The input summary follows the element description and the element input data.

The input summary for PLANE55 is shown in Figure 4.3. This element has four nodes, one degree of freedom (temperature), three possible real constants (thickness and mass transport velocity in x and y), seven possible material properties, three possible surface loads (convection, heat flux, and radiation), one possible body load (heat generation at each of the four nodes), and one special feature (birth and death).

PLANE55 Input Summary

Nodes

I, J, K, L

DOFs

TEMP

Real Constants

THK, VX, VY

THK = Thickness (used only if KEYOPT(3) = 3)

VX = Mass transport velocity in X (used only if KEYOPT(8) > 0)

VY = Mass transport velocity in Y (used only if KEYOPT(8) > 0)

Material Properties

MP command: KXX, KYY, DENS, C, ENTH, VISC, MU (VISC and MU used only if KEYOPT (9) = 1.

Do not use ENTH with KEYOPT(8) = 1 or 2.

Surface Loads

Convection or Heat Flux (but not both) and Radiation (using Lab = RDSF) --

face 1 (J-I), face 2 (K-J), face 3 (L-K), face 4 (I-L)

Body Loads

Heat Generations --

HG(I), HG(J), HG(K), HG(L)

Special Features

Birth and death

Figure 4.3 ANSYS Element Library Entry Excerpt for PLANE55: Input Summary.

4.3.1. Element Names

Each element in ANSYS has a name followed by a number. The combination of a name and a number cannot exceed 8 characters. The name indicates the element's family (FLUID, PLANE, SHELL, SOLID, etc.). The number is a unique identifier called the element routine number. For example, a PLANE182 element is a member of the PLANE element family and it is element routine number 182. When element types are defined using the command line or input files, only the routine number is required. The element family name is optional.

Element routine numbers were originally assigned sequentially and chronologically. Thus, elements with larger routine numbers are generally newer. The list of elements in the Mechanical APDL Element Reference (Figure 4.4, left) has a number of gaps in the element numbering. These gaps can occur for three reasons. First the element may be documented elsewhere. For example, it may be a legacy element (Figure 4.4, right) or a user element. Second, the element may have been undocumented. Finally, the element routine number may be reserved for an element that has not been created or documented yet.

Element Reference
1. About This Reference
2. Element Classifications
3. General Element Features
4. Element Support for Material Models
5. Elements for Stress Analysis
6. Elements for Multiphysics Analysis
7. Element Library
I. Element Library
SOLID5
LINK11
PLANE13
COMBIN14
MASS21
PLANE25
MATRIX27
FLUID29
FLUID30
LINK31
LINK33
LINK34
PLANE35

Feature Archive
About This Archive
I. Legacy Features
II. Legacy Commands
III. Legacy Material Properties
IV. Legacy Elements
BEAM4
CONTAC12
PIPE16
PIPE18
PLANE42
SOLID45
CONTAC52
PIPE59
SHELL63
FLUID79
FLUID80
FLUID81
PLANE82
SOLID92
SOLID95
V. Legacy Theory

Figure 4.4 Element Listings from the ANSYS 17.2 Element Library (Left) and Feature Archive (Right).

4.3.2. Element Shapes

There are eight possible element shapes in ANSYS: points (for point elements); lines (for line elements); triangles or quadrilaterals (for area elements); and tetrahedrons, pyramids, prisms, or bricks (for volume elements). These shapes are shown in Figure 4.5.

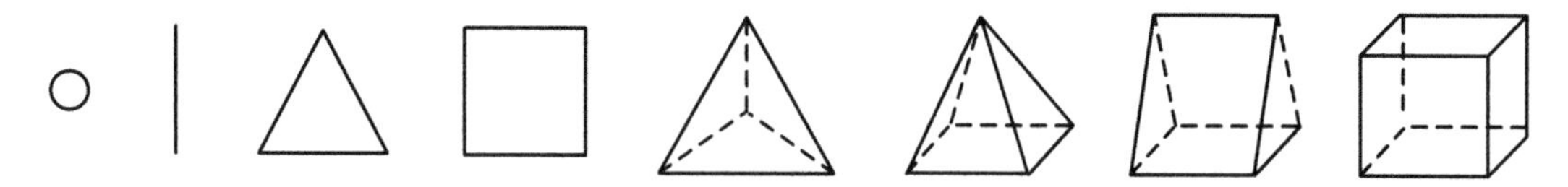

Figure 4.5 Possible Element Shapes in ANSYS.

Points, lines, triangles, quadrilaterals (quads), tetrahedrons (tets), and bricks are the characteristic shapes of most ANSYS elements. With few exceptions (e.g., HSFLD242), pyramids and prisms are degenerate shapes of brick elements. (See section 4.3.4 for more information on degenerate shapes.) In addition, some elements, like MATRIX27 and MATRIX50, do not have a shape.

The graphic pictorials in Section 2.2 of the Mechanical APDL Element Reference seem to indicate that additional element shapes are possible. For example, FLUID38, SOLID272, and SOLID273 appear to be cylinders. But a careful inspection of the element definition reveals that this is not the case. FLUID38 only has 2 nodes. Thus, its underlying geometry is a 1D line, not a 3D cylinder. Its cylindrical properties are included in the model as real constants. (See section 4.3.7 for more information on real constants.) Similarly, SOLID272 and SOLID273 are quadrilateral elements. Their cylindrical properties are included via axisymmetry.

4.3.3. Number of Nodes

All elements have nodes that define their location in space. For example, quadrilateral elements like PLANE55 have four nodes (I, J, K, and L)—one for each corner (Figure 4.6, left). Some elements also have mid-side nodes, or one node between each of the nodes located at the vertices. For example, PLANE77 is the counterpart to PLANE55 with mid-side nodes. It has a total of eight nodes (I, J, K, L, M, N, O and P) - one for each corner and one in the middle of each side (Figure 4.6, right). The node letters refer to the position of the nodes on a generic element (i.e., on the element type). The node lettering convention starts with the letter I and increments one letter per node.

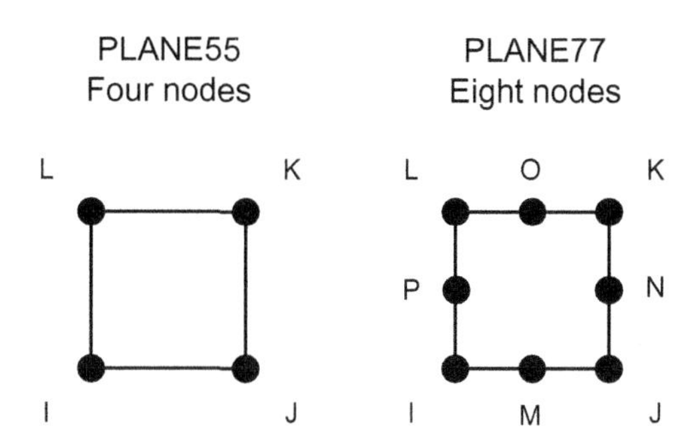

Figure 4.6 Quadrilateral Elements Without and With Mid-Side Nodes.

Some elements in ANSYS, including most triangular and tetrahedral elements and all electromagnetic elements, are only available with mid-side nodes. Triangular and tetrahedral elements require mid-side nodes to decrease their inherent stiffness. Electromagnetic elements require mid-side nodes because their results are only calculated at the mid-side nodes and not at the corner integration points.

Elements with mid-side nodes should be used in structural models where the geometry is irregular and stress gradients are high. Elements without mid-side nodes should be used in nonlinear analyses since the nonlinear results for these models are calculated at the integration points which are only associated with corner nodes. Otherwise, the benefits of using elements with mid-side nodes should be weighed against their cost. In ANSYS, the time required to solve a problem increases with the square of the number of degrees of freedom in the model. The use of elements with mid-side nodes in a 2D model doubles the number of nodes (and the degrees of freedom) and thus quadruples the solution time. The use of a 20-noded brick instead of its 8-noded equivalent increases the solution time by a factor of 6.25. In many cases, this additional cost adds no benefit to the solution. Thus, you should use elements without mid-side nodes by default.

4.3.4. Degenerate Shapes

Quadrilateral and brick elements can be forced into degenerate shapes with one or more triangular faces in order to mesh geometries that could not be meshed otherwise. For example, a quadrilateral element may collapse into a triangle (Figure 4.7, left), while a brick element may collapse into a prism, a pyramid, or a tetrahedron. This is achieved by defining the same node number for multiple nodes (Figure 4.7, right).

Degenerate element shapes are often found in irregular geometries and in the boundary between a fine and a coarse mesh. These elements lose terms in their element shape functions in the process of becoming degenerate and thus become stiffer than their nondegenerated counterparts. (See section 4.3.5 for more details about element shape functions.) For example, compressing an eight-noded brick element into a four-noded tetrahedron reduces the number of independent nodes by half. Thus, it effectively becomes a constant strain element. Degenerate shapes created from elements without mid-side nodes are significantly less accurate than their characteristic shapes and should be avoided in areas of interest (for example, where stress gradients are high).

Degenerate shapes created from elements with mid-side nodes should be used with caution. When degenerate elements are inserted, ANSYS will issue a warning.

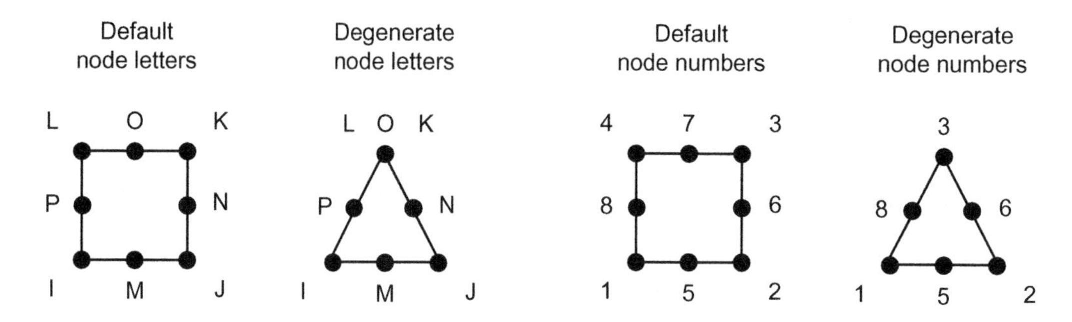

Figure 4.7 Quadrilateral Elements with Default and Degenerate Shapes: Node Letters for a Generic Element (left) and Node Numbers for a Specific Element in a FE Mesh (right).

When choosing an element shape to use, you must assess the risk of creating degenerate elements. If there is little or no risk of creating degenerate shapes, then you should use quadrilateral or brick elements because these elements historically produced more accurate results. However, three-dimensional models can be difficult to mesh using only bricks. For complex geometries, it is often better to use a 10-noded tetrahedral element rather than risk having degenerate brick elements in your mesh.

4.3.5. Element Shape Functions and Extra Displacement Shapes

Some element types allow extra displacement shapes which can be activated through the use of a key option. (See section 4.3.8 for more information on key options.) Extra displacement shapes grant additional flexibility to an element so you may receive some of the benefits of mid-side nodes without some of the disadvantages. To explain this, we will refer briefly to finite element theory.

The displacement U at any point within a four-noded element is calculated by summing some linear mathematical function f of each of the displacements of each of the four nodes I, J, K, and L. The function is not necessarily the same for each node. (The functions themselves are unimportant to this discussion. The details of how element shape functions are implemented in ANSYS can be found in the Mechanical APDL Theory Reference.)

$$U_{\text{element,4N}} = f_1(U_I) + f_2(U_J) + f_3(U_K) + f_4(U_L)$$

Similarly, the displacement U of an eight-noded element is calculated by summing some linear mathematical function f of each of the displacements for the eight nodes I, J, K, L, M, N, O, and P.

$$U_{\text{element,8N}} = f_1(U_I) + f_2(U_J) + f_3(U_K) + f_4(U_L) + f_5(U_M) + f_6(U_N) + f_7(U_O) + f_8(U_P)$$

The presence of more terms in the calculation of the displacement of each element is one of the reasons why it is inefficient to use mid-side nodes unless necessary.

However, the displacement U of a four-noded element with extra displacement shapes is calculated by summing some linear mathematical function f of each of the displacements for the four nodes I, J, K, and L plus some quadratic function g of two "nodeless variables" 1 and 2.

$$U_{\text{element,ESF}} = f_1(U_I) + f_2(U_J) + f_3(U_K) + f_4(U_L) + g_1(U_1) + g_2(U_2)$$

The addition of these two second-order terms gives some of the additional flexibility of mid-side nodes but at a lower computational cost since the calculations are done when the elements are formulated rather than during solution. You should consider using the extra displacement shapes key option when applicable.

4.3.6. Degrees of Freedom

The degrees of freedom (DOFs) are the primary unknowns in the equations that constitute a finite element model. Solving the equations determines the values of the DOFs for each node in the model. These values are referred to as the "primary data" in the documentation. The derived data (stresses, strains, gradients, fluxes, etc.) are calculated from the DOF solution. ANSYS DOFs include displacements (UX, UY, UZ), rotations (ROTX, ROTY, ROTZ), temperature (TEMP), fluid pressure (PRES), fluid kinetic energy (ENKE), magnetic vector potential (AX, AY, AZ), voltage (VOLT), and current (CURR).

The degrees of freedom included in each element type reflect the physics of the underlying problem. You should choose elements that offer only the degrees of freedom that you require since additional degrees of freedom increase computation time and provide no benefit. For example, structural analysis does not (usually) involve heat transfer or voltages. Therefore, the element(s) used to model structures should not have TEMP or VOLT DOFs.

4.3.7. Real Constants

Real constants are values that are required to calculate the element matrices but are not determined by the geometry, element type(s), or material properties. Examples of real constants include values for shell and plate thicknesses, pipe inner and outer diameters, hydraulic diameters, and stiffness scaling factors. Some elements (like continuum elements) do not require real constants.

The real constants that are required may depend on the key option settings so you should review the element documentation carefully before defining an element and its real constants.

4.3.8. Key Options

Elements in ANSYS can have alternate, and sometimes mutually exclusive, capabilities. These capabilities are specified and the behavior and output of the elements are fine-tuned using key options (KEYOPTs). For example, some key options allow you to specify the element coordinate system (global or element specific), the element behavior (plane strain, axisymmetric, etc.), the degrees of freedom included in the model, the use of extra shape functions, and additional results output.

There is a default state for some key options so you should check all default states before assigning an element type. In addition, the default state in an ANSYS multiphysics product may be different than the default state in other members of the ANSYS Mechanical APDL product family. If you might use your model in multiple Mechanical APDL family members then you should specify each key option rather than letting the program assign the default settings.

4.3.9. Required and Permitted Material Properties

The documentation for most elements in ANSYS includes a list of material properties (Figure 4.3). Some material properties are required for every analysis, while others are required only for certain analyses. For example, PLANE55 can accept thermal conductivity in x (KXX) and y (KYY), density (DENS), specific heat (C), enthalpy (ENTH), viscosity (VISC), and boundary admittance (MU). But for simple steady-state problems with isotropic conduction, only thermal conductivity (KXX) needs to be specified. The other material properties are used to simulate more complex phenomena such as thermal transients, mass transport heat flow from a constant velocity field, and nonlinear porous flow.

Some elements do not require any material properties to be specified. For example, contact elements require material properties only if you want to include friction, cohesion (for debonding analyses), or fluid damping. Finally, some elements, such as target elements, do not need and will not accept any material properties.

When the **SOLVE** command is issued, ANSYS checks the model input (including material properties) before beginning to calculate the solution. If an entity in the model is assigned a material number that does not satisfy the material properties of the assigned element type, ANSYS will issue an error and terminate the solution so you can define the required material properties. Defining material properties will be discussed in chapter 5. Assigning element types and material properties to geometric entities will be discussed in chapter 6.

4.3.10. Permitted Loads

Every element in ANSYS has permitted body and surface loads. No loads are required to solve a model, but failing to apply any loads will usually result in an uninteresting solution. Failing to apply any constraints will result in an error ("Check for an insufficiently constrained model") during solution.

The GUI filters the loading options based upon the degrees of freedom for the elements specified and will prevent you from applying inappropriate loads to elements. If you try to apply an inappropriate load via the command prompt, ANSYS will issue a warning saying that the current degree of freedom set does not support the load.

4.3.11. Special Features

Finally, some elements have special features. These are most common in elements that have nonlinear capabilities. Examples of special features include plasticity, swelling, large deflection, large strain, stress stiffening, adaptive descent, error estimation, birth and death, hyperelasticity, and viscoelasticity. Special features are beyond the scope of this book.

4.4. ANSYS Element Families*

Elements in ANSYS are organized in families. (The number and type of element families are subject to change with each new release of the program.) Each element family has a family name and a shared set of properties.

BEAM—These 1D elements have two or three nodes and represent structural beams in either two or three dimensions. Beam properties such as cross sectional area, area moment of inertia, and initial strain are defined as real constants or section properties. Beam elements typically include rotational degrees of freedom whereas continuum elements do not. Thus, connecting beam and continuum elements in a single model is not a simple task.

CIRCU—These 1D elements are used for circuit analysis. Elements in this family include a piezoelectric circuit and an electric circuit.

COMBIN—Combination elements are lumped parameter elements that are generally used in simulations of dynamic systems. Examples of combination elements include a spring-damper system and a nonlinear spring. Generally, these elements are not used in combination with continuum elements.

CONTA—Contact elements represent potentially deformable surfaces that may make, maintain, or break physical contact with one another or slide relative to each other.

CPT—Coupled pore pressure elements represent structural solids with a pressurized fluid located in the interstitial spaces. These elements have both structural and pressure degrees of freedom and can be subject to mechanical, pressure, and thermal loads.

FLUID—Fluid elements are used in analyses involving fluid mechanics and dynamics. They include capabilities for acoustics and wave propagation, fluids contained in vessels, and squeeze film fluids. There are no documented CFD elements in ANSYS Mechanical APDL.

FOLLW—There is only one follower element (FOLLW201). It is a one-noded element that can be overlaid onto an existing node to specify external forces and moments that follow a structure in a nonlinear analysis.

HSFLD—Hydrostatic fluid elements represent fluids that are fully contained within a solid body and can be used to model fluid–solid interaction. Hydrostatic fluids are assumed to have a uniform fluid pressure, temperature, and density. They may not have a free surface. Inertial effects (sloshing) and viscosity are not permitted. However, compressibility and fluid mass can be included.

INFIN—Infinite elements represent infinite boundaries and infinite solids for unbounded electric and magnetic field problems. These elements are intended to replace the large numbers of elements that otherwise would be required to represent a boundary far from the area of interest.

INTER—Interface elements represent magnetic interfaces to couple magnetic vector and scalar potentials, gaskets for problems involving structural assemblies, and cohesive elements for modeling interface failure and delamination.

LINK—Link elements have two or three nodes and represent spars, trusses, linear actuators, radiation links, conduction bars, convection links, and coupled thermal electric lines. For structural problems, links look like beam elements but do not have rotational degrees of freedom or beam bending capability.

MASS—Mass elements have one node and represent structural and thermal concentrated masses.

MATRIX—There are only two matrix elements. MATRIX27 provides an elastic kinematic response that is specified by stiffness, damping, or mass coefficients without the need to specify any underlying geometry. MATRIX50 is a superelement that is created using the AUX15 processor.

MESH—There is only one mesh element (MESH200). It is a "mesh-only" element and is ignored during solution if not replaced by another element type. It can be used for multistep meshing operations or as a placeholder for elements whose physics has not yet been defined.

MPC There is only one multipoint constraint element (MPC184). It allows kinematic constraints to be imposed on parts or bodies in the model using Lagrange multipliers. This can be used to model various types of joint elements in kinematic models.

PIPE—Pipe elements have 2 or 3 nodes and are used to represent straight and curved pipes, tees, and immersed pipes or cables.

PLANE—These are 2D elements with 4, 6, or 8 nodes that can be used in structural, dynamic, thermal, magnetic, or coupled (thermal solid, thermal electric, magnetic electric, and piezoelectric) analyses. They are very commonly used.

PRETS—There is only one pretension element (PRETS179). It is used to model a pretension section in a structure created from other structural elements (solids, beams, shells, pipes, or links).

REINF—These elements allow smeared and discrete fiber reinforcement to be added to solid and shell elements.

ROM—There is only one reduced order model element (ROM144). It represents a reduced order model of a coupled electrostatic–structural system for use in electromechanical circuit simulations.

SHELL—Shell elements represent thin walled structural or thermal systems. The thickness of the shell is defined as a real constant.

SOLID—Solid elements are the 3D equivalent of the plane elements. They have 8, 10, or 20 nodes and can be used for structural, dynamic, thermal, magnetic, or coupled (thermal solid, thermal electric, magnetic electric, and piezoelectric) analyses. There are also layered structural solids that allow layers of different materials to be "stacked" in the same element. They are very commonly used.

SOLSH—There is only one solid structural shell element (SOLSH190). It can represent shells with both thin and moderately thick walls. Structurally, this element behaves like a shell but is defined by the 8 nodes at the corners of the element instead of the 4 nodes at the element midplane. The inclusion of corner nodes makes it easier to connect continuum elements to SOLSH elements than to mid-plane shell elements.

SOURC—There is only one source element (SOURC36). It is a 3-noded magnetic electric current source.

SURF—Surface elements are used for structural or thermal loads and surface effect applications including radiosity. Surface elements are used for the application of extra or special loadings to the surface of a model. Surface elements do not contribute to the element stiffness matrix but they do contribute to the load vector.

TARGE—Target elements are used to define the "target" surfaces for the associated contact elements in a contact analysis.

TRANS—There is only one documented transducer element (TRANS126). It is a two-noded electromechanical transducer that converts energy from the electrostatic domain into the structural domain (and vice versa) while also allowing for energy storage.

Section 2.2 of the Mechanical APDL Element Reference contains a pictorial summary of most current-technology elements. This may help you to visualize each of the element families.

4.5. Product Codes and Product Restrictions

Your access to certain elements may be limited by your ANSYS product license. For example, users with ANSYS Mechanical Pro licenses cannot access any of the dynamics, acoustics, coupled-field, and hydrodynamic elements in the program. Similarly, users without an LS-DYNA license cannot access any of the explicit dynamics elements.

The documentation identifies ANSYS product variants by abbreviated product codes (Table 4.1). The product variants that can access a given element are listed in the upper right corner of each entry in the ANSYS Mechanical APDL Element Reference (see Figure 4.2). Restricted products are indicated by an empty (–) placeholder.

Table 4.1 ANSYS 17.2 Product Codes*

Code	Product
DesSpc	ANSYS Design Space
Pro	ANSYS Mechancial Pro
Premium	ANSYS Mechancial Premium
Enterprise	ANSYS Mechancial Enterprise
Ent PP	ANSYS Mechancial Enterprise PrepPost
Ent Solver	ANSYS Mechancial Enterprise Solver
DYNA	ANSYS LS-DYNA

The product codes for PLANE55 are shown in Figure 4.8. This indicates that users with licenses for ANSYS Mechanical Pro, Mechanical Premium, Mechanical Enterprise, Mechanical Enterprise PrepPost, and Mechanical Enterprise Solver can use the PLANE55 element.

Compatible Products: – | Pro | Premium | Enterprise | Ent PP | Ent Solver | –
Product Restrictions

Figure 4.8 Compatible Products for PLANE55*.

Even if an element is available to you, some of its options may not be. The restrictions for each element can be found at the bottom of its entry in the ANSYS Mechanical APDL Element Reference. You can also access the product restrictions by clicking on the hyperlink underneath the list of supported product codes. The product restrictions for the PLANE55 element are shown in Figure 4.9.

PLANE55 Product Restrictions

When used in the product(s) listed below, the stated product-specific restrictions apply to this element in addition to the general assumptions and restrictions given in the previous section.

ANSYS Mechanical Pro

- This element does not have the mass transport or fluid flow options. KEYOPT(8) and KEYOPT(9) default to 0 and cannot be changed.
- Birth and death is not available.

ANSYS Mechanical Premium

- This element does not have the mass transport or fluid flow options. KEYOPT(8) and KEYOPT(9) default to 0 and cannot be changed.
- Birth and death is not available.

Figure 4.9 ANSYS 17.2 Product Restrictions for PLANE55*.

4.6. Choosing an Element

Choosing which element(s) to use in a finite element analysis can be a daunting task for new users. Until you are familiar with the various ANSYS elements and their behavior, you may find it helpful to choose from a list of commonly used elements or to use the process of elimination.

4.6.1. Bottom-Up Element Selection: Commonly Used Elements

Most basic analyses use 2D or 3D continuum elements with structural or thermal degrees of freedom. These elements are shown in Tables 4.2 and 4.3. In general, the quad (2D) or brick (3D) elements without mid-side nodes (PLANE182 and SOLID185 for structural analyses and PLANE55 and SOLID70 for thermal analyses) should be used unless the geometry is very irregular.

Table 4.2 Commonly Used Structural Continuum Elements

Element	Order	Shape	Nodes	Physics	DOFs
PLANE182	2D	Quad	4	Structural	UX, UY
PLANE183	2D	Quad	8	Structural	UX, UY
SOLID185	3D	Brick	8	Structural	UX, UY, UZ
SOLID186	3D	Brick	20	Structural	UX, UY, UZ
SOLID187	3D	Tet	10	Structural	UX, UY, UZ

Table 4.3 Commonly Used Thermal Continuum Elements

Element	Order	Shape	Nodes	Physics	DOFs
PLANE55	2D	Quad	4	Thermal	TEMP
PLANE77	2D	Quad	8	Thermal	TEMP
SOLID70	3D	Brick	8	Thermal	TEMP
SOLID90	3D	Brick	20	Thermal	TEMP
SOLID87	3D	Tet	10	Thermal	TEMP

4.6.2. Top-Down Element Selection: Process of Elimination

If one of the commonly used elements will not suffice, you can use the process of elimination to choose your element(s). We suggest using the pictorial list of elements in Section 2.2 of the Mechanical APDL Element Reference to visualize and evaluate options. Begin by eliminating:

- All elements that do not address the physics environment(s) that you need
- All elements that do not support the material properties that you need
- All elements that do not permit the body/surface loads that you require
- All elements that do not have the special components, abilities, or features (hyperelasticity, viscoelasticity, birth and death, etc.) that you require
- All elements that are not permitted by your license.

All remaining elements could be used for your analysis. If you have not yet determined the dimensionality of your model, do so now.

- If you can model your system using lower dimension elements (1D solids to model beams, 2D solids to model plates, 1D and 2D shells to model thin walled structures, etc.) then you should eliminate all 3D elements.
- If a 3D model is required, you should eliminate all lower order elements.

Next, eliminate all elements which offer more features than you need, including:

- All elements with more physics environments than you need
- All elements with more degrees of freedom than you need

unless this would eliminate all possible choices. At this point, only a handful of elements should remain.

- If your analysis is nonlinear, eliminate options with mid-side nodes

unless this would eliminate all possible choices. Finally, choose the best element shape for the geometry (quadrilaterals or bricks for regular geometry and triangles or tetrahedrons for irregular geometry) and/or the element that offers the fewest unnecessary degrees of freedom.

4.7. Defining Element Types

Element types can be defined for use in an analysis with the **ET** command or the GUI path: **Main Menu > Preprocessor > Element Type > Add / Edit / Delete > Add. . ..**

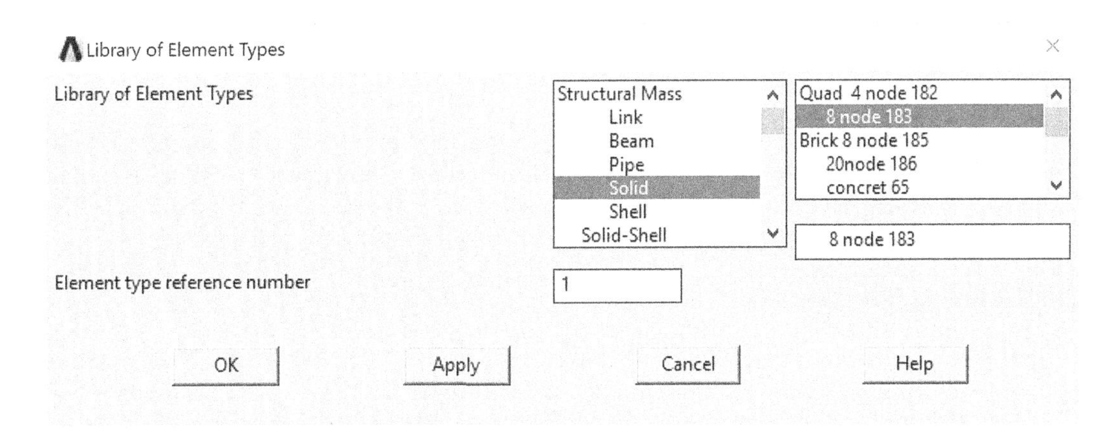

Figure 4.10 Element Library GUI.

Element types in the ANSYS Elements Library GUI are sorted by physics and then by element family in the left column. Clicking on an option in the left column reveals the individual element types in the right column. The elements in the right column are sorted by shape (quad, triangle, brick, tet), then by the number of nodes per element, and finally by the element routine number. To define an element type, select the desired element in the right column and click OK.

Like geometric entities, each element type is assigned a number when it is defined. By default, the element type reference number is assigned based on the order in which the elements were defined, starting with 1 and incrementing with each additional element. You can choose a specific element reference number using the text box in the middle of the bottom of the dialog box (Figure 4.10). You can also specify the element reference number in the second field of the **ET** command when using input files, batch files, and the command prompt.

For example, the command `ET,1,PLANE55` will define element type number 1 as a PLANE55 element and prepare it for use in the analysis. The command `ET,2,PLANE55` will assign an element reference number of 2 instead. Only the element routine number is required so the commands `ET,1,PLANE55` and `ET,1,55` will both result in PLANE55 being associated with element type reference number 1.

There is no limit to the number of times that an element routine can be assigned an element type number. For example, you can define element types #1, #2, and #3 to be PLANE55s. Defining multiple element types with the same routine number allows you specify different key options in different parts of the model. It can also make other operations, such as selecting, more efficient.

4.8. Deleting Element Types

Element types in ANSYS are deleted in much the same way they were created. To delete an element type, bring up the element library using the GUI path: **Main Menu > Preprocessor > Element Type > Add/Edit/Delete**, click on the desired element, and then click the "Delete" button. You can also use the **ETDELE** command.

4.9. Defining Real Constants

Real constants are defined using the **R** command or the GUI path: **Main Menu > Preprocessor > Real Constants > Add/Edit/Delete.** This will open the Real Constants dialog box. Clicking the "Add" option lists the defined element types and permits you to choose the element type for which the real constants will be defined. If the chosen element type requires real constants, a second element-specific dialog box will appear where you can enter these values.

If no element types have been defined when you try to define a real constant in the GUI, ANSYS will alert you with a note and display a generic real constant dialog box. Similarly, ANSYS will issue a note if you select an element for which no real constants are possible or if you need to set your element key options before you can define the real constants.

4.10. Defining Sections

The cross-sectional properties of some elements, such as BEAM188 and BEAM189, are not specified via real constants. Instead, they are defined as "sections." ANSYS offers a library of 11 commonly used cross-sectional shapes. You can use one of these sections by following the GUI path: **Main Menu > Preprocessor > Sections > Beam > Common Sections**. You can also use the **SECTYPE** command to define the type of cross section and the **SECDATA** command to define the dimensions associated with the cross section. It is possible to define custom cross sections and to create tapered beams. You can learn more about sections in the Mechanical APDL Structural Analysis Guide.

Exercise 4-1

Modeling a Simple 1D Cantilever Beam Using Beam Elements

Overview

In exercises 4-1, 4-2, and 4-3, you will perform a steady-state structural analysis of a cantilever beam using 1D, 2D, and 3D models meshed with beam, plane, and solid elements. The beam is made of aluminum with a Young's modulus of 73.1 GPa and a Poisson's ratio of 0.33. It is 1 m in length with a 10x10 cm cross section. It has a load of 5000 N applied to the unsupported end (Figure 4-1-1). The 1D model will be meshed with BEAM189 elements. The 2D model will be meshed with PLANE182 elements. And, the 3D model will be meshed with SOLID186 elements. The goal of each analysis is to determine the deflection at the end of the beam and the stresses throughout the beam. Once all three models have been solved, the differences and similarities among the results of the three models will be examined.

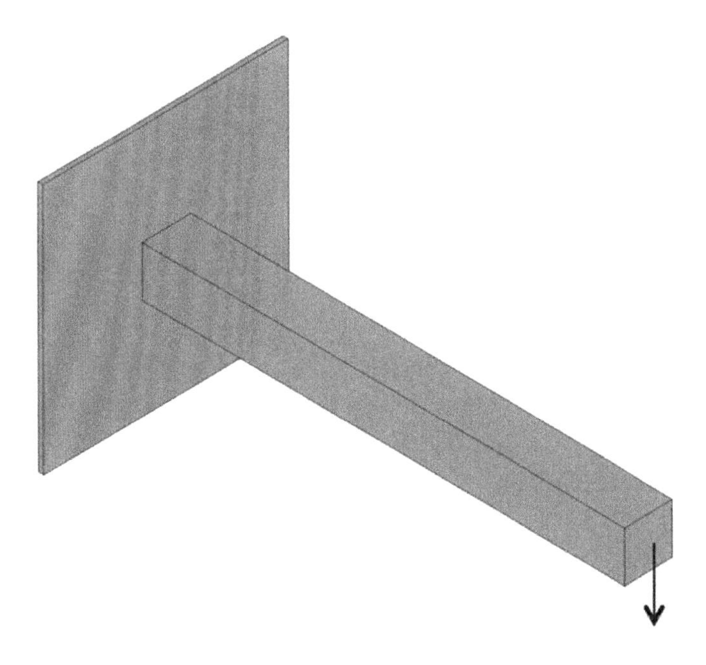

Figure 4-1-1 Schematic of Cantilever Beam with End Load.

ANSYS Mechanical APDL for Finite Element Analysis.
DOI: http://dx.doi.org/10.1016/B978-0-12-812981-4.00018-6

Model Attributes

Material Properties for 6061-T6 Aluminum

- Young's modulus—7.310e10 Pa
- Poisson's ratio—0.33

Loads

- 5000 N downward load applied to the center of the free end of the beam

Constraints

- The fixed end of the beam is fully constrained in *x*, *y*, and *z*

File Management

Create a new folder in your "Intro-to-ANSYS" folder named "Exercise4-1"

Open a new session of ANSYS using the Mechanical APDL Product Launcher

Change the Working Directory to the new "Exercise4-1" folder

Change the Jobname to "Exercise4-1"

Click Run to start ANSYS

Step 1: Define Geometry

1-1. Create keypoints to define the ends of the beam

- Preprocessor > Modeling > Create > Keypoints > In Active CS
- Supply (0,0,0) as the coordinates for the 1st KP
- Supply (1,0,0) as the coordinates for the 2nd KP

1-2. Create a line to connect the two keypoints

- Preprocessor > Modeling > Create > Lines > Lines > Straight Line

1-3. Save the model geometry

The solid model geometry specifies the location and length of the beam. The rest of the geometry is included via the beam's section properties including its moment of inertia.

Step 2: Define Element Types

2-1. Define the element type to use for this model

- Preprocessor > Element Type > Add/Edit/Delete
- Choose BEAM189 as the element type for this analysis

2-2. Define the section properties for your model

- Preprocessor > Sections > Beam > Common Sections
- Ensure that a rectangle is shown as the Sub-Type (Figure 4-1-2)
- Ensure that “Offset To” is set to “Centroid”
- Enter 0.1 for B (width)
- Enter 0.1 for H (height)
- Click “Preview” to view the defined cross section
- Click OK

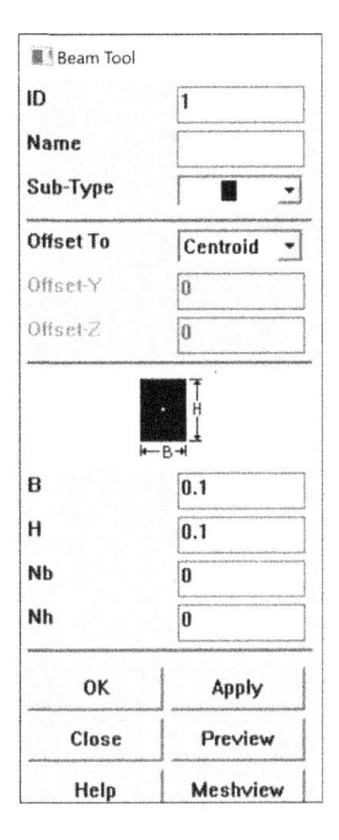

Figure 4-1-2 Defining Section Properties using the Beam Tool.

Step 3: Define Material Properties

3-1. Create a linear elastic material model for 6061-T6 aluminum

- Preprocessor > Material Props > Material Models
- Choose a structural, linear, elastic, isotropic material model
- Supply 7.310e10 as the value for Young’s modulus (EX)
- Supply 0.33 as the value for Poisson’s ratio (PRXY)

3-2. Save your progress

Step 4: Mesh

4-1. Create the mesh for the finite element model

- Preprocessor > Meshing > MeshTool
- Click the “Mesh” button
- Click the “Pick All” button in the Mesh Lines dialog box

4-2. Turn element numbering on

- Utility Menu > PlotCtrls > Numbering...
- Change "Elem/Attrib numbering" to "Element numbers"
- Click OK

4-3. Plot the finite element mesh

- Utility Menu > Plot > Elements

Note that this model only has two elements (element #1 and element #2).

4-4. Change to the isometric view

- Click the "Isometric View" button in the Pan Zoom Rotate menu

4-5. Turn element shape display on

- Utility Menu > PlotCtrls > Style > Size and Shape...
- Turn "[/ESHAPE] Display of element shapes based on real constant descriptions" on
- Click OK

The ***/ESHAPE*** *command displays elements and element results using shapes determined by the real constants or section properties that have been defined. This allows you to view the 3D representation of lower order models.*

4-6. Return to the front view

- Click the "Front View" button in the Pan Zoom Rotate menu

4-7. Turn element shape display off

- Utility Menu > PlotCtrls > Style > Size and Shape...
- Turn "[/ESHAPE] Display of element shapes based on real constant descriptions" off
- Click OK

4-8. Turn element numbering off

- Utility Menu > PlotCtrls > Numbering...
- Change "Elem/Attrib numbering" to "No numbering"
- Click OK

Step 5: Apply Constraint Boundary Conditions

5-1. Constrain the fixed end of the beam

- Solution > Define Loads > Apply > Structural > Displacement > On Keypoints
- Click on the keypoint at the origin or specify Keypoint 1 in the text box
- Click OK
- For "Lab2 DOFs to be constrained" choose "All DOF"
- For "VALUE Displacement value" enter 0
- Click OK

Two perpendicular little blue arrows should appear at the left end of the beam to indicate that the constraints have been successfully applied. In addition, two perpendicular orange double arrows should appear to indicate that rotations have also been constrained.

5-2. Save your constraints

Step 6: Apply Load Boundary Conditions

6-1. Apply a downward load to the free end of the beam

- Solution > Define Loads > Apply > Structural > Force/Moment > On Keypoints
- Click on the keypoint at the free (right) end of the beam or specify Keypoint 2 in the text box
- Click OK
- For "Lab Direction of force/mom" choose "FY"
- For "VALUE Force/moment value" enter −5000
- Click OK

A red arrow pointing downward should appear at the right end of the beam to indicate that the load has been applied successfully.

6-2. Save your loads

Step 7: Set the Solution Options

The default solution options can be used for this analysis.

Step 8: Solve

8-1. Select everything in your model

- Utility Menu > Select > Everything

8-2. Solve

- Solution > Solve > Current LS

8-3. Save your results

Step 9: Postprocess the Results

9-1. Change to the isometric view

- Click the "Isometric View" button in the Pan Zoom Rotate menu

9-2. Turn element shape display on

- Utility Menu > PlotCtrls > Style > Size and Shape...
- Turn "[/ESHAPE] Display of element shapes based on real constant descriptions" on
- Click OK

9-3. Plot the vertical deformation of the beam

- General Postproc > Plot Results > Contour Plot > Nodal Solu
- Choose "DOF Solution"
- Choose "Y-Component of displacement"
- Change "Scale Factor" to "True Scale"
- Click OK

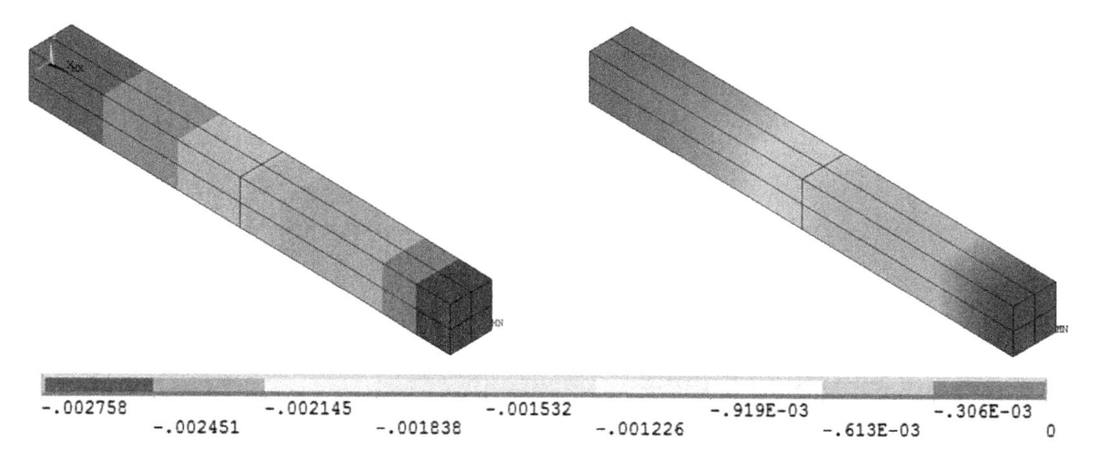

Figure 4-1-3 Plot of Displacement in *Y* (Isometric View, Element Shapes On): Win32 Graphics (left) and 3D Graphics (right).

Figure 4-1-3 shows that the deformations in the model are uniform through the cross section of the beam.

9-4. Change to the front view

- Click the "Front View" button in the Pan Zoom Rotate menu

9-5. Plot the x component of stress in the beam

- General Postproc > Plot Results > Contour Plot > Nodal Solu
- Choose "Stress"
- Scroll down and choose "X-Component of stress"
- Click OK

Figure 4-1-4 shows the longitudinal stress in the beam. You can see that the maximum stress occurs at the fixed end of the beam and has a value of 30 MPa. This is the same value as predicted by beam bending theory. The maximum stress on the top of the beam is equal and opposite in magnitude to the stress on the bottom of the beam, and there is no stress along the neutral axis or at the end of the beam where the load was applied. This is also consistent with our expectations.

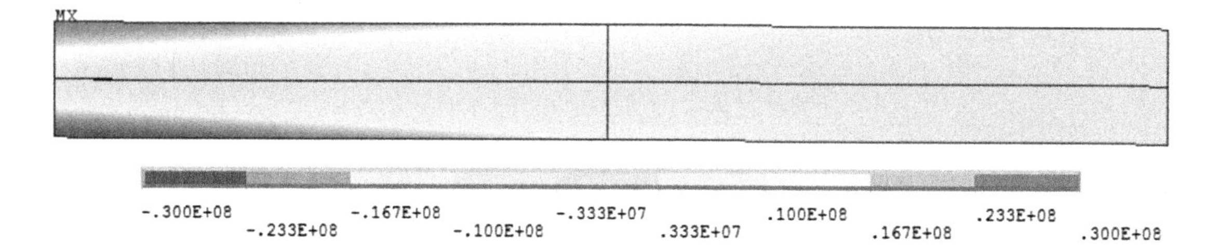

Figure 4-1-4 Plot of X-Component of Stress (Front View, Element Shapes On, 3D Graphics).

9-6. Change to the isometric view

- Click the "Isometric View" button in the Pan Zoom Rotate menu

9-7. Plot the equivalent stress in the beam

- General Postproc > Plot Results > Contour Plot > Nodal Solu
- Choose "Stress"
- Scroll down and choose "von Mises stress"
- Click OK

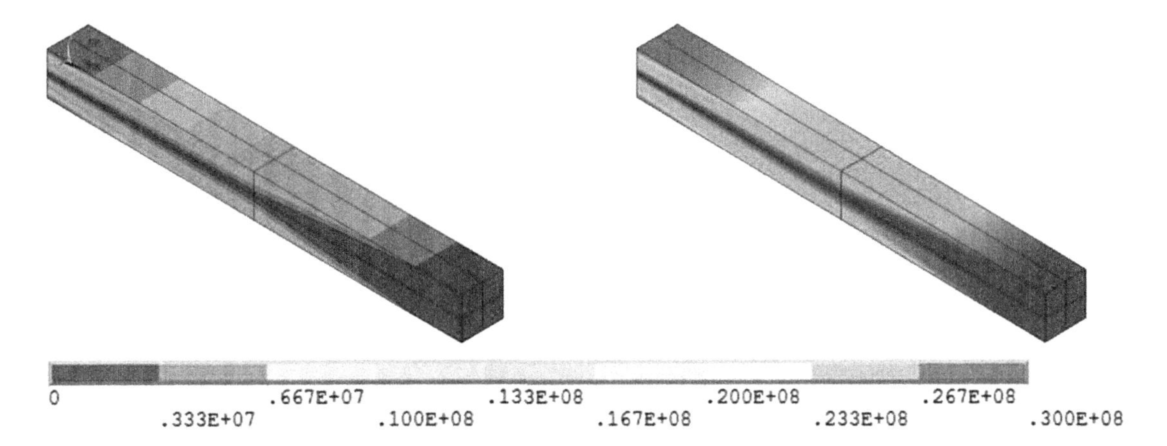

Figure 4-1-5 Plot of von Mises Stress (Isometric View, Element Shapes On): Win32 Graphics (left) and 3D Graphics (right).

Figure 4-1-5 shows that the stress at the free end of the beam is uniformly zero.

Step 10: Compare and Verify the Results

Figure 4-1-3 shows the vertical displacement in the beam. You can see that there is zero displacement at the fixed end of the beam and a maximum displacement of 2.758E-3 m at the free end of the beam. Beam bending theory predicts a maximum displacement of 2.736E-3 m. Thus, the results of this model are within 1% of the theory.

Figure 4-1-4 shows the longitudinal stress in the beam. You can see that the maximum stress occurs at the fixed end of the beam and has a value of 30 MPa. This is the same value as predicted by beam bending theory. The maximum stress on the top of the beam is equal and opposite in magnitude to the stress on the bottom of the beam, and there is no stress along the neutral axis or at the end of the beam where the load was applied. This is also consistent with our expectations.

Figure 4-1-5 shows the equivalent stress (von Mises stress) in the beam. The calculation of equivalent stress involves a square root so equivalent stress is always positive. In this plot, you can see that the equivalent stress is symmetric about the centerline of the beam. Equivalent stress is used in the Theory of Plasticity to determine if the material has yielded. This plot shows that the maximum stress is much lower than the yield stress of 6061-T6 aluminum (275 MPa). This validates our use of a linear elastic material model.

Close the Program

Utility Menu > File > Exit...

Sample Input File

```
/PREP7                  ! Enter the Preprocessor
K,,0,0,0                ! Create keypoint at (0,0,0)
K,,1,0,0                ! Create keypoint at (1,0,0)
L,1,2                   ! Create a line between Keypoints 1 and 2
ET,1,BEAM189            ! Use BEAM189 elements
SECTYPE,1,BEAM,RECT     ! Use a rectangular cross section for beam
SECOFFSET,CENT          ! Offset beam node to the centroid
SECDATA,0.1,0.1,        ! Use 0.1 x 0.1 cross section
MP,EX,1,7.31e10         ! Define Young's modulus for material #1
MP,PRXY,1,0.33          ! Define Poisson's ratio for material #1
LMESH,ALL               ! Mesh the line
FINISH                  ! Finish and Exit Preprocessor

/SOLU                   ! Enter the Solution Processor
DK,1,ALL,0              ! Constrain KP 1 in all DOFs
FK,2,FY,-5000           ! Apply -5000 N in y direction to KP 2
ALLSEL                  ! Select everything
SOLVE                   ! Solve the model
FINISH                  ! Finish and Exit Solution

/POST1                  ! Enter the General Postprocessor
/ESHAPE,1               ! Display element shapes using section data
/DSCALE,ALL,1           ! Plot using true scale
/VIEW,1,1,1,1           ! Change to isometric view
PLNSOL,U,Y,0,1          ! Plot displacement in y
/VIEW,1,,,1             ! Change to front view
PLNSOL,S,X,0,1          ! Plot stress in x
/VIEW,1,1,1,1           ! Change to isometric view
PLNSOL,S,EQV,0,1        ! Plot the equivalent stress
FINISH                  ! Finish and Exit Postprocessor

SAVE                    ! Save the database
!/EXIT                  ! Exit ANSYS
```

Exercise 4-2

Modeling a Simple 2D Cantilever Beam Using PLANE Elements

Overview

This exercise is a continuation of the cantilever beam analysis that was started in exercise 4-1. In exercise 4-2, you will take advantage of symmetry in the beam cross section to create a 2D model using PLANE182 elements.

File Management

Create a new folder in your "Intro-to-ANSYS" folder named "Exercise4-2"

Open a new session of ANSYS using the Mechanical APDL Product Launcher

Change the Working Directory to the new "Exercise4-2" folder

Change the Jobname to "Exercise4-2"

Click Run to start ANSYS

Step 1: Define Geometry

1-1. Create an area to serve as the center plane of the beam

- Preprocessor > Modeling > Create > Areas > Rectangle > By Dimensions
- Enter 0 for X1
- Enter 1 for X2
- Enter −0.05 for Y1
- Enter 0.05 for Y2
- Click OK

1-2. Save the model geometry

Step 2: Define Element Types

2-1. Define the element type to use for this model

- Preprocessor > Element Type > Add/Edit/Delete
- Choose PLANE182 as the element type for this analysis
- Do not close the Element Types dialog box

ANSYS Mechanical APDL for Finite Element Analysis.
DOI: http://dx.doi.org/10.1016/B978-0-12-812981-4.00019-8

2-2. Specify plane stress behavior for your model

- If you have closed the Element Types dialog box, reopen it
- Click the "Options..." button
- For "Element behavior K3" choose "Plane stress w/thk" (Figure 4-2-1)
- Click OK
- Close the Element Types dialog box

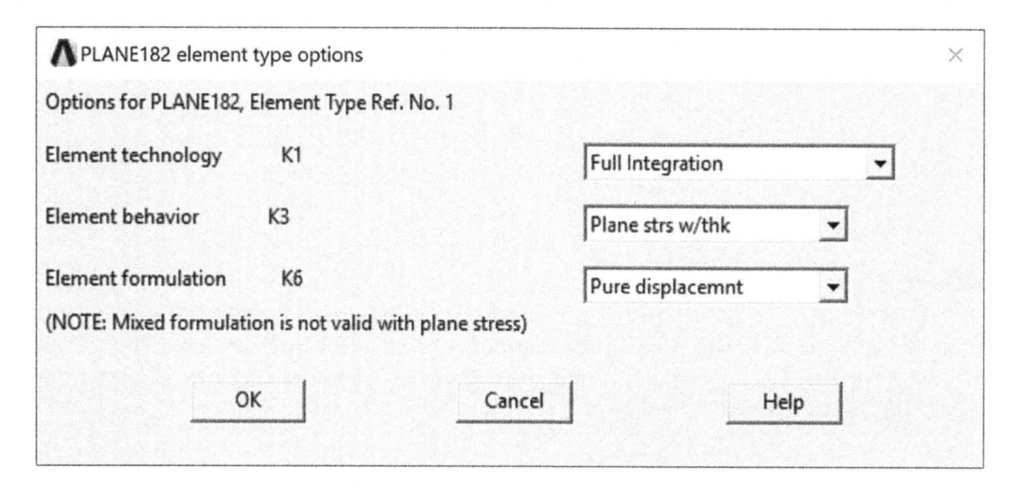

Figure 4-2-1 Element Options Dialog Box.

In general, thickness is not required for 2D analyses. There are two exceptions to this rule: (1) when force loads are applied and (2) when you want to plot the full 3D results using the element shapes option.

As noted in exercise 2-1, pressure loads in ANSYS are applied per unit area. When pressure loads are applied to lines, ANSYS assumes a unit depth of 1 unless otherwise specified. Usually this default value is ok. (For example, it was used in exercise 2-1 with no problem.) However, if the default value for depth will not result in the correct value for the applied load, the model thickness must be specified using a real constant. When forces are applied, the magnitude of the force must be adjusted for the thickness.

Element shapes allow you to visualize the full 3D results of a lower order model. You took advantage of this option in exercise 4-1 to visualize the results of the first cantilever beam. Element shapes do not affect the model results. They only affect postprocessing.

The model in this exercise can be created with or without a specified thickness. The results will be the same assuming that the correct value of force is applied. However, without the thickness value, you will not be able to generate the 3D plot shown in Figure 4-2-5.

2-3. Define the real constants for your model

- Preprocessor > Real Constants > Add/Edit/Delete
- Click the "Add..." button
- Click OK
- For "Thickness THK" enter 0.1 (Figure 4-2-2)
- Click OK
- Close the Real Constants dialog box

In this model, the beam thickness has not been specified via the solid model geometry. Thus, it must be defined using a real constant.

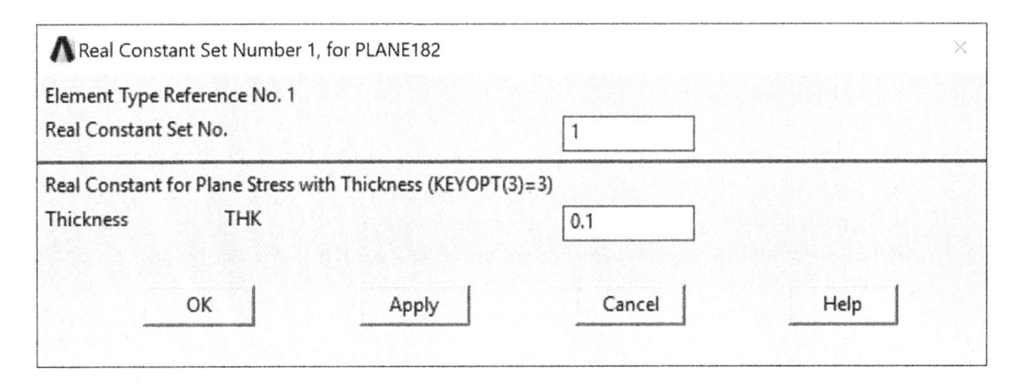

Figure 4-2-2 Real Constant Dialog Box.

Step 3: Define Material Properties

3-1. Create a linear elastic material model for 6061-T6 aluminum

- Preprocessor > Material Props > Material Models
- Choose a structural, linear, elastic, isotropic material model
- Enter 7.310e10 as the value for Young's modulus (EX)
- Enter 0.33 as the value for Poisson's ratio (PRXY)

3-2. Save your progress

Step 4: Mesh

4-1. Create the mesh for the finite element model

- Preprocessor > Meshing > MeshTool
- In the third area of the Mesh Tool, click the "Set" button associated with "Areas"
- Click "Pick All"
- Set the "SIZE Element edge length" to 0.0125
- Click OK
- In the fourth area of the Mesh Tool, click the "Mesh" button
- Click "Pick All"
- Close the Mesh Tool

This procedure specifies that all elements in the model should be 0.0125x0.0125 m long. It results in a coarse mesh with 8 elements through the thickness of the beam and 80 elements along the length of the beam.

4-2. Change to the isometric view

- Click the "Isometric View" button in the Pan Zoom Rotate menu

The isometric view clearly shows that this is a 2D model and that the finite element mesh is only one element thick.

4-3. Return to the front view

- Click the "Front View" button in the Pan Zoom Rotate menu

Step 5: Apply Constraint Boundary Conditions

5-1. Select the nodes at $x = 0$

- Utility menu > Select > Entities...
- Ensure that "Nodes" are the entity to be selected
- Choose "By Location" as the selection method
- Ensure that "X coordinates" is set as the coordinate to select by
- Enter 0 in the "Min,Max" box
- Ensure that "From Full" is selected
- Click OK

5-2. Plot the selected nodes to confirm that the selection operation was successful

- Utility menu > Plot > Nodes

The plot shows that a single vertical line of nodes has been selected. These nodes are generally indistinguishable from any other line of vertical nodes. However, the presence of the global triad (i.e., the origin) at the center of the line shows that this is the correct line of nodes.

5-3. Constrain the fixed end of the beam in x

- Solution > Define Loads > Apply > Structural > Displacement > On Nodes
- Click "Pick All"
- For "Lab2 DOFs to be constrained" choose "UX"
- For "VALUE Displacement value" enter 0
- Click OK

5-4. Select the node at (0,0)

- Utility menu > Select > Entities...
- Ensure that "Nodes" are the entity to be selected
- Ensure that "By Location" is the selection method
- Change the coordinate to select by to "Y coordinates"
- Enter 0 in the "Min,Max" box
- Change "From Full" to "Reselect"
- Click OK

Reselecting entities from the currently selected set creates a new subset. Selecting "From Full" creates a new set from the set of all entities defined in the model. "Also Select" adds to the current set or creates a union between the old set and the new set. "Unselect" removes entities from the current set. See chapter 7 *for more details.*

5-5. List the nodes to ensure that the correct node was selected

- Utility menu > List > Nodes...
- Click OK
- Close the NLIST window when you are finished reviewing the nodes

You can confirm that the correct node was selected according to its (x,y,z) coordinates.

5-6. Constrain the fixed end of the beam in *y*

- Solution > Define Loads > Apply > Structural > Displacement > On Nodes
- Click "Pick All"
- For "Lab2 DOFs to be constrained" unselect "UX" if it is still selected
- For "Lab2 DOFs to be constrained" choose "UY"
- For "VALUE Displacement value" enter 0
- Click OK

Constraining only a single point in y allows the beam to expand and contract in the y-direction and thus does not overconstrain the model.

5-7. Save your constraints

Step 6: Apply Load Boundary Conditions

6-1. Select the node at (1,0)

- Utility menu > Select > Entities...
- Ensure that "Nodes" are the entity to be selected
- Ensure that "By Location" is the selection method
- Change the coordinate to select by to "X coordinates"
- Enter 1 in the "Min,Max" box
- Change "Reselect" to "From Full"
- Click Apply

This selects all of the nodes at $x = 1$.

- Change the coordinate to select by to "Y coordinates"
- Enter 0 in the "Min,Max" box
- Change "From Full" to "Reselect"
- Click OK

6-2. List the nodes to ensure that the correct node was selected

- Utility Menu > List > Nodes...
- Click OK
- Close the NLIST window when you are finished reviewing the nodes

6-3. Apply a downward load to the center of the free end of the beam

- Solution > Define Loads > Apply > Structural > Force/Moment > On Nodes
- Click "Pick All"
- For "Lab Direction of force/mom" choose "FY"
- For "VALUE Force/moment value" enter −5000
- Click OK

6-4. List the forces to ensure that the load was correctly applied

- Utility Menu > List > Loads > Forces > On All Nodes
- Close the FLIST window when you are finished reviewing the loads

6-5. Save your loads

Step 7: Set the Solution Options

The default solution options can be used for this analysis.

Step 8: Solve

8-1. Select everything in the model

- Utility Menu > Select > Everything

8-2. Solve

- Solution > Solve > Current LS

8-3. Save your results

Step 9: Postprocess the Results

9-1. Plot the vertical deformation of the beam

- General Postproc > Plot Results > Contour Plot > Nodal Solu
- Choose "DOF Solution"
- Choose "Y-Component of displacement"
- Change the "Scale Factor" to "True Scale"
- Click OK

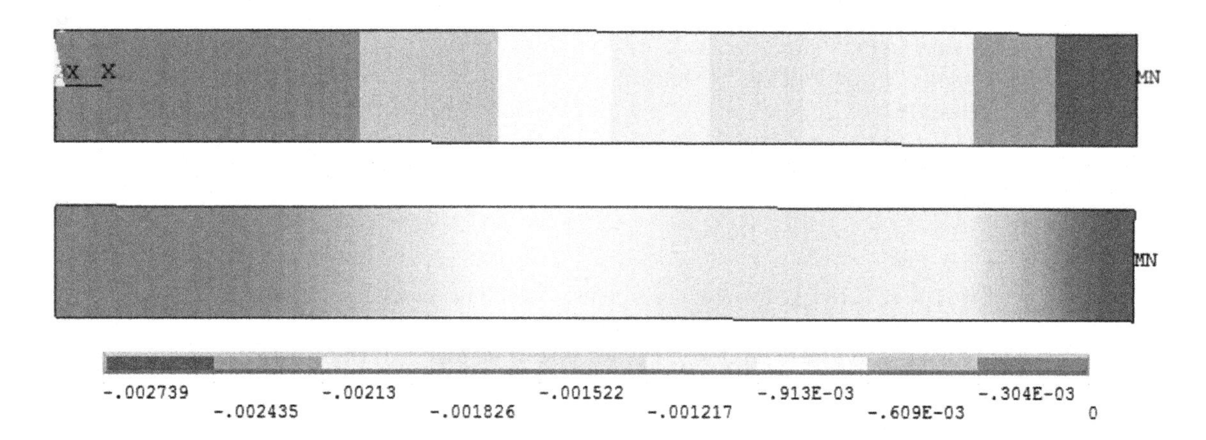

Figure 4-2-3 Plot of Displacement in *Y*: Win32 Graphics (top) and 3D Graphics (bottom).

9-2. Plot the *x* component of stress in the beam

- General Postproc > Plot Results > Contour Plot > Nodal Solu
- Choose "Stress"
- Scroll down and choose "X-Component of stress"
- Click OK

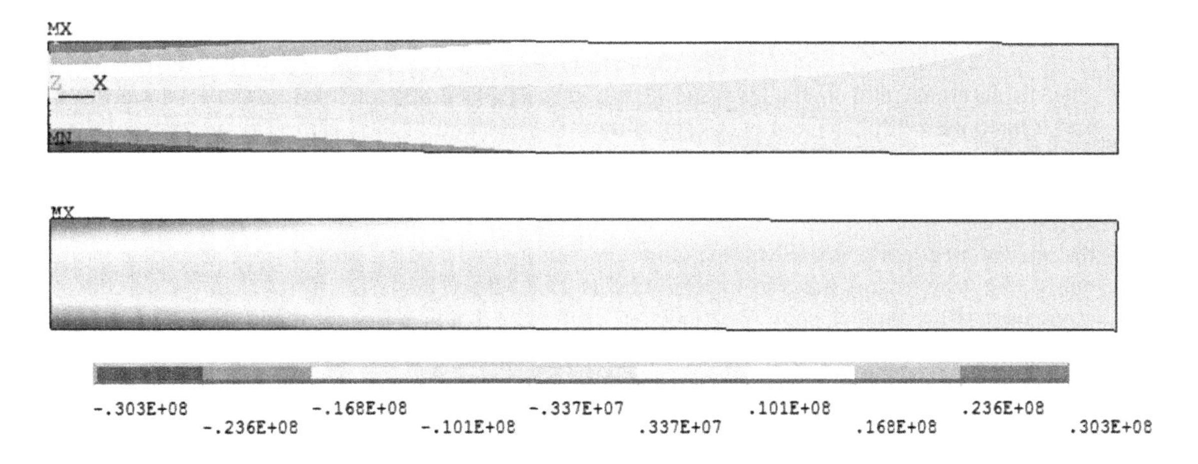

Figure 4-2-4 Plot of X-Component of Stress: Win32 Graphics (top) and 3D Graphics (bottom).

9-3. Change to the isometric view

- Click the "Isometric View" button in the Pan Zoom Rotate menu

9-4. Turn element shape display on

- Utility Menu > PlotCtrls > Style > Size and Shape...
- Turn "[/ESHAPE] Display of element shapes based on real constant descriptions" on
- Click OK

9-5. Plot the equivalent stress in the beam

- General Postproc > Plot Results > Contour Plot > Nodal Solu
- Choose "Stress"
- Scroll down and choose "von Mises stress"
- Click OK

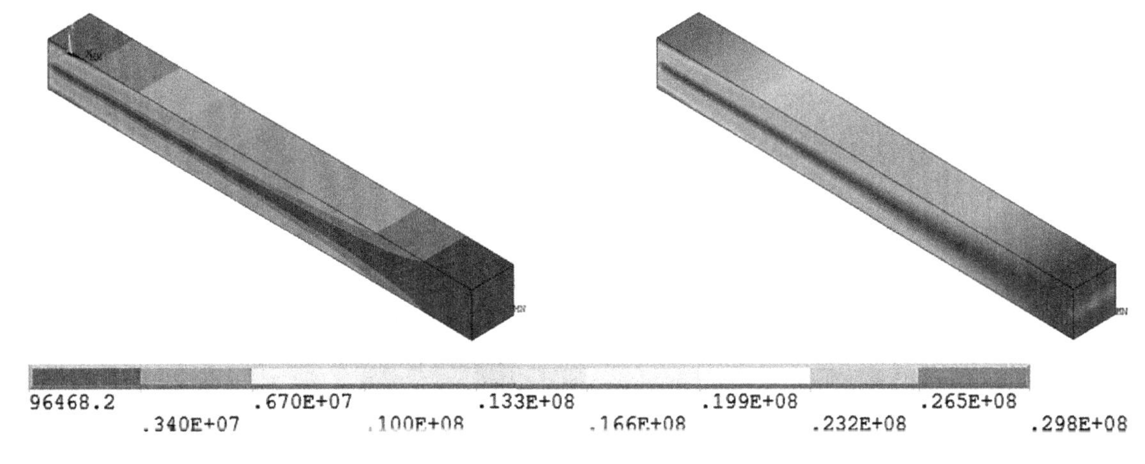

Figure 4-2-5 Plot of von Mises Stress (Element Shapes On): Win32 Graphics (left) and 3D Graphics (right).

Step 10: Compare and Verify the Results

The displacement plot in Figure 4-2-3 shows zero displacement at the fixed end of the beam and a maximum displacement of 2.739E-3 m at the free end of the beam. This is within 1% of the result from exercise 4-1 (2.758E-3 m) and very similar to the value predicted by beam theory (2.736E-3 m). The plot of the longitudinal stress in Figure 4-2-4 shows that the maximum stress occurs at the fixed end of the beam and has a value of 30.3 MPa. This is 1% higher than the maximum stress predicted by the model in exercise 4-1 and by beam theory. The plot of the equivalent (von Mises) stress in Figure 4-2-5 is 29.8 MPa. This is 1% lower than the maximum equivalent stress predicted by the model in exercise 4-1 and by beam theory. Based on these comparisons, we can conclude that the model is in excellent agreement with the theory and can be used for engineering design and analysis. We can also conclude that there are no major differences between the solutions of the 1D and 2D models.

Close the Program

- Utility Menu > File > Exit...

Sample Input File

```
/PREP7                  ! Enter the Preprocessor
RECT,0,1,-0.05,0.05     ! Create a rectangle from x(0:1), y(-0.5:0.5)
ET,1,PLANE182           ! Use PLANE183 Elements
KEYOPT,1,3,3            ! Use plane stress w/thickness
R,1,0.1,                ! Set plane stress thickness to 0.1 m
MP,EX,1,7.31e10         ! Define Young's modulus for material #1
MP,PRXY,1,0.33          ! Define Poisson's ratio for material #1
ESIZE,0.0125            ! Set element size to 0.0125
AMESH,ALL               ! Mesh all areas
FINISH                  ! Finish and Exit Preprocessor

/SOLU                   ! Enter the Solution Processor
NSEL,S,LOC,X,0          ! Select all nodes at x=0
D,ALL,UX,0              ! Constrain those nodes in x
NSEL,R,LOC,Y,0          ! Reselect the node at y=0
D,ALL,UY,0              ! Constrain that node in y
NSEL,S,LOC,X,1          ! Select all nodes at x=1
NSEL,R,LOC,Y,0          ! Reselect the node at y=0
F,ALL,FY,-5000          ! Apply a downward load of -5000 N
ALLSEL                  ! Select everything
SOLVE                   ! Solve the model
FINISH                  ! Finish and Exit Solution

/POST1                  ! Enter the General Postprocessor
/DSCALE,ALL,1           ! Plot using true scale
/VIEW,1,,,1             ! Change to front view
PLNSOL,U,Y,0,1          ! Plot displacement in y
PLNSOL,S,X,0,1          ! Plot stress in x
/VIEW,1,1,1,1           ! Change to isometric view
/ESHAPE,1               ! Display element shapes using section data
PLNSOL,S,EQV,0,1        ! Plot the equivalent stress
FINISH                  ! Finish and Exit Postprocessor
SAVE                    ! Save the database
!/EXIT                  ! Exit ANSYS
```

Exercise 4-3

Modeling a Simple 3D Cantilever Beam Using SOLID Elements

Overview

This exercise is a continuation of the cantilever beam analysis that was started in exercises 4-1 and 4-2. In exercise 4-3, you will create a full 3D model using SOLID186 elements.

File Management

Create a new folder in your "Intro-to-ANSYS" folder named "Exercise4-3"

Open a new session of ANSYS using the Mechanical APDL Product Launcher

Change the Working Directory to the new "Exercise4-3" folder

Change the Jobname to "Exercise4-3"

Click Run to start ANSYS

Step 1: Define Geometry

1.1. Create the beam volume

- Preprocessor > Modeling > Create > Volumes > Block > By Dimensions
- Enter 0 for X1
- Enter 1 for X2
- Enter −0.05 for Y1 and Z1
- Enter 0.05 for Y2 and Z2
- Click OK

1.2. Save the model geometry

Step 2: Define Element Types

2.1. Define the element type to use for this model

- Preprocessor > Element Type > Add/Edit/Delete
- Choose SOLID186 as the element type for this analysis

ANSYS Mechanical APDL for Finite Element Analysis.
DOI: http://dx.doi.org/10.1016/B978-0-12-812981-4.00020-4

This model uses an element with mid-side nodes (SOLID186 instead of SOLID185) because it substantially reduces the number of elements required to produce good results.

Step 3: Define Material Properties

3.1. Create a linear elastic material model for 6061-T6 aluminum

- Preprocessor > Material Props > Material Models
- Choose a structural, linear, elastic, isotropic material model
- Supply 7.310e10 as the value for Young's modulus (EX)
- Supply 0.33 as the value for Poisson's ratio (PRXY)

3.2. Save your progress

Step 4: Mesh

4.1. Create the mesh for the finite element model

- Preprocessor > Meshing > MeshTool
- In the third area of the Mesh Tool, click the "Set" button associated with "Areas"
- Click "Pick All"
- Set the "SIZE Element edge length" to 0.025
- Click OK
- In the fourth area of the Mesh Tool, change "Shape" to Hex
- Click the "Mesh" button
- Click "Pick All"
- Click OK

4.2. Change to the isometric view

- Click the "Isometric View" button in the Pan Zoom Rotate menu

This plot clearly shows that you have generated 3D geometry and a 3D finite element mesh.

4.3. Return to the front view

- Click the "Front View" button in the Pan Zoom Rotate menu

Step 5: Apply Constraint Boundary Conditions

5.1. Select the nodes at $x = 0$

- Utility menu > Select > Entities...
- Ensure that "Nodes" are the entity to be selected
- Choose "By Location" as the selection method
- Ensure that "X coordinates" is set as the coordinate to select by
- Enter 0 in the "Min,Max" box
- Ensure that "From Full" is selected
- Click OK

5.2. Constrain the fixed end of the beam in *x*

- Solution > Define Loads > Apply > Structural > Displacement > On Nodes
- Click "Pick All"
- Click OK
- For "Lab2 DOFs to be constrained" choose "UX"
- For "VALUE Displacement value" enter 0
- Click OK

5.3. Select the nodes at (0,0)

- Utility menu > Select > Entities...
- Ensure that "Nodes" are the entity to be selected
- Ensure that "By Location" is the selection method
- Change the coordinate to select by to "Y coordinates"
- Enter 0 in the "Min,Max" box
- Change "From Full" to "Reselect"
- Click OK

5.4. Constrain the fixed end of the beam in *y*

- Solution > Define Loads > Apply > Structural > Displacement > On Nodes
- Click "Pick All"
- Click OK
- For "Lab2 DOFs to be constrained" choose "UY"
- For "VALUE Displacement value" enter 0
- Click OK

5.5. Select the node at (0,0,0)

- Utility menu > Select > Entities...
- Ensure that "Nodes" are the entity to be selected
- Ensure that "By Location" is the selection method
- Change the coordinate to select by to "Z coordinates"
- Enter 0 in the "Min,Max" box
- Ensure that you will "Reselect" from the current set
- Click OK

5.6. Constrain the fixed end of the beam in *z*

- Solution > Define Loads > Apply > Structural > Displacement > On Nodes
- Click "Pick All"
- Click OK
- For "Lab2 DOFs to be constrained" choose "UZ"
- For "VALUE Displacement value" enter 0
- Click OK

5.7. Save your constraints

Step 6: Apply Load Boundary Conditions

6.1. Select the node at (1,0,0)

- Utility menu > Select > Entities...
- Ensure that "Nodes" are the entity to be selected
- Ensure that "By Location" is the selection method
- Change the coordinate to select by to "X coordinates"
- Enter 1 in the "Min,Max" box
- Change "Reselect" to "From Full"
- Click Apply

This selects the plane of nodes at x = 1.

- Change the coordinate to select by to "Y coordinates"
- Enter 0 in the "Min,Max" box
- Change "From Full" to "Reselect"
- Click Apply

This selects the line of nodes at (1,0).

- Change the coordinate to select by to "Z coordinates"
- Enter 0 in the "Min,Max" box
- Ensure that you will "Reselect" entities from the currently active set
- Click OK

This selects the node at (1,0,0).

6.2. List the nodes to ensure that the correct node was selected

- Utility menu > List > Nodes...
- Click OK
- Close the NLIST window when you are finished reviewing the nodes

6.3. Apply a downward load to the center of the free end of the beam

- Solution > Define Loads > Apply > Structural > Force/Moment > On Nodes
- Click "Pick All"
- For "Lab Direction of force/mom" choose "FY"
- For "VALUE Force/moment value" enter −5000
- Click OK

6.4. Save your loads

Step 7: Set the Solution Options

The default solution options can be used for this analysis.

Step 8: Solve

8.1. Select everything in the model

- Utility Menu > Select > Everything

8.2. Solve

- Solution > Solve > Current LS

8.3. Save your results

Step 9: Postprocess the Results

9.1. Plot the vertical deformation of the beam

- General Postproc > Plot Results > Contour Plot > Nodal Solu
- Choose "DOF Solution"
- Choose "Y-Component of displacement"
- Change "Scale Factor" to "True Scale"
- Click OK

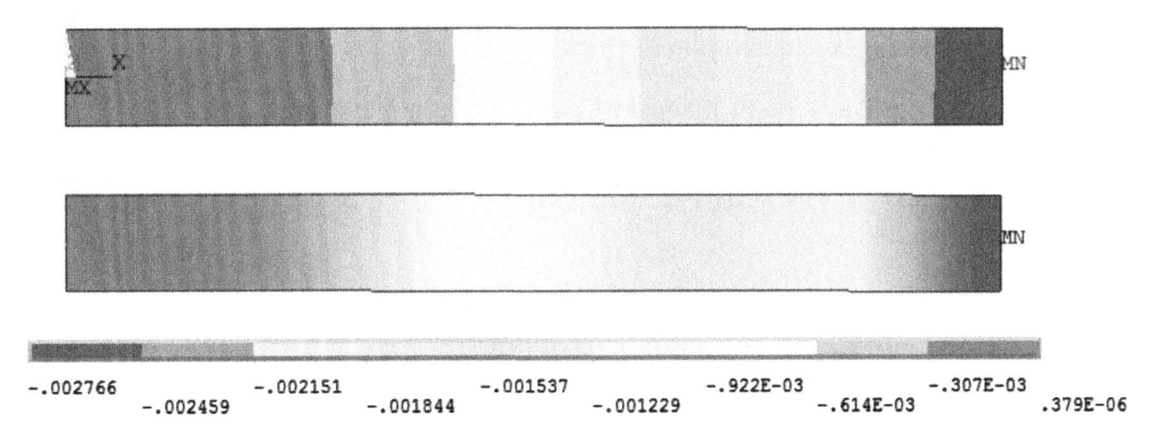

Figure 4-3-1 Plot of Displacement in *Y*: Win32 Graphics (top) and 3D Graphics (bottom).

9.2. Change to the isometric view

- Click the "Isometric View" button in the Pan Zoom Rotate menu

9.3. Plot the x component of stress in the beam

- General Postproc > Plot Results > Contour Plot > Nodal Solu
- Choose "Stress"
- Scroll down and choose "X-Component of stress"
- Click OK

9.4. Turn Power Graphics off

- ANSYS Toolbar > POWRGRPH
- Check the "Off" radio button
- Click OK

Power Graphics is the default graphics display method in ANSYS. It displays results more quickly by showing results only for the model surface. In this model, the deformation in the y direction and the component stress in the x direction are uniform through the beam cross section. Therefore, the results for the surface of the model are representative for the entire model. However, the equivalent (von Mises) stress is not uniform through the beam cross section. To determine the maximum and minimum values for the entire 3D model from the plot legends, we must first turn Power Graphics off.

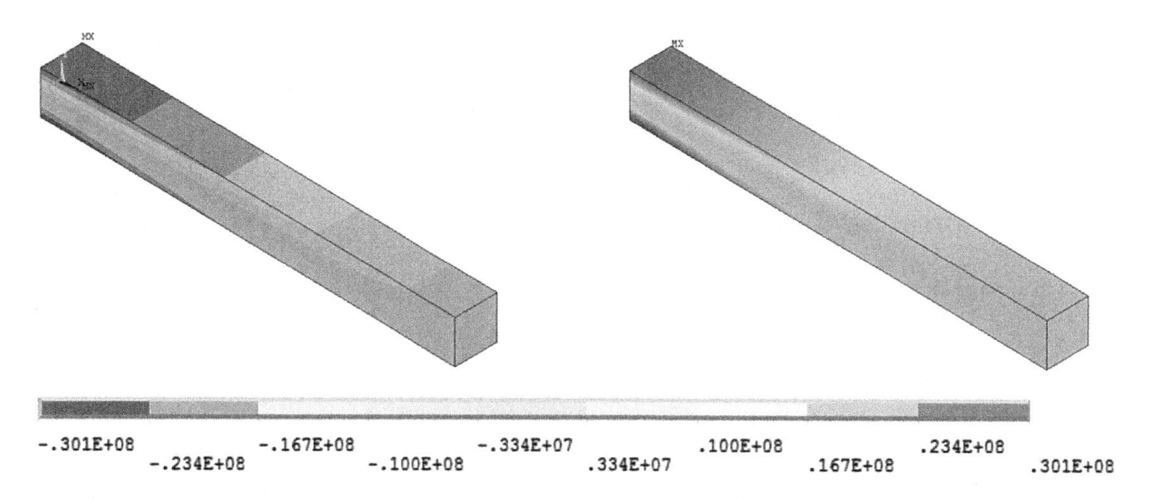

Figure 4-3-2 Plot of X-Component of Stress: Win32 Graphics (left) and 3D Graphics (right)

9.5. Plot the equivalent stress in the beam

- General Postproc > Plot Results > Contour Plot > Nodal Solu
- Choose “Stress”
- Scroll down and choose “von Mises stress”
- Change the “Scale Factor” to “True Scale”
- Click OK

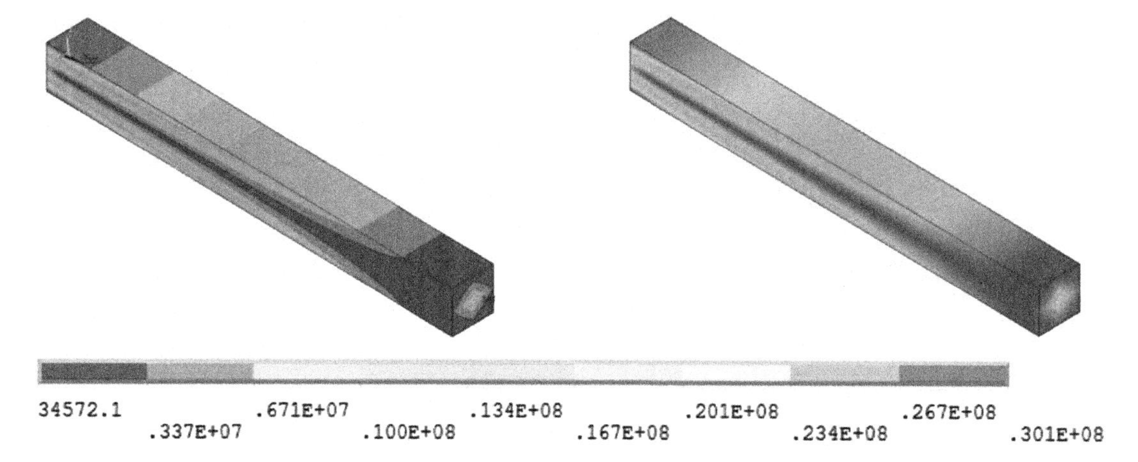

Figure 4-3-3 Plot of von Mises Stress (Power Graphics Off): Win32 Graphics (left) and 3D Graphics (right).

9.6. Change to the right view

- Click the “Right View” button in the Pan Zoom Rotate menu

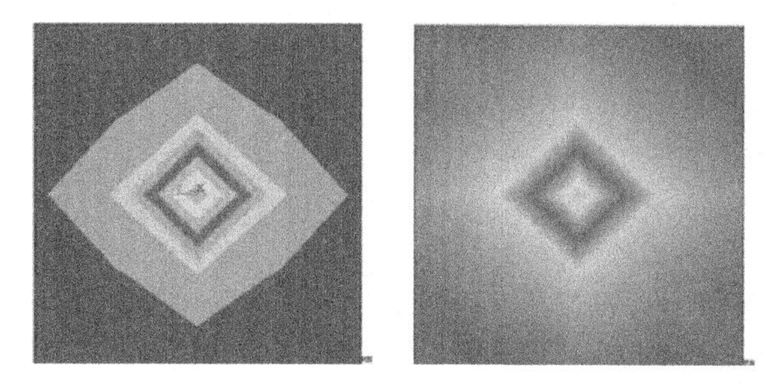

Figure 4-3-4 Plot of von Mises Stress (Right View): Win32 Graphics (left) and 3D Graphics (right).

Figure 4-3-4 shows that there is a stress concentration around the node where the downward force was applied. This occurs because the force was applied to a single node instead of being distributed over an area. This stress anomaly could be eliminated by distributing the force over the corner nodes of the element at the end of the beam. (Force type loads should not be applied to mid-side nodes.) But it is not necessary since this stress concentration does not affect the deflection of the beam or the peak stress at the beam support.

Step 10: Compare and Verify the Results

The displacement plot in Figure 4-3-1 shows zero displacement at the fixed end of the beam and a maximum displacement of 2.766e-3 m at the free end of the beam. This is within 1% of the results from exercise 4-1 (2.758e-3 m) and exercise 4-2 (2.739e-3 m), and is very similar to the theoretical value (2.736e-3 m). The plot of the longitudinal stress in Figure 4-3-2 shows that the maximum stress occurs at the fixed end of the beam and has a value of 30.1 MPa. This is within 1% of the values obtained in exercise 4-1 (30 MPa), exercise 4-2 (30.3 MPa), and from beam theory (30 MPa). Finally, Figure 4-3-3 shows that the maximum equivalent stress is also 30.1 MPa. This is also within 1% of the values obtained in exercise 4-1 (30 MPa), exercise 4-2 (29.8 MPa), and from beam theory (30 MPa).

Table 4-3-1 provides a comparison of the values for maximum deflection and maximum stress predicted by the theory and by the 1D, 2D, and 3D models. The results from the finite element models are all within 1.1% of the theoretical values and within 1% of each other. This confirms that each of these options represents a valid way to create a finite element model.

Table 4-3-1 Comparison of Results for a Simple Cantilever Beam

	Maximum Stress	Maximum Displacement
Theory	3.00e7 Pa	2.736E-3
Beam189	3.00e7 Pa	2.758E-3
Plane182	3.03e7 Pa	2.739E-3
Solid186	3.01e7 Pa	2.766E-3

Although their results are similar, the three models are not completely equivalent. The 1D model using beam elements required section data to be defined. Similarly, the 2D model using plane elements required a key option to define the element behavior and a real constant to supply the beam thickness during preprocessing. Thus, these models were more complicated to build than the full 3D model. In addition, both the 1D and 2D models benefited from the use of element shapes for the display of results. As a result, these models were also more complicated to postprocess than the full 3D model.

From this perspective, the 3D model seems to be the preferred choice. However, both the 2D and 3D models are very sensitive to the finite element mesh. Unlike most finite element models, a finer mesh in exercises 4-2 and 4-3 makes the results worse instead of better. This is because the boundary conditions in these models are applied at the locations of maximum stress. In contrast, the beam elements are very insensitive to mesh density. Increasing the mesh density by a factor of 5 or 10 does not impact the results at all.

You should draw two conclusions from this discussion. First, different element types are intended for different applications. Beam elements were designed to model structures that are composed of beams (i.e., structures with long, thin members). Thus, it is reasonable to expect that beam elements will work well when the structure satisfies the underlying assumptions of beam theory. Second, different element types have different properties and input requirements. It is important to understand how an element functions and what key options, real constants, and section data can and must be supplied before using it. Your choice of elements should be dictated by the problem that you are trying to solve, not your comfort and familiarity (or lack thereof) with a given element.

Close the Program

- Utility Menu > File > Exit...

Sample Input File

```
/PREP7                              ! Enter the Preprocessor
BLOCK,0,1,-0.05,0.05,-0.05,0.05 ! Create a block that is 1 x 0.05 x 0.05
ET,1,SOLID186                       ! Use SOLID186 Elements
MP,EX,1,7.31e10                     ! Define Young's modulus for material #1
MP,PRXY,1,0.33                      ! Define Poisson's ratio for material #1
ESIZE,0.025                         ! Set element size to 0.025
VMESH,ALL                           ! Mesh all areas
FINISH                              ! Finish and Exit Preprocessor

/SOLU                               ! Enter the Solution Processor
NSEL,S,LOC,X,0                      ! Select all nodes at x=0
D,ALL,UX,0                          ! Constrain those nodes in x
NSEL,R,LOC,Y,0                      ! Reselect the node at y=0
D,ALL,UY,0                          ! Constrain that node in y
NSEL,R,LOC,Z,0                      ! Reselect the node at z=0
D,ALL,UZ,0                          ! Constrain that node in z
NSEL,S,LOC,X,1                      ! Select all nodes at x=1
NSEL,R,LOC,Y,0                      ! Reselect the node at y=0
NSEL,R,LOC,Z,0                      ! Reselect the node at z=0
F,ALL,FY,-5000                      ! Apply a downward load of -5000 N
ALLSEL                              ! Select everything
SOLVE                               ! Solve the model
FINISH                              ! Finish and Exit Solution

/POST1                              ! Enter the General Postprocessor
/GRAPHICS,OFF                       ! Turn power graphics off
/ESHAPE,1                           ! Display element shapes using section data
/DSCALE,ALL,1                       ! Plot using true scale
/VIEW,1,,,1                         ! Change to front view
PLNSOL,U,Y,0,1                      ! Plot displacement in y
/VIEW,1,1,1,1                       ! Change to isometric view
PLNSOL,S,X,0,1                      ! Plot stress in x
PLNSOL,S,EQV,0,1                    ! Plot the equivalent stress
FINISH                              ! Finish and Exit Postprocessor

SAVE                                ! Save the database
!/EXIT                              ! Exit ANSYS
```

Chapter

5

Defining Material Properties

Suggested Reading Assignments:
Mechanical APDL Basic Analysis Guide: Chapter 1 Section 1.1.4
Mechanical APDL Material Reference: Chapters 3 and 4

CHAPTER OUTLINE

In this chapter, you will learn about material models and how they are implemented in ANSYS. You will also learn how to specify the material model that you wish to use, how to supply material property values, where to find material property data, and how to save your material data for future use.

5.1. What are Material Models?

A material, or constitutive, model is a mathematical representation of the expected behavior of a given material in response to an applied load. For example, we expect the deformation (or strain ε) of many materials in the elastic regime to be linearly proportional to a coaxial applied force (or stress σ) according to Hooke's Law:

$$\varepsilon_{axial} = \sigma_{axial}/E$$

In this case, the constant of proportionality is Young's modulus E. We further expect that the resulting strain in the transverse directions will be negative and linearly proportional to the axial strain through Poisson's ratio υ:

$$d\varepsilon_{trans} = -\upsilon d\varepsilon_{axial}$$

ANSYS Mechanical APDL for Finite Element Analysis.
DOI: http://dx.doi.org/10.1016/B978-0-12-812981-4.00005-8

Similarly, we often expect that the heat flux (q) in a body will be proportional to the temperature gradient (∇T) present through the thermal conductivity (k) of the material according to Fourier's Law:

$$q = -k\nabla T$$

Unlike analytical models, finite element models do not require you to derive and solve these equations. But you do need to identify the materials that will be used in the model (copper, glass, polyethylene, etc.) and determine how they will behave under the expected conditions. Based on that information, you must identify the material model(s) (linear elasticity, kinematic hardening plasticity, etc.) that you wish to use and then supply the material property values (E, υ, k, etc.) for those models, so the program can set up and solve the equations for you.

5.2. Material Models in ANSYS

ANSYS currently offers more than 60* predefined material models. These can be used to model most engineering materials under most conditions. Like elements, each material model in ANSYS has a number of properties associated with it. These properties include the material model name and label, the material model linearity, the material property values that can or must be supplied, the material property labels and constants used to define the material property values, the spatial and temperature dependencies of the material model, the permissible material model combinations, the elements that support the material model, and the product restrictions associated with the material model.

5.2.1. Material Model Name

Each material model has a formal name that is used in the ANSYS documentation. Some names describe the behavior of the material model (anisotropic elasticity, swelling, etc.). Other names identify the material that is most commonly associated with that behavior (cast iron, concrete, shape memory alloy, etc.). Material model names can also refer to the internal structure of the material (honeycomb, foam, etc.) or describe how the material is used (gasket, geological cap, etc.). Finally, some material models are named for the researchers who developed them (Bergstrom-Boyce, Chaboche, Drucker-Prager, etc.). The material model names used in the documentation generally reflect the nomenclature from engineering textbooks and the research literature.

5.2.2. Linearity

In ANSYS, material model linearity is defined based on solution iteration. Linear material models require only a single solution pass to solve the system of equations while nonlinear material models require an iterative solution. Although this may not seem like an important distinction, it affects how the program implements and evaluates the material properties. This, in turn, affects how you must supply material information to the program. For example, material properties for linear material models are defined using the **MP** family of commands while nonlinear material properties are defined using the **TB** family of commands. The material documentation reflects this distinction, so it is important that you are aware that it exists.

5.2.3. Material Property and Material Model Labels

Each linear material property has a 1 to 4 character abbreviation of its name that is used as its material property label (**MP**, *Lab*). For example, EX is the label for Young's modulus in the x direction, C is the label for specific heat, and DENS is the label for mass density. These labels specify which linear material property to define using the **MP** and **MPDATA** commands. A list of all linear material property labels can be found in Table 3.1 in the Mechanical APDL Material Reference.

Each nonlinear material model has a 2 to 8 character abbreviation of its name that is used as its material model label (**TB**, *Lab*). For example, AHYPER is the label for anisotropic hyperelasticity, while BB is the label for the Bergstrom-Boyce model. The material model label identifies which material model to use in conjunction with the **TB** command. A full list of the material models and their labels can be found in Chapter 2 of the Mechanical APDL Material Reference.

5.2.4. Material Property Values

Each material model represents a general type of behavior that is applicable to a particular class of materials (metals, thermoplastics, etc.). Material property values (in the form of constants, experimental curves, or both) are required to customize the material model for a specific material. For example, at low applied stresses, aluminum 6061-T6 can be modeled using the same linear elastic structural material model as AISI 1060 steel. However, these two materials will undergo very different deformations for the same applied load. Thus, different values for the Young's modulus and Poisson's ratio must be supplied.

Linear material properties values are supplied as constants (*C0*) using the **MP** command. Material property values for some nonlinear material models are supplied as constants (*C1, C2, C3*, etc.) using the **TBDATA** command. For other nonlinear material models, material property values are supplied as curves using the **TBPT** command. You must refer to the description of each material model hyperlinked from the list in Chapter 2 of the Mechanical APDL Material Reference to determine which command (**TBDATA** or **TBPT**) to use and for details on how to use it. The syntax and usage of these commands are discussed in more detail in Section 5.3.

Most nonlinear material models require linear material properties to be defined in addition to the nonlinear material properties.

5.2.5. Spatial Dependence

Many real materials have a spatial or orientation dependence. For example, natural materials like wood are stronger when loaded parallel to the grain and weaker when loaded perpendicular to the grain. Similarly, fiber reinforced materials have different properties depending on the orientation of the fibers. Most material models in ANSYS can be spatially dependent.

If a material has properties that are the same in every direction, the material is isotropic. In this case, only one value per material property is needed. For example, isotropic thermal analyses require only 1 value for thermal conductivity (KXX). Linear elastic isotropic structural analyses require 1 value for each of 2 material properties to calculate a solution: Young's modulus (EX) and Poisson's ratio (PRXY). (The shear modulus GXY is automatically calculated based on the Young's modulus and Poisson's ratio.).

If the material has properties that are different in each of the three coordinate directions, the material is orthotropic. For example, orthotropic thermal analyses with heat conduction require 3 thermal conductivities (KXX, KYY, KZZ): one for each coordinate direction. Similarly, 3 values must be supplied for each of the 3 material properties in an orthotropic structural analysis: Young's modulus in the three coordinate directions (EX, EY, EZ), the three major Poisson's ratios (PRXY, PRYZ, PRXZ) or the three minor Poisson's ratios (NUYX, NUYZ, NUXZ) but not both, and the three shear moduli (GXY, GYZ, GXZ).

Some materials have properties that are the same in two of the three coordinate directions. In ANSYS, these materials are still considered to be orthotropic materials and values for all three coordinate directions still need to be supplied.

Finally, material properties that do not exhibit any symmetry are anisotropic. For anisotropic electromagnetic materials, 6 values must be supplied to populate the permittivity matrix. For structural analyses, 10 values must be supplied to populate the elastic coefficient matrix for 2D analyses. 21 values are needed for the 3D case.

In general, spatial dependence does not affect the linearity of the material model.

5.2.6. Temperature Dependence

The properties of all real materials change with temperature. Materials generally become softer when heated and become brittle when cooled. Most materials expand when heated and contract when cooled. They also tend to conduct heat better when warm and conduct electricity better when cold. For this reason, most material models in ANSYS can also be temperature dependent.

Material properties can be treated as temperature independent if there is no temperature variation during the course of the analysis or if the temperature variation is small enough to have a negligible impact on the results. When an analysis is performed at a constant temperature, only one value per material property needs to be supplied.

If a finite element model will be solved at a variety of different temperatures or if the model will experience a significant change in temperature during the solution process, temperature-dependent material properties need to be defined. The details for how to input temperature-dependent material properties are given in section 5.3.

Temperature-dependent analyses where temperature is a degree of freedom require an iterative solution and thus are classified as nonlinear analyses. However, the **MPTEMP** and **MPDATA** commands are still used to define temperature-dependent linear material properties.

5.2.7. Supported Material Model Combinations

Some material models can be combined to simulate more complex behaviors. This most commonly occurs when a plasticity and/or creep model is combined with a hardening model. Chapter 7* of the Mechanical APDL Material Reference lists the material model combinations that are currently possible in ANSYS. If you access the interactive documentation through the program, you will also find links to examples of these material combinations in other parts of the documentation. Material model combinations are an advanced topic and are not addressed in this book.

5.2.8. Supported Elements

As noted in chapter 4, material models can only be used with elements that support them. Chapter 2 of the Mechanical APDL Material Reference lists the supported elements for each material model in the program, and Chapter 4 of the Mechanical APDL Element Reference lists the supported material models that can be used with each element.

5.2.9. Product Restrictions

Although all material models are listed in the Material Reference, you can only use the ones associated with the product(s) that you have a license to run. The product restrictions for linear material properties are listed at the bottom of the documentation for the **MP** command in the Mechanical APDL Command Reference. The product restrictions for all nonlinear material models are listed at the bottom of the document for the **TB** command in the Mechanical APDL Command Reference.

5.3. Defining Material Properties

The most convenient way to choose a material model and to define material model parameters in ANSYS is to use the Material Models GUI. It can be accessed via the GUI path: **Main Menu > Preprocessor > Material Props > Material Models**.

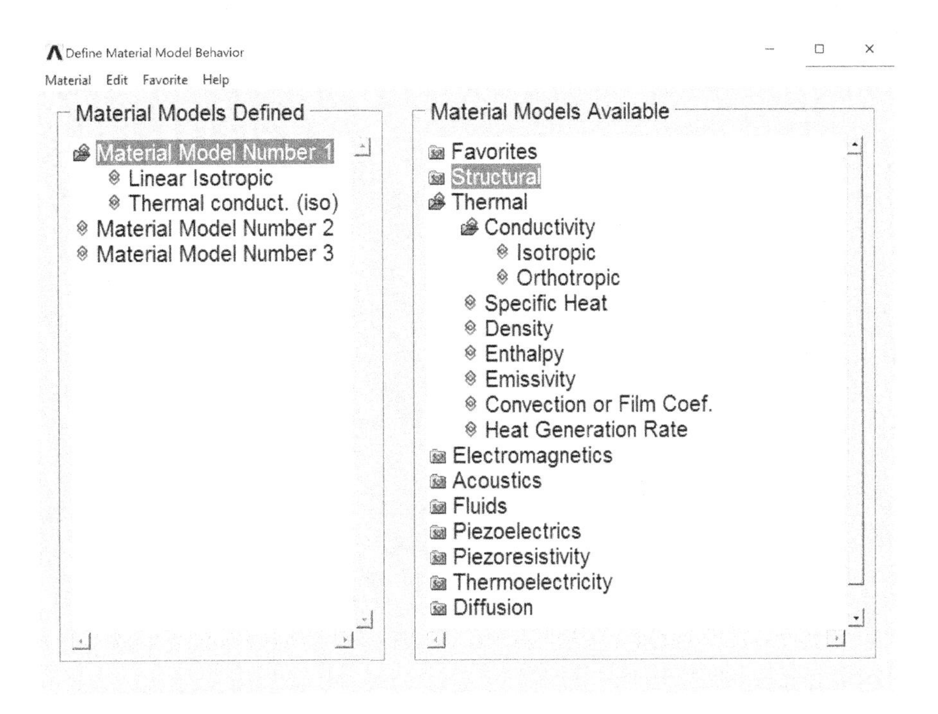

Figure 5.1 The Material Models GUI.

The Material Models GUI (Figure 5.1) is divided into two parts. The left side lists the material models that have already been defined for the analysis. You can add a new material model by following the path **Material > New Model**... in the dialog box menu. The default reference number for each new material model will be the maximum currently defined material model reference number plus one unless you specify a different number. You can change a material reference number by using the GUI path: **Main Menu > Preprocessor > Material Props > Change Mat Num** or by using the **MPCHG** command.

Once at least one material property per model has been defined, a folder symbol will be added to the material model icon in the left column of the Material Models GUI. This folder can then be expanded to show its contents if desired.

The right side of the Material Models GUI is a menu tree that lists the available material models organized by physics, linearity, and spatial dependence. Clicking on a material folder icon will expand the menu and list the material model options inside. Clicking on a double diamond icon will open a data input dialog box for the selected material model.

5.3.1. Inputting Linear Material Property Values Using the GUI

Selecting a material model in the right column will open a data input dialog box where you can supply material property values. The input boxes for linear material properties are labeled with the material property labels used with the **MP** command. Figure 5.2 shows the data input dialog box for an isotropic linear elastic structural material (aluminum 6061-T6).

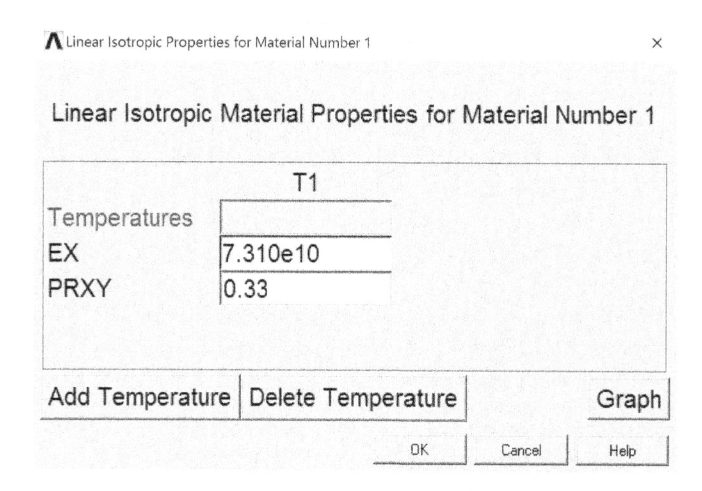

Figure 5.2 Material Model Data Input Dialog Box for a Linear Elastic Structural Material (Aluminum 6061-T6).

5.3.2. Inputting Linear Material Property Values Using Commands

As noted in chapter 2, every action in the GUI issues a command, or a group of commands, with complete syntax to the program. The command structure associated with material property definition is more complex than usual. There are multiple combinations of commands that can be used to define a given set of material properties. And, the command combinations used by the GUI are not the most intuitive or streamlined. This will become important when writing or editing input files. To prepare you for that discussion, the commands needed to define material properties are introduced in this chapter.

Linear material properties are usually defined using the **MP** command. The full syntax for the command is: **MP**, *Lab, MAT, C0, C1, C2, C3, C4*. The *Lab* argument specifies the label of the material property being defined (EX for Young's modulus, PRXY for Poisson's ratio, etc.). The *MAT* argument specifies the material reference number for the analysis. The remaining arguments are used to supply the material property value (*C0*) or the coefficients for a temperature-dependent polynomial equation (see section 5.3.8 for more details).

Example 5.1 lists the commands to define the Young's modulus and Poisson's ratio for aluminum 6061-T6 (material model reference #1).

Example 5.1

```
/PREP7                    ! Enter the Preprocessor
MP,EX,1,7.31e10           ! Define Young's modulus (E=7.31e10): material #1
MP,PRXY,1,0.33            ! Define Poisson's ratio (v=0.33): material #1
```

5.3.3. Inputting Nonlinear Material Property Values Using the GUI

The process for inputting nonlinear material property values using the GUI is similar to the one used for linear material models. The input boxes for nonlinear material properties are labeled either with abbreviations of the required properties (Figure 5.3) or with numbered constants (Figure 5.4). The material properties associated with each constant for each nonlinear material model are listed in a table (Figure 5.5) along with the material model description in Chapter 4 of the Mechanical APDL Material Reference.

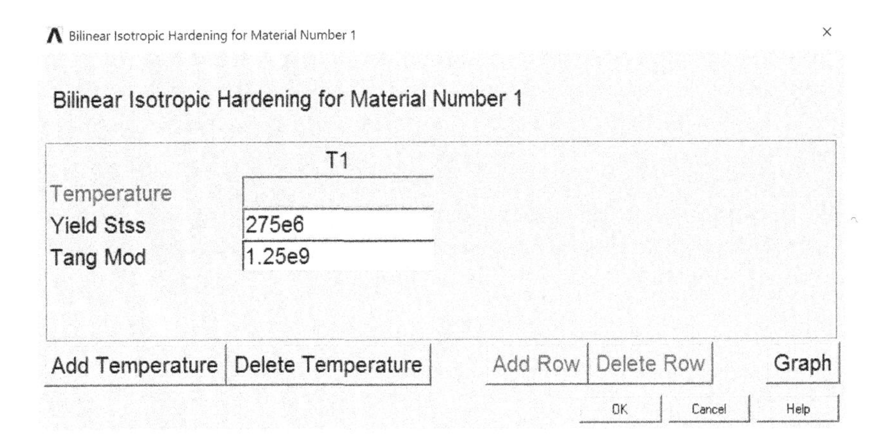

Figure 5.3 Material Model Data Input Dialog Box for a Bilinear Plastic Material (Aluminum 6061-T6).

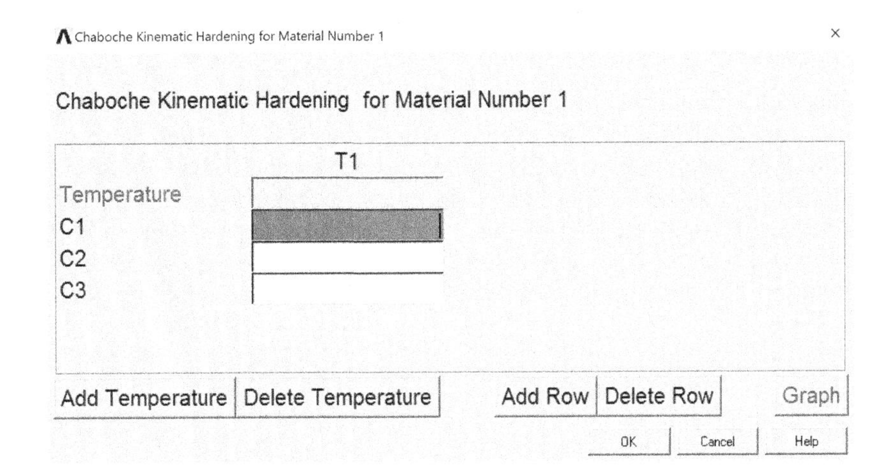

Figure 5.4 Material Model Data Input Dialog Box for Chaboche Plasticity.

Constant	Meaning	Property
C1	σ_0	Yield stress
C2	E_T	Tangent modulus

Figure 5.5 Material Property Constants from the Mechanical APDL Material Reference (Bilinear Isotropic Hardening Model).

5.3.4. Inputting Nonlinear Material Property Values Using Commands

Defining nonlinear material property values requires a minimum of two commands: the **TB** command and the **TBDATA** command. The **TB** command specifies the material model to use, the material reference number to use, the number of temperatures that will be supplied, and the number of data points that will be supplied (if applicable). The full syntax of the command is: **TB**, *Lab, MAT, NTEMP, NPTS, TBOPT, EOSOPT, FuncName*. The *Lab* argument is the material model label. The *MAT* argument specifies the material reference number for the analysis. *NTEMP* is the number of temperatures for which the data will be supplied. For a non-temperature dependent analysis, *NTEMP* is set to 1. *NPTS* is the number of data points that will be supplied (if applicable). The final three arguments depend on the chosen material model.

The **TBDATA** command is then used to supply the material property data values. The full syntax of the command is: **TBDATA**, *STLOC, C1, C2, C3, C4, C5, C6. STLOC* is the starting location in the table where the data will be entered. If the default value (the last filled location plus one) is ok, that field can be left blank. The remaining arguments are for the material property constants.

Example 5.2 lists the commands to define the properties of aluminum 6061-T6 using a temperature-independent bilinear isotropic hardening plasticity model (BISO).

Example 5.2

```
/PREP7                  ! Enter the Preprocessor
MP,EX,1,7.31e10         ! Define Young's modulus (E=7.31e10): material #1
MP,PRXY,1,0.33          ! Define Poisson's ratio (v=0.33): material #1
TB,BISO,1,1,2           ! Activate data table for BISO: material #1
                        ! 1 temperature, 2 data points per temperature
TBDATA,,275e6,1.25e9    ! C1 (yield stress)=275e6, C2 (tan modulus)=1.25e9
```

5.3.5. Inputting Nonlinear Material Properties Curves Using the GUI

The properties of some materials cannot be defined with a few constants. In these cases, multiple data points (experimental curves) are needed to define the material behavior. The Material Models GUI for these material models initially presents input boxes for a single data point. You can add input boxes by clicking the "Add Point" button until you have entered all the data points needed for that material (Figure 5.6). The maximum number of points that can be included depends on the particular material model. Input boxes can be deleted using the "Delete Point" button. You can also plot the data to confirm that it is correct by clicking the "Graph" button (Figure 5.7).

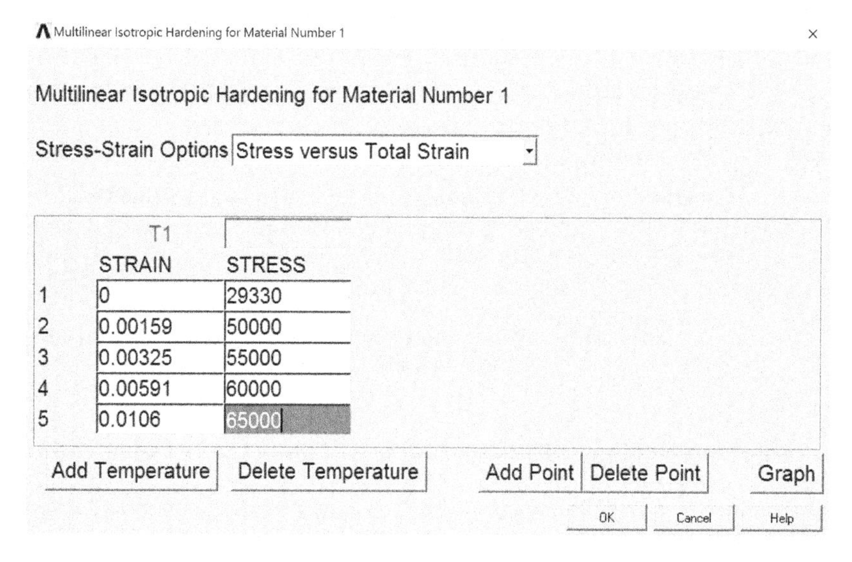

Figure 5.6 Material Model Data Input Dialog Box for Multilinear Isotropic Hardening Material.

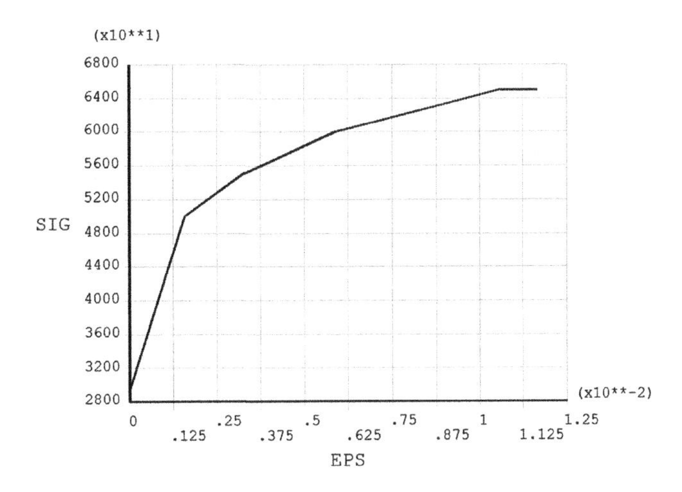

Figure 5.7 Graph of Stress Strain Curve from Figure 5.6.

5.3.6. Inputting Nonlinear Material Property Curves Using Commands

To define nonlinear material property curves, the **TB** command is again used to specify the material model, the material reference number, the number of temperatures that will be supplied, and the number of data points that will be supplied. The **TBPT** command is then used to define new values for those data points. The full syntax for the command is: **TBPT**, *Oper, X1, X2, X3, . . ., XN*. The *OPER* argument must be either DEFI (define a new data point) or DELE (delete an existing data point). (The DEFI option is the default for the first argument so this field can be left blank during material data definition if desired.) The remaining arguments are the material property values to be supplied. Example 5.3 lists the commands to produce the material model shown in Figures 5.6 and 5.7.

Example 5.3

```
/PREP7                    ! Enter the Preprocessor
MP,EX,1,14.665E6          ! Define Young's modulus (E=14.665E6): material #1
MP,PRXY,1,0.3             ! Define Poisson's ratio (v=0.3): material #1
TB,MISO,1,1,5,            ! MISO data table, mat #1, 1 temp, 5 data points
TBPT,DEFI,0,29.33E3       ! New data points: Strain=0; Stress=29.33e3
TBPT,DEFI,1.59E-3,50E3    ! New data points: Strain=1.59e-3; Stress=50e3
TBPT,DEFI,3.25E-3,55E3    ! New data points: Strain=3.25e-3; Stress=55e3
TBPT,DEFI,5.91E-3,60E3    ! New data points: Strain=5.91e-3; Stress=60e3
TBPT,DEFI,1.06E-2,65E3    ! New data points: Strain=1.06e-2; Stress=65e3
```

Nonlinear materials may have a minimum number of data points required and/or a maximum number of data points permitted. These limits are specified in the documentation for the **TB** command in the Mechanical APDL Command Reference.

5.3.7. Inputting Temperature-Dependent Material Property Data Using the GUI

Material property data is assumed to be temperature independent, so the data input dialog boxes for all material models that support temperature-dependent properties initially request properties for a single temperature (T1). Clicking the "Add Temperature" button activates the Temperature row that was previously grayed out and allows you to specify a temperature for the first column of material property data. It also adds a new column where you can specify a new temperature and the material properties for that temperature. Figure 5.8 shows selected thermal conductivities for calcium silicate. After the temperature data points have been entered for a material, they can be graphed by clicking the "Graph" button in the lower right-hand corner of the dialog box. This procedure is the same for both linear and nonlinear material models.

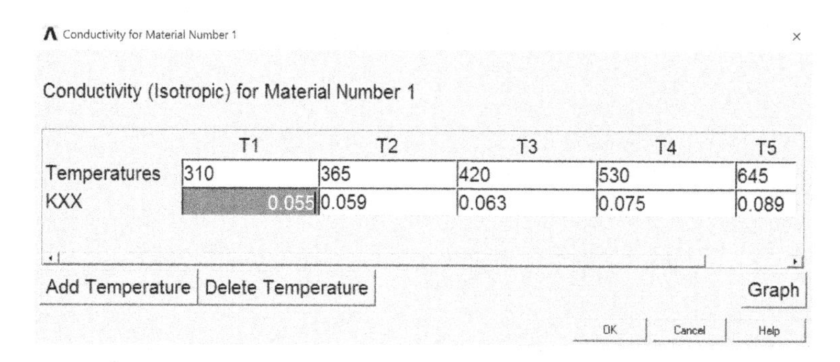

Figure 5.8 Defining Temperature Dependent Material Properties (Thermal Conductivity for Calcium Silicate).

5.3.8. Inputting Linear Temperature-Dependent Material Property Data Using Commands: Polynomial Equations

You can define linear, quadratic, cubic, or quartic temperature-dependent material properties with the form:

$$\text{Property} = C0 + C1\,T + C2\,T^2 + C3\,T^3 + C4\,T^4$$

where *C0, C1, C2, C3*, and *C4* are coefficients whose values are supplied by the user and *T* is temperature. The five coefficients are supplied as the final 5 arguments of the **MP** command.

If only *C0* is supplied by the user, only *C0* is stored in the database. In this case, the other coefficients are assumed to be zero and the material is treated as temperature-independent. However, if any value for *C1* though *C4* is supplied, ANSYS creates a table using the temperatures defined by the current **MPTEMP** command and calculates the associated properties using the polynomial equation. (See section 5.3.9 for more details.) If the **MPTEMP** command has not been issued, ANSYS assumes a range of temperatures instead. Once the initial values in the table are defined, the polynomial equation is ignored. At solution time, ANSYS will do a linear interpolation on the table values to determine the property value at the solution temperature.

This option is not available through the GUI.

5.3.9. Specifying Temperature Dependence Using Commands: Temperature Tables

Temperature-dependent material property data also can be input using the **MPTEMP** and **MPDATA** commands. The **MPTEMP** command is used to define a temperature table for linear material properties. The full syntax for the **MPTEMP** command is: **MPTEMP**, *STLOC, T1, T2, T3, T4, T5, T6*. Again, *STLOC* is the starting location in the table where the data will be entered. The remaining arguments are the temperatures to be defined.

The **MPDATA** command is then used to supply the material property data associated with those temperatures. The full syntax for the **MPDATA** command is: **MPDATA**, *Lab, MAT, STLOC, C1, C2, C3, C4, C5, C6*. The *Lab* argument is the material property label (EX, PRXY, etc.). *STLOC* is the starting location in the table where the data will be entered. The remaining arguments are the data to be supplied. Up to 6 temperatures may be defined with a single **MPTEMP** command and up to 6 data points may be defined with a single **MPDATA** command. The commands may be repeated as often as needed.

Example 5.4 lists the commands issued by the GUI to define the temperature-dependent thermal conductivity from Figure 5.8.

Example 5.4

```
/PREP7                                          ! Enter the Preprocessor
MPTEMP,1,310,365,420,530,645,750                ! Define temps for mat #1
MPDATA,KXX,1,,0.055,0.059,0.063,0.075,0.089,0.104 ! Define KXXs for mat #1
```

The temperature table defined by the **MPTEMP** command remains valid for all subsequent properties until it is changed. To change the temperature table, you must issue a blank **MPTEMP** command (`MPTEMP,,,,,,,,`).

If new temperatures are supplied without erasing the previous table, the new temperatures will replace the old temperatures in the current table. Erasing the temperature table does not erase or delete temperature data for previously defined material properties.

5.3.10. How the GUI Specifies Linear Material Properties Using Commands

The Material Models GUI writes all linear material properties to the log file using the **MPTEMP** and **MPDATA** commands regardless of temperature dependence. Example 5.5 lists the commands issued by the GUI for the material model defined in Figure 5.2 (non-temperature-dependent Young's modulus and Poisson's ratio for aluminum 6061-T6).

Example 5.5

```
/PREP7                  ! Enter the Preprocessor
MPTEMP,,,,,,,,          ! Clear existing material temperature table
MPTEMP,1,0              ! New material temperature table: temp=0
MPDATA,EX,1,,7.31e10    ! Define Young's modulus (E-7.31e10): material #1
MPDATA,PRXY,1,,0.33     ! Define Poisson's ratio (v=0.33): material #1
```

The GUI never issues the **MP** command.

5.3.11. Inputting Nonlinear Temperature-Dependent Material Property Data Using Commands

While inputting temperature-dependent linear material data is relatively straightforward, inputting temperature-dependent nonlinear material data is not. This is due to the amount of data needed to implement a nonlinear material model and because there are multiple ways to define that data.

To define temperature-dependent nonlinear material properties, you must use the **TB, TBTEMP**, and **TBDATA** or **TBPT** commands. The **TB** command specifies both the intended material model and the number of temperatures for which the data will be supplied. The **TBTEMP** command specifies the current temperature for which the data will be defined. Each **TBTEMP** command is then followed by a series of **TBDATA** or **TBPT** commands to supply the material property values for that temperature. Example 5.6 lists the commands to define the material model from section 5.3.6 for two temperatures: 0°C and 500°C.

Example 5.6

```
/PREP7                      ! Enter the Preprocessor
MPTEMP,1,0,500              ! New temperature data table, temp1=0, temp2=500
MPDATA,EX,1,,14.665E6,12.423e6 ! Define Young's modulus for both temperatures
MPDATA,PRXY,1,,0.3          ! Define Poisson's ratio for both temperatures
TB,PLASTIC,1,2,5,MISO       ! Plastic MISO data table, mat #1, 2 temps, 5 pts
TBTEMP,0                    ! Define temperature = 0 for TBPT command
TBPT,DEFI,0,29.33E3         ! New data points: Strain=0; Stress=29.33e3
TBPT,DEFI,1.59E-3,50E3      ! New data points: Strain=1.59e-3; Stress=50e3
TBPT,DEFI,3.25E-3,55E3      ! New data points: Strain=3.25e-3; Stress=55e3
TBPT,DEFI,5.91E-3,60E3      ! New data points: Strain=5.91e-3; Stress=60e3
TBPT,DEFI,1.06E-2,65E3      ! New data points: Strain=1.06e-2; Stress=65e3
TBTEMP,500                  ! Define temperature = 500 for TBPT command
TBPT,DEFI,0,27.33E3         ! New data points: Strain=0; Stress=27.33e3
TBPT,DEFI,2.02E-3,37E3      ! New data points: Strain=2.02e-3; Stress=37e3
TBPT,DEFI,3.76E-3,40.3E3    ! New data points: Strain=3.76e-3, Stress=40.3e3
TBPT,DEFI,6.48E-3,43.7E3    ! New data points: Strain=6.48e-3; Stress=43.7e3
TBPT,DEFI,1.12E-2,47E3      ! New data points: Strain=1.12e-2; Stress=47e3
```

5.3.12. GUI Inaccessible Materials

Some of the material models available in ANSYS are not accessible through the GUI. A complete list of these materials is contained in Chapter 9 of the Mechanical APDL Material Reference. GUI inaccessible material models are more sophisticated material models that require an understanding of the material model and how it is implemented in ANSYS. GUI inaccessible materials are beyond the scope of this book.

5.3.13. User Materials

Although the material models provided by ANSYS are both extensive and highly adaptable, they may not be sufficient for all users—especially individuals involved in material model research. You can define your own material model (a "user material") in ANSYS. However, this requires programming at the Fortran level of the program and is beyond the scope of this book.

5.4. Choosing Which Material Properties to Define

The Material Models GUI provides no guidance about which material model to choose or which material properties are required for a given analysis. However, there are rules of thumb to follow.

For structural analyses:

- Young's modulus (EX) is required for all analyses.
- Poisson's ratio (PRXY) is required for all analyses.
- Density (DENS) is required when gravity loads are applied and for all dynamic analyses including transient analyses, modal analyses, and harmonic analyses.
- Nonlinear material properties are required if the maximum stress in the model is, or is expected to be, greater than the yield stress of the material.

For thermal analyses:

- Thermal conductivity (KXX) is required for all analyses.
- Density (DENS) and specific heat (C) are required for all transient thermal analyses.

- Enthalpy (ENTH) is required for all analyses that involve phase change.
- Viscosity (VISC) and fiction coefficient (MU) are needed for analyses that involve mass transport.

5.5. Finding Material Property Data

One of the most difficult and time-consuming parts of finite element modeling can be finding good material property values for your analysis. The properties of many common materials can be found in engineering handbooks and in material property databases. You can often find information about more exotic materials in the academic literature (journal articles) and in government reports. However, it may be necessary for you to perform experiments to obtain the information that you need.

It is your responsibility as an engineer to determine whether or not a material model can be used for a given application. It is also your responsibility to ensure that the material constants are from a reliable source and can be verified by multiple references. All material property values provided in this book and in the ANSYS Material Library are for demonstration purposes only. They should not be used in real analyses without confirmation from external sources.

It is critical that the values supplied for the material properties match the nominal temperature of the analysis. Using material properties determined at room temperature for high or low temperature applications is guaranteed to give bad results. For biomedical analyses, it is also important to determine if your material properties were measured in vitro or in vivo and which set of values you need.

5.6. Potential Pitfalls Associated with Material Property Evaluation in ANSYS

There are a number of potential pitfalls associated with the definition and evaluation of material properties in ANSYS.

5.6.1. Too Few Material Properties

When the **SOLVE** command is issued, ANSYS checks the model data to ensure that it has all of the information necessary to assemble the system of equations. If some or all of the material properties required to generate those equations are missing, the program will issue an error message and terminate checking. You will be unable to solve the model until the minimum material properties are supplied.

5.6.2. Too Many Material Properties

ANSYS makes very few assumptions about what material properties are needed and what values those properties should have. Thus, you can supply more material properties than are needed for a given analysis. In general, the program will use only the material properties required to satisfy the physics of the problem and will ignore the rest. However, providing extraneous material properties can sometimes lead to unexpected and unintended results. For example, if a plasticity material model is defined, the program will calculate plastic strains when stresses exceed yield even if this was not your intention. Similarly, if a thermal expansion coefficient is supplied without temperature loads, thermal strains will not be calculated even if you expected them to be. We strongly recommend supplying only the material properties required by your analysis and no more.

5.6.3. Insufficient Number of Points on a Material Data Curve

ANSYS uses linear interpolation to evaluate all data stored in tables, including material data curves. As the number of data points increases, this technique approximates a smooth curve. However, if an insufficient number of data points are supplied, the material property values used by the program may not adequately reflect the behavior of the real material.

Fortunately, the program also graphs temperature-dependent material properties and material property data using a linear interpolation between data points, so you can visualize your material data curves when they are defined. You should check the approximation before issuing the **SOLVE** command. If it is not a good representation of the material behavior, you should increase the number of material property data points until it is.

5.7. Saving Material Properties

There are two good ways to save material properties for future use in ANSYS. You can export your material properties using the ANSYS Material Library or you can save the commands used to define the material properties and reissue them at a later time.

The Material Models GUI has an option to name material models and save them as "Favorites." However, this option will save only the material model and not the properties contained within it. All material property values specified within a "Favorite" material model will be lost when you clear the database. For this reason, saving material models as "Favorites" is not recommended.

5.7.1. Exporting Material Properties as ANSYS Material Library Files

Once you have defined the properties for at least one material model, you can export that information as an ANSYS material library file using the GUI path: **Main Menu > Preprocessor > Material Props > Material Library > Export Library** or the **MPWRITE** command. This will display a dialog box (Figure 5.9) that prompts you to specify the system of units that the materials have been defined in. Once this has been done, it opens a second dialog box (Figure 5.10) where you can specify the reference number of the material model to export and the name of the file to which you will export. The file will be saved in your working directory as filename.unit_MPL unless you specify another location.

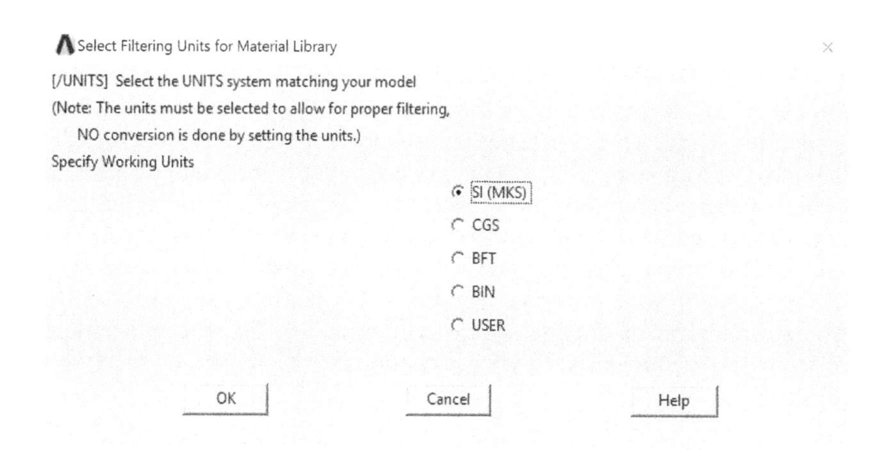

Figure 5.9 Select Units for Export Material Library Dialog Box.

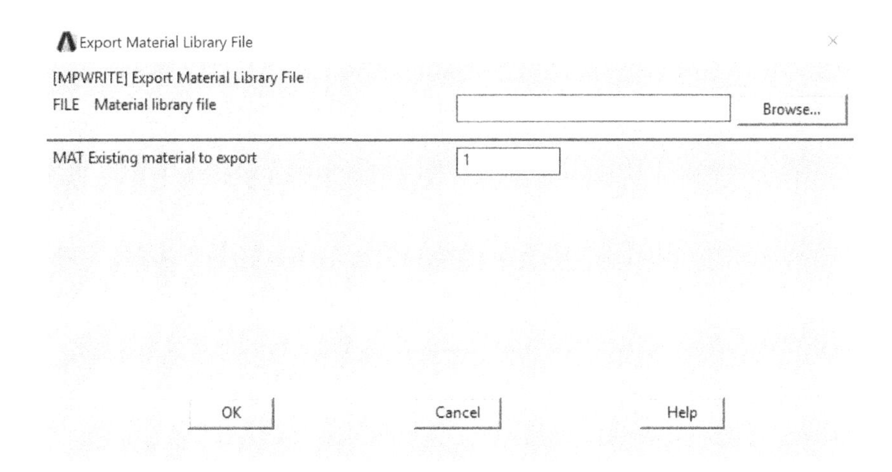

Figure 5.10 Export Material Library File Dialog Box.

To import a material library file that you have created or whose content you have verified, use the GUI path: **Main Menu > Preprocessor > Material Props > Material Library > Import Library** or the **MPREAD** command.

The material library will attempt to import material models with a new material model reference number. However, if you specify an existing reference number, it will overwrite the existing data for that model.

5.7.2. Saving Material Properties as Commands

Finally, you can save the commands used to define a specific material model in a plain text file. You can read this file into the program to re-create the material model at a later time using the GUI path: **Utility Menu > File > Read Input From**... or using the **/INPUT** command. You can also copy the commands and paste them into the command prompt. We recommend this method for saving material properties because it is simple and robust. Input files will be addressed in more detail in chapter 10.

Exercise 5-1

Temperature-Dependent Plasticity Analysis of a Plate with a Central Hole

Overview

In this exercise, you will analyze the tensile behavior of a plate with a central hole (Figure 5-1-1) using a linear elastic material model at room temperature and an elastic-plastic material model at a variety of temperatures. The plate is made of 2024-T6 aluminum. Its material properties are given below. The plate is 15 cm wide and 10 cm tall. The hole is 2 cm in diameter. The plate is thin, so this can be treated as a 2D problem. A load of 200 MPa will be placed on the right end of the plate. The left end of the plate will be fixed. The reaction force from this constraint will effectively apply a second 200 MPa load on the left. The value of this load is relatively high and will cause the area around the hole to deform plastically. Although the plate geometry is the same as in exercise 3-1, symmetry will not be used in this model because the maximum plastic strains are expected to occur near the planes of symmetry. As noted in exercise 4-3, it is generally advisable to avoid applying boundary conditions to the area(s) of maximum stress and strain. This rule of thumb is more important in plastic analyses than in elastic analyses.

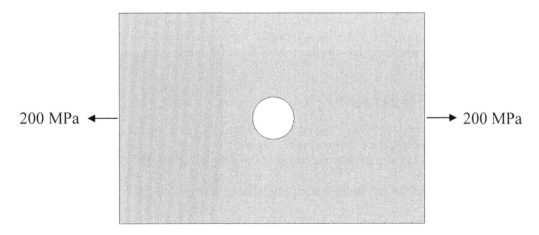

Figure 5-1-1 Plate with a Central Hole.

ANSYS Mechanical APDL for Finite Element Analysis.
DOI: http://dx.doi.org/10.1016/B978-0-12-812981-4.00022-8

Model Attributes

Material Properties for 2024-T6 Aluminum

Temperature	Young's Modulus	Poisson's Ratio	Yield Stress	Plastic Modulus
25°C	73.3 GPa	0.33	413 MPa	4.56 GPa
100°C	72.2 GPa	0.33	394 MPa	4.15 GPa
149°C	65.7 GPa	0.33	337 MPa	3.90 GPa
204°C	60.4 GPa	0.33	282 MPa	3.50 GPa
260°C	55.0 GPa	0.33	149 MPa	3.00 GPa
316°C	50.2 GPa	0.33	67 MPa	0

Loads

- Tensile load of 200 MPa on the right edge of the plate.

Constraints

- No displacement of the left edge of the plate in the horizontal (x) direction.
- No displacement of the lower left corner of the plate in the vertical (y) direction.

File Management

Create a new folder in your "Intro-to-ANSYS" folder for Exercise5-1

Change the Working Directory and the Jobname

Start ANSYS

Step 1: Define Geometry

1-1. Create the plate

- Preprocessor > Modeling > Create > Areas > Rectangle > By 2 Corners
- Place the lower left corner of the plate at the origin (0,0) and the upper right corner at (0.15,0.1)

1-2. Create the hole in the plate

- Preprocessor > Modeling > Create > Area > Circle > Solid Circle
- The hole has a radius of 1 cm (0.01 m)
- It should be placed in the center of the plate (0.075,0.05)

1-3. Subtract the circle from the rectangle to create one area

- Preprocessor > Modeling > Operate > Booleans > Subtract > Areas
- Remember to select the plate first and the hole second

1-4. Save your progress

Step 2: Define Element Types

2-1. Define the element type to use for this model

- Preprocessor > Element Type > Add/Edit/Delete
- Use PLANE182 elements

This exercise uses an element without mid-side nodes because plastic strains are only calculated for the corner nodes. In this case, mid-side nodes would add computation cost but provide no benefit.

Step 3: Define Material Properties

3-1. Create a linear elastic material model for aluminum 2024-T6

Since the applied loads will exceed the yield stress of the material, an elastic analysis is only possible as long as the inelastic material properties are not supplied. For this reason, the plastic material properties will not be entered until after the first elastic solution.

3-1-1. Define the elastic material properties at room temperature

- Preprocessor > Material Props > Material Models
- Structural > Linear > Elastic > Isotropic
- Enter 73.3e9 for the Young's modulus (EX)
- Enter 0.33 for the Poisson's ratio (PRXY)
- Click Add Temperature

Clicking the "Add Temperature" button does two things: it activates the temperature row in the dialog box and it adds a second column to the dialog box for a second set of data. Repeating this process will add additional columns to the dialog box (Figure 5-1-2).

3-1-2. Specify the value for room temperature

- Enter 25 as the temperature for the first column

3-1-3. Define the elastic material properties at 100°C

- Enter 100 as the temperature for the second column
- Enter 72.2e9 for the Young's modulus (EX)
- Enter 0.33 for the Poisson's ratio (PRXY)

3-1-4. Define the elastic material properties at 149°C

- Click Add Temperature
- Enter 149 as the temperature for the third column
- Enter 65.7e9 for the Young's modulus (EX)
- Enter 0.33 for the Poisson's ratio (PRXY)

3-1-5. Define the elastic material properties at 204°C using the table above

3-1-6. Define the elastic material properties at 260°C using the table above

3-1-7. Define the elastic material properties at 316°C using the table above

3-1-8. Click OK

Clicking OK saves your material properties to the database and closes the dialog box. This prevents you from losing your work when you graph the material properties.

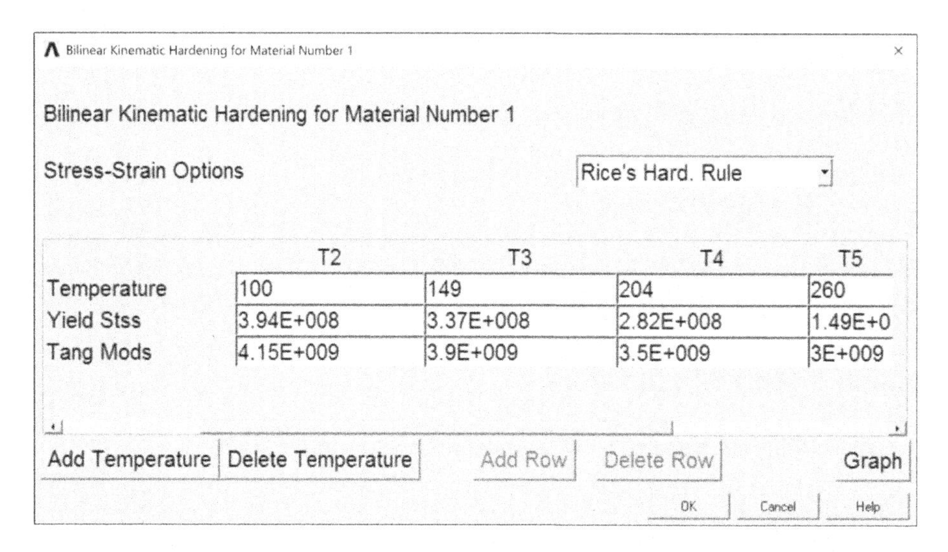

Figure 5-1-2 Defining Temperature-Dependent Elastic Material Properties.

3-1-9. Click on your new material model to reopen the dialog box

- Material Model Number 1 > Linear Isotropic

3-1-10. Graph the temperature-dependent Young's modulus

- Click the "Graph" button in the Material Models dialog box
- Choose "EX" as the material property to plot.

The graph of the temperature-dependent Young's modulus is shown in Figure 5-1-3.

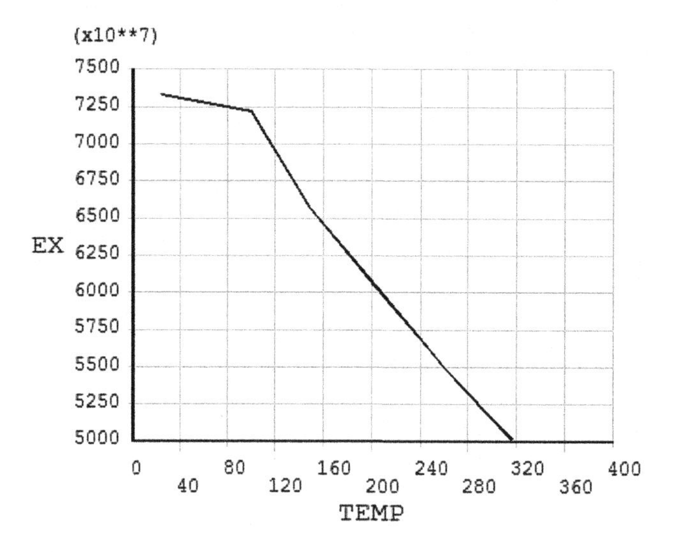

Figure 5-1-3 Plot of Temperature-Dependent Young's Modulus.

3-1-11. Complete the elastic material property definition

- Click OK
- Close the material model dialog box

3-2. Save your progress

Step 4: Mesh

4-1. Create the initial mesh for the finite element model

- Preprocessor > Meshing > MeshTool
- Specify a Smart Size of 3
- Mesh all of the areas in the model

A large stress gradient is expected near the hole for the elastic analysis and a large plastic strain gradient is expected near the hole for the inelastic analysis. Therefore, the mesh must be refined.

4-2. Refine the mesh around the hole

- Preprocessor > Meshing > MeshTool
- Refine the mesh along the four lines that comprise the central hole
- Use a level of refinement of 1

4-3. Refine the entire mesh

- Preprocessor > Meshing > MeshTool
- Refine the area mesh for the entire model
- Use a level of refinement of 3.

The final finite element mesh is shown in Figure 5-1-4.

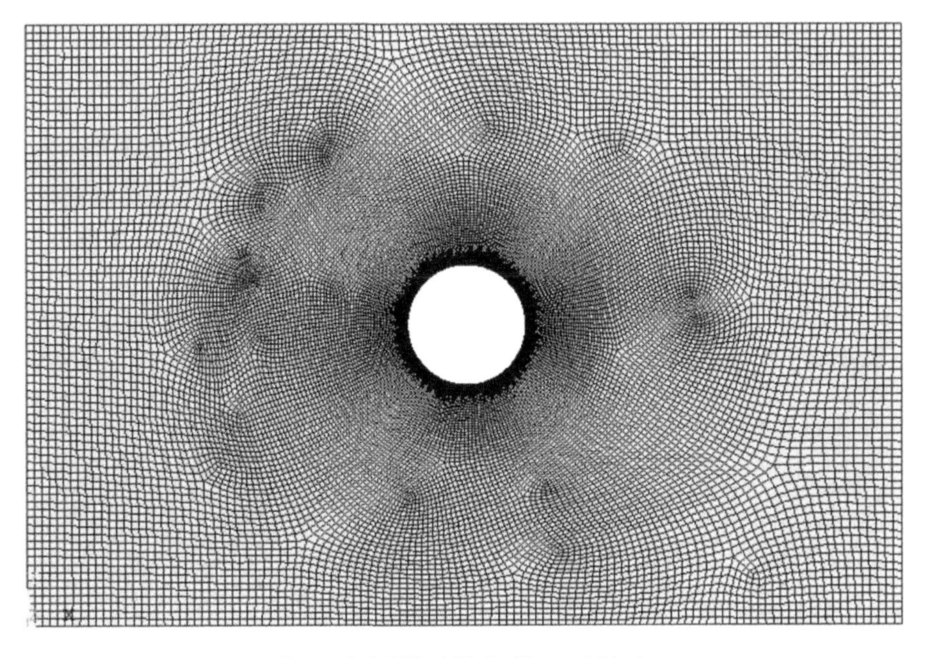

Figure 5-1-4 Final Finite Element Mesh.

4-4. Save your finite element mesh

Step 5: Apply Constraint Boundary Conditions

5-1. Constrain the left side of the plate in *x*

- Solution > Define Loads > Apply > Structural > Displacement > On Lines
- Pick the line at the left end of the model
- Choose UX as the DOF to be constrained
- Set the displacement value to 0

5-2. Constrain the lower left corner of the plate in *y*

- Solution > Define Loads > Apply > Structural > Displacement > On Keypoints
- Pick the keypoint at the lower left corner of the model (0,0,0)
- Choose UY as the DOF to be constrained
- Set the displacement value to 0

5-3. Save your constraints

Step 6: Apply Load Boundary Conditions

6-1. Apply load boundary conditions to the right end of the plate

- Solution > Define Loads > Apply > Structural > Pressure > On Lines
- Pick the line at the right end of the model
- Specify −200e6 for the applied load

6-2. Define a uniform temperature for the first load step

- Solution > Define Loads > Apply > Structural > Temperature > Uniform Temp
- Enter 25 for "TUNIF Uniform Temperature"
- Click OK

6-3. Save your loads

Step 7: Set the Solution Options

The default solution options can be used for this analysis.

Step 8: Solve

8-1. Select everything in the model

8-2. Solve

- Solution > Solve > Current LS

8-3. Save your results

Step 9: Postprocess the Results

9-1. Plot the elastic equivalent (von Mises) stress distribution in the model

- General Postproc > Plot Results > Contour Plot > Nodal Solu

The yield stress of 2024-T6 aluminum at room temperature is 413 MPa. However, Figure 5-1-5 shows that the maximum equivalent stress for this model is 632 MPa. This value is well above the yield stress and confirms that a nonlinear solution is necessary. To accurately model the deformation of and stresses within the plate, you must supply nonlinear material properties and perform a plasticity analysis.

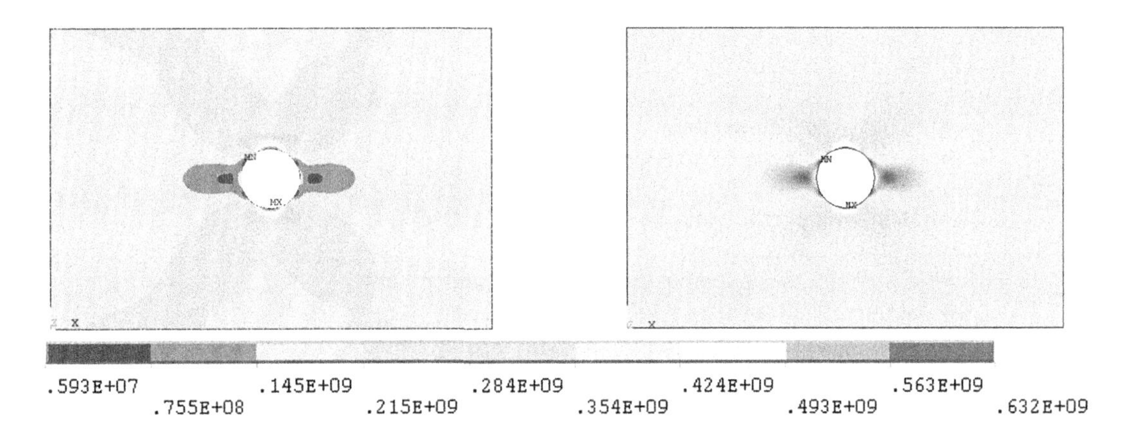

Figure 5-1-5 Plot of Elastic Equivalent Stress: Win32 Graphics (left), 3D Graphics (right).

Step 10: Set Up the Plasticity Analysis

To perform a plasticity analysis, nonlinear material properties must be included. You could return to the Preprocessor to enter the remaining material properties. However, the Material Models GUI is also available in the Solution processor. So you will return to the Solution processor, enter the nonlinear material properties, and solve the model again.

10-1. Change the jobname to Exercise5-1-plastic

- Utility Menu > File > Change Jobname ...
- Change the jobname to Exercise5-1-plastic
- Check the box to create new log and error files
- Click OK

*By default, ANSYS assumes that a new analysis should be started each time the Solution processor is entered. So when you issue the **SOLVE** command for the plasticity analysis, the program will overwrite the elastic results. By changing the jobname, you will automatically create a new results file and protect your old results. For more information, see* chapter 8.

10-2. Enter the plastic material properties

You will use a bilinear kinematic hardening plasticity model for this analysis. This is a relatively simple plasticity model. Only the yield stress and the tangent modulus (i.e., the slope of the plastic part of the stress-strain curve) are required as inputs.

10-2-1. Define the plastic material properties at room temperature

- Solution > Load Step Opts > Other > Change Mat Props > Material Models
- Material Model Number 1 should be highlighted
- Click on: Structural > Nonlinear > Inelastic > Rate Independent > Kinematic Hardening Plasticity > Mises Plasticity > Bilinear
- Enter 4.13e8 for the yield stress ("Yield Stss") (Figure 5-1-6)
- Enter 4.56e9 for the tangent modulus ("Tang Mods")

10-2-2. Specify the value for room temperature

- Click Add Temperature
- Enter 25 for the T1 temperature

10-2-3. Define the plastic material properties at 100°C

- Enter 100 as the temperature for the second column
- Enter 3.94e8 for the yield stress ("Yield Stss")
- Enter 4.15e9 for the tangent modulus ("Tang Mods")

10-2-4. Define the plastic material properties at 149°C using the table above

10-2-5. Define the plastic material properties at 204°C using the table above

10-2-6. Define the plastic material properties at 260°C using the table above

10-2-7. Define the plastic material properties at 316°C using the table above.

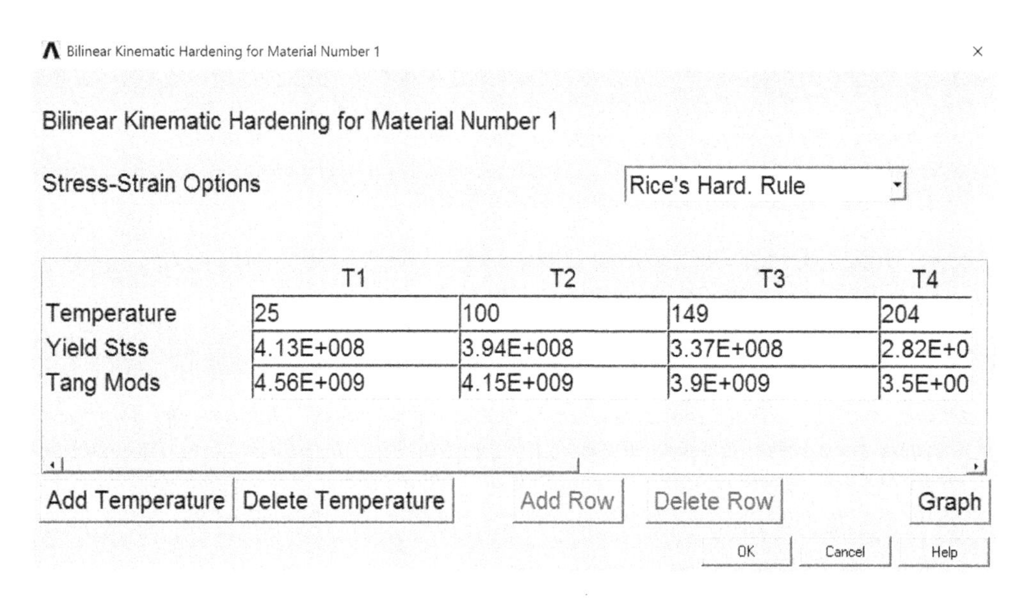

Figure 5-1-6 Defining Temperature-Dependent Plastic Material Properties.

10-3. Set the time for the end of the first load step

- Solution > Analysis Type > Sol'n Controls
- Enter 1 for "Time at end of loadstep" (Figure 5-1-7)
- Click OK

Nonlinear analyses in ANSYS are solved using the Newton-Raphson procedure. This procedure requires a load parameter to track the increments in the iterations. ANSYS uses time as that parameter since transient analyses also usually require an iterative solution and time is a convenient variable. If you do not supply a value for time in an analysis that requires Newton-Raphson iterations, ANSYS issues a warning and uses a default value of the previous time plus 1. For more information, see chapter 8.

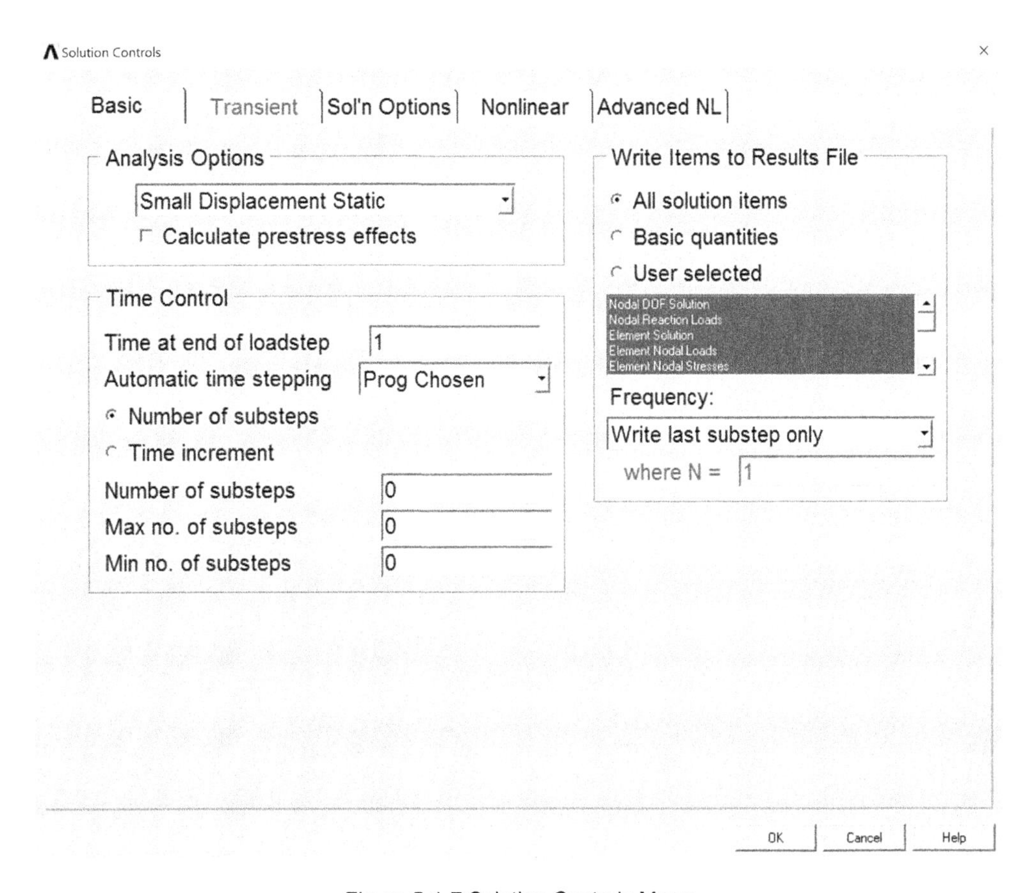

Figure 5-1-7 Solution Controls Menu.

10-4. Solve the first load step

- Solution > Solve > Current LS.

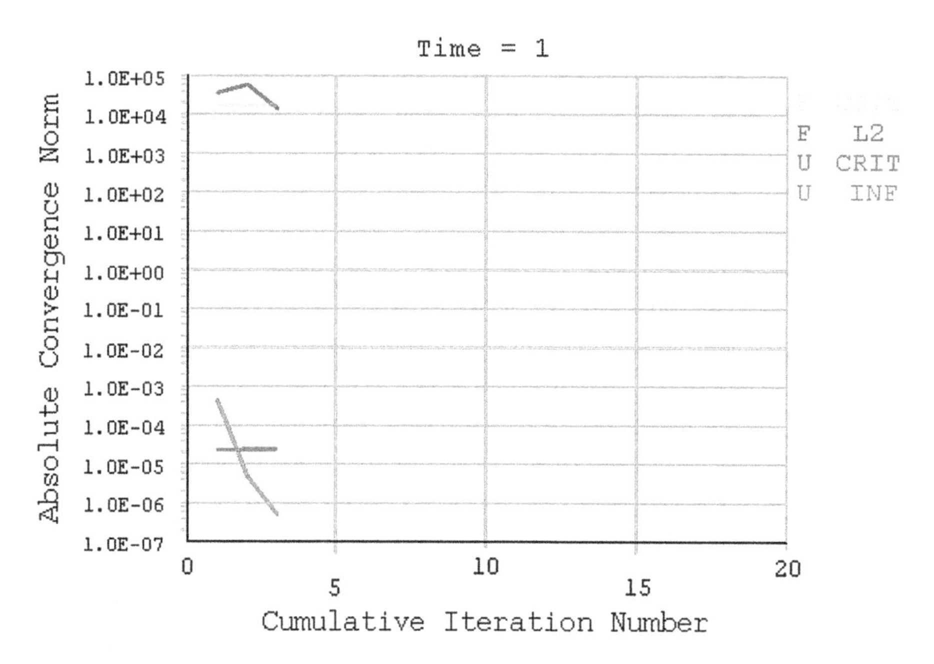

Figure 5-1-8 Graphical Solution Tracking.

```
   DISP CONVERGENCE VALUE   = 0.4887E-05  CRITERION= 0.2308E-04 <<< CONVERGED
  EQUIL ITER   2 COMPLETED.  NEW TRIANG MATRIX.  MAX DOF INC=  0.4887E-05
   FORCE CONVERGENCE VALUE  = 0.5791E+05  CRITERION= 0.1719E+05
curEqn=  31101  totEqn=  57586 Job CP sec=     40.422
     Factor Done=  50% Factor Wall sec=     0.049 rate=    6731.4 Mflops
curEqn=  57586  totEqn=  57586 Job CP sec=     40.531
     Factor Done= 100% Factor Wall sec=     0.103 rate=    6436.3 Mflops
   DISP CONVERGENCE VALUE   = 0.5135E-06  CRITERION= 0.2356E-04 <<< CONVERGED
  EQUIL ITER   3 COMPLETED.  NEW TRIANG MATRIX.  MAX DOF INC=  0.5135E-06
   FORCE CONVERGENCE VALUE  = 0.1389E+05  CRITERION= 0.1754E+05 <<< CONVERGED
  >>> SOLUTION CONVERGED AFTER EQUILIBRIUM ITERATION   3
```

Figure 5-1-9 Convergence Tracking in the Output Window.

Nonlinear analyses are iterative and can take several iterations before the solution converges. As the solution is being calculated, ANSYS displays the convergence norms and the corresponding convergence criteria for each iteration graphically in the Graphical Solution Tracking feature (Figure 5-1-8) and as text in the Output Window (Figure 5-1-9).

In this analysis, the four lines being plotted are F L2 (the L2 norm of the reaction force variable), F Crit (the convergence criterion for the reaction force), U L2 (the L2 norm of the displacement), and U Crit (the convergence criterion for the displacement). The equations that define the convergence norms can be found in the Mechanical APDL Theory Reference. The analysis convergences when both L2 norms (referred to as "convergence values" in the Output Window) are below their respective convergence criteria. See chapter 8 *for more details.*

10-5. Define the temperature for the second load step

- Solution > Define Loads > Apply > Structural > Temperature > Uniform Temp
- Enter 100 for "TUNIF Uniform Temperature"
- Click OK

This will replace the previously specified temperature.

10-6. Set the time for the end of the second load step

- Solution > Analysis Type > Sol'n Controls
- Enter 2 for "Time at end of loadstep"
- Click OK

10-7. Solve the second load step

- Solution > Solve > Current LS

10-8. Define the temperature for the third load step

- Solution > Define Loads > Apply > Structural > Temperature > Uniform Temp
- Enter 200 for TUNIF
- Click OK

10-9. Set the time for the end of the third load step

- Solution > Analysis Type > Sol'n Controls
- Enter 3 for "Time at end of loadstep"
- Allow all other values to default
- Click OK

10-10. Solve the third load step

- Solution > Solve > Current LS

10-11. Define the temperature for the fourth load step

- Solution > Define Loads > Apply > Structural > Temperature > Uniform Temp
- Enter 250 for TUNIF
- Click OK

10-12. Set the time for the end of the fourth load step

- Solution > Analysis Type > Sol'n Controls
- Enter 4 for "Time at end of loadstep"
- Allow all other values to default
- Click OK

10-13. Solve the fourth load step

- Solution > Solve > Current LS

In the first three solutions, the solution converged quickly. For example, the first load step solved in three iterations. But for the fourth solution, 31 cumulative iterations were required. Longer solution times are common in nonlinear analyses.

10-14. Save your results

Step 11: Postprocess the Results of the Plastic Analysis

11-1. List the results sets available for this analysis

- General Postproc > Results Summary

The results summary in Figure 5-1-10 shows that there are four solution sets on the results file for this exercise.

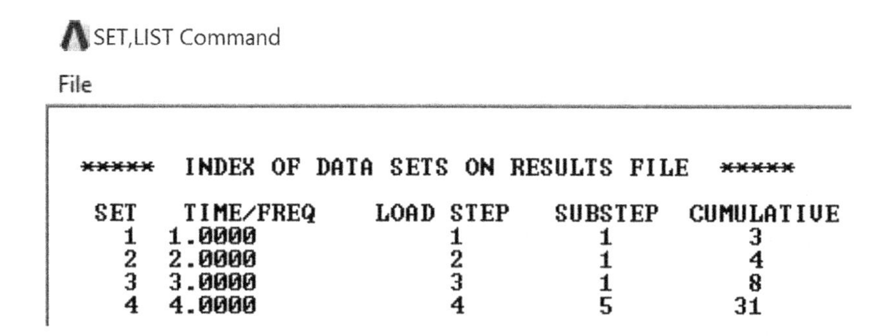
SET,LIST Command

File

***** INDEX OF DATA SETS ON RESULTS FILE *****

SET	TIME/FREQ	LOAD STEP	SUBSTEP	CUMULATIVE
1	1.0000	1	1	3
2	2.0000	2	1	4
3	3.0000	3	1	8
4	4.0000	4	5	31

Figure 5-1-10 Results Summary for Plastic Analysis.

These results are saved in the file Exercise5-1-plastic.rst. The elastic solution is on a different results file Exercise5-1.rst. You can find both results files in your working directory.

11-2. Read the first set of results

- General Postproc > Read Results > First Set

*When ANSYS solves a problem, it writes the solution to the results file and to the database. However, only one set of results can reside in the database at any given time. When a new **SOLVE** command is issued, the results in the database from the previous solution are replaced. Most postprocessing commands only operate on the results that are currently stored in the database. Thus, the first set of results must be restored to the database before they can be postprocessed.*

11-3. Plot the deformation of the plate at 25°C

- General Postproc > Plot Results > Contour Plot > Nodal Solu
- Choose "DOF Solution"
- Choose "Displacement vector sum"
- Set the "Undisplaced shape key" to "Deformed shape with undeformed edge"
- Leave the "Scale Factor" set to "Auto Calculated"
- Click OK

Figure 5-1-11 shows the deformed plate with an overlay of the edges of the undeformed model. You can see that the central hole has been elongated and that the plate has become both longer and narrower. The maximum total displacement in this model is 0.446 mm. As in exercise 2-1, you can see that the asymmetry in the boundary conditions also results in a slight asymmetry in the results.

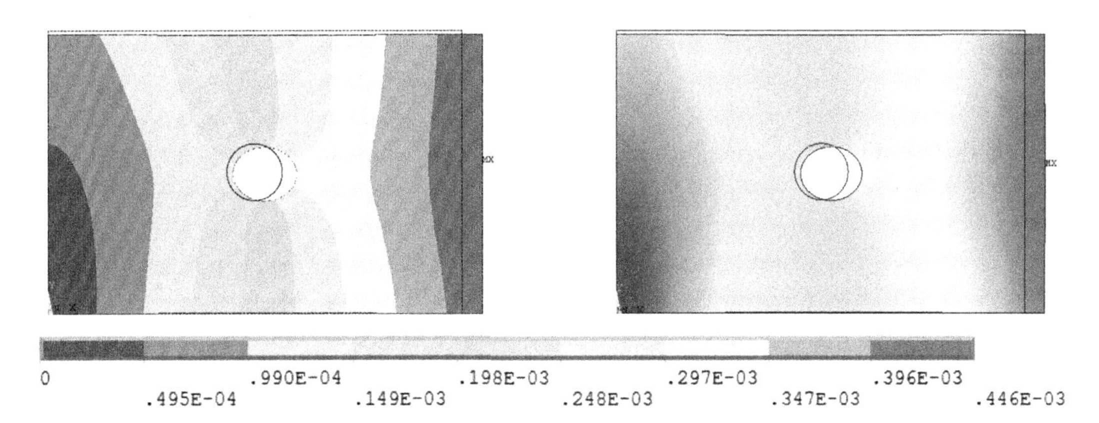

Figure 5-1-11 Plot of Total Displacement with Plasticity at 25°C: Win32 Graphics (left), 3D Graphics (right).

11-4. Plot the equivalent stress distribution in the model at 25°C

- General Postproc > Plot Results > Contour Plot > Nodal Solu
- Choose "Stress"
- Choose "von Mises stress"
- Click OK

Figure 5-1-12 shows that the maximum equivalent stress for this model is 437 MPa. This is just above the yield stress (413 MPa) and much lower than the value predicted by the elasticity analysis (632 MPa).

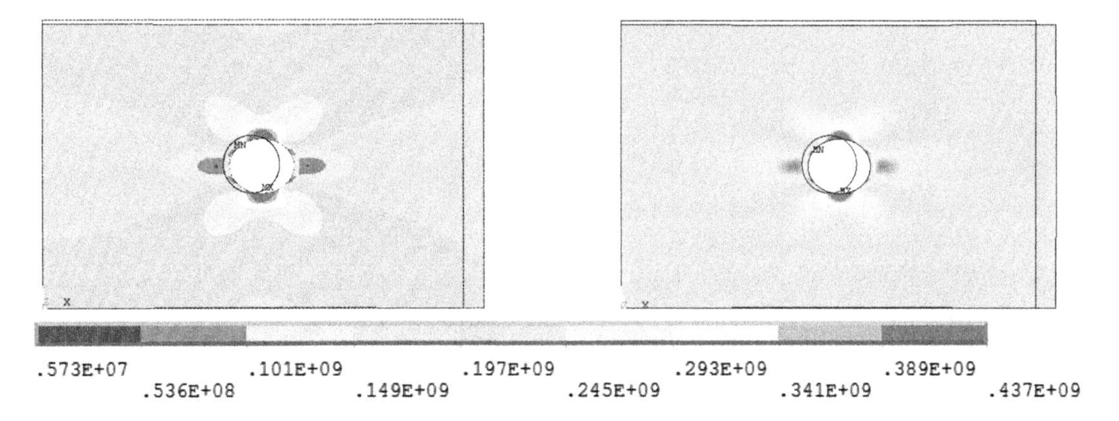

Figure 5-1-12 Plot of Equivalent Stress with Plasticity at 25°C: Win32 Graphics (left), 3D Graphics (right).

11-5. Plot the von Mises plastic strain at 25°C

- General Postproc > Plot Results > Nodal Solu
- Choose "Plastic Strain"
- Choose "von Mises plastic strain"
- Click OK

Figure 5-1-13 shows that the maximum von Mises plastic strain for this model is 0.487%. All plastic deformation is occurring just above and below the central hole.

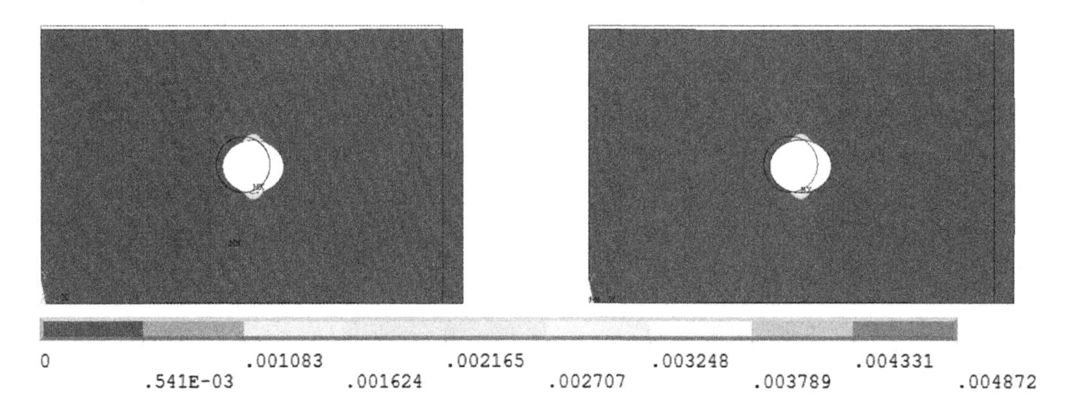

Figure 5-1-13 Plot of von Mises Plastic Strain at 25°C: Win32 Graphics (left), 3D Graphics (right).

11-6. Read the second set of results

- General Postproc > Read Results > By Pick
- Click on the line for Data Set 2 (Figure 5-1-14)
- Click "Read"
- Click "Close"

Results File: Exercise5-1-plasti).T9

Available Data Sets:

Set	Time	Load Step	Substep	Cumulative
1	1.0000	1	1	3
2	2.0000	2	1	4
3	3.0000	3	1	8
4	4.0000	4	5	31

Read Next Previous

Close Help

Figure 5-1-14 Choosing a Results Set by Picking.

11-7. Plot the equivalent stress distribution in the model at 100°C

- General Postproc > Plot Results > Contour Plot > Nodal Solu

This plot shows that the maximum equivalent stress for this model is 417 MPa. This is just above the yield stress (394 MPa) and lower than the maximum equivalent stress at 25°C (437 MPa).

11-8. Plot the von Mises plastic strain at 100°C

- General Postproc > Plot Results > Contour Plot > Nodal Solu

This plot shows that the maximum equivalent plastic strain for this model is 0.528%. All plastic deformation occurs just above and below the central hole. This is higher than the equivalent plastic strain at 25°C (0.487%).

11-9. Read the third set of results

- General Postproc > Read Results > By Time/Freq
- Enter 3 for "TIME Value of time or freq" (Figure 5-1-15)
- Click OK.

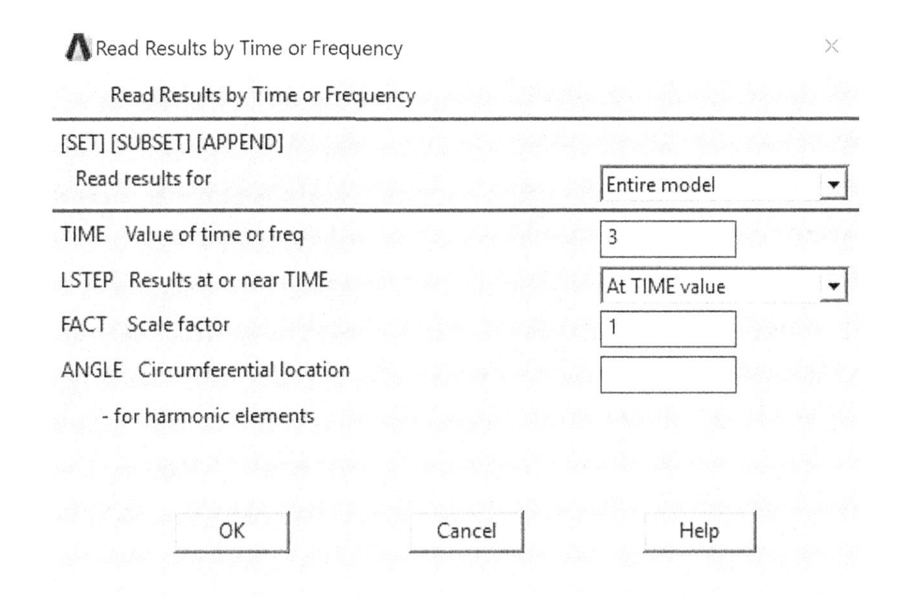

Figure 5-1-15 Choosing a Results Set by Time.

11-10. Plot the equivalent stress distribution in the model at 200°C

- General Postproc > Plot Results > Contour Plot > Nodal Solu

This plot shows that the maximum equivalent stress for this model is 332 MPa. This is above the yield stress (~282 MPa) and substantially lower than the maximum equivalent stress at 25°C (437 MPa) and at 100°C (417 MPa).

11-11. Plot the von Mises plastic strain at 200°C

- General Postproc > Plot Results > Contour Plot > Nodal Solu

This plot shows that the maximum equivalent plastic strain for this model is 1.229%. All plastic deformation occurs just above and below the central hole. This is higher than the equivalent plastic strain at 25°C (0.487%) and at 100°C (0.528%).

11-12. Read the fourth set of results

- General Postproc > Read Results > Last Set

11-13. Plot the equivalent stress distribution in the model at 250°C

- General Postproc > Plot Results > Contour Plot > Nodal Solu

This plot shows that the maximum equivalent stress for this model is 471 MPa. This is well above the yield stress (~149 MPa) and substantially higher than the maximum equivalent stress at 25°C (437 MPa), 100°C (417 MPa), and 200°C (332 MPa).

11-14. Plot the von Mises plastic strain at 250°C

- General Postproc > Plot Results > Contour Plot > Nodal Solu

Figure 5-1-16 shows that the maximum equivalent plastic strain for this model is 9.121%. This is a very large value for plastic strain. You can see from this plot that the entire plate is starting to yield.

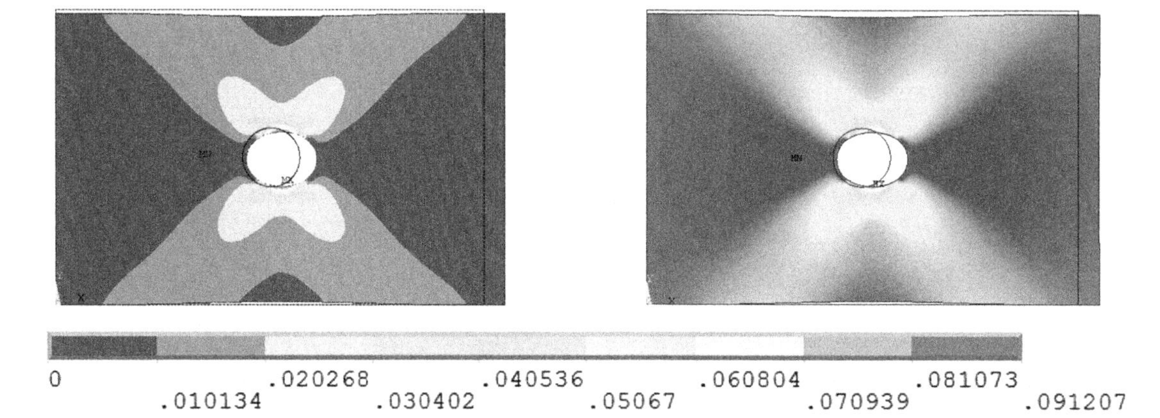

Figure 5-1-16 Plot of von Mises Plastic Strain at 250°C: Win32 Graphics (left), 3D Graphics (right).

Close the Program

- Utility Menu > File > Exit ...
- Choose "Quit - No save!"
- Click OK

Sample Input File

```
/PREP7                              ! Enter the Preprocessor
RECTNG,0,0.15,0,0.1                 ! Create a rectangle that is 0.15 x 0.1 m
CYL4,0.075,0.05,0.01                ! Create a circle centered at (0.075,0.05)
ASBA,1,2                            ! Subtract area 2 (circle) from area 1
                                    ! (rect.)
ET,1,PLANE182                       ! Use PLANE182 elements
MPTEMP,,,,,,,,                      ! Clear any existing temperature tables
MPTEMP,1,25,100,149,204,260,316     ! Define temperatures for data table
MPDATA,EX,1,1,7.33e10,7.22e10,6.57e10,6.04e10,5.50e10,5.02e10
                                    ! Define Young's modulus for 2024-T6
                                    ! aluminum
MPDATA,PRXY,1,,.33,.33,.33,.33,.33,.33
                                    ! Define Poisson's ratio for 2024-T6
                                    ! aluminum
SMRT,3                              ! Element size determined by Smart Size of 3
AMESH,all                           ! Mesh all areas
LSEL,S,LOC,X,0.05,0.1               ! Select the lines in the plate center by x
LSEL,R,LOC,Y,0.025,0.075            ! Select the lines in the plate center by y
LREFINE,ALL,,,1,1,1,1               ! Refine the mesh attached to the selected
                                    ! lines
AREFINE,ALL,,,3,0,1,1               ! Refine the mesh for all areas in the plate
FINISH                              ! Finish and Exit Preprocessor
```

```
/SOLU                               ! Enter the Solution Processor
LSEL,S,LOC,X,0                      ! Select the line at x=0
DL,ALL,,UX,0                        ! Constrain the line from movement in x
KSEL,S,LOC,X,0                      ! Select the keypoints at x=;0
KSEL,R,LOC,Y,0                      ! Reselect the keypoint at (0,0)
DK,ALL,UY,0                         ! Constrain the keypoint from movement in y
LSEL,S,LOC,X,0.15                   ! Select the line at x=0.15
SFL,ALL,PRES,-200e6                 ! Apply a pressure of -200e6 Pa to the line
ALLSEL,ALL                          ! Select everything in the model
SOLVE                               ! Solve problem
FINISH                              ! Finish and Exit Solution

/POST1                              ! Enter the General Postprocessor
PLNSOL,S,EQV,                       ! Plot equivalent (von Mises) stress
FINISH                              ! Finish and Exit the Postprocessor
SAVE                                ! Save the database

/FILNAME,Exercise5-1-plastic,0      ! Change the jobname for the analysis

/SOLU                               ! Return to Solution for plastic analysis
TB,BKIN,1,6,2,1                     ! Bilinear plasticity with kinematic
                                    ! hardening
                                    ! Use 6 temperatures
TBTEMP,25                           ! First temperature is 25 C
TBDATA,,4.13e8,4.56e9               ! Yield stress and plastic modulus at 25 C
TBTEMP,100                          ! Second temperature is 100 C
TBDATA,,394e6,4.15e9                ! Yield stress and plastic modulus at 100 C
TBTEMP,149                          ! Third temperature is 149 C
TBDATA,,337e6,3.90e9                ! Yield stress and plastic modulus at 149 C
TBTEMP,204                          ! Fourth temperature is 204 C
TBDATA,,282e6,3.5e9                 ! Yield stress and plastic modulus at 204 C
TBTEMP,260                          ! Fifth temperature is 260 C
TBDATA,,149e6,3.0e9                 ! Yield stress and plastic modulus at 260 C
TBTEMP,316                          ! Sixth temperature is 316 C
TBDATA,,67e6,0                      ! Yield stress and plastic modulus at 316 C
TIME,1                              ! Set time=1
SOLVE                               ! Solve the current model (25 C)

TIME,2                              ! Set time=2
TUNIF,100                           ! Set model temperature=100 C
SOLVE                               ! Solve the current model (100 C)

TIME,3                              ! Set time=3
TUNIF,200                           ! Set model temperature=200 C
SOLVE                               ! Solve the current model (200 C)

TIME,4                              ! Set time=4
TUNIF,250                           ! Set model temperature=250 C
SOLVE                               ! Solve the current model (250 C)

/POST1                              ! Enter the General Post Processor
SET,1                               ! Choose results set 1
PLNSOL,U,SUM,2,1                    ! Plot deformation w/undeformed edge
PLNSOL,S,EQV,2,1                    ! Plot von Mises stress
PLNSOL,EPPL,EQV,2,1                 ! Plot von Mises plastic strain
```

```
SET,2                         ! Choose results set 2
PLNSOL,S,EQV,2,1              ! Plot von Mises stress
PLNSOL,EPPL,EQV,2,1           ! Plot von Mises plastic strain

SET,3                         ! Choose results set 3
PLNSOL,S,EQV,2,1              ! Plot von Mises stress
PLNSOL,EPPL,EQV,2,1           ! Plot von Mises plastic strain

SET,4                         ! Choose results set 4
PLNSOL,S,EQV,2,1              ! Plot von Mises stress
PLNSOL,EPPL,EQV,2,1           ! Plot von Mises plastic strain

FINISH                        ! Finish and Exit Postprocessor

SAVE                          ! Save the database
!/EXIT                        ! Exit ANSYS
```

Chapter

6

Meshing

Suggested Reading Assignments:
Mechanical APDL Modeling and Meshing Guide: Chapter 2 Section 2.6
Mechanical APDL Modeling and Meshing Guide: Chapter 7
Mechanical APDL Modeling and Meshing Guide: Chapter 8 Sections 8.1 and 8.4

CHAPTER OUTLINE

There are two ways to create a finite element mesh in ANSYS. You can either create the nodes and elements manually as described in chapter 3 or you can use the program's built-in meshing capabilities to mesh the solid model geometry. Depending on which method you choose and which meshing options you specify, you can have complete control over the resulting mesh, almost no control over the resulting mesh, and every level of control in between.

This chapter provides an overview of the meshing process in ANSYS. You will learn how to specify the attributes of the elements to be generated and how to set the mesh controls. You will learn about free and mapped meshes and how to generate each one. You will learn about the factors that affect mesh quality and how to determine the quality of a given finite element mesh. Finally, you will learn how to modify, improve, and delete a finite element mesh. These are the final operations that will be performed in the Preprocessor.

6.1. Meshing Overview

There are six basic steps in the meshing process. First, element attributes (element type, material type, real constants, etc.) are assigned to the solid model geometry. Next, the mesh controls are set. These dictate the element shape to use, the size of the elements to create, the type of mesh

ANSYS Mechanical APDL for Finite Element Analysis.
DOI: http://dx.doi.org/10.1016/B978-0-12-812981-4.00006-X

to generate, etc. Once these tasks are complete, some, or all, of the model can be meshed. This initial mesh can be copied or extruded to mesh other parts of the model if desired. After all meshing operations are completed, the quality of the resulting mesh should be evaluated. If the mesh quality is insufficient or additional refinement is desired, the mesh can be revised, refined, or cleared and regenerated. These steps are summarized below:

1. Assign element (mesh) attributes
2. Set the mesh controls
3. Mesh the model
4. Copy or extrude the mesh if desired
5. Evaluate the mesh quality
6. Revise, refine, or regenerate the mesh if necessary.

Several of these steps can be done out of order. For example, some element attributes can be assigned or adjusted after meshing. In addition, it can be more efficient to refine a mesh before evaluating its quality. It can also be more efficient to evaluate and improve a mesh before operating on it. If different attributes and/or mesh controls will be used for different parts of the model, this procedure can be repeated until the entire model has been meshed.

6.2. Element Attributes

Thus far, the exercises and examples in this book have been limited to simple models with a single set of element attributes. However, more complicated analyses may require different attributes for different parts of the model. For example, Figure 6.1 shows a cantilever beam composed of a thick piece of acrylic sandwiched between two thin pieces of aluminum. If this system is modeled in two dimensions, the solid model will consist of three areas. All three layers of the beam can be meshed using continuum elements, so only one element type is needed. However, the model will require two material types: material #1 (aluminum) for Areas 1 and 3 and material #2 (acrylic) for Area 2.

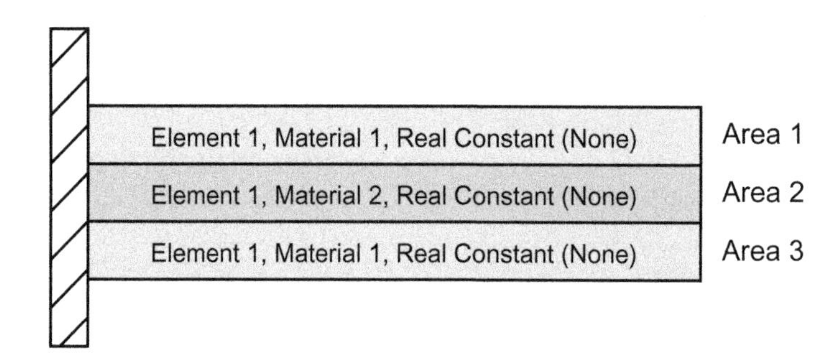

Figure 6.1 Cantilever Sandwich Beam Composed of Two Materials (Not to Scale).

Each element generated by ANSYS is assigned a total of 10 attributes. They are:

1. The element number	Defaults to the next available value
2. The element type number	Defaults to currently defined or 1
3. The material model reference number	Defaults to currently defined or 1
4. The real constant set number	Defaults to currently defined or 1
5. The element coordinate system number	Defaults to currently defined or 0
6. The section number	Defaults to currently defined or 1
7. The birth and death state	Defaults to alive = 1
8. The solid model reference number	Based on the lowest order entity
9. The shape identifier	Automatically set by the program
10. The P-method exclusion key	No longer documented

All 10 element attributes are defined for all generated elements even if the attributes are not used. For example, sections are not usually needed for continuum elements. Similarly, birth and death generally only applies to structural and thermal elements and must be activated using the **EKILL** command. ANSYS assigns the element number, the solid model reference number, and the shape identifier automatically during meshing. It also controls the birth and death state during solution. The remaining attributes can be specified by the user. If no element attributes are defined, ANSYS will mesh the entire model using the default settings for element attributes: element type #1, material model reference #1, real constant set #1 (if applicable), and element coordinate system #0. If no element type has been defined (i.e., if there is no element type #1), an error will be generated and the meshing operation will fail.

6.2.1. Setting Global Element Attributes

You can set the element attributes for the entire model (i.e., the global element attributes) using the GUI path: **Main Menu > Preprocessor > Meshing > Mesh Attributes > Default Attribs** or **Main Menu > Preprocessor > Modeling > Create > Elements > Elem Attributes**. You can also use commands to define individual global element attributes.

- The **TYPE** command specifies the element type number to use during meshing.
- The **MAT** command specifies the material model reference number to use.
- The **REAL** command specifies the real constant set number to use.
- The **ESYS** command specifies the element coordinate system number to use.
- The **SECNUM** command specifies the section number to use.

Once an element attribute has been set, it continues to be the default until changed. Thus, all parts of the model that are meshed after the element attributes are set will share these attributes.

Example 6.1 lists the commands to create and mesh the cantilever sandwich beam shown in Figure 6.1 using global element attributes. First, the model is built, and the element type and material properties are defined. Next, Areas 1 and 3 are selected using **ASEL** commands, the active material model reference number is set using the **MAT** command, and the selected areas are meshed using the **AMESH** command. Finally, Area 2 is selected using the **ASEL** command, the active material model reference number is changed using the **MAT** command, and the newly selected area is meshed using the **AMESH** command.

Example 6.1

```
/PREP7                    ! Enter the Preprocessor
RECTNG,0,10,0.9,1         ! Create the top beam
RECTNG,0,10,0.1,0.9       ! Create the middle beam
RECTNG,0,10,0,0.1         ! Create the bottom beam
NUMMRG,ALL                ! Merge all solid model entities
ET,1,182                  ! Define Plane182 as element #1
MP,EX,1,7.3e10            ! Define Young's modulus for aluminum: material #1
MP,PRXY,1,0.33            ! Define Poisson's ratio for aluminum: material #1
MP,EX,2,3e9               ! Define Young's modulus for acrylic: material #2
MP,PRXY,2,0.4             ! Define Poisson's ratio for acrylic: material #2
ASEL,S,,,1                ! New set: select Area 1
ASEL,A,,,3                ! Also select Area 3
MAT,1                     ! Use material #1
AMESH,ALL                 ! Mesh selected areas
ASEL,S,,,2                ! New set: select Area 2
MAT,2                     ! Use material #2
AMESH,ALL                 ! Mesh selected areas
```

In this example, the element attributes could also be set before selecting the areas to mesh. The order of these two operations is not important.

6.2.2. Setting Local Element Attributes

You can define local element attributes for some or for all solid model entities (keypoints, lines, areas, or volumes) of a given type. The GUI path to define the element attributes for selected solid model entities is: **Main Menu > Preprocessor > Meshing > Mesh Attributes > Picked [Entities]**. The GUI path to define the element attributes for all solid model entities of a given type is: **Main Menu > Preprocessor > Meshing > Mesh Attributes > All [Entities]**. Both GUI paths issue one of the following four commands depending on the entity type in question.

- The **KATT** command specifies the element attributes for the selected keypoints.
- The **LATT** command specifies the element attributes for the selected lines.
- The **AATT** command specifies the element attributes for the selected areas.
- The **VATT** command specifies the element attributes for the selected volumes.

Each of these commands takes 5 arguments: *MAT*, *REAL*, *TYPE*, *ESYS*, and *SECNUM*.

Example 6.2 lists the commands to mesh the cantilever sandwich beam shown in Figure 6.1 using local element attributes. First, the model is built, and the elements and material properties are defined. Next, Areas 1 and 3 are selected using **ASEL** command, and their element attributes are set using the **AATT** command. Then, Area 2 is selected using the **ASEL** command, and its element attributes are set using the **AATT** command. Finally, all areas in the model are selected using the **ASEL** command and are meshed using the **AMESH** command. In this case, the order of operations is very important. You must select the solid model geometry before assigning the element attributes.

Example 6.2

```
/PREP7                   ! Enter the Preprocessor
RECTNG,0,10,0.9,1        ! Create the top beam
RECTNG,0,10,0.1,0.9      ! Create the middle beam
RECTNG,0,10,0,0.1        ! Create the bottom beam
NUMMRG,ALL               ! Merge all solid model entities
ET,1,182                 ! Define Plane182 as element #1
MP,EX,1,7.3e10           ! Define Young's modulus for aluminum: material #1
MP,PRXY,1,0.33           ! Define Poisson's ratio for aluminum: material #1
MP,EX,2,3e9              ! Define Young's modulus for acrylic: material #2
MP,PRXY,2,0.4            ! Define Poisson's ratio for acrylic: material #2
ASEL,S,,,1               ! New set: select Area 1
ASEL,A,,,3               ! Also select Area 3
AATT,1,,1                ! Set element attributes: material #1, element #1
ASEL,S,,,2               ! New set: select Area 2
AATT,2,,1                ! Set elements attributes: material #2, element #1
ASEL,ALL                 ! Select all areas in the model
AMESH,ALL                ! Mesh all areas
```

6.2.3. Modifying Element Attributes

You can modify the element attributes after meshing using the GUI path: **Main Menu > Preprocessor > Modeling > Move/Modify > Elements > Modify Attrib** or the **EMODIF** command.

6.3. Mesh Controls

Although many meshing options are available, usually only two options need to be set: whether to generate a free or a mapped mesh and what element sizing rule(s) to use.

6.3.1. Free Versus Mapped Meshing

ANSYS can create two types of meshes: free meshes and mapped meshes (Figure 6.2). A free mesh has no restrictions on the organization of its elements or on the type of geometry that can be meshed. The program makes almost all decisions regarding the mesh and may generate a slightly different mesh each time. Both triangular and quadrilateral elements can be used to free mesh an area, but only tetrahedral elements can be used to free mesh a volume.

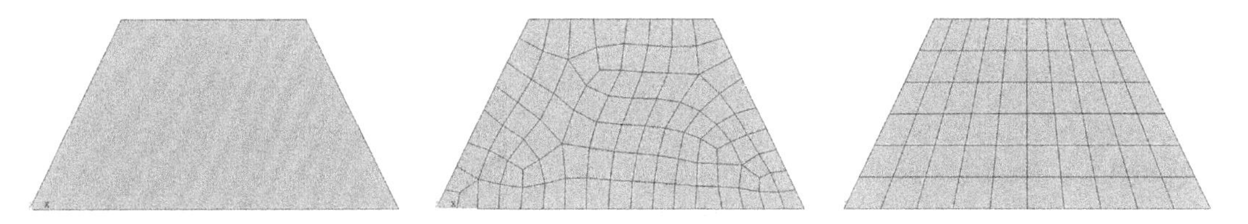

Figure 6.2 Trapezoidal Area: Unmeshed (left), Free Meshed (center), and Map Meshed (right).

A mapped mesh contains elements organized in a regular pattern. As a result, you can only map mesh areas that are bounded by three or four lines and volumes that are bounded by four, five, or six areas. (Several methods to work around this limitation are presented in Section 6.5.) Triangular and quadrilateral elements can be used to map mesh an area, but only brick elements can be used to map mesh a volume.

Free meshing is easy and robust. For this reason, ANSYS free meshes all models by default. However, map meshing is useful for some applications. To specify a free or a mapped mesh, you must set the mesh key. This is automatically set by your choice of GUI path for area meshes (Figure 6.3, left) and volume meshes (Figure 6.3, center). The mesh key can be set by issuing the **MSHKEY** command via the command prompt (`MSHKEY,0` for free meshing and `MSHKEY,1` for mapped meshing). It can be set via the Basic SmartSize Settings menu (GUI path: **Main Menu > Preprocessor > Meshing > Size Contrls > SmartSize > Basic**) (Figure 6.4). And, it can be set by specifying the type of mesh to create in the fourth area of the Mesh Tool (Figure 6.3, right).

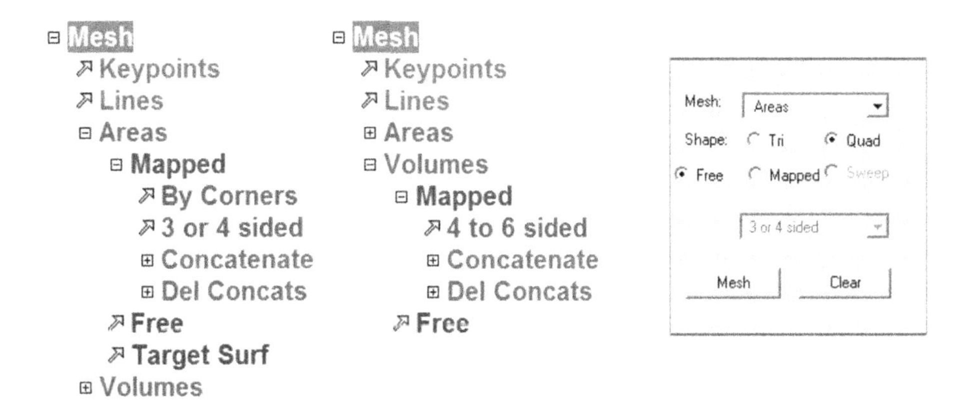

Figure 6.3 Specifying a Free or Mapped Mesh in the GUI Main Menu (left and center) and in the GUI Mesh Tool (right).

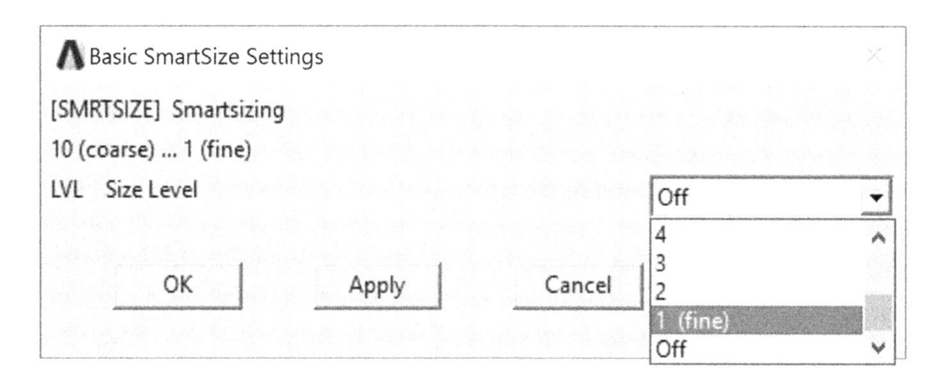

Figure 6.4 Specifying a Free (LVL 1 through 10) or Mapped Mesh (Off) in the Basic SmartSize Settings Menu.

6.3.2. Element Sizing

There are three ways to specify the element sizing in a model: you can use Smart Sizing, you can specify the element edge length, or you can specify the number of element divisions along a line. For global element sizing, these three options are mutually exclusive. You can also specify the element size locally. If you do not specify any element sizing options, your mesh will be generated using the default values associated with the **DESIZE** command.

6.3.2.1. Smart Sizing the Mesh

Smart Sizing allows you to specify the density of a free mesh on a scale from 1 (fine) to 10 (coarse). The default setting is 6. Once the desired density is set, the program automatically generates the mesh. The biggest advantage of Smart Sizing (aside from convenience) is that all element edge lengths are estimated and then refined based on the curvature of the geometry and the proximity to other features before meshing. This improves the quality of the mesh and leads to more accurate results. Smart Sizing can be activated by following the GUI path: **Main Menu > Preprocessor > Meshing > Size Cntrls > SmartSize > Basic**, by using the **SMRT (SIZE)** command, or by using the Mesh Tool.

Figure 6.5 demonstrates the effect of Smart Sizing on the meshing of a plate with a central hole. The plate on the left was free meshed using a Smart Size of 1. The plate in the center was meshed using a Smart Size of 6. The plate on the right was meshed using a Smart Size of 10. Example 6.3 lists the commands to create the three meshes shown in Figure 6.5.

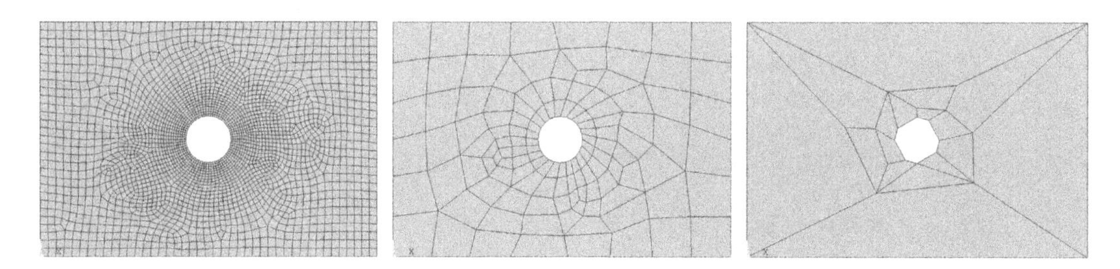

Figure 6.5 Free Meshing of a Plate with a Central Hole using the Smart Size Command: Smart Size of 1 (left), Smart Size of 6 (center), and Smart Size of 10 (right).

Example 6.3

```
/PREP7                        /PREP7                        /PREP7
RECTNG,0,0.15,0,0.1           RECTNG,0,0.15,0,0.1           RECTNG,0,0.15,0,0.1
CYL4,0.075,0.05,0.01          CYL4,0.075,0.05,0.01          CYL4,0.075,0.05,0.01
ASBA,1,2                      ASBA,1,2                      ASBA,1,2
ET,1,PLANE182                 ET,1,PLANE182                 ET,1,PLANE182
SMRT,1                        SMRT,6                        SMRT,10
AMESH,ALL                     AMESH,ALL                     AMESH,ALL
```

6.3.2.2. Specifying the Element Edge Length

You can specify the nominal edge length for the elements in your mesh by following the GUI path: **Main Menu > Preprocessor > Meshing > Size Cntrls > ManualSize > Global > Size** or by using the **ESIZE** command. The syntax is **ESIZE**, *SIZE, NDIV* where *SIZE* is the desired element edge length and *NDIV* is the desired number of element divisions per boundary. These two arguments are mutually exclusive. If a non-zero value is specified for the element *SIZE*, the *NDIV* value will be ignored. You can also set the element edge length in the Mesh Tool.

Figure 6.6 demonstrates the effect of element edge length on the meshing of a 0.15x0.1 plate with a 0.02 diameter central hole. The plate on the left was free meshed using an element edge length of 0.0025. This resulted in a relatively uniform mesh with 40 elements along the short edges of the plate and 60 elements along the long edges of the plate. The central hole is composed of four lines, each with a length of 0.0158. Thus, seven elements are needed per line, for a total of 28 elements around the central hole. The plate in the center was meshed with an element edge length of 0.01. This resulted in 10 elements along the short edges of the plate and 15 elements along the long edges of the plate. Each of the four segments on the central hole was meshed with two elements, causing the central hole to become an octagon. Finally, the plate on the right was meshed with an element edge length of 0.1. This specified a single element along each of the four segments on the central hole, forcing the central hole to become a diamond. Example 6.4 lists the commands to create the three meshes shown in Figure 6.6.

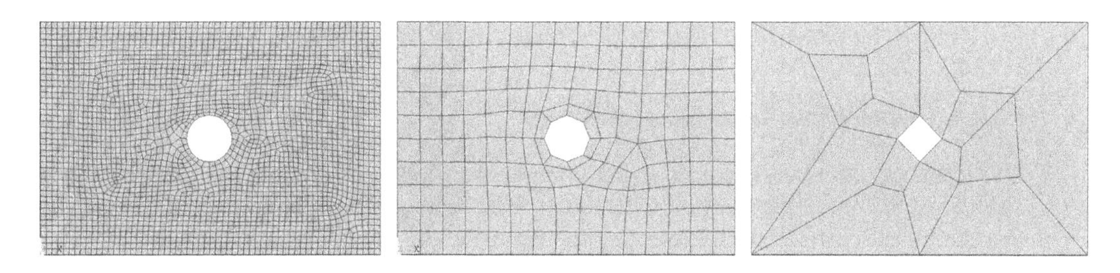

Figure 6.6 Free Meshing of a Plate with a Central Hole with Fixed Element Edge Lengths: ESIZE of 0.0025 (left), ESIZE of 0.01 (center), and ESIZE of 0.1 (right).

Example 6.4

```
/PREP7                        /PREP7                        /PREP7
RECTNG,0,0.15,0,0.1           RECTNG,0,0.15,0,0.1           RECTNG,0,0.15,0,0.1
CYL4,0.075,0.05,0.01          CYL4,0.075,0.05,0.01          CYL4,0.075,0.05,0.01
ASBA,1,2                      ASBA,1,2                      ASBA,1,2
ET,1,PLANE182                 ET,1,PLANE182                 ET,1,PLANE182
ESIZE,0.0025                  ESIZE,0.01                    ESIZE,0.1
AMESH,ALL                     AMESH,ALL                     AMESH,ALL
```

6.3.2.3. Specifying the Number of Divisions

Finally, you can specify the number of element divisions along a line by using the GUI path: **Main Menu > Preprocessor > Meshing > Size Cntrls > ManualSize > Global > Size**, by using the **ESIZE** command, or by setting the global element size preferences in the Mesh Tool. For example, Figure 6.7 shows a regular trapezoidal area that has been meshed with 10 element divisions per line (left), 5 element divisions per line (center), and 2 element divisions per line (right). Example 6.5 lists the commands to create the three meshes shown in Figure 6.7.

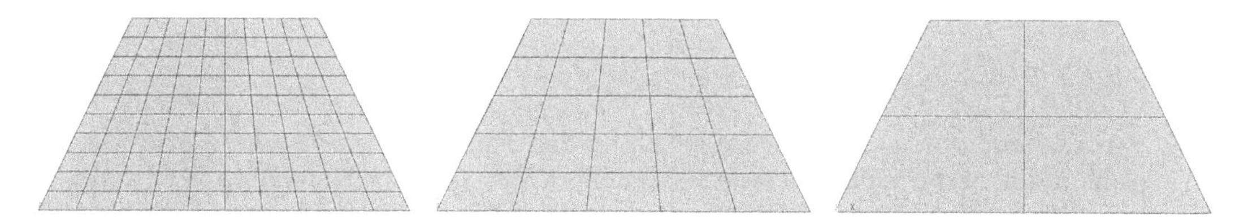

Figure 6.7 Mapped Meshing of a Trapezoidal Area: 10 Element Divisions Per Line (left), 5 Element Divisions Per Line (center), and 2 Element Divisions Per Line (right).

Example 6.5

```
/PREP7              /PREP7              /PREP7
K,1,0,0,0           K,1,0,0,0           K,1,0,0,0
K,2,1,0,0           K,2,1,0,0           K,2,1,0,0
K,3,0.25,0.5        K,3,0.25,0.5        K,3,0.25,0.5
K,4,0.75,0.5        K,4,0.75,0.5        K,4,0.75,0.5
A,1,3,4,2           A,1,3,4,2           A,1,3,4,2
ET,1,182            ET,1,182            ET,1,182
MSHKEY,1            MSHKEY,1            MSHKEY,1
ESIZE,,10           ESIZE,,5            ESIZE,,2
AMESH,ALL           AMESH,ALL           AMESH,ALL
```

6.3.2.4. Global Versus Local Element Sizing

Like element attributes, element sizing can be specified both globally and locally. There are three ways to control the local element sizing. First, you can go through several iterations of setting the global element sizing, selecting the entities to be meshed, and meshing. This method is not recommended because the resulting mesh is dependent on the order in which the model entities are meshed.

The second and better option is to define the element edge length or the number of element divisions near selected keypoints or on selected lines. This can be done using the GUI Paths:

- **Main Menu > Preprocessor > Meshing > Size Cntrls > Manual Size > Keypoints > Picked KPs**
- **Main Menu > Preprocessor > Meshing > Size Cntrls > Manual Size > Lines > Picked Lines**
- **Main Menu > Preprocessor > Meshing > Size Cntrls > Manual Size > Areas > Picked Areas**

Alternatively, you can use commands to define local element sizing.

- The **KESIZE** command specifies the element sizing for the selected keypoints.
- The **LESIZE** command specifies the element sizing for the selected lines.
- The **AESIZE** command specifies the element sizing for the selected areas.

Finally, you can use global element sizing to create a preliminary mesh and then later refine the mesh in the regions of interest.

These options are all available in the Mesh Tool.

6.3.3. Always Save Before Meshing

Meshing, like Boolean operations, should be considered a 'risky' procedure in ANSYS. There are two reasons for this. First, some ANSYS products have a maximum number of nodes and elements that can be created. If the licensing limits are exceeded, ANSYS will issue an error and close the program without giving you a chance to save. Second, complex meshing operations can cause the program to crash or freeze. Because of this, you should always save your database before meshing. Saving before meshing allows you to restart the program, resume the database, adjust the mesh parameters, and retry the meshing operation without losing your work.

6.4. Generating a Mesh

Meshing operations in ANSYS depend on the dimension of the geometry to be meshed, the element type to be used, the type of mesh to create (free or mapped mesh), and the dimension of the meshing command. All four must match in order for meshing to be performed. Otherwise, an error will be generated and the meshing operation will fail.

6.4.1. The Mesh Tool

The Mesh Tool (Figure 6.8) is an interactive dialog box that allows you to set the meshing options and to perform all major meshing operations including meshing, refining the mesh, and clearing the mesh. Like the Pan Zoom Rotate Menu, the Mesh Tool will remain open until it is closed or until you exit the Preprocessor. To open the Mesh Tool, use the GUI path: **Main Menu > Preprocessor > Meshing > MeshTool**.

The Mesh Tool is divided into six sections. The first section allows you to set the global and local element attributes. The second section allows you to activate and control Smart Sizing. The third section allows you to set the element size and the number of element divisions per line for the entire model or for specific entity types. The fourth section allows you to choose the type of entity to mesh, whether to generate a free or a mapped mesh, and the element shape to use. This section also allows you to mesh the model and to clear a mesh. The fifth section of the Mesh Tool allows you to refine a mesh. The last section provides general controls for the dialog box.

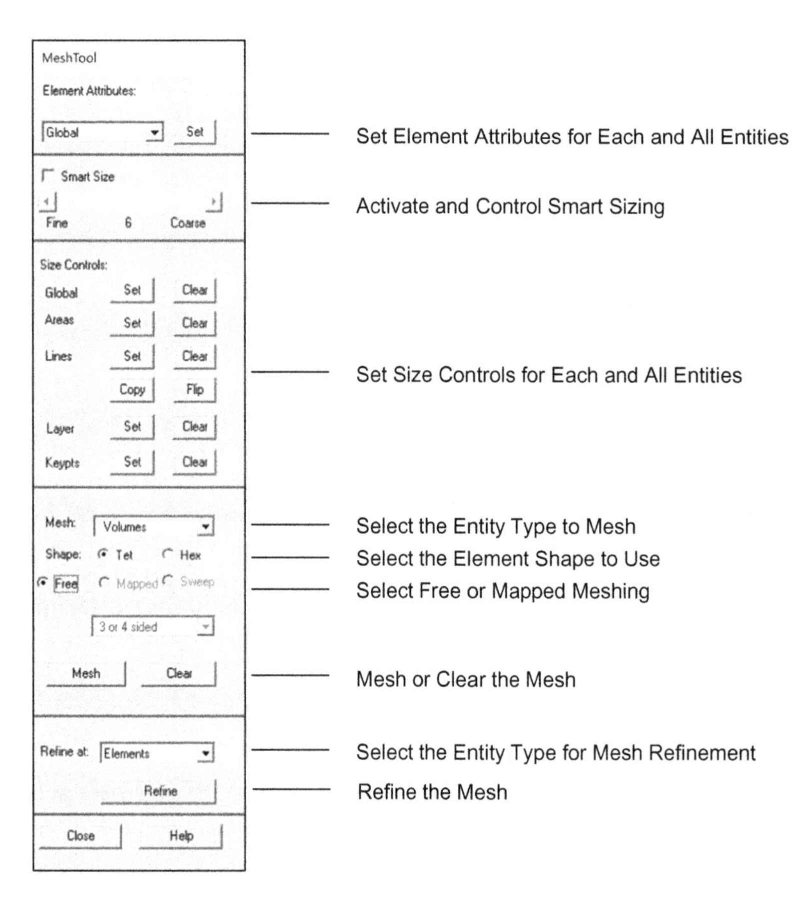

Figure 6.8 The Mesh Tool.

6.4.2. Meshing Commands

You can also issue individual meshing commands directly to the program. There is one meshing command for each type of solid model entity:

- **KMESH** for keypoints
- **LMESH** for lines
- **AMESH** for areas
- **VMESH** for volumes.

The GUI paths for these commands can be found under: **Main Menu > Preprocessor > Meshing > Mesh**.

6.4.3. Meshing Order

Meshing in ANSYS starts from the outside of the model and proceeds inward, beginning with the keypoints, and progressing to the lines, then to the areas, and finally to the volumes. For example, meshing will begin with the first line on the first area of a volume, starting with the keypoints and then moving towards the interior of the line. Once the first line is meshed, the meshing operation will proceed to the second line, and this process will continue until all lines on that area are meshed. Once the lines have been meshed, meshing will continue with the interior of the first area. Then the next area will be selected and the same procedure will begin with the next unmeshed line. Once all of the areas have been meshed, the interior of the volume will be meshed. This process is automatic and cannot be controlled by the user.

6.5. Mapped Meshing

As noted previously, a mapped mesh can be generated only for areas that are bounded by three or four lines and for volumes that are bounded by four, five, or six areas. If you try to map mesh bodies that are composed of more lines or areas, ANSYS will issue an error and ignore the command. For example, if you try to map mesh the hexagonal area shown in Figure 6.9 using PLANE182 elements, you will receive the following error: "Area 1 is irregular. Cannot be map meshed with quadrilaterals." However, it is sometimes desirable to map mesh more complex geometry. In these cases, you have three options: you can use Boolean operations, you can concatenate some of the extra model entities, or you can map mesh by corners.

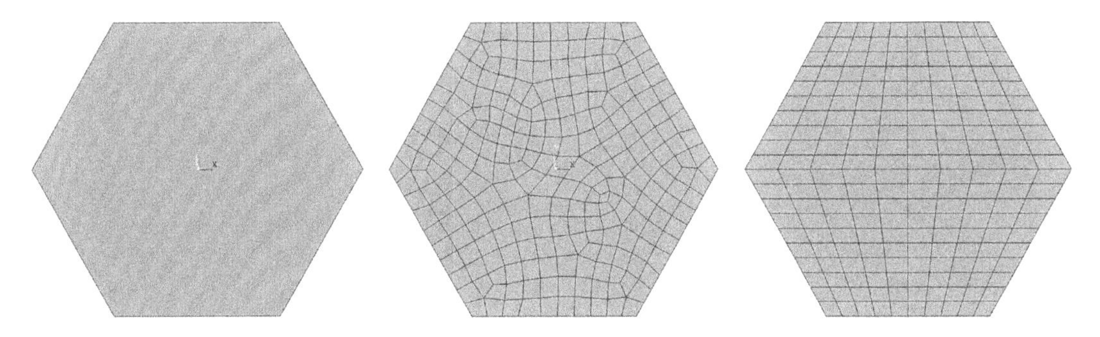

Figure 6.9 Hexagonal Areas: Unmeshed (left), Free Meshed (center), and Map Meshed using Concatenation (right).

6.5.1. Map Meshing Using Boolean Operations

Boolean add operations can be used to reduce the number of lines or areas in a model by permanently combining them. This approach is reasonably robust and allows additional operations to be performed on the solid model. The main disadvantage is that you can no longer select the original entities independently.

Alternatively, Boolean divide operations can be used to create a larger number of entities with fewer sides. For example, the hexagon shown in Figure 6.9 could be divided into two four-sided trapezoids. The resulting trapezoids could be map meshed.

6.5.2. Map Meshing Using Concatenation

Concatenation is a less permanent combination of solid model entities for the sole purpose of meshing. Solid modeling operations cannot be performed on concatenated entities, and solid model boundary conditions cannot be applied to them. For this reason, concatenation should be done just before meshing. To concatenate lines, use the GUI path: **Main Menu > Preprocessor > Meshing > Concatenate > Lines** or the **LCCAT** command. To concatenate areas, use the GUI path: **Main Menu > Preprocessor > Meshing > Concatenate > Areas** or the **ACCAT** command. Concatenating areas automatically concatenates any required lines.

For example, consider the octahedron shown in Figure 6.10. Initially, the volume has 10 areas: one on the top, one on the bottom, and one for each of the 8 sides. These 10 areas can be reduced to 6 through four concatenations: one for Areas 3 and 4, one for Areas 5 and 6, one for Areas 7 and 8, and one for Areas 9 and 10. These operations create four new areas: Areas 11–14. (Unlike a Boolean operation, concatenation does *not* delete Areas 3–10.) Once this is accomplished, the volume can be mapped meshed. Example 6.6 lists the commands to create the map meshed octahedron as shown in Figure 6.10.

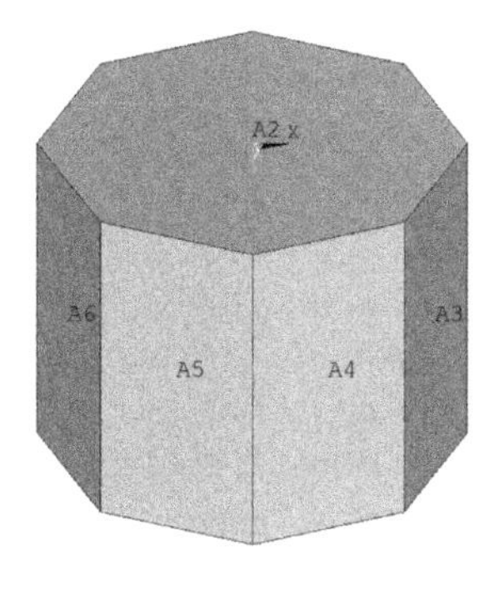

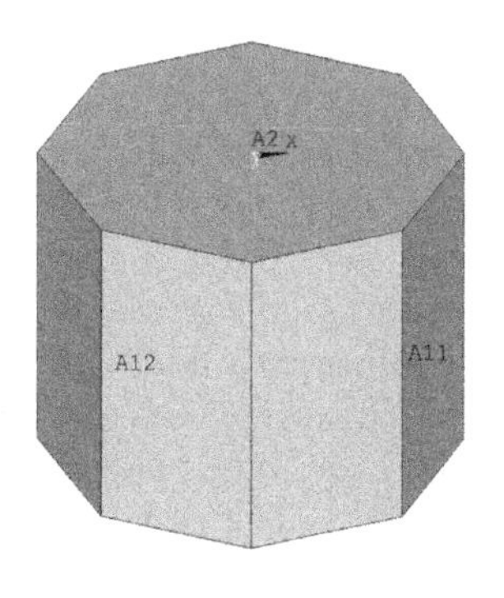

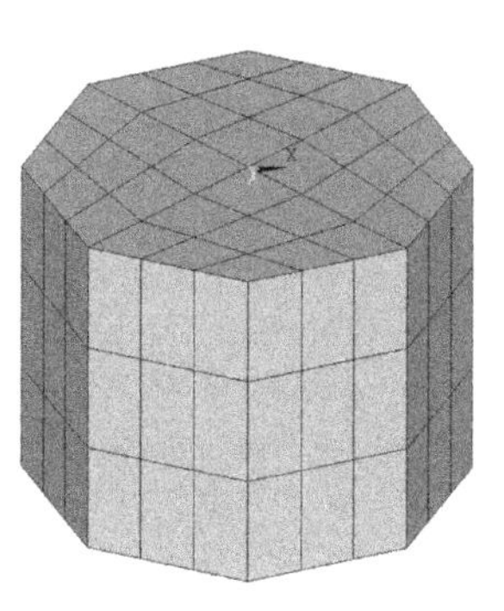

Figure 6.10 Octagonal Volume with Area Numbering Turned On (left), After Concatenation (center), and After Meshing (right).

Example 6.6

```
/PREP7                     ! Enter the preprocessor
RPR4,8,0,0,2,,3            ! Create an octahedron at (0,0),radius=2, height=3
/PNUM,AREA,1               ! Turn on area numbering
ACCAT,3,4                  ! Concatenate area 3 and area 4
ACCAT,5,6                  ! Concatenate area 5 and area 6
ACCAT,7,8                  ! Concatenate area 7 and area 8
ACCAT,9,10                 ! Concatenate area 9 and area 10
ET,1,185                   ! Use SOLID185 elements
MSHKEY,1                   ! Use mapped meshing
VMESH,ALL                  ! Mesh all volumes
```

If additional operations on the solid model are needed, the existing concatenations must be deleted. To delete concatenations, use the GUI path: **Main Menu > Preprocessor > Modeling > Delete > Del Concats** or **Main Menu > Preprocessor > Meshing > Concatenate > Del Concats.** This will delete all line or area concatenations in the model.

6.5.3. Map Meshing Areas By Corners

Finally, you can create a mapped area mesh by using GUI path: **Main Menu > Preprocessor > Meshing > Mesh > Areas > Mapped > By Corners** or the **AMAP** command. The program will automatically concatenate the lines between the chosen keypoints, mesh the area, and then delete the concatenations. A similar option is not available for volume map meshes.

6.6. Copying and Extruding a Mesh

Once a finite element mesh has been created, you can operate on it to mesh other parts of the model. For example, you can copy a mesh to another part of the model or reflect an existing mesh to create a new part of the model. You can also extrude or sweep an area mesh to create a volume mesh. You can find information on these operations in the Mechanical APDL Modeling and Meshing Guide.

6.7. Defining the Quality of a Mesh

The quality of a finite element model is strongly dependent on the quality of its mesh. There are two functions that the mesh must perform in order to produce good quality results: the mesh must represent the system being analyzed and the mesh must generate accurate results.

6.7.1. The Mesh Must Accurately Represent the System

As noted in chapter 1, the finite element method divides a system whose behavior cannot be predicted using closed form equations into small pieces, or elements, whose behavior is known. A finite element model approaches a perfect representation of the original system as the number of elements becomes infinite. Since it is impossible to generate an infinite number of elements, all finite element models are approximations. Still, some approximations are better than others.

When an insufficient number of elements is used, a finite element model may not represent the original geometry. For example, the center and right most plates in Figure 6.6 no longer have circular holes after meshing. Since the finite element meshes in these models do not correspond to the original solid model geometry, they should not be used in an analysis.

Having enough elements to accurately represent the original geometry does not guarantee that your model will produce good results, but it will guarantee that you produce results for the right model.

6.7.2. The Mesh Must Generate Accurate Results

The accuracy of your results will depend on two factors: the number of elements in the model and the shape of those elements. To explain how the finite element mesh impacts the accuracy of the solution, we will first explain how the results are calculated.

During solution, ANSYS first calculates the "primary" or "degree of freedom" results (displacements, temperatures, etc.) at the nodes using the original set of simultaneous equations. This solution is continuous throughout the model. Next, the program calculates the "derived," "secondary," or "element" results (stresses, gradients, fluxes, etc.) at the element integration points using the primary results and the element shape functions. The secondary values associated with any given node may be different for each element attached to that node. For example, Node 9 in Figure 6.11 is attached to four different elements, so it will be associated with four secondary result values. ANSYS then extrapolates the secondary (element) results from the integration points back to the nodes using the element shape functions. Finally, the secondary result coordinate values (stresses, strains, gradients, fluxes, etc., associated with x, y, z, xy, yz, and xz coordinates) are averaged at the nodes, while the derived secondary results values (principle stresses and strains, equivalent stresses and strains, stress and strain intensities, etc.) are recalculated at the nodes. These averaged (or recalculated) "nodal" results are the best estimate of the true values at each location, so these are the results that are used for plotting and other postprocessing operations.

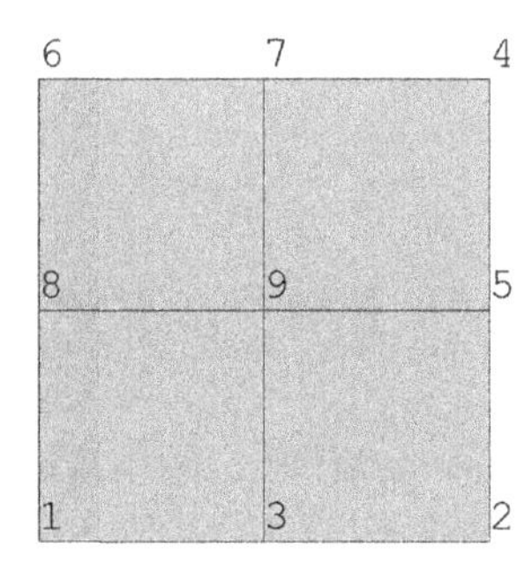

Figure 6.11 Four Elements with Node Numbering Turned On.

As the number of elements in a model increases, the individual elements become smaller and the integration points move closer together. This causes the differences between the secondary results at the various nodes to decrease and the solution to become more accurate, detailed, and continuous. As the elements become infinitely small, the integration point values will converge to a single final value.

It is neither necessary nor economical to refine the mesh beyond the point where reasonable results are obtained. But it is necessary to ensure that the maximum difference between the individual element results and the nodal averaged results is small. The required accuracy for any given model will depend on the problem, the location within the model being considered, and how the results will be used. But as a rule of thumb, differences on the order of 1−3% are usually acceptable.

Figure 6.12 demonstrates the effect of mesh density on the continuity and accuracy of the results for a notched plate in tension. The model on the left was meshed using the parameters from Exercise 2.1—a Smart Size of 1 and an element refinement level of 3. The model on the right was meshed using a Smart Size of 6. Both meshes are sufficient to model the original plate geometry. However, the maximum equivalent stress in these two models differs by over 14% (3.01 vs 2.58 MPa). The maximum equivalent stress in the coarser model is well below the theoretical maximum equivalent stress (2.58 vs 3.03 MPa). The stress distributions in the two plates are also noticeably different. This is because the mesh on the right does not have enough detail to accurately predict the stress fields near the notches.

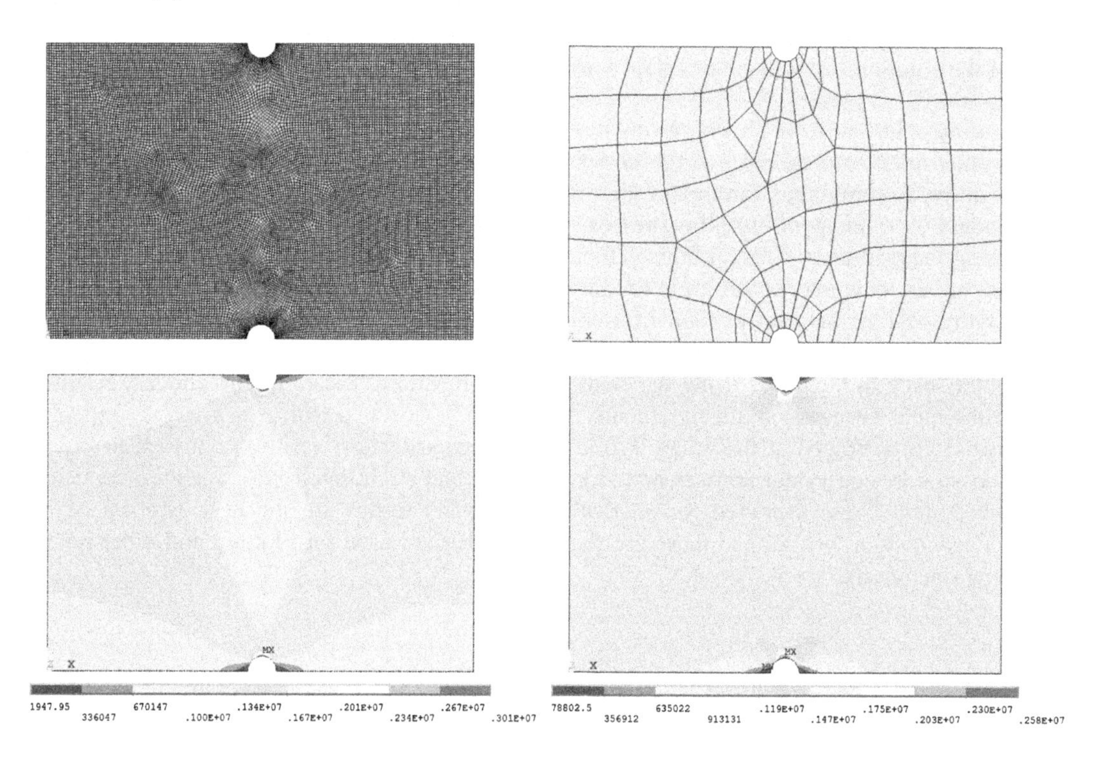

Figure 6.12 Effect of Mesh Density on a Finite Element Model of a Notched Plate: Adequate (top left) and Inadequate (top right) Meshes and von Mises Stress Plots from those Models (bottom) (Win32 Graphics).

The shape of the elements in your mesh can also affect the quality of the solution. The mathematical functions that estimate the solution values between the nodes are called "shape functions" because they depend on the base shape of the element that was assumed when the element was formulated. When an element is closer to its base shape (equilateral triangle, square, equilateral tetrahedron, or cube), the interpolation by the shape function is more accurate. Similarly, as the element becomes more distorted (long and skinny), the interpolation becomes less accurate. Badly distorted elements can produce noticeably poor results and should be avoided if possible.

6.8. Determining the Quality of a Mesh

As an analyst, it is your responsibility to evaluate the quality of your finite element mesh and to determine if its results can be used to make engineering decisions. This section introduces two of the tools that ANSYS provides for this purpose: element shape testing and energy error estimates. It also presents some guidelines that can be used to evaluate the quality of a mesh qualitatively by visual inspection and quantitatively through mesh convergence.

6.8.1. Evaluating Element Shapes with Element Shape Testing

After meshing, newly created continuum elements (2D solids, 3D solids, and 3D shells) are tested to determine if their shapes have been excessively distorted. Shape testing (or checking) looks at characteristics such as the element aspect ratio, the deviation from the element's optimal angles (either 60 or 90 degrees), the deviation from a parallel state (for quads and bricks), the maximum angle in the element, the element Jacobian ratio, and the element warping factor. Different elements are checked for different characteristics.

Two threshold values are defined for each characteristic: the value to trigger a warning and the value to trigger an error. The warning values are chosen to protect the accuracy of the solution while minimizing the frequency of warnings. In general, warning values have not been proven to adversely affect the solution. In contrast, error values usually indicate a noticeable solution error. When an error value is exceeded, the program displays an error message and causes the meshing operation to fail. Details for these thresholds can be found in Section 12.1 of the Mechanical APDL Theory Reference. If shape testing triggers any warning or error, a summary of the shape testing results is printed in the Output Window (Figure 6.13) and a pop-up dialog box is presented to notify you (Figure 6.14).

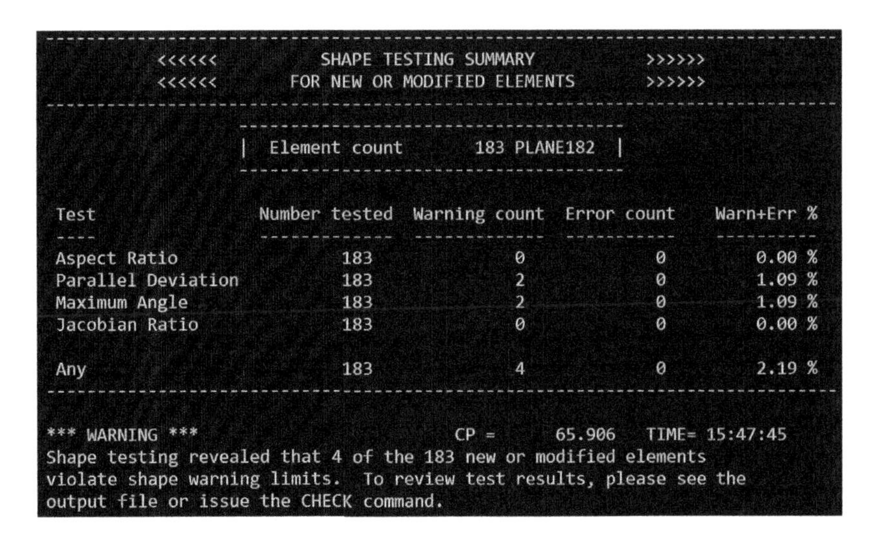

```
 --------------------------------------------------------------------------------
          <<<<<<          SHAPE TESTING SUMMARY           >>>>>>
          <<<<<<        FOR NEW OR MODIFIED ELEMENTS      >>>>>>
 --------------------------------------------------------------------------------

                   --------------------------------------
                   |  Element count        183 PLANE182  |
                   --------------------------------------

 Test                Number tested  Warning count  Error count    Warn+Err %
 ----                -------------  -------------  -----------    ----------
 Aspect Ratio               183              0            0         0.00 %
 Parallel Deviation         183              2            0         1.09 %
 Maximum Angle              183              2            0         1.09 %
 Jacobian Ratio             183              0            0         0.00 %

 Any                        183              4            0         2.19 %
 --------------------------------------------------------------------------------

*** WARNING ***                         CP =      65.906   TIME= 15:47:45
Shape testing revealed that 4 of the 183 new or modified elements
violate shape warning limits.  To review test results, please see the
output file or issue the CHECK command.
```

Figure 6.13 Typical Meshing Output with Element Shape Testing Warnings or Errors.

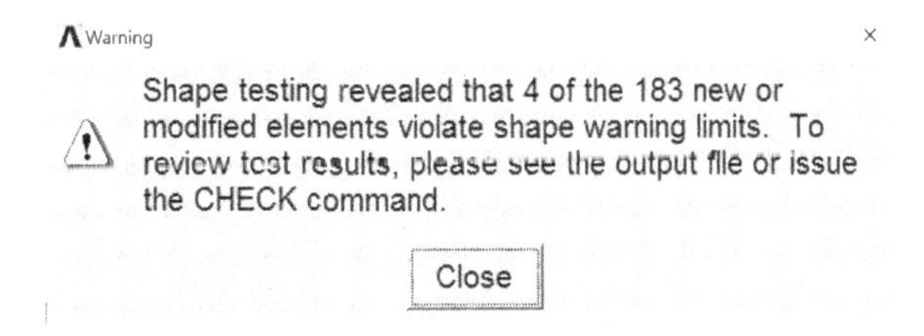

Figure 6.14 Element Shape Testing Pop Up Dialog Box.

Poorly shaped elements should be avoided in areas with high solution gradients and in areas of interest. You can plot the location of elements that have failed element shape checking by using the GUI path: **Main Menu > Preprocessor > Meshing > Check Mesh > Individual Elm > Plot Warning/Error Elements**. By default, element shape checking plots do not display good elements. You can set an option to display them in blue. Elements that generate warnings are shown in yellow. Elements that generate errors are shown in red. Figure 6.15 shows a detailed view of the finite element model from Figure 6.13 on the left and the associated element shape checking plot on the right.

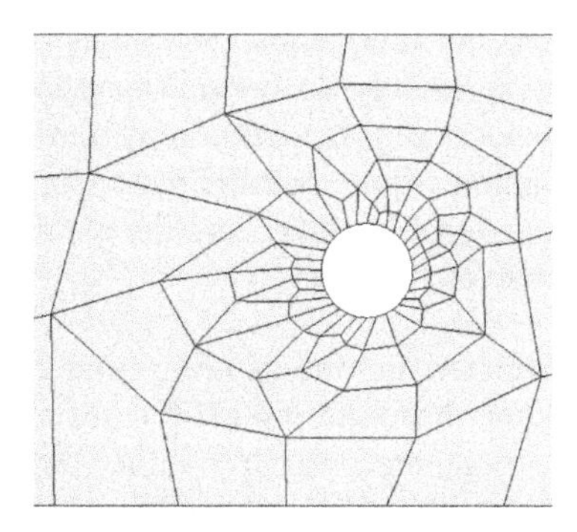
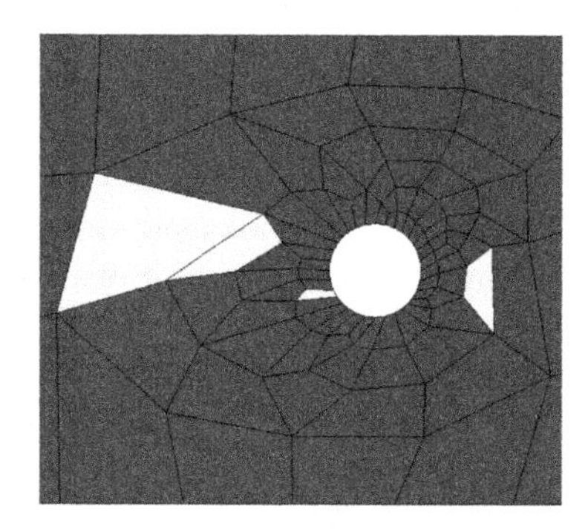

Figure 6.15 Element Plot (left) and Shape Checking Plot (right).

The **CHECK** command can be used to select elements that produce warning and error messages (`CHECK,ESEL,WARN`) or to select elements that produce only error messages (`CHECK,ESEL, ERR`). It will also produce up to five warning and/or error messages in the Output Window and will write all messages to the error file.

The Mechanical APDL Modeling and Meshing Guide notes that the element shape testing criteria are somewhat arbitrary and strongly depend on the analysis. Thus, a mesh with hundreds of warnings may still produce good results. Similarly, since the shape of the generated elements is only one factor that affects the quality of a mesh, a warning-free mesh does not guarantee good results. You should think of element shape testing as a way to obtain some, but not all, of the information needed to evaluate the quality of your mesh.

6.8.2. Evaluating Mesh Density by Visual Inspection

Most analysts automatically evaluate each finite element mesh that is generated by visual inspection. Over time, you will develop a sense of what is (and is not) a good mesh. Until then, you should use the following guidelines:

- The size of your elements should be small compared to the body that has been meshed.
- There should always be at least two elements through the thickness of a feature.
- More elements are needed in areas of interest and where the solution gradients will be large (e.g., where point loads and constraints are applied, in areas with stress concentrations, in regions with flow constrictions, etc.).
- Fewer elements are needed in areas where the solution gradients will be small and the geometry is uniform.

It is not always obvious where the large solution gradients in a model will be. You may need to solve your model in order to determine where they are, adjust the mesh accordingly, and solve the model again.

6.8.3. Evaluating Mesh Density through Energy Error Estimation

A more rigorous evaluation of the mesh density can be done using energy error estimation. Structural energy error (SERR) is a measure of the discontinuity of the stress field from element to element, while the thermal (TERR) and magnetic energy errors (BERR) are measures of the discontinuity of the heat and magnetic flux from element to element. All three energy errors are calculated by comparing the difference between the final averaged nodal value and the element value of stress or flux for each node on each element throughout the model. This error is then multiplied by the stress-strain matrix (or the conductivity matrix) for each node and integrated over each element to determine the energy error for the element. The energy error for the entire model is the sum of the energy errors for the individual elements.

The energy error for the entire model can also be normalized by the strain energy, the thermal dissipation energy, or the magnetic energy to determine the structural (SEPC), thermal (TEPC), or magnetic percentage error (BEPC). The details for the calculation of the energy errors can be found in Section 17.6 of the Mechanical APDL Theory Reference.

Once your model has been solved, the energy error may be plotted using the GUI path: **Main Menu > General Postproc > Plot Results > Contour Plot > Element Solu** and then choosing "Error Estimation" as the item to be contoured or by issuing the **PLESOL** command (`PLESOL, SERR; PLESOL,TERR; or PLESOL,BERR`). The percentage errors can be listed by turning Power Graphics off (`/GRAPHICS,OFF`) and then by following the path: **Main Menu > General Postproc > List Results > Percent Error** or by issuing the **PRERR** command.

Large values of energy error imply that the stress field is very discontinuous, and small values of energy error imply a more continuous stress field. While there are no absolutes regarding permissible values for energy error, a maximum percentage error of 5% was the default value used by the old ANSYS **ADAPT** macro and the largest value recommended by the literature.

Figure 6.16 shows the structural energy error (SERR) plots for the notched plate shown in Figure 6.12. The maximum energy errors for the plate are 1.3e-7 (left) and 8.73e-4 (right). The structural percentage errors (SEPC) for the plate are 0.046% (left) and 3.1093% (right). Both values are below the 5% limit. However, the differences between the maximum energy errors and the maximum equivalent stresses in the two plots are approximately two orders of magnitude. In addition, the structural energy error plots show that the higher energy errors are grouped around the two notches. This suggests that the mesh around the notches should be refined in the coarser model on the right at a minimum.

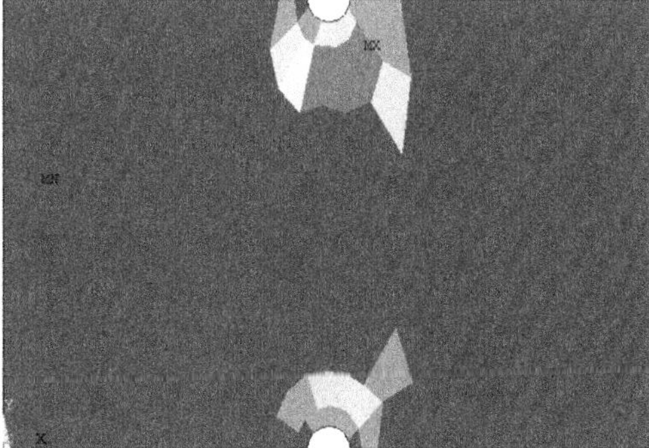

Figure 6.16 Plot of Structural Energy Error for a Notched Plate with Different Mesh Densities: Adequate Fine Mesh (left) and Inadequate Coarse Mesh (right).

6.8.4. Evaluating the Mesh Quality through Mesh Convergence

Finally, you can check the quality of a mesh by performing a simple mesh convergence experiment. As the number of elements in a model increases, the approximation of the system being evaluated improves, and the solution results will approach the asymptote of the 'true' system behavior. The mesh has "converged" when the solution becomes asymptotic. Once this occurs, large changes in mesh density will result in very small changes in the solution.

To test the mesh convergence of a model, build the model, review its solution, and note the value of one or more output parameters such as the maximum displacement in the model or the stress at a location of interest in the model. Next, substantially refine the mesh (e.g., use twice as many elements in the entire model if possible or twice as many elements in areas with high solution gradients if it is not). Then solve the model again and compare the results of the new model to the old one. If the output parameters being compared do not change significantly after mesh refinement, then the mesh has probably converged and the previous mesh density was sufficient to generate good results. If the output parameters change significantly, repeat the process until the mesh converges.

6.9. Modifying and Regenerating a Mesh

If the mesh quality is insufficient or if you are dissatisfied with a mesh, you can either refine the existing mesh or clear the mesh and generate a new one. Other mesh modifications such as adding or removing mid-side nodes, changing some of the element attributes, changing the node and element numbers, changing the outward normal direction of the elements, and detaching the mesh from the solid model for manual manipulation are beyond the scope of this book.

6.9.1. Refining a Mesh

Mesh refinement essentially divides the existing elements in half. As a result, mesh refinement operations approximately double the number of line elements, quadruple the number of area elements, and increase the number of volume elements by a factor of eight. Since the cost of the solution increases with the square of the number of DOFs, mesh refinement usually cannot be done indiscriminately. In addition, global refinement of the entire mesh is often unnecessary. For many problems, the mesh only needs to be refined in the areas of interest and/or in areas with high gradients. Thus, local mesh refinement is usually more economical and as effective as global mesh refinement. Figure 6.17 shows the effect of global versus local mesh refinement.

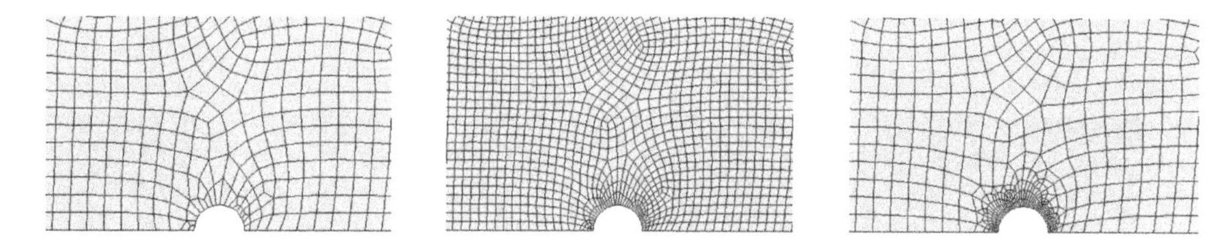

Figure 6.17 Global Versus Local Mesh Refinement: Original Mesh (left), Global Mesh Refinement (center), and Local Mesh Refinement (right).

The easiest way to refine a mesh is to use the "Refine" button at the bottom of the Mesh Tool. This will give you the option of refining the mesh locally at the selected elements, nodes, keypoints, lines, or areas, or to globally refine all of the elements in the model. After making this selection, another dialog box will appear that allows you to specify the level of refinement on a scale from 1 (Minimal) to 5 (Maximum). The amount of refinement is significant, so it is best to refine your mesh one level at a time.

Mesh refinement can also be done using the GUI path: **Main Menu > Preprocessor > Meshing > Modify Mesh > Refine At**. Or, you can use commands to refine the mesh.

- **KREFINE** refines the mesh around the specified keypoints.
- **LREFINE** refines the mesh around the specified lines.
- **AREFINE** refines the mesh around the specified areas.
- **EREFINE** refines the mesh around the specified elements.
- **NREFINE** refines the mesh around the specified nodes.

Mapped meshes cannot be refined locally. If local refinement is applied to a mapped mesh, free meshing will be used to perform the refinement. If refinement of a mapped mesh is required, it is best to clear the original mesh, modify the element size or the number of divisions on the edges, and remesh the model.

6.9.2. Clearing a Mesh

If you are dissatisfied with your current mesh or if you want to modify your solid model after it has been meshed, you can clear (i.e., delete) it by clicking the "CLEAR" button in the Mesh Tool or by following the path: **Main Menu > Preprocessor > Meshing > Clear**. You can also clear the mesh using one of the mesh clearing commands:

- **KCLEAR** to clear meshed keypoints.
- **LCLEAR** to clear meshed lines.
- **ACLEAR** to clear meshed areas.
- **VCLEAR** to clear meshed volumes.

Mesh clearing commands clear all nodes and elements associated with the selected solid model entities regardless of the selection status of those nodes and elements. However, they do not delete the nodes shared by adjacent entities that are to remain meshed. Because of this, you cannot clear the mesh for an entity that is of a lower order than the overall mesh. For example, you cannot clear the mesh for a line that is attached to a meshed area.

Exercise 6-1

Determining the Mesh Convergence of a Heated Plate With a Central Hole

Overview

In this exercise, you will perform a series of steady-state thermal analyses of a rectangular plate with a central hole to explore the impact of the mesh on the solution quality. The plate is made of high carbon steel with a thermal conductivity of 40 W/mK. It is 15 cm wide and 10 cm tall. The hole is 4 cm wide and 4 cm tall. It has 0.3 cm fillets on each corner. The plate is thin, so it can be modeled in two dimensions (Figure 6-1-1). One side of the plate will be held at 400 K, and the other side will be at 300 K. This will create a temperature gradient in the plate. The goal of each analysis is to visualize the temperature distribution in the plate and to determine the maximum heat flux.

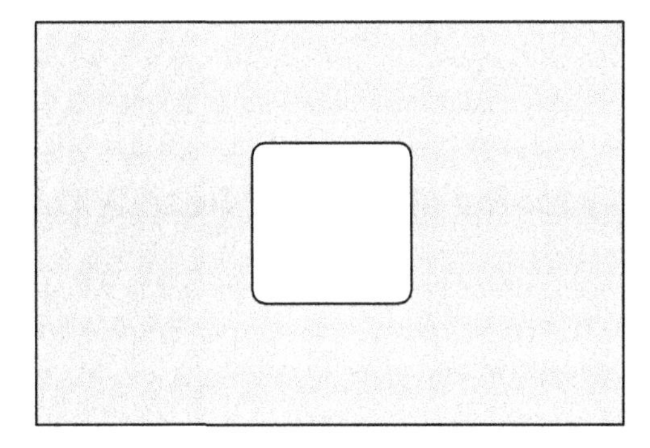

Figure 6-1-1 Schematic of the Rectangular Plate with a Central Hole.

The mesh will be refined after each analysis to generate a total of six solutions. The six meshes will be created using:

- No element size specified (**DESIZE** defaults)
- Smart Size of 5
- Smart Size of 3
- Smart Size of 1
- Smart Size of 1 with refinement of the lines around the hole
- Smart Size of 1 with global element refinement

After all six models have been solved, their results will be compared to determine how the mesh affects the quality of the solution.

ANSYS Mechanical APDL for Finite Element Analysis.
DOI: http://dx.doi.org/10.1016/B978-0-12-812981-4.00023-X

Model Attributes

Material Properties for High Carbon Steel

- Thermal conductivity—40 W/mK

Boundary Conditions

- Temperature of 400 K on the left side of the plate.
- Temperature of 300 K on the right side of the plate.

File Management

Create a new folder in your Intro-to-ANSYS directory for Exercise6-1

Change the Working Directory to the new Exercise6-1 folder

Change the Jobname to Exercise6-1-mesh1

Start ANSYS

*In this exercise, we will issue the **AMESH** and **SOLVE** commands six times. Each subsequent **AMESH** command will replace the previous mesh and each **SOLVE** command will replace the old results in the database unless you prevent this. We could create, solve, and postprocess each model in order, making sure to get all of the information needed before moving on to the next step and overwriting the model. However, this could make it difficult to take breaks or correct mistakes. Instead, we will save each model and solution using a different jobname. This means that each mesh will exist in a separate database file and each set of results will be contained in its own results file. We will postprocess the models after they have all been solved. More information about protecting your results can be found in* chapter 8.

Define a Storage Array for the Model and Results Data

You will need to store some of the results from each of the six models in this exercise to evaluate the mesh convergence. You could record this data externally (in a separate file or on a piece of paper), but it is more convenient and more accurate to use an array.

The array will be named "ex61results" because it will store the results from Exercise 6-1. The array will be six rows long to store the results from the six solutions. It will be five columns wide to store the five parameters of interest: the number of elements in each model (first column), the maximum heat flux from the averaged nodal solution (second column), the maximum heat flux from the secondary element solution (third column), the thermal percentage error (TEPC) (fourth column), and the maximum thermal energy error (TERR) (fifth column).

Arrays are a part of the ANSYS Parametric Design Language.

- Utility Menu > Parameters > Array Parameters > Define/Edit . . .
- Click "Add . . ."
- Enter "ex61results" for the Par Parameter Name (Figure 6-1-2)
- Leave the parameter type set to "Array"
- Enter 6, 5, and 1 for the I, J, and K number of rows, cols, and planes
- Click OK
- Close the Array Parameter dialog box

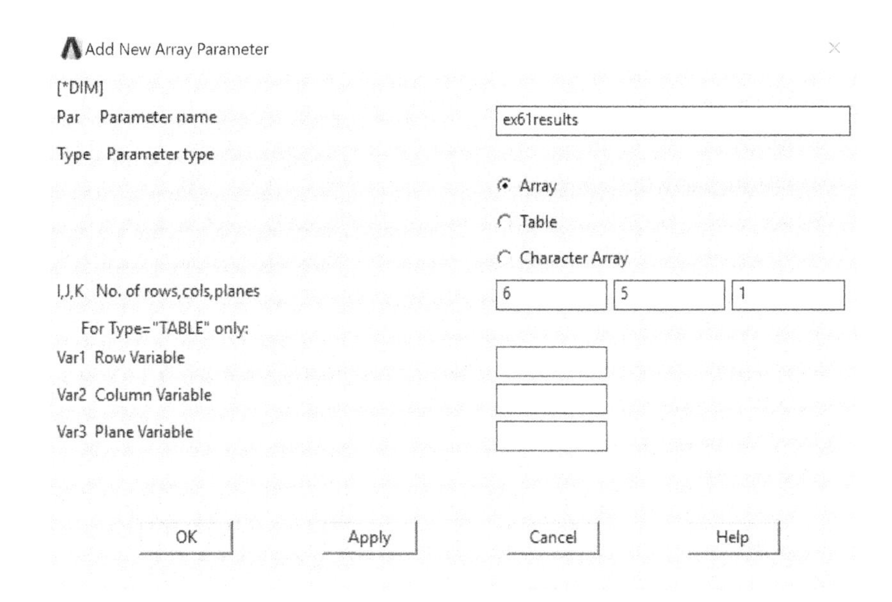

Figure 6-1-2 Array Parameter Dialog Box.

Create the Model and Run the Analysis (Model #1)

Step 1: Define Geometry

1-1. Create a rectangle to represent the plate

- Preprocessor > Modeling > Create > Areas > Rectangle > By 2 Corners
- Place the lower left corner of the plate at the origin (0,0) and the upper right corner at (0.15,0.1).

1-2. Create a square to represent the central hole

- Preprocessor > Modeling > Create > Areas > Rectangle > By Centr & Cornr
- Place the center of the square at (0.075,0.05). It should be 0.04 m tall and wide.

Perfectly square corners on internal features are rarely seen in practice because they are difficult to manufacture and they are a source of undesirable stress concentrations. Therefore, fillets will be added to the internal corners in the model. Even if square corners were present in the real system, fillets might still be needed in the finite element model because perfectly square corners introduce numerical singularities that can cause nonconvergence of the solution and unrealistic secondary results (stresses, fluxes, etc.).

Fillets can be added to the lines in the model, but this cannot be done while the lines are attached to an area. In the next step, you will delete the areas but keep the keypoints and lines in place. Then, you will create the fillets and create the final area.

1-3. Delete all areas in the model, leaving the lines and keypoints

- Preprocessor > Modeling > Delete > Areas Only
- Click Pick All

1-4. Turn line numbering on

- Utility Menu > PlotCtrls > Numbering ...
- Turn line numbering on

1-5. Plot the lines in the model

- Utility Menu > Plot > Lines

1-6. Create the fillets for the square hole

- Preprocessor > Modeling > Create > Lines > Line Fillet
- Click on Line 5 (L5)
- Click on Line 6 (L6)
- Click OK
- Enter the fillet radius of 0.003
- Click Apply
- Repeat for Lines 6 and 7
- Repeat for Lines 7 and 8
- Repeat for Lines 8 and 5
- Click OK

There should now be a total of 12 lines in the model: 4 for the outer rectangle, 4 for the inner rectangle, and 4 for the fillets.

1-7. Create a new area from the lines

- Preprocessor > Modeling > Create > Areas > Arbitrary > By Lines
- Check the Min,Max,Inc radio button (Figure 6-1-3)
- In the text box enter "1,12,1"
- Click OK

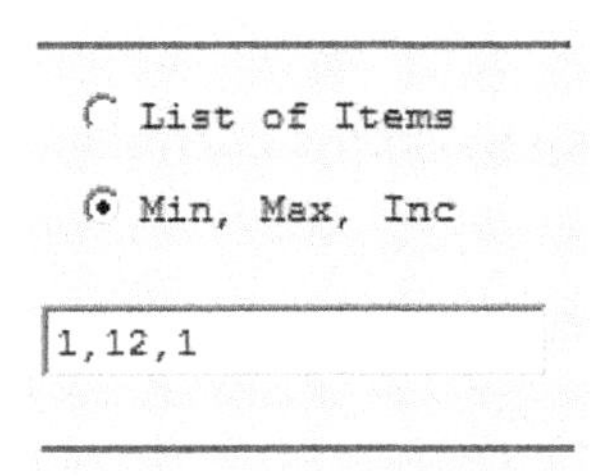

Figure 6-1-3 Entering Item Numbers into the Picker.

The Min,Max,Inc option in the Picker allows you to select all of the entities between the minimum (Min) and maximum (Max) entity number specified in the active set in increments of Inc. Inc defaults to 1, therefore both 1,12 and 1,12,1 will select lines 1 through 12 in the model. The Picker will be discussed in chapter 7.

1-8. Turn line numbering off

- Utility Menu > PlotCtrls > Numbering ...
- Turn line numbering off

1-9. Save your progress

Step 2: Define Element Types

2-1. Define the element type to use for this model

- Preprocessor > Element Type > Add/Edit/Delete
- Use PLANE55 elements

PLANE55 is a 2D thermal solid that can calculate heat transfer in a plane. It has four nodes with one degree of freedom (temperature) at each node.

Step 3: Define Material Properties

3-1. Create a thermal conductivity material model for high carbon steel

- Preprocessor > Material Props > Material Models
- Thermal > Conductivity > Isotropic
- Enter 40 for the thermal conductivity (KXX)

3-2. Save your progress

Step 4: Mesh

4-1. Create the initial mesh for the finite element model

- Preprocessor > Meshing > MeshTool
- Click the "Mesh" button
- Click the "Pick All" button in the Mesh Areas dialog box

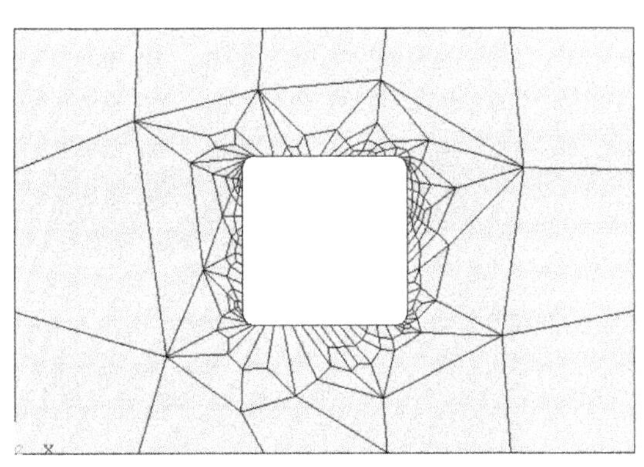
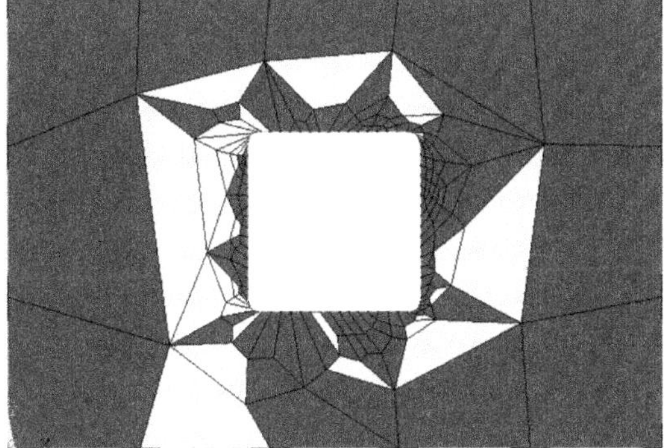

Figure 6-1-4 Element Plot (left) and Shape Checking Plot (right) for the First Mesh (DESIZE).

Because no mesh options were set, ANSYS used the default values associated with the ***DESIZE*** *command. These include a minimum of three elements per line for elements without mid-side nodes (such as PLANE55) and a maximum of 15 elements per line.*

The resulting mesh (Figure 6-1-4, left) is relatively poor. Although there are at least two elements through the thickness of all parts, many of the elements are large relative to the body being meshed. High temperature gradients are expected near the corners of the hole, but the mesh density is still low in those areas. Finally, almost 20% of the elements in this model (36 of 183 in the authors' model at Release 17.2) violate the shape checking warning limits. You should expect this solution to have significant errors. It is unlikely that the results of this model will adequately capture the behavior of the system.

*Using the **DESIZE** defaults decreases the repeatability of the meshing operation. Therefore, the number, shape, and distribution of the elements in your model may differ substantially from the ones shown here.*

4-2. Plot the elements with warning and error messages

- Preprocessor > Meshing > Check Mesh > Individual Elm > Plot Warning/Error Elements
- Check the radio button to show "Good Elements (blue)"
- Leave all other radio buttons checked
- Click OK

Figure 6-1-4 (right) shows that many of the elements that generate shape checking warnings are located around the central hole. Because this will be where the maximum heat flux occurs, these poorly shaped elements may negatively affect the results.

4-3. Retrieve and store the maximum number of elements in model #1

These operations will retrieve the maximum element number and store it in position (1,1) in the ex61results array. Data is stored in the array by (row, column). The rows represent the various meshes for this problem. The number of elements is stored in the first column.

- Utility Menu > Parameters > Get Scalar Data . . .
- In the left window, pick "Model data" (Figure 6-1-5)
- In the right window, scroll down and pick "For selected set"
- Click OK

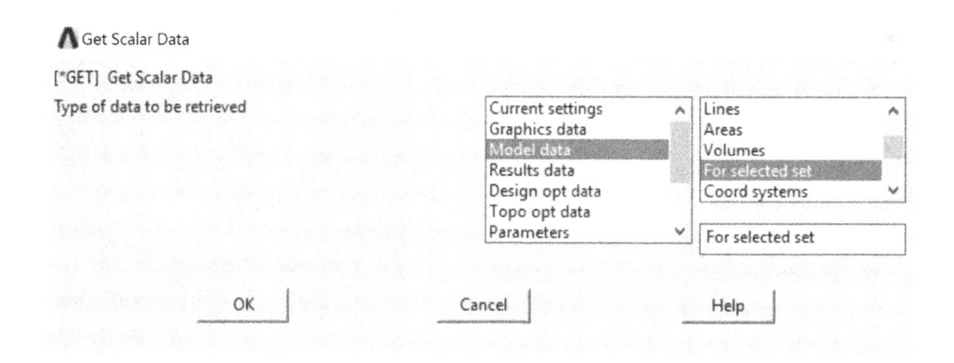

Figure 6-1-5 Get Scalar Data Dialog Box.

- For Name of parameter to be defined enter "ex61results(1,1)" (Figure 6-1-6)
- In left window, pick "Current elem set"
- In the right window, pick "Highest elem num"
- Click OK

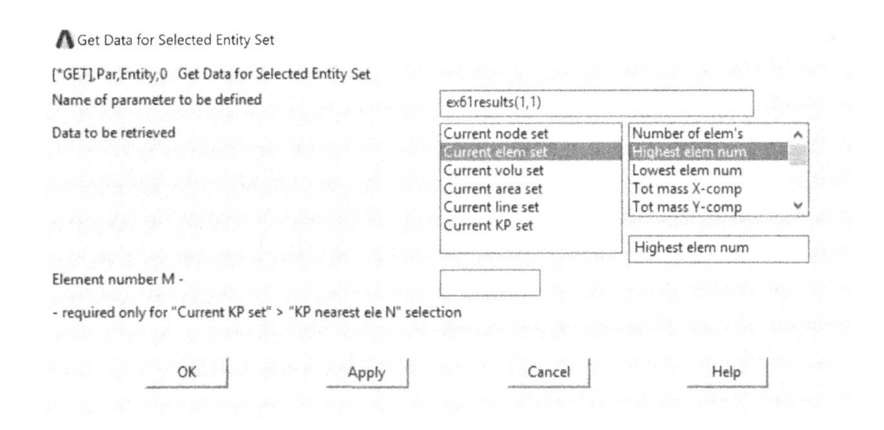

Figure 6-1-6 Get Scalar Data for Selected Entity Set Dialog Box.

4-4. Save your progress

Step 5: Apply Constraint Boundary Conditions

5-1. Apply the first temperature constraint to the model

- Solution > Define Loads > Apply > Thermal > Temperature > On Lines
- Pick the line on the left edge of the plate (vertical line at $x = 0$)
- Choose temperature as the degree of freedom to constrain
- Set the temperature to 400
- Click Apply

5-2. Apply the second temperature constraint to the model

- Pick the line on the right edge of the plate (vertical line at $x = 0.15$)
- Choose temperature as the degree of freedom to constrain
- Set the temperature to 300
- Click OK

Because the boundary conditions were applied to the solid model and not the finite element model, you can remesh the model without needing to delete and reapply the boundary conditions.

5-3. Save your progress

Step 6: Apply Load Boundary Conditions

Thermal loads in ANSYS include surface convections, heat fluxes, and heat generation. There are no applied loads for this exercise.

Step 7: Set the Solution Options

The default solution options can be used for this analysis.

Step 8: Solve Model #1

8-1. Select everything in the model

8-2. Solve

8-3. Save your results

Refine the Mesh and Rerun the Analysis (Model #2)

Step 9: Change the jobname

- Utility Menu > File > Change Jobname . . .
- Change the jobname to Exercise6-1-mesh2
- Click OK

Step 10: Mesh Model #2

10-1. Remesh the model

- Preprocessor > Meshing > Mesh Tool
- Activate Smart Sizing and choose a Smart Size of 5
- Click the Mesh button
- Click Pick All
- Confirm that you want to remesh the model by clicking OK (Figure 6-1-7)

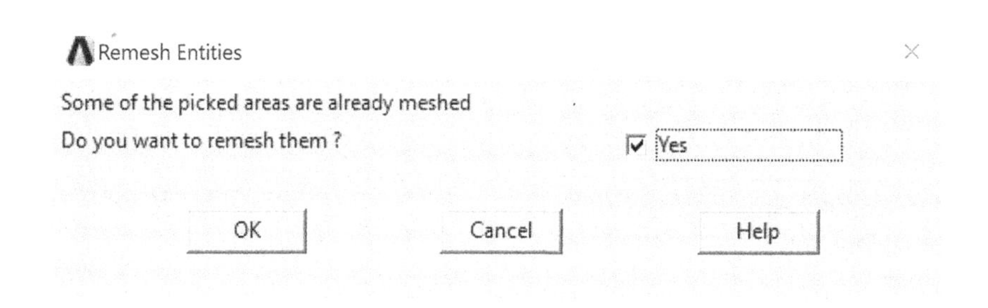

Figure 6-1-7 Confirm Remeshing.

10-2. Plot the elements with warning and error messages

- Preprocessor > Meshing > Check Mesh > Individual Elm > Plot Warning/Error Elements
- Check the radio button to show "Good Elements (blue)"
- Click OK

Figure 6-1-8 (left) shows that this mesh has more than twice as many elements as the first mesh. In addition, these elements are better formed and generated fewer shape checking warnings (3 vs 36) (Figure 6-1-8, right). You should expect this mesh to produce results that are more continuous in the bulk of the plate. However, there are still relatively few elements in the areas of interest, so this mesh may not accurately predict the maximum heat flux in the model.

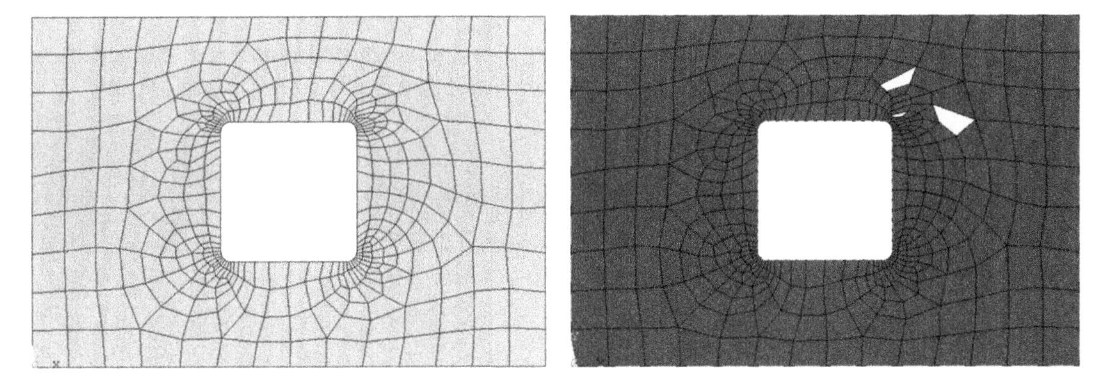

Figure 6-1-8 Element Plot (left) and Shape Checking Plot (right) for the Second Mesh (SMRT,5).

10-3. Retrieve and store the maximum number of elements in model #2

- Utility Menu > Parameters > Get Scalar Data . . .
- In the left window, "Model data" should still be selected
- In the right window, "For selected set" should still be selected
- Click OK
- For Name of parameter to be defined enter "ex61results(2,1)"
- In left window, "Current elem set" should still be selected
- In the right window, "Highest elem num" should still be selected
- Click OK

Step 11: Solve Model #2

11-1. Select everything

11-2. Solve the model

11-3. Save your results

11-4. Turn off Boundary Condition Symbols

- Utility Menu > PlotCtrls > Symbols . . .
- In the first section [/PBC] Boundary condition symbol check the radio button for "None"
- Click OK

This step prevents applied and reaction forces from being displayed.

11-5. Confirm that you have one set of files per jobname

- Open your working directory
- Sort the files by type
- Confirm that you have an Exercise6-1-mesh1.db
- Confirm that you have an Exercise6-1-mesh1.rth
- Confirm that you have an Exercise6-1-mesh2.db
- Confirm that you have an Exercise6-1-mesh2.rth
- Ensure that all of these files have a non-zero size

If Exercise6-1-mesh2.rth is shown as having a file size of 0 KB, move on to Step 12 and then check the file sizes again after the jobname has been updated. If this does not result in non-zero file sizes, you will need to go back and re-run the affected solution(s).

Refine the Mesh and Rerun the Analysis (Model #3)

Step 12: Change the jobname

- Utility Menu > File > Change Jobname . . .
- Change the jobname to Exercise6-1-mesh3

Step 13: Mesh Model #3

13-1. Remesh the model using a Smart Size of 3

- Preprocessor > Meshing > Mesh Tool

13-2. Plot the elements with warning and error messages if desired

- Preprocessor > Meshing > Check Mesh > Individual Elm > Plot Warning/Error Elements

This model has the same features as the previous mesh but has more than twice as many elements. The mesh is now relatively fine in the corners near the location of maximum heat flux and may produce reasonable results (*Figure 6-1-9*).

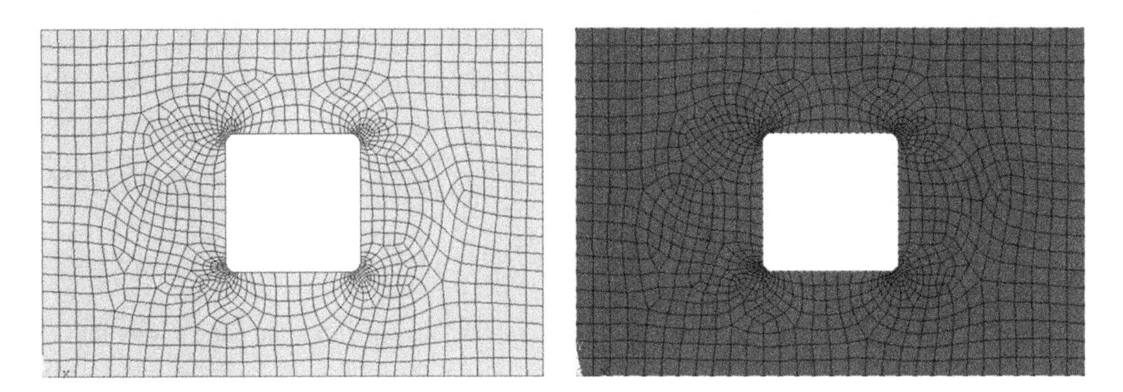

Figure 6-1-9 Element Plot (left) and Shape Checking Plot (right) for the Third Mesh (SMRT,3).

13-3. Retrieve and store the maximum number of elements in model #3

- Utility Menu > Parameters > Get Scalar Data . . .

Step 14: Solve Model #3

14-1. Select everything

14-2. Solve the model

14-3. Save your results

Refine the Mesh and Rerun the Analysis (Model #4)

Step 15: Change the jobname to Exercise6-1-mesh4

Step 16: Mesh Model #4

16-1. Remesh the model using a Smart Size of 1

- Preprocessor > Meshing > MeshTool

16-2. Plot the elements with warning and error messages if desired

- Preprocessor > Meshing > Check Mesh > Individual Elm > Plot Warning/Error Elements

  The resulting mesh and the shape checking plot are shown in Figure 6-1-10.

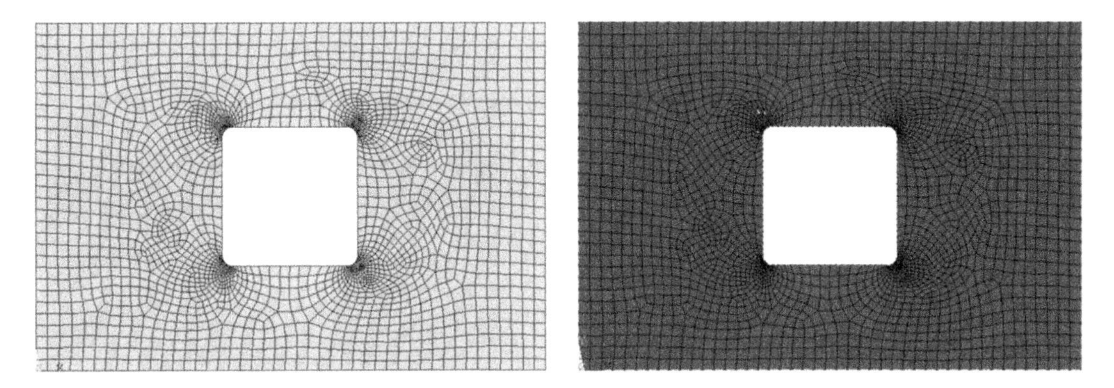

Figure 6-1-10 Element Plot (left) and Shape Checking Plot (right) for the Fourth Mesh (SMRT,1).

16-3. Retrieve and store the maximum number of elements in model #4

- Utility Menu > Parameters > Get Scalar Data . . .

Step 17: Solve Model #4

17-1. Select everything

17-2. Solve the model

17-3. Save your results

Refine the Mesh and Rerun the Analysis (Model #5)

Step 18: Change the jobname to Exercise6-1-mesh5

Step 19: Mesh Model #5

19-1. Refine the mesh along the lines around the central hole

- Preprocessor > Meshing > MeshTool
- Change the Refine at: option from "Elements" to "Lines"
- Click the Refine button
- Check the Min,Max,Inc radio button
- In the text box, type "5,12"

This selects lines 5 through 12 in the model, i.e., all of the interior lines and fillets.

- Click OK
- Leave the "Level of Refinement" set to "1 (Minimal)"
- Click OK

19-2. Plot the elements with warning and error messages if desired

- Preprocessor > Meshing > Check Mesh > Individual Elm > Plot Warning/Error Elements

Figure 6-1-11 shows that many of the elements that generate shape checking warnings are located at the boundary between the refined mesh and the coarse mesh.

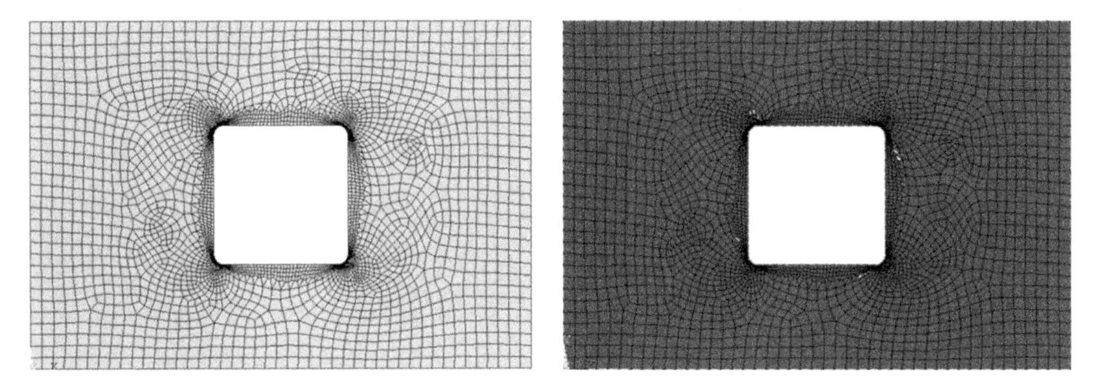

Figure 6-1-11 Element Plot (left) and Shape Checking Plot (right) for the Fifth Mesh (SMRT,1 and LREF of 1).

19-3. Retrieve and store the maximum number of elements in model #5

- Utility Menu > Parameters > Get Scalar Data . . .

Step 20: Solve Model #5

20-1. Select everything

20-2. Solve the model

20-3. Save your results

Refine the Mesh and Rerun the Analysis (Model #6)

Step 21: Change the jobname to Exercise6-1-mesh6

Step 22: Mesh Model #6

22-1. Remesh the model using a Smart Size of 1

- Preprocessor > Meshing > Mesh Tool

22-2. Refine the entire mesh using an element refinement level of 1

- Preprocessor > Meshing > MeshTool
- Change the Refine at: option from "Lines" back to "Elements"
- Click the Refine button
- Pick All
- Leave the "LEVEL Level of Refinement" set to "1 (Minimal)"
- Click OK

22-3. Plot the elements with warning and error messages if desired

- Preprocessor > Meshing > Check Mesh > Individual Elm > Plot Warning/Error Elements

  The resulting mesh and the shape checking plot are shown in Figure 6-1-12.

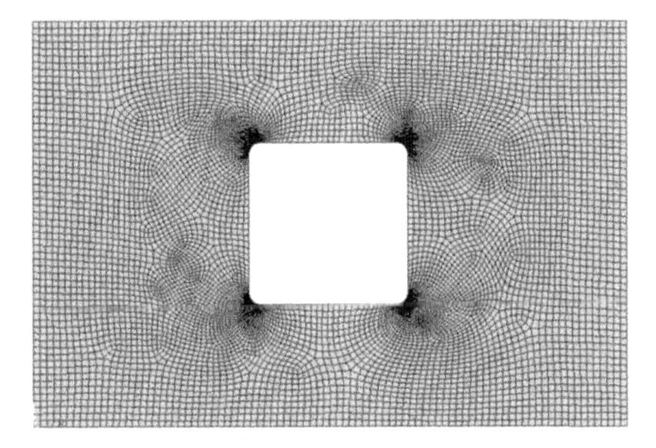

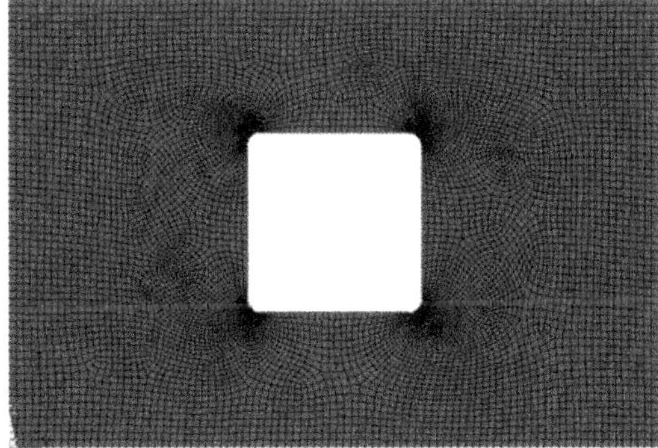

Figure 6-1-12 Element Plot (left) and Shape Checking Plot (right) for the Sixth Mesh (SMRT,1 and EREF of 1).

22-4. Retrieve and store the maximum number of elements in model #6

- Utility Menu > Parameters > Get Scalar Data . . .

Step 23: Solve Model #6

23-1. Select everything

23-2. Solve the model

23-3. Save your results

Verify and Save Your Model and Results Data Array

Step 24: Confirm that the element number array values have been stored correctly

24-1. Open the Array Parameters dialog box

- Utility Menu > Parameters > Array Parameters > Define/Edit . . .
- Click "Edit . . ."

The values in your array may be noticeably different from the values in the array shown in Figure 6-1-13 because of differences between your mesh and the meshes shown in this exercise.

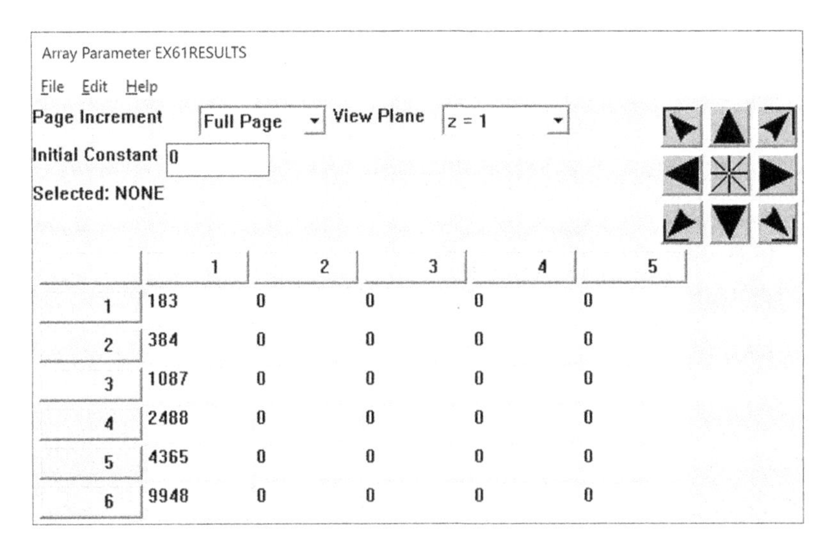

Figure 6-1-13 Array Filled with Maximum Element Numbers (Counts).

24-2. Close the Array Parameters dialog boxes

- In the Array Parameter EX61RESULTS dialog box, follow the path: File > Quit
- In the Array Parameters dialog box click Close

24-3. Save the Array Parameter

- Utility Menu > Parameters > Save Parameters ...
- In the Lab Parameters to be written drop down box, choose "Scalar and Array" (Figure 6-1-14)
- For Fname Write to file, enter "Exercise6-1-Parameters"
- Click OK

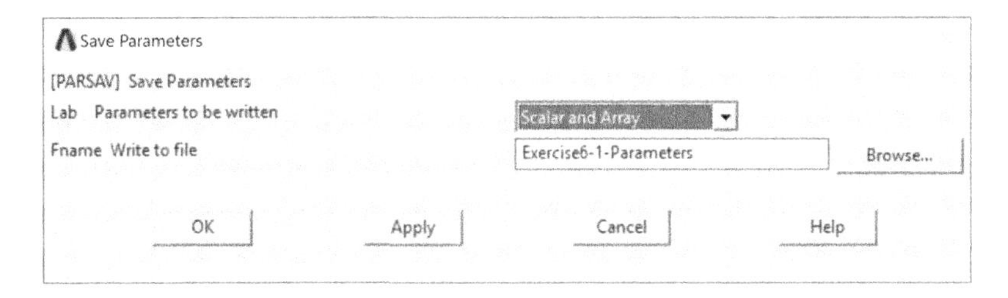

Figure 6-1-14 Save Parameters Dialog Box.

This is a convenient stopping point in the exercise if you need a break. All of your work so far has been saved. All remaining work uses the saved files.

Postprocess All Models

Step 25: Postprocess the Results of Model #1

25-1. Clear the database

- Utility Menu > File > Clear & Start New ...
- Click OK
- Click Yes

25-2. Change the jobname back to Exercise6-1-mesh1

- Utility Menu > File > Change Jobname ...

25-3. Resume the database for Model #1

- ANSYS Toolbar > RESUM_DB

Or

- Utility Menu > File > Resume from ...
- Select the file Exercise6-1mesh1.db
- Click OK

25-4. Load the saved parameters

- Utility Menu > Parameters > Restore Parameters ...
- In the "Lab Existing paramcters will be" drop down box, choose "Merged with new"
- For "Fname Restore from file", enter "Exercise6-1-Parameters" (or browse for the appropriate file)
- Click Open
- Click OK
- Click Yes

25-5. Confirm that the array values have been loaded correctly

- Utility Menu > Parameters > Array Parameters > Define/Edit . . .

25-6. Plot the temperature distribution in the plate

- General Postproc > Plot Results > Contour Plot > Nodal Solu
- Choose DOF Solution > Nodal Temperature
- Click OK

The temperature plot in Figure 6-1-15 shows that the maximum temperature (400 K) is located on the left side of the plate and that the minimum temperature (300 K) is located on the right side of the plate as expected. The temperature contours are spaced more closely near the square hole. This implies that the temperature gradients are steeper in this area. However, the temperature contours are neither smooth nor perfectly symmetric because there are too few elements to ensure a continuous nodal solution.

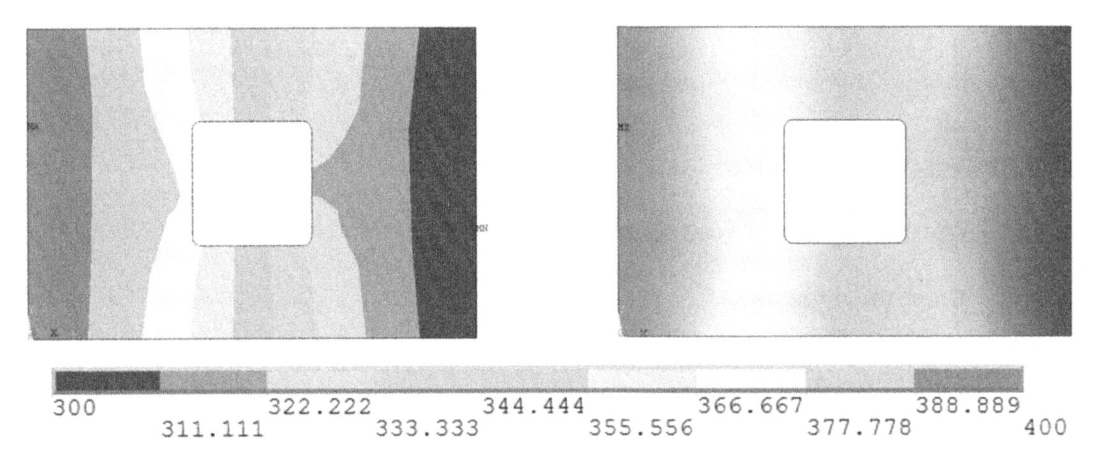

Figure 6-1-15 Temperature Contour Plot for the First Mesh: Win32 Graphics (left), 3D Graphics (right).

25-7. Plot the nodal (averaged) thermal flux distribution in the plate

- General Postproc > Plot Results > Contour Plot > Nodal Solu
- Choose Nodal Solution > Thermal Flux > Thermal flux vector sum
- Click OK

The thermal flux plot shows the heat flow through the plate. Because the hole acts as a flow constriction, there is a higher overall heat flux in the center of the plate. The nodal heat flux contours are neither smooth nor perfectly symmetric because there are too few elements in the mesh. The maximum heat flux is found near the corners where the thermal constriction begins.

25-8. Retrieve and store the maximum nodal thermal flux in the plate

- Enter the following command into the command prompt (Figure 6-1-16):
  `*GET,EX61RESULTS(1,2),PLNSOL,,MAX`

```
*GET, Par, Entity, ENTNUM, Item1, IT1NUM, Item2, IT2NUM
*GET,ex61results(1,2),PLNSOL,,MAX
```

Figure 6-1-16 Command Prompt with *GET Command.

Remember to include the extra comma before "max". "Max" is the fourth argument for this command and must be supplied in the fourth field. The double comma tells the program to use the default value for the third field (if any) and to ignore that field if no value is needed.

Commands in ANSYS are not case-sensitive. The capitalization has been added for clarity.

The ****GET*** *command (pronounced: Star Get) retrieves and stores values from the database. In Step 4-3, the* ****GET*** *command was used to retrieve the maximum element number in the model via the GUI. This set of arguments stores the maximum contour value from the current (thermal flux) plot. This option is not available through the GUI, so you must issue the command directly to the program.*

25-9. Plot the element thermal flux distribution in the plate

- General Postproc > Plot Results > Contour Plot > Element Solu
- Choose Element Solution > Thermal Flux > Thermal flux vector sum
- Choose Deformed shape with undeformed model
- Click OK

The plots in Figure 6-1-17 (right) show the unaveraged secondary (element) results for the first model. As the mesh is refined, the discontinuities across the elements will become smaller and the element solution will approach the nodal solution (Figure 6-1-17, left).

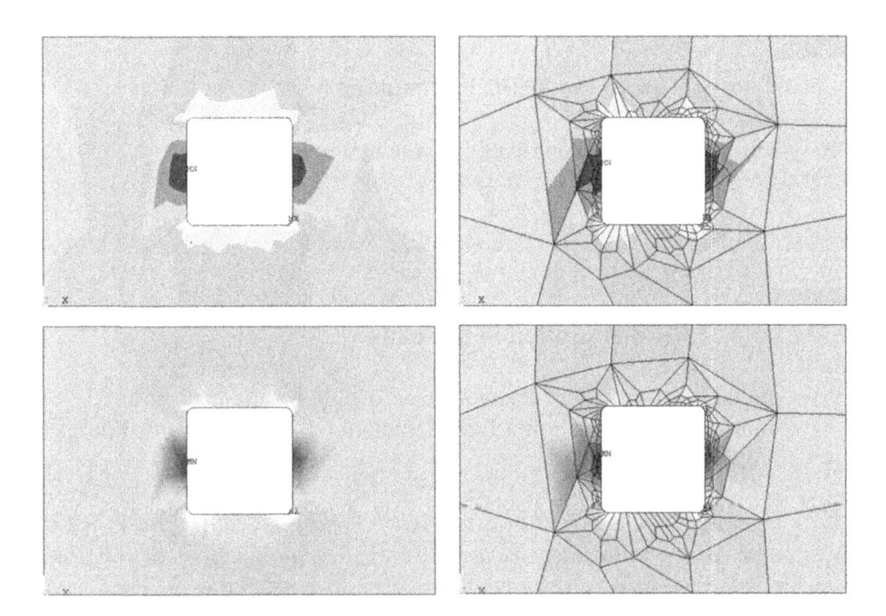

Figure 6-1-17 Nodal (left) and Element (right) Thermal Flux Contour Plot for the First Mesh: Win32 Graphics (top), 3D Graphics (bottom).

25-10. Retrieve and store the maximum element thermal flux in the plate

- Enter the following command into the command prompt:
  ```
  *GET,EX61RESULTS(1,3),PLNSOL,,MAX
  ```

Or

- Click the downward arrow to the right of the command prompt to display the recently used commands
- Select the previous ***GET** command
- Update the array indices to say (1,3) instead of (1,2)
- Press “enter” or the return key to execute the command

25-11. Turn off Power Graphics

- Utility Menu > PlotCtrls > Style > Hidden Line Options . . .
- Under “[/GRAPHICS] Used to control the way a model is displayed Graphic display method is:”
- Change from “PowerGraphics” to “Full model”
- Click OK

Or

- ANSYS Toolbar > POWRGRPH
- Check the “Off” radio button
- Click OK

Some results, including energy errors, are not available while PowerGraphics is in use. PowerGraphics must be turned off to calculate the thermal energy error for this model.

25-12. List the TEPC in the plate

- General Postproc > List Results > Percent Error

Note that this value is well above the 5% limit discussed in chapter 6.

25-13. Retrieve and store the TEPC in the plate

- Enter the following command into the command prompt:
  ```
  *GET,EX61RESULTS(1,4),PRERR,,TEPC
  ```

You could also copy the value listed in the PRERR window and paste it into the array. However, this will not retrieve all of 14 digits of the result.

25-14. Plot the TERR distribution in the plate

- General Postproc > Plot Results > Contour Plot > Element Solu
- Choose Element Solution > Error Estimation > Thermal Error Energy
- Click OK

The TERR for the first model is shown in Figure 6-1-18. Recall that the TERR is a measure of the discontinuity of the secondary (element) results including the heat flux from element to element. Large values of energy error indicate large discontinuities in the secondary results and small values of energy error indicate a continuous secondary solution. Small values are preferred.

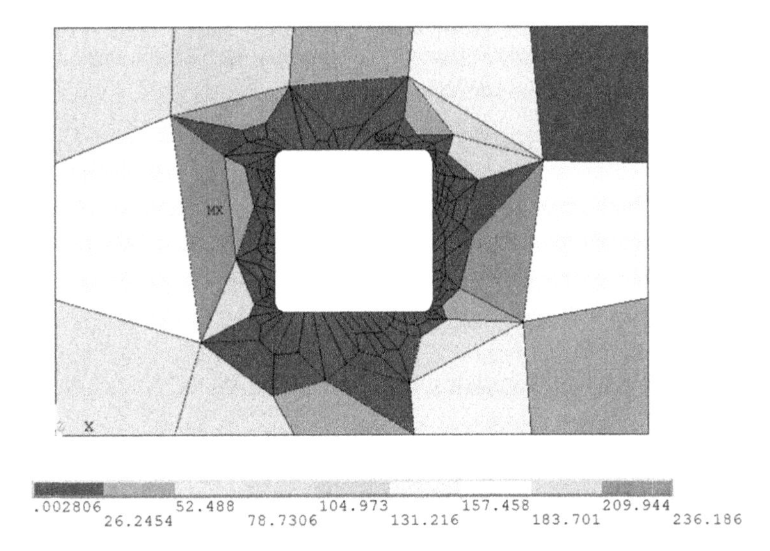

Figure 6-1-18 Plot of Thermal Energy Error for the First Mesh (Win32 Graphics).

25-15. Retrieve and store the maximum TERR in the plate

- Enter the following command into the command prompt:
  `*GET,EX61RESULTS(1,5),PLNSOL,,MAX`

25-16. Confirm that all array values have been stored correctly

- Utility Menu > Parameters > Array Parameters > Define/Edit ...
- Click "Edit ..."
- Review the stored data (Figure 6-1-19)
- Close the Array Parameters dialog boxes when finished

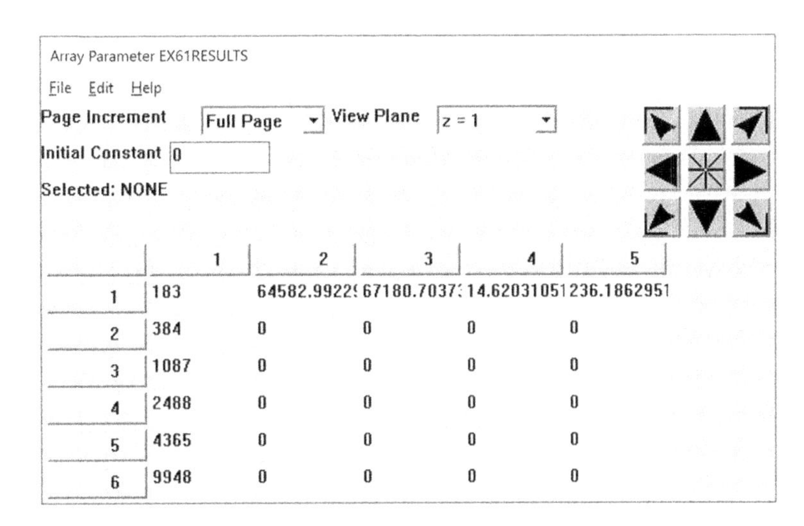

	1	2	3	4	5
1	183	64582.9922!	67180.7037:	14.62031051	236.1862951
2	384	0	0	0	0
3	1087	0	0	0	0
4	2488	0	0	0	0
5	4365	0	0	0	0
6	9948	0	0	0	0

Figure 6-1-19 Array with Postprocessing Data from the First Mesh.

If you double click on the value in ex61results(1,2), the significant figures that were not visible before will be displayed. ANSYS reports all noninteger values with 14 digits. However, the results are not accurate to 14 significant figures.

25-17. Save the Array Parameter

- Utility Menu > Parameters > Save Parameters . . .
- In the "Lab Parameters to be written" drop down box, choose "Scalar and Array"
- For "Fname Write to file", enter "Exercise6-1-Parameters" or browse for the file
- Click OK

Step 26: Postprocess the Results of Model #2

26-1. Clear the database

- Utility Menu > File > Clear & Start New . . .

26-2. Change the jobname back to Exercise6-1-mesh2

- Utility Menu > File > Change Jobname . . .

26-3. Resume the database for Model #2

- ANSYS Toolbar > RESUM_DB

26-4. Load the saved parameters

- Utility Menu > Parameters > Restore Parameters . . .
- Remember to choose "Merged with new."

26-5. Confirm that the array values have been loaded correctly

- Utility Menu > Parameters > Array Parameters > Define/Edit . . .

26-6. Plot and review the temperature distribution in the plate

- General Postproc > Plot Results > Contour Plot > Nodal Solu
- DOF Solution > Nodal Temperature

Notice that the contours on this temperature plot are much smoother than in the first model.

26-7. Plot and review the nodal thermal flux distribution in the plate

- General Postproc > Plot Results > Contour Plot > Nodal Solu
- Nodal Solution > Thermal Flux > Thermal flux vector sum

Notice that the nodal thermal flux plot shown in Figure 6-1-20 (left) is more symmetric than in the first model.

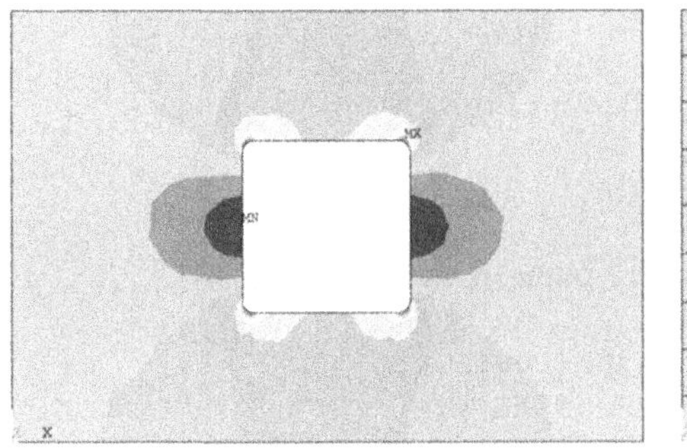 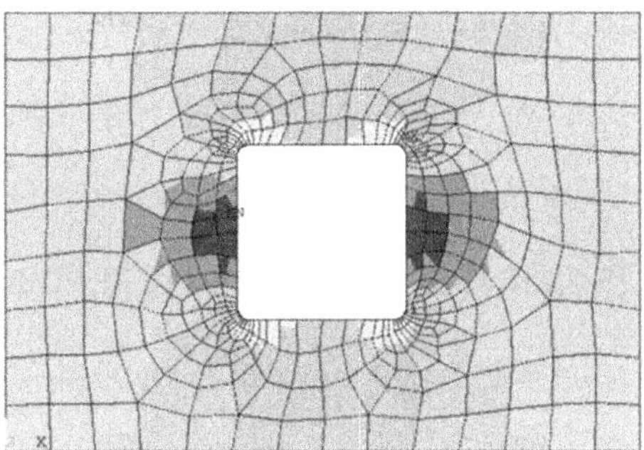

Figure 6-1-20 Nodal (left) and Element (right) Thermal Flux Contour Plot for the Second Mesh (Win32 Graphics).

26-8. Retrieve and store the maximum nodal thermal flux in the plate in ex61results(2,2)

- Enter the following command into the command prompt:
  `*GET,EX61RESULTS(2,2),PLNSOL,,MAX`

Remember that you can reuse previously issued commands by clicking on the down arrow at the right side of the command prompt. While the previously commands are displayed, click on the desired command with the mouse. This will place that text into the command prompt. Update the information in the command and press enter to execute (Figure 6-1-21).

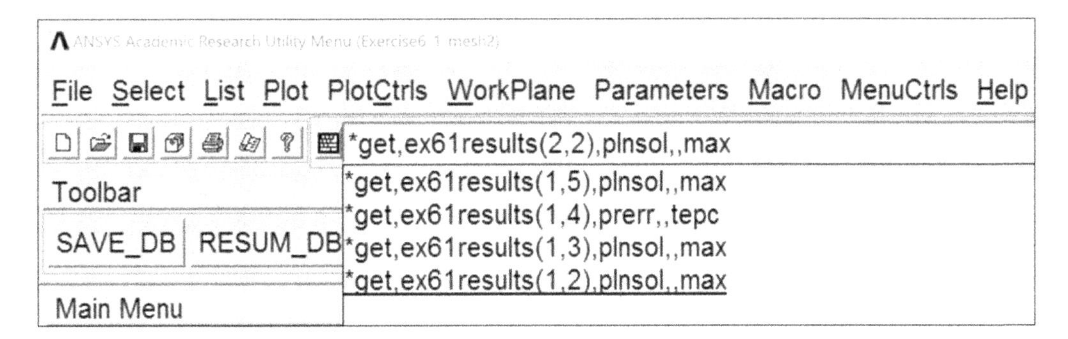

Figure 6-1-21 Previously Issued Commands in the Command Prompt.

26-9. Plot and review the element thermal flux distribution in the plate

- General Postproc > Plot Results > Contour Plot > Element Solu
- Element Solution > Thermal Flux > Thermal flux vector sum

Although the element thermal flux plot (Figure 6-1-20, right) is more symmetric and the nodal and element heat flux plots are more similar than in the first model, the discontinuities in the solution are still clearly present.

26-10. Retrieve and store the maximum element thermal flux in the plate in ex61results(2,3)

- Enter the following command into the command prompt:
  `*GET,EX61RESULTS(2,3),PLNSOL,,MAX`

26-11. List the TEPC in the plate

- General Postproc > List Results > Percent Error

Notice that this value is still above the 5% limit.

26-12. Retrieve and store the TEPC in the plate in ex61results(2,4)

- Enter the following command into the command prompt:
  `*GET,EX61RESULTS(2,4),PRERR,,TEPC`

26-13. Plot and review the TERR distribution in the plate

- General Postproc > Plot Results > Contour Plot > Element Solu
- Choose Element Solution > Error Estimation > Thermal Error Energy
- Click OK

Notice that the maximum TERR has decreased by roughly two orders of magnitude because of the mesh refinement (Figure 6-1-22).

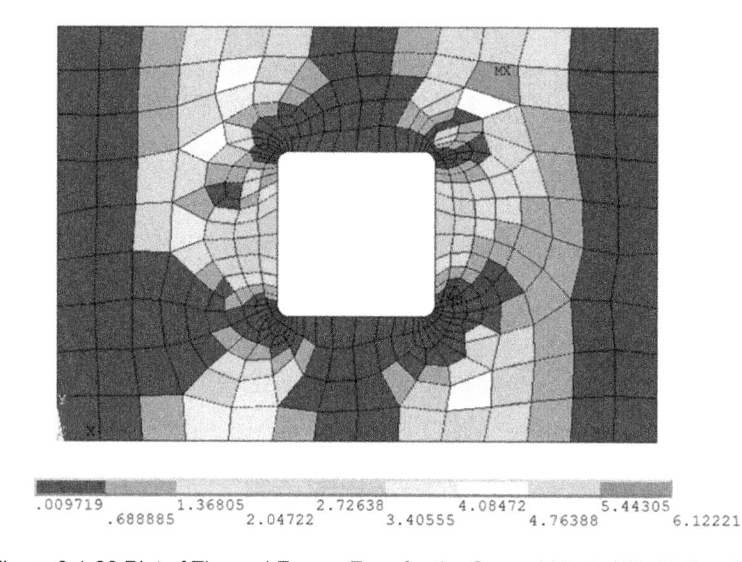

Figure 6-1-22 Plot of Thermal Energy Error for the Second Mesh (Win32 Graphics).

26-14. Retrieve and store the maximum TERR in the plate in ex61results(2,5)

- Enter the following command into the command prompt:
  `*GET,EX61RESULTS(2,5),PLNSOL,,MAX`

26-15. Confirm that all array values have been stored correctly

- Utility Menu > Parameters > Array Parameters > Define/Edit ...

26-16. Save the Array Parameter

- Utility Menu > Parameters > Save Parameters ...
- Choose "Scalar and Array"
- Save the file as "Exercise6-1-Parameters" or browse for the file.

Step 27: Postprocess the Results of Model #3

27-1. Clear the database

- Utility Menu > File > Clear & Start New ...

27-2. Change the jobname back to Exercise6-1-mesh3

- Utility Menu > File > Change Jobname ...

27-3. Resume the database for Model #3

- ANSYS Toolbar > RESUM_DB

27-4. Load the saved parameters

- Utility Menu > Parameters > Restore Parameters ...
- Remember to choose "Merged with new"

27-5. Confirm that the array values have been loaded correctly

- Utility Menu > Parameters > Array Parameters > Define/Edit ...

27-6. Plot and review the temperature distribution in the plate

- General Postproc > Plot Results > Contour Plot > Nodal Solu
- DOF Solution > Nodal Temperature

27-7. Plot and review the nodal thermal flux distribution in the plate

- General Postproc > Plot Results > Contour Plot > Nodal Solu
- Nodal Solution > Thermal Flux > Thermal flux vector sum

Notice that the nodal thermal flux plot is much smoother than in the second model (Figure 6-1-23, left).

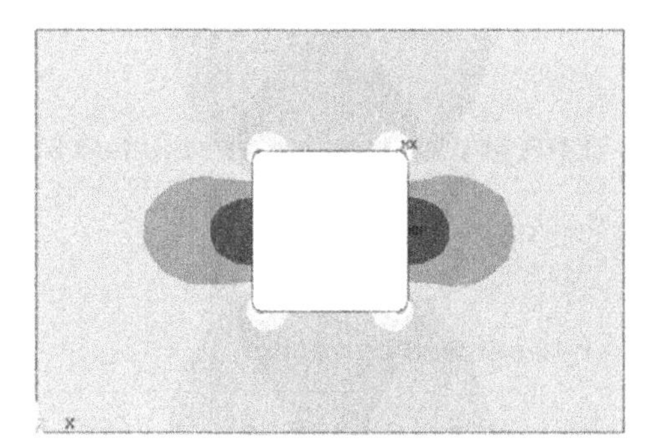
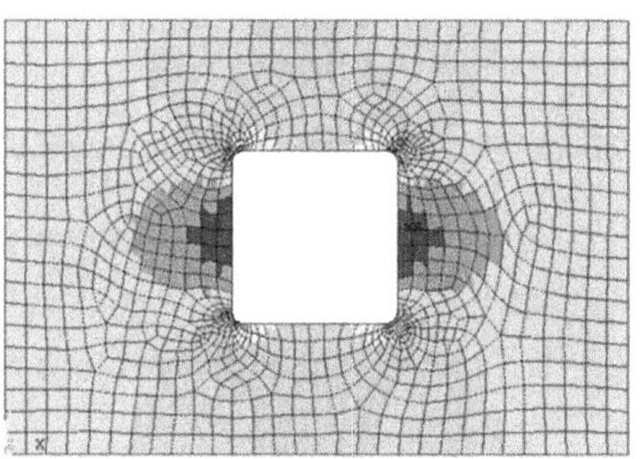

Figure 6-1-23 Nodal (left) and Element (right) Thermal Flux Contour Plot for the Third Mesh (Win32 Graphics).

27-8. Retrieve and store the maximum nodal thermal flux in the plate in ex61results (3,2)

- Enter the following command into the command prompt:
  `*GET,EX61RESULTS(3,2),PLNSOL,,MAX`

27-9. Plot and review the element thermal flux distribution in the plate

- General Postproc > Plot Results > Contour Plot > Element Solu
- Element Solution > Thermal Flux > Thermal flux vector sum

Notice that the element thermal flux plot (Figure 6-1-23, right) is more similar to its corresponding nodal solution than in the second model.

27-10. Retrieve and store the maximum element thermal flux in the plate in ex61results(3,3)

- Enter the following command into the command prompt:
  `*GET,EX61RESULTS(3,3),PLNSOL,,MAX`

27-11. List the TEPC in the plate

- General Postproc > List Results > Percent Error

Notice that this value is now below the 5% limit.

27-12. Retrieve and store the TEPC in the plate in ex61results(3,4)

- Enter the following command into the command prompt:
  `*GET,EX61RESULTS(3,4),PRERR,,TEPC`

27-13. Plot and review the TERR distribution in the plate

- General Postproc > Plot Results > Contour Plot > Element Solu
- Choose Element Solution > Error Estimation > Thermal Error Energy
- Click OK

Notice that the maximum TERR has decreased by half an order of magnitude because of the mesh refinement (Figure 6-1-24).

27-14. Retrieve and store the maximum TERR in the plate in ex61results(3,5)

- Enter the following command into the command prompt:
  `*GET,EX61RESULTS(3,5),PLNSOL,,MAX`

27-15. Confirm that all array values have been stored correctly

- Utility Menu > Parameters > Array Parameters > Define/Edit ...

27-16. Save the Array Parameter

- Utility Menu > Parameters > Save Parameters ...
- Choose "Scalar and Array"
- Save the file as "Exercise6-1-Parameters" or browse for the file.

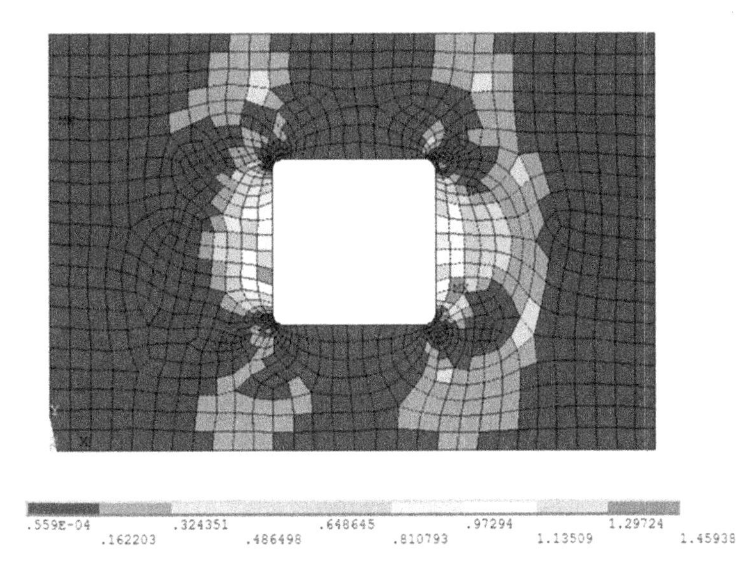

Figure 6-1-24 Plot of Thermal Energy Error for the Third Mesh (Win32 Graphics).

Step 28: Postprocess the Results of Model #4

28-1. Clear the database

28-2. Change the jobname back to Exercise6-1-mesh4

28-3. Resume the database for Model #4

28-4. Load the saved parameters

28-5. Confirm that the array values have been loaded correctly

28-6. Plot and review the temperature distribution in the plate

28-7. Plot and review the nodal thermal flux distribution in the plate

The nodal thermal flux distribution for the fourth model is shown in Figure 6-1-25 (left).

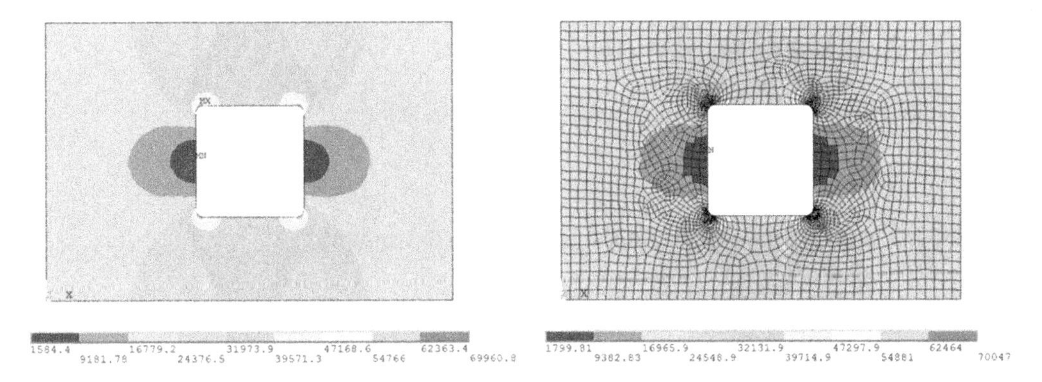

Figure 6-1-25 Nodal (left) and Element (right) Thermal Flux Contour Plot for the Fourth Mesh (Win32 Graphics).

28-8. Retrieve and store the maximum nodal thermal flux in the plate in ex61results(4,2)

28-9. Plot and review the element thermal flux distribution in the plate

The element thermal flux distribution for the fourth model is shown in Figure 6-1-25 (right).

28-10. Retrieve and store the maximum element thermal flux in the plate in ex61results(4,3)

28-11. List the TEPC in the plate

28-12. Retrieve and store the TEPC in the plate in ex61results(4,4)

28-13. Plot and review the TERR distribution in the plate

The thermal energy error distribution for the fourth model is shown in Figure 6-1-26.

28-14. Retrieve and store the maximum TERR in the plate in ex61results(4,5)

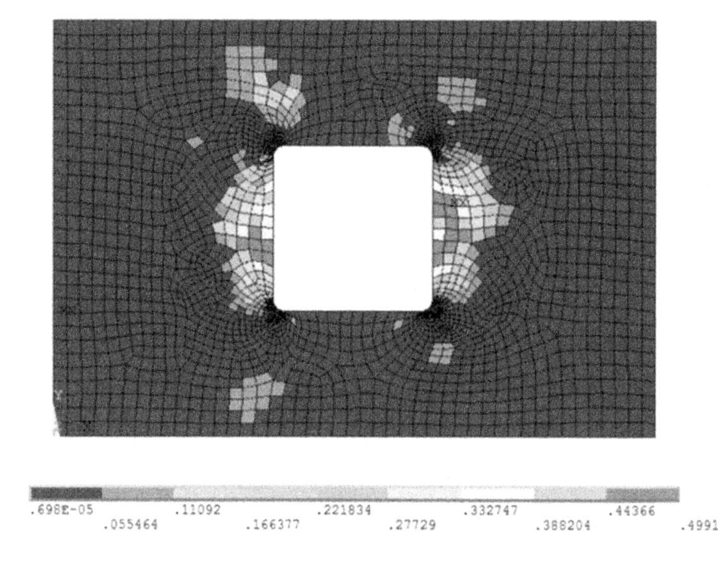

Figure 6-1-26 Plot of Thermal Energy Error for the Fourth Mesh (Win32 Graphics).

28-15. Confirm that all array values have been stored correctly.

28-16. Save the Array Parameter.

Step 29: Postprocess the Results of Model #5

29-1. Clear the database.

29-2. Change the jobname back to Exercise6-1-mesh5.

29-3. Resume the database for Model #5.

29-4. Load the saved parameters.

29-5. Confirm that the array values have been loaded correctly.

29-6. Plot and review the temperature distribution in the plate.

29-7. Plot and review the nodal thermal flux distribution in the plate.

The nodal thermal flux distribution for the fifth model is shown in Figure 6-1-27 (left).

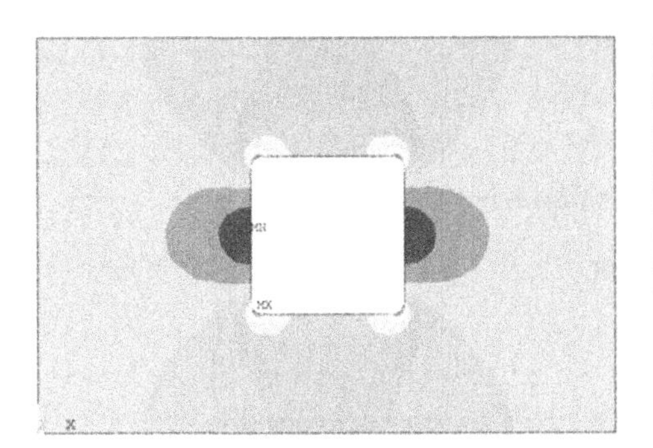

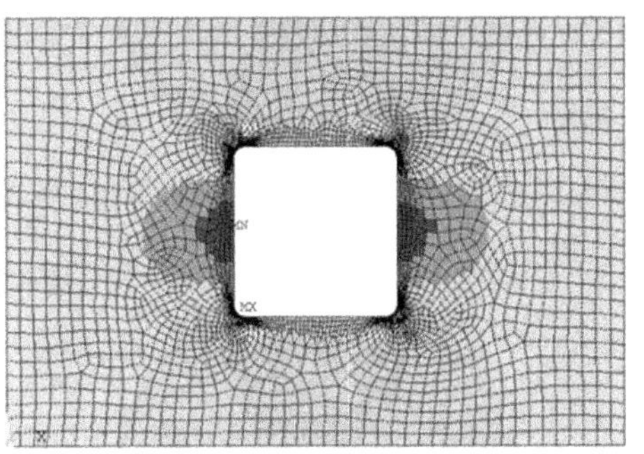

Figure 6-1-27 Nodal (left) and Element (right) Thermal Flux Contour Plot for the Fifth Mesh (Win32 Graphics).

29-8. Retrieve and store the maximum nodal thermal flux in the plate in ex61results(5,2).

29-9. Plot and review the element thermal flux distribution in the plate.

The element thermal flux distribution for the fifth model is shown in Figure 6-1-27 (right).

29-10. Retrieve and store the maximum element thermal flux in the plate in ex61results(5,3).

29-11. List the TEPC in the plate.

29-12. Retrieve and store the TEPC in the plate in ex61results(5,4).

29-13. Plot and review the TERR distribution in the plate.

The thermal energy error distribution for the fifth model is shown in Figure 6-1-28.

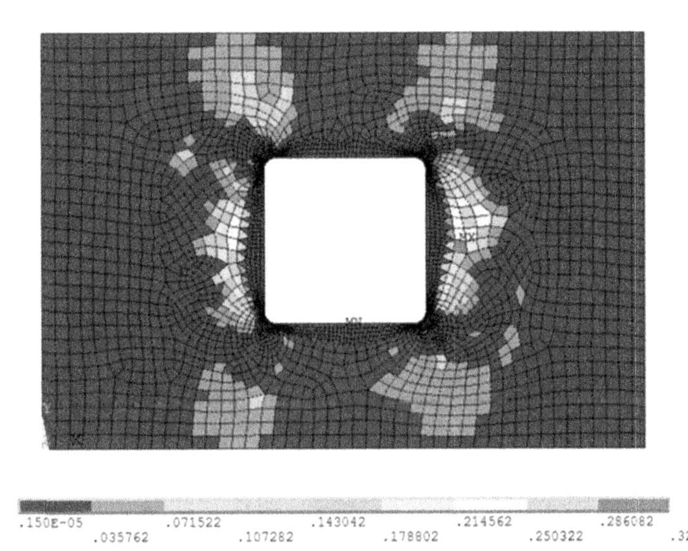

Figure 6-1-28 Plot of Thermal Energy Error for the Fifth Mesh (Win32 Graphics).

29-14. Retrieve and store the maximum TERR in the plate in ex61results(5,5).

29-15. Confirm that all array values have been stored correctly.

29-16. Save the Array Parameter.

Step 30: Postprocess the Results of Model #6

30-1. Clear the database.

30-2. Change the jobname back to Exercise6-1-mesh6.

30-3. Resume the database for Model #6.

30-4. Load the saved parameters.

30-5. Confirm that the array values have been loaded correctly.

30-6. Plot and review the temperature distribution in the plate.

30-7. Plot and review the nodal thermal flux distribution in the plate.

The nodal thermal flux distribution for the sixth model is shown in Figure 6-1-29 (left).

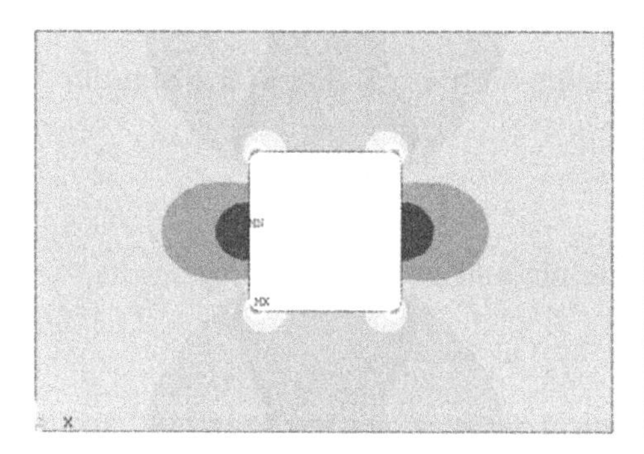

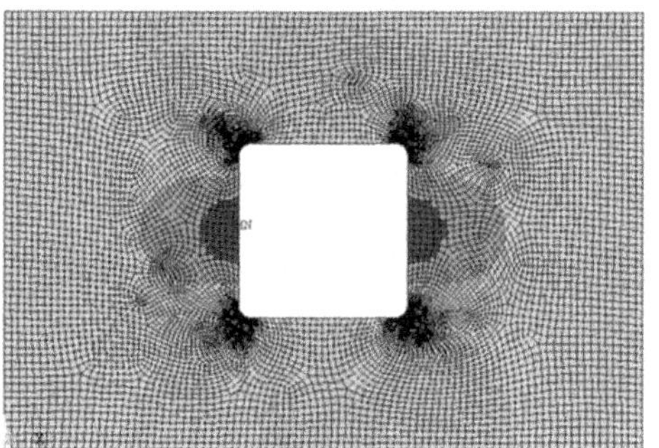

Figure 6-1-29 Nodal (left) and Element (right) Thermal Flux Contour Plot for the Fifth Mesh (Win32 Graphics).

30-8. Retrieve and store the maximum nodal thermal flux in the plate in ex61results(6,2).

30-9. Plot and review the element thermal flux distribution in the plate.

The element thermal flux distribution for the sixth model is shown in Figure 6-1-29 (right). Notice that even for a very fine mesh, the nodal and element solutions do not correspond exactly.

30-10. Retrieve and store the maximum element thermal flux in the plate in ex61results(6,3).

30-11. List the TEPC in the plate.

30-12. Retrieve and store the TEPC in the plate in ex61results(6,4).

30-13. Plot and review the TERR distribution in the plate.

The thermal energy error distribution for the sixth model is shown in Figure 6-1-30.

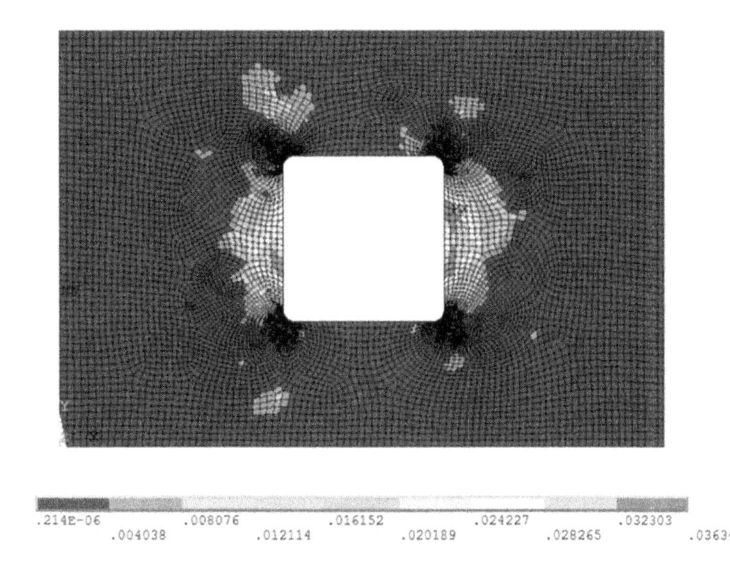

Figure 6-1-30 Plot of Thermal Energy Error for the Sixth Mesh (Win32 Graphics).

30-14. Retrieve and store the maximum TERR in the plate in ex61results(6,5).

30-15. Confirm that all array values have been stored correctly.

30-16. Save the Array Parameter.

This is another convenient stopping point in the exercise if you need a break. All remaining work uses the saved array parameter.

The following operations are also very advanced. These instructions are provided only for your information.

Export the Data from the ex61results Array

This step saves the data in your array to a plain text file in your working directory. The ANSYS Parametric Design Language follows Fortran conventions. Because Fortran is column major, array data is stored in linear memory by incrementing the row index first (down the column) and then by incrementing the column index (across the row). When used as arguments in ANSYS commands, APDL arrays are usually called by specifying only the first entry in the column (see below).

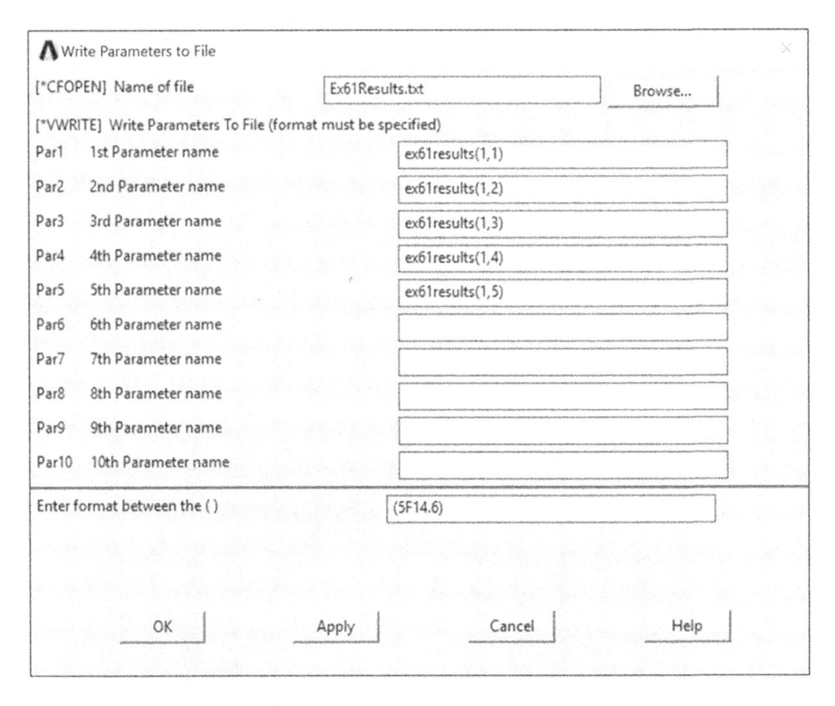

Figure 6-1-31 Write Parameters to File Dialog Box.

- Utility Menu > Parameters > Array Parameters > Write to File . . .
- In the "[*CFOPEN] Name of file" box enter "Ex61Result.txt" (Figure 6-1-31)
- For "Par1 1st Parameter name" enter "ex61results(1,1)"
- For "Par2 2nd Parameter name" enter "ex61results(1,2)"
- For "Par3 3rd Parameter name" enter "ex61results(1,3)"
- For "Par4 4th Parameter name" enter "ex61results(1,4)"
- For "Par5 5th Parameter name" enter "ex61results(1,5)"
- Where it says "Enter format between the ()" enter "(5F14.6)"
- Click OK

The file format for export is specified using Fortran syntax. The leading "5" specifies the number of times to repeat the operation. The "F" specifies real numbers in decimal form. The 14 specifies input up to 14 total digits including the decimal point. The "6" specifies output with 6 decimal places.

Figure 6-1-32 shows the contents of the newly created file (Ex61Results.txt).

Ex61Results - Notepad

File Edit Format View Help

```
   183.000000  64582.992299  67180.703737     14.620311    236.186295
   384.000000  66477.408092  67115.224671      5.629633      6.122214
  1087.000000  69281.490740  69554.336086      3.730326      1.459382
  2488.000000  69960.771703  70047.006543      2.683391      0.499117
  4365.000000  69972.478015  69982.362163      2.242958      0.321842
  9948.000000  69968.679320  69997.803372      1.353811      0.036341
```

Figure 6-1-32 Plain Text File From the Array Export.

Plot and Compare the Mesh Refinement Results

You can plot the results that were stored in the ex61results array using the Time History Postprocessor (POST26) or you can export the data and plot it using another program. This section provides instructions for both options.

Step 31: Plot the nodal and element heat flux versus the number of elements in the model

31-1. Enter the Time History Postprocessor

- Main Menu > TimeHist Postpro
- Close any dialog boxes that open

31-2. Allocate storage for the data to be plotted

- Main Menu > TimeHist Postpro > Store Data
- Change "Lab New data will" to "Allocation only"
- Set the "NPTS For allocation option" to 6
- Click OK

This step tells the program to expect six data points in each new variable that will be defined.

31-3. Move the data from the ex61results array to POST26 storage

- TimeHist Postpro > Table Operations > Parameter to Var
- In the "Par Array parameter" box enter "ex61results(1,1)"
- Enter 1 in the "IR Variable containing data" box
- Click Apply
- In the "Par Array parameter" box enter "ex61results(1,2)"
- Enter 2 in the "IR Variable containing data" box
- Click Apply
- In the "Par Array parameter" box enter "ex61results(1,3)"
- Enter 3 in the "IR Variable containing data" box
- Click Apply
- In the "Par Array parameter box" enter "ex61results(1,5)"
- Enter 5 in the "IR Variable containing data" box
- Click OK

This step stores each of the columns from the ex61results array as a variable that can be used as the x- or y-axis variable in a plot.

31-4. Set the number of elements (IR = 1) as the independent (*x*) variable for the plot

- Main Menu > TimeHist Postpro > Settings > Graph
- In the "[XVAR] X-axis variable" section, change the radio button from "Time (or freq)" to "Single variable"
- Set the "Single variable" number to 1
- Click OK

This tells the program to use variable 1 (the number of elements) as the x-axis variable in the next plot.

31-5. Plot the maximum nodal and element thermal flux versus the number of elements in the model

- Main Menu > TimeHist Postpro > Graph Variables
- Enter 2 in the "NVAR1 1st variable to graph" box
- Enter 3 in the "NVAR2 2nd variable" box
- Click OK

This tells the program to create a graph that uses variables 2 (maximum nodal thermal flux) and 3 (maximum element thermal flux) as the y-axis variables.

The plot of the maximum nodal and element thermal fluxes versus the number of elements in the model is shown in Figure 6-1-33 (left).

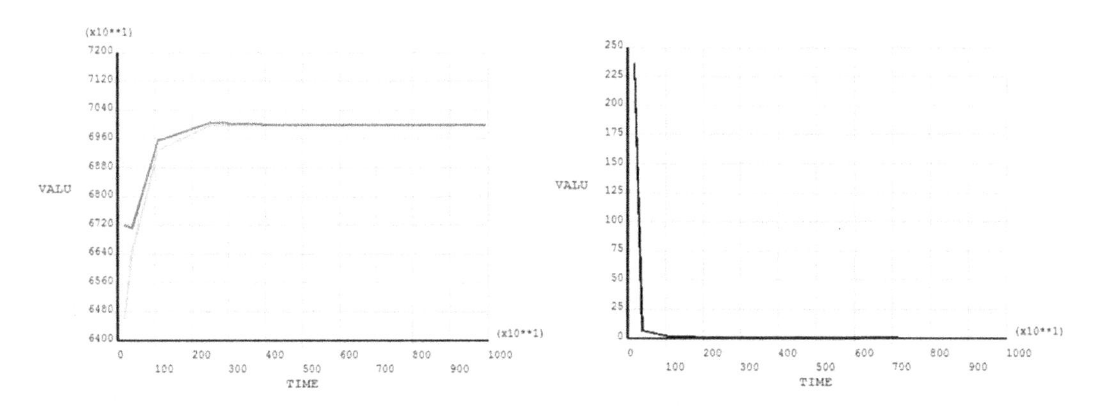

Figure 6-1-33 Plot of the Maximum Nodal and Element Thermal Flux (left) and Plot of the Maximum Thermal Energy Error (right) vs. the Number of Elements in the Model.

31-6. Plot the maximum TERR versus the number of elements in the model

- Main Menu > TimeHist Postpro > Graph Variables
- Enter 5 in the "NVAR1 1st variable to graph" box
- Ensure that the "NVAR 2nd variable" box is empty
- Click OK

The plot of the maximum thermal energy error versus the number of elements in the model is shown in Figure 6-1-33 (right).

Discussion

Table 6-1-1 shows the five parameters of interest from the six models created for this exercise. Because each meshing operation creates a slightly different mesh, your models may have slightly different results. Older versions of ANSYS may produce substantially different results, especially for the coarser meshes.

These results can be evaluated from four different perspectives. First, you can calculate the percentage difference between the nodal and element results for each of the six models (using the nodal solution as the basis) as a measure of solution continuity: ((nodal-element)/nodal). These calculations are shown on the right side of in Table 6-1-1. Second, if you assume that the nodal results from the model with the finest mesh represent the 'true' solution, you can calculate the percentage difference between the results of the finest model and the other models as a measure of solution quality ((fine-coarse)/fine). These calculations for both the nodal and element results are shown in Table 6-1-2. Third, you can compare the TEPC to the 5% limit proposed in the literature. Finally, you can see if the maximum TERRs are tending towards zero (Table 6-1-1 and Figure 6-1-33, right).

Table 6-1-1 Results Parameters From the Six Meshes

Mesh	Number of Elements	Maximum Nodal Thermal Flux	Maximum Element Thermal Flux	Thermal Energy Percent Error	Maximum TERR	% Difference Between Nodal and Element Solutions
1	183	64,583	67,181	14.62	236.19	− 4.02
2	384	66,477	67,115	5.63	6.12	− 0.96
3	1087	69,281	69,554	3.73	1.46	− 0.39
4	2488	69,961	70,047	2.68	0.50	− 0.12
5	4365	69,972	69,982	2.24	0.32	− 0.01
6	9948	69,969	69,998	1.35	0.036	− 0.04

Table 6-1-2 Comparison of Results From the Six Meshes

Comparison Between Meshes	% Difference Between Nodal Solutions	% Difference Between Element Solutions
6 and 1	7.70	4.02
6 and 2	4.99	4.12
6 and 3	0.98	0.63
6 and 4	0.01	− 0.070
6 and 5	− 0.004	0.023

These tables show that the first model did not produce good results. The maximum thermal flux in the nodal and element solutions differed by more than 4%, indicating that the mesh was insufficient to generate a continuous solution. In addition, the maximum thermal flux for the nodal solution for the first mesh was almost 8% less than the solution from final mesh, whereas the element solution differed from the final element solution by more than 4%. These differences are clearly shown in the plot in Figure 6-1-33, left. The thermal energy percent error (14.6%) was well over the 5% limit from the literature. And the maximum TERR was four orders of magnitude greater than the energy error for the final model.

The second model that was produced using a Smart Size of 5 performed better. The difference between the nodal and element solution was less than 1%. The nodal solution differed from the final solution by just under 5%, whereas the element solution differed from the final solution by just over 4%. And, the maximum thermal energy percent error was just over the 5% limit.

The models produced using a Smart Size of 3 and smaller all demonstrate reasonable mesh convergence. The differences between the nodal and element solutions are all less than 1%. The differences between the solutions for the coarser models and the finest model are on the order of 1% or less. The thermal energy percent errors are all well below the 5% limit. And, the maximum thermal fluxes and the maximum TERR have all begun to approach their asymptotic final values. Below a Smart Size of 3, refining the mesh for this model adds cost but no value.

Close the Program

- Utility Menu > File > Exit . . .

Sample Input File

```
*DIM,EX61RESULTS,ARRAY,6,5,1    ! Create a 6x5x1 array named "ex61results"
/FILNAME,Exercise6-1-mesh1      ! Change the jobname to Exercise6-1-mesh1
/PREP7                          ! Enter the Preprocessor
BLC4,0,0,0.15,0.10              ! Create 0.15x0.1 m rectangle w/corner at (0,0)
BLC5,0.075,0.05,0.04,0.04       ! Create 0.04x0.04 m rect. center at (0.075,0.05)
ADELE,ALL                       ! Delete areas only; leave lines and keypoints
LFILLT,5,6,0.003                ! Create 0.003 m fillet between Lines 5 and 6
LFILLT,6,7,0.003                ! Create 0.003 m fillet between Lines 6 and 7
LFILLT,7,8,0.003                ! Create 0.003 m fillet between Lines 7 and 8
LFILLT,8,5,0.003                ! Create 0.003 m fillet between Lines 8 and 5
AL,ALL                          ! Create an area from all lines in the model
ET,1,PLANE55                    ! Use PLANE55 elements
MP,KXX,1,40                     ! Define thermal conductivity for material #1
AMESH,ALL                       ! Mesh all areas in the model
*GET,ex61results(1,1),ELEM,,NUM,MAX  ! Retrieve the maximum element number
/SOLU                           ! Enter the Solution processor
LSEL,S,LOC,X,0                  ! Select line at x=0
DL,ALL,,TEMP,400,               ! Apply Temperature constraint of 400 C to line
LSEL,S,LOC,X,0.15               ! Select line at x=0.15
DL,ALL,,TEMP,300,               ! Apply Temperature constraint of 300 C to line
ALLSEL                          ! Select everything
SOLVE                           ! Solve the model
SAVE                            ! Save the database
FINISH                          ! Exit the Solution Processor


/FILNAME,Exercise6-1-mesh2      ! Change the jobname to Exercise6-1-mesh2
/PREP7                          ! Enter the Preprocessor
ACLEAR,ALL                      ! Clear all area meshes
SMRT,5                          ! Use a Smart Size of 5
AMESH,ALL                       ! Mesh all areas in the model
*GET,ex61results(2,1),ELEM,,NUM,MAX  ! Retrieve the maximum element number
/SOL                            ! Enter the Solution processor
ALLSEL,ALL                      ! Select everything
SOLVE                           ! Solve the model
SAVE                            ! Save the database
FINISH                          ! Exit the Solution Processor
```

```
/FILNAME,Exercise6-1-mesh3     ! Change the jobname to Exercise6-1-mesh3
/PREP7                         ! Enter the Preprocessor
ACLEAR,ALL                     ! Clear all area meshes
SMRT,3                         ! Use a Smart Size of 3
AMESH,ALL                      ! Mesh all areas in the model
*GET,ex61results(3,1),ELEM,,NUM,MAX  ! Retrieve the maximum element number
/SOL                           ! Enter the Solution processor
ALLSEL,ALL                     ! Select everything
SOLVE                          ! Solve the model
SAVE                           ! Save the database
FINISH                         ! Exit the Solution Processor


/FILNAME,Exercise6-1-mesh4     ! Change the jobname to Exercise6-1-mesh4
/PREP7                         ! Enter the Preprocessor
ACLEAR,ALL                     ! Clear all area meshes
SMRT,1                         ! Use a Smart Size of 1
AMESH,ALL                      ! Mesh all areas in the model
*GET,ex61results(4,1),ELEM,,NUM,MAX  ! Retrieve the maximum element number


/SOL                           ! Enter the Solution processor
ALLSEL,ALL                     ! Select everything
SOLVE                          ! Solve the model
SAVE                           ! Save the database
FINISH                         ! Exit the Solution Processor

/FILNAME,Exercise6-1-mesh5     ! Change the jobname to Exercise6-1-mesh5
/PREP7                         ! Enter the Preprocessor
LSEL,S,LOC,X,0.05,0.10         ! Select lines between x=0.05 and x=0.10
LSEL,R,LOC,Y,0.025,0.075       ! Also select lines between y=0.025 and 0.075
LREFINE,ALL,,,1                ! Level 1 refinement for all selected lines
*GET,ex61results(5,1),ELEM,,NUM,MAX  ! Retrieve the maximum element number
/SOL                           ! Enter the Solution processor
ALLSEL,ALL                     ! Select everything
SOLVE                          ! Solve the model
SAVE                           ! Save the database
FINISH                         ! Exit the Solution Processor

/FILNAME,Exercise6-1-mesh6     ! Change the jobname to Exercise6-1-mesh6
/PREP7                         ! Enter the Preprocessor
ACLEAR,ALL                     ! Clear all area meshes
SMRT,1                         ! Use a Smart Size of 1
AMESH,ALL                      ! Mesh all areas in the model
EREF,ALL,,,1                   ! Level 1 refinement for all elements
*GET,ex61results(6,1),ELEM,,NUM,MAX  ! Retrieve the maximum element number
/SOL                           ! Enter the Solution processor
ALLSEL,ALL                     ! Select everything
SOLVE                          ! Solve the model
SAVE                           ! Save the database
FINISH                         ! Exit the Solution processor
PARSAVE,ALL, 'Exercise6-1-Parameters'  ! Save all parameters in the model
```

```
/CLEAR
/FILNAME,Exercise6-1-mesh1      ! Change the jobname to Exercise6-1-mesh1
RESUME                          ! Resume the database
PARRES,CHANGE,'Exercise6-1-Parameters'  ! Load the array parameters
/POST1                          ! Enter the General Postprocessor
PLNSOL,TEMP,,0                  ! Plot the nodal temperature
PLNSOL,TF,SUM,0                 ! Plot the nodal thermal flux
*GET,ex61results(1,2),PLNSOL,,MAX  ! Retrieve the max nodal thermal flux
PLESOL,TF,SUM,0                 ! Plot the element thermal flux
*GET,ex61results(1,3),PLNSOL,,MAX  ! Retrieve the max element thermal flux
*GET,ex61results(1,4),PRERR,,TEPC  ! Retrieve thermal energy percent error
/GRAPHICS,OFF                   ! Turn Power Graphics off
PLESOL,TERR,,0                  ! Plot the thermal energy error
*GET,ex61results(1,5),PLNSOL,,MAX  ! Retrieve the max thermal energy error
PARSAVE,ALL,'Exercise6-1-Parameters'  ! Save all parameters in the model
FINISH                          ! Exit the Postprocessor

/CLEAR
/FILNAME,Exercise6-1-mesh2      ! Change the jobname to Exercise6-1-mesh2
RESUME                          ! Resume the database
PARRES,CHANGE,'Exercise6-1-Parameters'  ! Load the array parameters
/POST1                          ! Enter the General Postprocessor
PLNSOL,TEMP,,0                  ! Plot the nodal temperature
PLNSOL,TF,SUM,0                 ! Plot the nodal thermal flux
*GET,ex61results(2,2),PLNSOL,,MAX  ! Retrieve the max nodal thermal flux
PLESOL,TF,SUM,0                 ! Plot the element thermal flux
*GET,ex61results(2,3),PLNSOL,,MAX  ! Retrieve the max element thermal flux
*GET,ex61results(2,4),PRERR,,TEPC  ! Retrieve thermal energy percent error
/GRAPHICS,OFF                   ! Turn Power Graphics off
PLESOL,TERR,,0                  ! Plot the thermal energy error
*GET,ex61results(2,5),PLNSOL,,MAX  ! Retrieve the max thermal energy error
PARSAVE,ALL,'Exercise6-1-Parameters'  ! Save all parameters in the model
FINISH                          ! Exit the Postprocessor

/CLEAR
/FILNAME,Exercise6-1-mesh3      ! Change the jobname to Exercise6-1-mesh3
RESUME                          ! Resume the database
PARRES,CHANGE,'Exercise6-1-Parameters'  ! Load the array parameters
/POST1                          ! Enter the General Postprocessor
PLNSOL,TEMP,,0                  ! Plot the nodal temperature
PLNSOL,TF,SUM,0                 ! Plot the nodal thermal flux
*GET,ex61results(3,2),PLNSOL,,MAX  ! Retrieve the max nodal thermal flux
PLESOL,TF,SUM,0                 ! Plot the element thermal flux
*GET,ex61results(3,3),PLNSOL,,MAX  ! Retrieve the max element thermal flux
*GET,ex61results(3,4),PRERR,,TEPC  ! Retrieve thermal energy percent error
/GRAPHICS,OFF                   ! Turn Power Graphics off
PLESOL,TERR,,0                  ! Plot the thermal energy error
*GET,ex61results(3,5),PLNSOL,,MAX  ! Retrieve the max thermal energy error
PARSAVE,ALL,'Exercise6-1-Parameters'  ! Save all parameters in the model
FINISH                          ! Exit the Postprocessor

/CLEAR
/FILNAME,Exercise6-1-mesh4      ! Change the jobname to Exercise6-1-mesh4
RESUME                          ! Resume the database
PARRES,CHANGE,'Exercise6-1-Parameters'  ! Load the array parameters
```

```
/POST1                          ! Enter the General Postprocessor
PLNSOL,TEMP,,0                  ! Plot the nodal temperature
PLNSOL,TF,SUM,0                 ! Plot the nodal thermal flux
*GET,ex61results(4,2),PLNSOL,,MAX  ! Retrieve the max nodal thermal flux
PLESOL,TF,SUM,0                 ! Plot the element thermal flux
*GET,ex61results(4,3),PLNSOL,,MAX  ! Retrieve the max element thermal flux
*GET,ex61results(4,4),PRERR,,TEPC  ! Retrieve thermal energy percent error
/GRAPHICS,OFF                   ! Turn Power Graphics off
PLESOL,TERR,,0                  ! Plot the thermal energy error
*GET,ex61results(4,5),PLNSOL,,MAX  ! Retrieve the max thermal energy error
PARSAVE,ALL,'Exercise6-1-Parameters'  ! Save all parameters in the model
FINISH                          ! Exit the Postprocessor

/CLEAR
/FILNAME,Exercise6-1-mesh5      ! Change the jobname to Exercise6-1-mesh5
RESUME                          ! Resume the database
PARRES,CHANGE,'Exercise6-1-Parameters'  ! Load the array parameters
/POST1                          ! Enter the General Postprocessor
PLNSOL,TEMP,,0                  ! Plot the nodal temperature
PLNSOL,TF,SUM,0                 ! Plot the nodal thermal flux
*GET,ex61results(5,2),PLNSOL,,MAX  ! Retrieve the max nodal thermal flux
PLESOL,TF,SUM,0                 ! Plot the element thermal flux
*GET,ex61results(5,3),PLNSOL,,MAX  ! Retrieve the max element thermal flux
*GET,ex61results(5,4),PRERR,,TEPC  ! Retrieve thermal energy percent error
/GRAPHICS,OFF                   ! Turn Power Graphics off
PLESOL,TERR,,0                  ! Plot the thermal energy error
*GET,ex61results(5,5),PLNSOL,,MAX  ! Retrieve the max thermal energy error
PARSAVE,ALL,'Exercise6-1-Parameters'  ! Save all parameters in the model
FINISH                          ! Exit the Postprocessor

/CLEAR
/FILNAME,Exercise6-1-mesh6      ! Change the jobname to Exercise6-1-mesh6
RESUME                          ! Resume the database
PARRES,CHANGE,'Exercise6-1-Parameters'  ! Load the array parameters
/POST1                          ! Enter the General Postprocessor
PLNSOL,TEMP,,0                  ! Plot the nodal temperature
PLNSOL,TF,SUM,0                 ! Plot the nodal thermal flux
*GET,ex61results(6,2),PLNSOL,,MAX  ! Retrieve the max nodal thermal flux
PLESOL,TF,SUM,0                 ! Plot the element thermal flux
*GET,ex61results(6,3),PLNSOL,,MAX  ! Retrieve the max element thermal flux
*GET,ex61results(6,4),PRERR,,TEPC  ! Retrieve thermal energy percent error
/GRAPHICS,OFF                   ! Turn Power Graphics off
PLESOL,TERR,,0                  ! Plot the thermal energy error
*GET,ex61results(6,5),PLNSOL,,MAX  ! Retrieve the max thermal energy error
PARSAVE,ALL,'Exercise6-1-Parameters'  ! Save all parameters in the model
FINISH                          ! Exit the Postprocessor

/POST26                         ! Enter the Time History Postprocessor
STORE,ALLOC,6                   ! Store data for 6 variables
VPUT,ex61results(1,1),1         ! Move vector 1 into the variable
VPUT,ex61results(1,2),2         ! Move vector 2 into the variable
VPUT,ex61results(1,3),3         ! Move vector 3 into the variable
```

```
VPUT,ex61results(1,5),5        ! Move vector 5 into the variable
XVAR,1                         ! Specify the x variable for the graph
PLVAR,2,3                      ! Specify the y variables and graph
XVAR,1                         ! Specify the x variable for the graph
PLVAR,5                        ! Specify the y variable and graph

! Write the parameter array to an external file
*CREATE,ansuitmp               ! Create a command macro file
*CFOPEN,'Ex61Results','txt',   ! Open a command file
*VWRITE,ex61results(1,1),ex61results(1,2),ex61results(1,3),ex61results(1,4),
  ex61results(1,5),,,,         ! Write the data to a file
(5F14.6)                       ! With this format sequence
*CFCLOS                        ! Close the command file
*END                           ! Close the macro file
/INPUT,ansuitmp                ! Run the macro file

FINISH                         ! Finish and Exit Postprocessor
!/EXIT                         ! Exit ANSYS
```

Chapter

7

Selecting Entities

Suggested Reading Assignments:
Mechanical APDL Basic Analysis Guide: Chapter 8
Mechanical APDL Operations Guide: Chapter 5

CHAPTER OUTLINE

This chapter provides an overview of entity specification, selection, and picking in ANSYS. You will learn about the selection methods and options available in ANSYS, how to use the Select menu and select commands, and how to save sets of selected entities for future use. You will also learn how to use the Picker, how the Picker operates, and how to interpret the commands issued by the Picker.

7.1. Specifying Entities in ANSYS

Most ANSYS commands operate on and/or associate various types of information with solid model or finite element model entities. These entities can be specified in one of three ways.

First, the entity numbers can be specified as arguments in a command. For example, the command `L,1,2` creates a line between Keypoints 1 and 2. Similarly, `ASBA,1,2` subtracts Area 2 from Area 1. Specifying entity numbers directly does not change the active set of entities in the model (see section 7.2.3 for more details).

Second, entity numbers can be supplied as command arguments via graphical picking. When entity-related arguments are required for a command that is issued through the GUI, ANSYS opens a menu called the Picker. The Picker allows you to click on the desired entities with the mouse or to enter their entity numbers (Keypoint 2, Lines 3 and 5, Nodes 5 through 10, etc.) in a text box. The numbers of the picked entities are collected and supplied to the command as a single argument. The effect is the same as supplying the entity number(s) directly to the

ANSYS Mechanical APDL for Finite Element Analysis.
DOI: http://dx.doi.org/10.1016/B978-0-12-812981-4.00007-1

command. However, this method does not require you to know or use the command syntax. It does not require you to know the numbers of the desired entities. And, it does not require the entity numbers to be sequential. Picking does not change the active set of entities in the model.

Finally, you can select the desired entities to create a new active set of entities in the model and then operate on that set. Some commands only operate on the active set. These commands do not require and will not accept any entity-related arguments. For example, the **AATT** command assigns the specified element attributes to all areas in the active set. For commands that require entity-related arguments, such as the **AMESH** command, the argument supplied to the command is often "ALL." For example, `AMESH,1` will mesh only Area 1, while `AMESH,ALL` meshes all areas in the active set.

Although entity specification is required for many aspects of finite element modeling including solid modeling, meshing, and postprocessing, it is most critical for the application of boundary conditions and initial conditions. Since boundary conditions and initial conditions are applied in the Solution processor, selection is discussed in chapter 7 before Solution is discussed in chapter 8.

7.2. Selection Overview

Selection in ANSYS requires three pieces of information:

- The type of entity to select (keypoint, line, area, volume, node, element, or component).
- The selection method (by num/pick, by location, by attributes, attached to, etc.).
- The set operation (from full, reselect, also select, unselect, select all, select none, or invert).

7.2.1. Type of Entity

Selection requires you to identify the entity type to be selected. The selection status of each entity type in ANSYS is independent of the other entity types. Selecting an entity has no effect on the other entities that are attached to or contained within it.

7.2.2. Selection Methods

The selection method specifies how the desired entity or entities should be selected. Each method requires, or will accept, different arguments to fine-tune its behavior.

7.2.2.1. Selecting "By Num/Pick"

The "By Num/Pick" option allows you to select entities based on their entity number or to "pick" entities using the mouse. This is the most intuitive and interactive selection method. However, it is also the least preferred for two reasons. First, as you have seen in the exercises from the previous chapters, graphical picking is imprecise and can be difficult in 3D models. Second, entity numbers are not stable. Many modeling and meshing operations change the entity numbering within the model. In addition, the original entity numbers assigned can depend on the version of ANSYS being used.

7.2.2.2. Selecting "By Location"

The "By Location" option allows you to select entities based on their location in the active coordinate system. You can only specify a value or a range of values for one coordinate (*x, y, z,* or equivalent) per command. If there are no entities at a specified location, nothing will be selected. For example, if you specify a location that falls between two nodes, a new empty set of nodes will be created. Tolerances exist for selecting by location (and can be adjusted using the **SELTOL** command), but these tolerances are small. You should ensure that your selection coordinate values are sufficient to select all of the entities required. Selecting by location is one of the most robust selection methods and is strongly recommended.

7.2.2.3. Selecting "By Attributes"

The "By Attributes" option allows you to select entities based on their element (mesh) attributes. The available attributes depend on the entity type to be selected but generally include material reference number, element type number, real constant set number, element coordinate system number, and section number. This is a robust selection method and is strongly recommended for models whose attributes have been assigned.

7.2.2.4. Selecting Entities "Attached To" Other Entities

The "Attached To" option allows you to select entities that are attached to other entities. For example, you can select all of the nodes attached to a given element or a set of elements. This is a nested type of selection. It is only meaningful if it follows a selection operation to create an active set of the entity type to which the new set is attached. Each entity type has a limited number of other entity types to which it can be attached for selection purposes. For example:

- Keypoints can be attached only to nodes and lines.
- Lines can be attached only to keypoints and areas.
- Areas can be attached only to lines and volumes.
- Volumes can be attached only to areas.
- Nodes can be attached to keypoints, lines, areas, volumes, and elements.
- Elements can be attached only to lines, areas, volumes, and/or nodes depending on the dimensions of the entities involved. (For example, 3D elements cannot be selected based on the lines they are "attached to".)

Each combination of entity, the entity to which it is attached, and the selection option chosen (interior, exterior, any, all, etc.) produces a different result. You should check the documentation carefully before using one of these select commands. You should also list or plot the selected entities after using one of these commands to ensure that the selected set is as intended. The "Attached To" selection option is recommended when selecting small entities that are difficult to pick with the mouse and for large sets of entities that cannot be conveniently selected by location.

7.2.2.5. Other Selection Methods

Additional selection methods exist for specific entity types. For example, lines and areas can be selected based on their concatenation status. Similarly, nodes and elements can be selected based on their results after a solution has been calculated. The full list of available selection methods can be found in the documentation for each select command.

7.2.3. Set Operations

The selection process in ANSYS is built upon set theory. When entities are created, they become part of the full set of that entity type. For example, if Keypoints 1, 2, 3, and 4 exist, the full set of keypoints in the model consists of those four keypoints. Creating Keypoints 5 and 6 adds those keypoints to the full set of keypoints defined for that analysis. Selecting entities creates an active subset of that entity type (i.e., sets the selection status of those entities to +1 or leaves them at +1) and inactivates all other entities from the full set (i.e., sets the selection status of the remaining entities to −1). For example, selecting Keypoints 1 through 4 will set the selection status of Keypoints 5 and 6 to −1. By default, the full set of each entity type is active.

There are seven set operations available in ANSYS:

- From Full—Create a new active subset from the full set of entities in the model
- Reselect—Select from the currently active subset to create a smaller active subset
- Also Select—Add new entities to the currently active subset
- Unselect—Remove entities from the currently active set
- Select All—Select all entities in the model (the full set becomes the active set)
- Select None—Unselect all entities in the model (create an empty active set)
- Invert—Select all currently unselected entities and unselect all currently selected entities

These operations are illustrated in Figure 8.1 of the Mechanical APDL Basic Analysis Guide.

7.3. Selecting Entities in ANSYS

There are two ways to select entities in ANSYS: you can use the Select menu located within the Utility Menu or you can issue select commands directly to the program.

7.3.1. The Select Menu

The Select menu can be accessed using the GUI path: **Utility Menu > Select > Entities**. This brings up a dialog box with four sections (Figure 7.1).

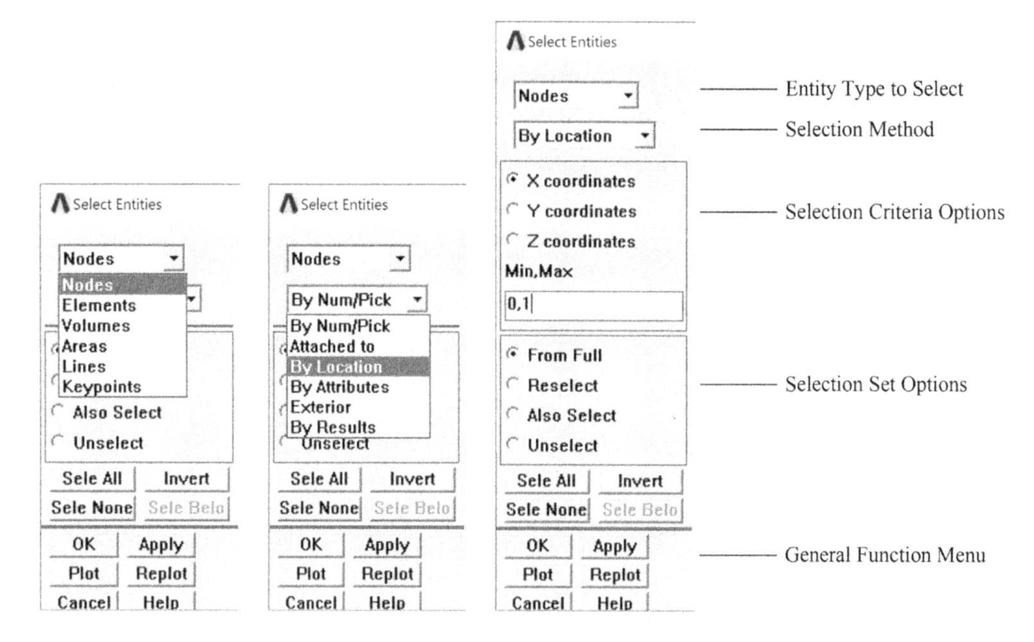

Figure 7.1 The Select Entities Dialog Box.

The first section contains two drop down menus. The upper drop down menu allows you to choose the entity type to select. The lower drop down menu allows you to choose the selection method to use. The second section allows you to refine the choices made in the first section by setting the selection criteria options. The specific options available depend on the entity type and the selection method chosen. The third section allows you to specify the selection set operation to perform (select from full, reselect, unselect, etc.). The fourth section is the general function menu that allows you to accept the selection options, to plot or replot the selected entities, to cancel the selection process, and to display the documentation.

If "By Num/Pick" is chosen in the second drop down menu, clicking "Apply" or "OK" will open the Picker. Otherwise, the selection operation will be performed immediately. The Graphics Window will not update to show the newly selected entities. You will need to (re)plot or list the selected entity type to see the changes.

7.3.2. Select Commands

Selection also can be performed using commands. All select commands are global commands. They can be issued in any processor and at the Begin Level. As noted above, each select command performs one operation. If you need to select multiple entity types, use multiple selection methods, use multiple selection criteria, or perform multiple set operations, you must issue multiple select commands or their GUI equivalents.

There are three types of select commands: basic select commands, crossover select commands, and the Select All command.

7.3.2.1. Basic Select Commands

There is one basic select command for each entity type. They are:

- **KSEL** — Select keypoints
- **LSEL** — Select lines
- **ASEL** — Select areas
- **VSEL** — Select volumes
- **NSEL** — Select nodes
- **ESEL** — Select elements

Each of these commands has the form: **KSEL**, *TYPE, ITEM, COMP, VMIN, VMAX, VINC, [].* The first argument (*TYPE)* specifies the set operation. The available set operations are:

- S—Select a new set (default)
- R—Reselect from the current set (create a subset)
- A—Additionally select (add to the current subset)
- U—Unselect (remove from the current set)
- ALL—Restore the full set
- NONE—Create an empty set
- INVE—Invert the current set
- STAT—Display the current selection status

The second argument (*ITEM*) specifies the selection method. The most common items include:

- KP—Select by keypoint number (**KSEL** only)
- LINE—Select by line number (**LSEL** only)
- AREA—Select by area number (**ASEL** only)
- VOLU—Select by volume number (**VSEL** only)
- NODE—Select by node number (**NSEL** only)
- ELEM—Select by element number (**ESEL** only)
- LOC—Select by location
- MAT—Select by material reference number
- TYPE—Select by element type number
- REAL—Select by real constant set number

The third argument (*COMP*) fine-tunes the selection method and depends on both the command and the *ITEM* argument. If selecting by location, this is where you specify the coordinate to use (*X, Y,* or *Z*). The next three arguments specify the minimum value of the range to select (*VMIN*), the maximum value of the range to select (*VMAX*), and the increment to use (*VINC*). If selecting a single entity, only *VMIN* is required. *VINC* defaults to 1 so every entity within the range (*VMIN* to *VMAX*) is selected unless you specify otherwise. The final argument is specific to the individual command. The full list of available arguments for each command can be found in the Mechanical APDL Command Reference.

7.3.2.2. Crossover Select Commands

In addition, there is one crossover select command for each entity-to-entity combination associated with the "Attached To" selection method. For example, **ASLL** selects the areas attached to the currently selected lines, while **NSLE** selects the nodes attached to the currently selected elements. The full list of crossover select commands can be found in Table 2.6 of the Mechanical APDL Command Reference. Each crossover select command takes one or two arguments. The first argument (*TYPE*) specifies the set operation to perform. The second argument (*LSKEY, ARKEY, VKEY, NKEY,* or *EKEY*) fine-tunes the behavior of the command.

7.3.2.3. The Select All Command

The Select All command (syntax: **ALLSEL**, *LabT, Entity*) can be used to perform three types of selection operations. It can select all entities in the model, select all entities of a given type and "below," or select all entities below the selected entities of a given type. The hierarchy used to determine which entities are "below" is volumes, areas, lines, keypoints, elements, and nodes.

By default, the **ALLSEL** command selects everything in the model. If this is the desired behavior, both arguments are optional. Typing `ALLSEL,ALL,ALL` or `ALLSEL,ALL` or simply `ALLSEL` into the command prompt will set the selection status of every entity in the model to +1.

Supplying "ALL" as the *LabT* argument selects all entities of the type specified and all entities of the same kind (either solid model or finite element model) below it. For example, `ALLSEL,ALL,AREA` selects all areas, lines, and keypoints in the model. `ALLSEL,ALL,ELEM` selects all elements and nodes in the model.

Supplying "BELOW" as the *LabT* argument selects all solid model and finite element model entity types below the selected entities of the type specified. For example, `ALLSEL,BELOW,AREA` selects all lines, keypoints, elements, and nodes associated with the areas in the active set. The "BELOW" option can be thought of as a reselect command since it is only meaningful when selecting from an active subset of entities.

7.3.3. Examples Using Select Commands

This section demonstrates some of the select options and commands discussed above. Example 7.1 recreates the solid model geometry and defines the material properties for the sandwich beam from Example 6.1.

Example 7.1

```
/PREP7                     ! Enter the Preprocessor
RECTNG,0,10,0.9,1          ! Create the top beam
RECTNG,0,10,0.1,0.9        ! Create the middle beam
RECTNG,0,10,0,0.1          ! Create the bottom beam
NUMMRG,ALL                 ! Merge all entities in the model
ET,1,182                   ! Define Plane182 as element #1
MP,EX,1,7.3e10             ! Define Young's modulus for aluminum: material #1
MP,PRXY,1,0.33             ! Define Poisson's ratio for aluminum: material #1
MP,EX,2,3e9                ! Define Young's modulus for acrylic: material #2
MP,PRXY,2,0.4              ! Define Poisson's ratio for acrylic: material #2
```

Example 7.2 presents three methods for assigning the element attributes to this model. Each column contains a series of commands to select Areas 1 and 3, plot the areas in the active set to confirm that the selection operation was successful, and assign material model reference #1. Each set of commands then selects Area 2, plots the areas in the active set to confirm that the selection operation was successful, and assigns material model reference #2. Next, all areas in the model are selected and meshed. Element numbering is turned on and set to material numbers. Finally, the elements in the active (full) set are plotted to ensure that the material properties were correctly assigned.

In the first column, all three areas are selected by area number (*ITEM*=AREA). In the second column, Areas 1 and 3 are selected by selecting the full set of areas in the model (*TYPE*=ALL) and then unselecting Area 2 (*TYPE*=U). Once the material model number is assigned, the active set is inverted to select Area 2 and to unselect Areas 1 and 3 (*ITEM*=INVE). The `ASEL,ALL` command in the first line is unnecessary since all entities in a model are included in the active set by default. However, it is good practice to reset the status of all model entities before performing selection operations. In the third column, all areas between $y=0.1$ and $y=0.9$ are selected (*ITEM*=LOC, *COMP*=Y). Because Area 2 is the only area that meets those criteria, this creates a new active set that contains Area 2. Areas 1 and 3 are selected by inverting the active set. Once the material model number is assigned, the active set is inverted again and the material model number for Area 2 is assigned.

Example 7.2

```
ASEL,S,AREA,,1
ASEL,A,AREA,,3
APLOT
AATT,1,,1
ASEL,S,AREA,,2
APLOT
AATT,2,,1
ASEL,ALL
AMESH,ALL
/PNUM,MAT,1
EPLOT
```

```
ASEL,ALL
ASEL,U,,,2
APLOT
AATT,1,,1
ASEL,INVE
APLOT
AATT,2,,1
ASEL,ALL
AMESH,ALL
/PNUM,MAT,1
EPLOT
```

```
ASEL,S,LOC,Y,0.1,0.9
ASEL,INVE
APLOT
AATT,1,,1
ASEL,INVE
APLOT
AATT,2,,1
ASEL,ALL
AMESH,ALL
/PNUM,MAT,1
EPLOT
```

The resulting mesh is adequate for this model, and refinement is not necessary. However, if we wanted to refine the mesh in Area 2, the elements associated with Area 2 would need to be selected. Example 7.3 shows three options for doing this. In the first column, Area 2 is selected by entity number. All elements attached to the areas in the active set are selected and the elements in the active set are plotted to confirm that the select operation was successful. Finally, these elements are refined using a refinement level of 1. In the second column, the nodes in the active set (all nodes in the model) are plotted. Next, all nodes between $y=0.1$ and $y=0.9$ are selected, and the nodes in the active set are plotted again to show the difference. Then, the elements attached to the nodes in the active set are selected and plotted. Finally, the selected elements are refined. In the third column, all elements associated with material model reference #2 are selected, plotted, and then refined.

Example 7.3

```
ASEL,S,,,2
ESLA,S
EPLOT
EREF,ALL,,,1
```

```
NPLOT
NSEL,S,LOC,Y,0.1,0.9
NPLOT
ESLN,S,1
EPLOT
EREF,ALL,,,1
```

```
ESEL,S,MAT,,2
EPLOT
EREF,ALL,,,1
```

By default, the **ESLN** command selects an element if any of its nodes are contained in the active set. Thus, the command `ESLN,S` or `ESLN,S,0` would select all elements associated with Area 2 and the elements from Areas 1 and 3 that border Area 2 (Figure 7.2, left). Because only the elements associated with Area 2 were to be refined, the *EKEY* argument was set to 1 (`ESLN,S,1`). This ensured that only elements with all of their nodes contained in the active set were selected (Figure 7.2, right).

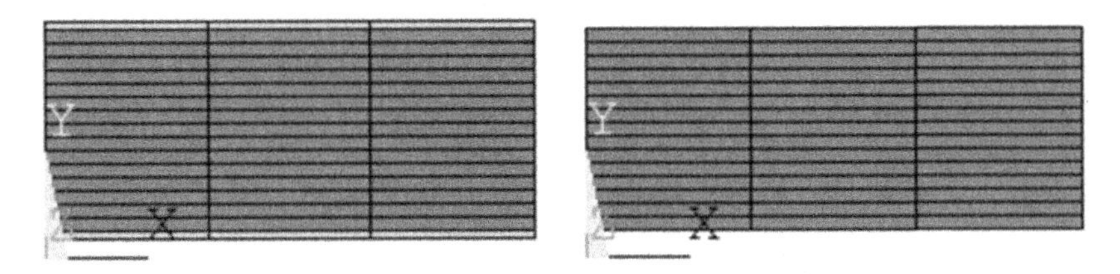

Figure 7.2 Select Elements Attached to Nodes, ANY (`ESLN,S,0`) (left) and Attached to Nodes, ALL (`ESLN,S,1`) (right).

As seen in the examples above, there is usually more than one selection operation or set of selection operations that can be used to accomplish a goal. Some of these options may be more convenient or more robust than the others. Each option may also require a different number of commands. But in general, there is no right or wrong selection method or combination of selection methods to use as long as the selection operation can be performed.

7.3.4. Saving Selection Sets Using Components and Assemblies

Once an active set of entities has been created, it can be named and saved as a component by using the GUI Path: **Utility Menu > Select > Comp/Assembly > Create Component** or by using the **CM** command. Components can be created for any entity type. However, each component may contain only one entity type. Component names must be alphanumeric and can contain up to 32 characters. There is no limit to the number of items that can be included in a component. There is no limit to the number of components that can be created in a model. And, there is no limit to the number of components to which a given entity can belong.

Example 7.4 creates components for the areas and elements associated with the aluminum parts of the sandwich beam. The first two commands select Areas 1 and 3. The third command creates a new area component named "ALUAREA." The fourth command selects the elements attached to the areas in the active set. The final command creates a new element component named "ALUELEM."

Example 7.4

```
ASEL,S,AREA,,1            ! Create a new set that contains Area 1
ASEL,A,AREA,,3            ! Also select Area 3 (add to the active set)
CM,ALUAREA,AREA           ! Create an area component named "ALUAREA"
ESLA,S                    ! Select the elements attached to the areas
CM,ALUELEM,ELEM           ! Create an element component named "ALUELEM"
```

You can group multiple components into an assembly using the GUI path: **Utility Menu > Select > Comp/Assembly > Create Assembly** or the **CMGRP** command. Unlike components, assemblies are not limited to a single entity type. Assemblies can be nested up to five levels deep.

Components and assemblies can be selected like other entities using the GUI path: **Utility Menu > Select > Comp/Assembly > Select Comp/Assembly** or the **CMSEL** command. This makes components and assemblies a convenient way to save an active set for future use.

Example 7.5 demonstrates the selection of previously defined components. First, all entities in the model are selected and the areas in the active set are plotted. Next, the newly created component "ALUAREA" is selected and the areas in the active set are plotted again to show the difference. Then, the model elements are plotted. Finally, the newly created component "ALUELEM" is selected and the elements in the active set are plotted again to show the difference.

Example 7.5

```
ALLSEL,ALL                ! Select all entities in the model
APLOT                     ! Plot the areas in the active (full) set
CMSEL,S,ALUAREA           ! Select the component "ALUAREA"
APLOT                     ! Plot the areas in the new active set
EPLOT                     ! Plot the elements in the active (full) set
CMSEL,S,ALUELEM           ! Select the component "ALUELEM"
EPLOT                     ! Plot the elements in the new active set
```

7.4. The Picker

The Picker can be accessed in three ways: you can issue a command through the GUI that requires entity-related arguments, you can enter "Pick" or "P" in the *ITEM* field of a select command issued through the command prompt, or you can choose the "By Num/Pick" option in the Select menu. The first option initiates a picking operation (i.e., identifies the entities to supply as an argument for a given command). The second and third options initiate a selection operation (i.e., change the active set of entities in the model).

The Picker is composed of 5 sections (Figure 7.3). The top section allows you to specify whether to pick (select) entities to add to the new set or to unpick (unselect) entities from the current set. The second section allows you to specify how to pick (select) the entities.

- The "Single" option allows you to pick one entity at a time.
- The "Box" option allows you to pick all entities within a box drawn by the mouse.
- The "Polygon" option allows you to pick all entities within a polygon drawn by the mouse.
- The "Circle" option allows you to pick all entities within a circle drawn by the mouse.
- The "Loop" option allows you to pick all entities within a loop of edges. This option is only available for entities associated with loops (i.e., lines and areas).

The third section provides feedback about the picking (selection) operation.

- The "Count" field tells you how many entities have been picked.
- The "Maximum" field tells you how many entities may be picked.
- The "Minimum" field tells you how many entities must be picked.
- The "[Entity] No." field tells you the number of the entity that has been picked most recently. In Figure 7.3, the entity number shown is the keypoint number (KeyP No.).

The fourth section allows you to pick (select) by entity number. The "List of Items" option allows you to enter a comma separated list of the entity numbers to pick. The "Min, Max, Inc" option allows you to specify a minimum entity number, a maximum entity number, and an increment value. This will pick all of the entities within that range using that increment. (The increment still defaults to 1.) The fifth section offers general picking controls.

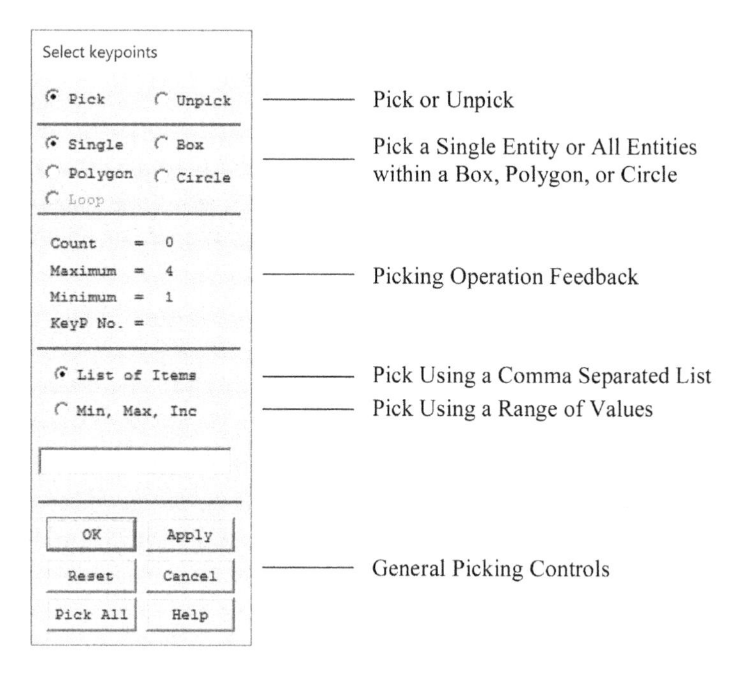

Figure 7.3 The Picker.

If the Picker is used for picking, you will only be able to choose entities in the active set. The "Pick All" option will choose all entities in the active set. After the picking operation, the active set will be the same as it was before picking was initiated.

If the Picker is used for selecting, it will create a new set of entities or modify the active set of entities depending on the set operation chosen. If a new set has been specified in the Select menu, all entities in the model will be available for graphical picking regardless of their previous state. In this case, the "Pick All" option will pick all entities in the full set, making it equivalent to an **ALLSEL** command. If a reselect operation is chosen, only the active set will be available. In this case, "Pick All" will select all of the entities in the active set (i.e., do nothing). After the selection process is complete, the new set will remain active until changed by another selection operation.

7.5. Picker Commands

The internal behavior of the Picker depends on whether it is used for picking or selecting. It also depends on the number of entities to be selected. If a single entity is selected, the Picker issues a single select command with the number of the specified entity supplied as an argument.

Otherwise, a series of commands, called a Picker command block, is issued to produce the desired operation.

You will never need to issue Picker commands directly to the program. Picker command blocks are discussed here so you can identify a Picker block in your log file and replace the Picker commands with select commands when editing input files.

7.5.1. The FLST Helper Command

FLST is always the first command in a picker block. It has the syntax: **FLST**, *NFIELD, NARG, TYPE, Otype, LENG.* The goal of a picking operation is to obtain information to be used in some other command, called the intended command. **FLST** specifies the data required for the picking operation and is essentially a helper command. *NFIELD* indicates the field number (or location in the command syntax) for the picked data in the intended command. *NARG* is the number of items that were picked and *TYPE* refers to the entity type (node, element, keypoint, etc.) that was picked. The most common *TYPE* of items to be picked are as follows:

- 1—Node numbers
- 2—Element numbers
- 3—Keypoint numbers
- 4—Line numbers
- 5—Area numbers
- 6—Volume numbers

The *Otype* field specifies whether the data supplied will be ordered (ORDE) or not (NOOR). Finally, *LENG* specifies the number of items in the list. This number should be equal to *NARG*.

For example, consider a picking or selection operation that selects two areas (Areas 1 and 3) with the **ASEL** command (syntax: ASEL, *Type, Item, Comp, VMIN, VMAX, VINC, KSWP*). The area numbers to be selected are supplied in the fifth field of the **ASEL** commands (*VMIN*), so the value of *NFIELD* would be 5. Two items were picked, so *NARG* would be 2. Since areas were the entity type picked, the *TYPE* would also be 5. The full command would read `FLST,5,2,5,ORDE,2`.

7.5.2. The FITEM Helper Command

The **FLST** command is followed by one or more **FITEM** commands. The syntax for the **FITEM** command is **FITEM**, *NFIELD, ITEM, ITEMY, ITEMZ*, where *NFIELD* is the field of the command for which picking was invoked (the same as above) and *ITEM* is the entity number of the object that was picked. To pick Areas 1 and 3, *NFIELD* would still be 5 for both **FITEM** commands, the *ITEM* argument would be 1 for the first **FITEM** command (e.g., `FITEM,5,1`), and the *ITEM* argument would be 3 for the second **FITEM** command (e.g., `FITEM,5,3`).

If the value provided in the *ITEM* field of the second **FITEM** command is negative, all entity numbers between the first and second **FITEM** command are also picked. For example, `FITEM,2,10` and `FITEM,2,20` will select Element 10 and Element 20 (2 elements), while `FITEM,2,10` and `FITEM,2,-20` will select Elements 10 through 20 (11 elements).

7.5.3. Picker Command Blocks for Selecting Operations

If the Picker is used for selecting, the **FLST** and **FITEM** commands will be followed by a single select command. A “P51X” will take the place of the argument which would usually specify the entity number to select.

Example 7.6 lists the Picker commands for an operation to select Areas 1 and 2. The **FLST** command specifies that two entities will be chosen and they will be specified in the fifth field of the intended command. The first **FITEM** command specifies that entity number 1 will occupy the fifth field of the intended command. The second **FITEM** command specifies that entity number 2 will also occupy the fifth field of the intended command. The negative sign in the second **FITEM** command indicates a range of areas (1 through 2) instead of a list. Finally, the **ASEL** command selects the areas whose entity numbers are supplied by the **FITEM** commands.

Example 7.6

```
FLST,5,2,5,ORDE,2          ! Pick 2 areas for the 5th field of a command
FITEM,5,1                  ! Identify Area 1
FITEM,5,-2                 ! Identify (through) Area 2
ASEL,S,,,P51X              ! Select areas associated with FITEM commands
```

Because this Picker block was used for selection, these operations changed the active set of entities in the model.

7.5.4. Picker Command Blocks for Picking Operations

Graphical picking operations do not change the selection status of the model entities (unless the intended command changes the entities themselves). Therefore, Picker command blocks often include component commands (**CM, CMSEL, CMDELE**, etc.) to save the selection status of the model entities before picking begins and to return the program status to its original condition after the picking operation has finished.

Example 7.7 shows a Picker command block taken from a log file that selects Areas 1 and 3 and sets their element attributes to material #1 and element coordinate system #1. **ASEL** is the intended command in this example. It is needed because **AATT** does not have any entity-related arguments.

Example 7.7

```
FLST,5,2,5,ORDE,2          ! Pick 2 areas for the 5th field of a command
FITEM,5,1                  ! Identify Area 1
FITEM,5,3                  ! Identify Area 3
CM,_Y,AREA                 ! Create a component for the active set of areas
ASEL,,,,P51X               ! Select areas associated with FITEM commands
CM,_Y1,AREA                ! Create a component of the picked areas
CMSEL,S,_Y                 ! Select the areas in the original active set
!*
CMSEL,S,_Y1                ! Select the picked areas
AATT,1,,1,0,               ! Set the element attributes of the picked areas
CMSEL,S,_Y                 ! Select the areas in the original active set
CMDELE,_Y                  ! Delete the component for the original areas
CMDELE,_Y1                 ! Delete the component for the picked areas
```

If all areas were originally in the active set, the same result could have been achieved with the commands in Example 7.8.

Example 7.8

```
ASEL,,,,1,3,2            ! Select areas 1 and 3
AATT,1,,1,0,             ! Set the attributes of the selected areas
ALLSEL,ALL               ! Select everything (reset the select status)
```

Since entity numbers are not stable from release to release, you should replace picking command blocks with selection commands that are independent of entity number when editing a log file to create an input file. This topic will be discussed again in chapter 10.

Chapter

8

Solution

Suggested Reading Assignments:
Mechanical APDL Basic Analysis Guide: Chapters 2 and 4

CHAPTER OUTLINE

This chapter provides an overview of the activities performed in the Solution Processor. You will learn how the term "solution" is used in ANSYS. You will learn how to apply boundary conditions (BCs) and initial conditions (ICs) to your model. You will learn how to set the solution options, initiate a solution, interpret the solution feedback provided by ANSYS, and terminate the solution if necessary. Finally, you will learn how to protect the results of your analysis and append new results to the results file.

8.1. Defining "Solution"

In ANSYS, the word "solution" has four different meanings depending on the context.

1. When capitalized, "Solution" is shorthand for the Solution Processor. For example, "boundary conditions are applied in Solution."
2. When lowercase and used without an article, "solution" refers to the time that begins when the **SOLVE** command is issued and continues until a solution has been found or the attempt to find one has been abandoned. For example, "during solution, detailed feedback is provided in the Output Window."

ANSYS Mechanical APDL for Finite Element Analysis.
DOI: http://dx.doi.org/10.1016/B978-0-12-812981-4.00008-3

3. When lowercase and used with an article, the "solution" may refer to the process of assembling and solving the model's system of equations. For example, "you may need to terminate the solution if your computer has insufficient memory."
4. When lowercase and used with an article, a "solution" may also refer to a set of results produced by the equation solving process. For example, "you should review the steady-state solution to ensure that the model is behaving as expected before solving the transient case."

This chapter uses "solution" in all of these contexts and clarifies the meaning when possible. The ANSYS documentation and members of the ANSYS community also use "solution" in all of these ways. However, they usually do so without clarification and with the expectation that you will understand the meaning from the context.

8.2. Boundary Condition Overview

The activities performed in the Preprocessor specify the data needed to assemble a system of equations that define the degrees of freedom and their interactions within the model. However, these equations cannot be solved unless values for some of the degrees of freedom are supplied. In the classic literature, these are referred to as "initial values" or "initial conditions" if the associated DOFs can vary over the course of the analysis and as "boundary values" or "boundary conditions" if they are fixed for all time. Boundary conditions are usually divided into two categories: "geometric" or "essential" boundary conditions (i.e., constraints) and "natural" or "force" boundary conditions (i.e., loads).

8.2.1. Constraints

A constraint is an input/output value pair that defines the value for a given degree of freedom. It says that the solution will have a specific, known value for a given independent variable at a particular location. Constraints are based on the physical situation. For example, they can specify a zero displacement for the clamped end of a fixed beam or a known deflection at the free end of a fixed beam. They can also specify a zero flow velocity at the wall of a fluid channel or the temperature at a given location in a thermal analysis. The values of constraints are unaffected by other factors such as external loads.

All models must have at least one constraint for each order of the differential equation involved. For example, if a system can be modeled using a second order differential equation, then two constraints must be supplied. It is important to recognize that the degrees of freedom in this context are the degrees of freedom associated with the physical situation and not the element degrees of freedom. For example, a 3D structural problem that uses a SOLID185 element with three nodal degrees of freedom (UX, UY, and UZ) will have six actual degrees of freedom: three translational degrees of freedom and three rotational degrees of freedom. Therefore, a minimum of six constraints will be required for a general 3D structural analysis.

8.2.2. Loads

Loads are perturbations (forces, pressures, voltages, etc.) applied to the system from an external source. The purpose of most simulations is to determine how the model will behave in response to these loads. External loads that are applied in a time-variable analysis before time $t=0$ and do not change with time are called preloads. External loads that are applied in a time-variable analysis before time $t=0$ and change with time are called initial conditions.

8.3. Boundary Conditions in ANSYS

The ANSYS documentation classifies boundary conditions based on their physics, the type of entity to which they are applied, and how they interact with the model.

8.3.1. Boundary Conditions in ANSYS—Organized by Physics

Section 2.1 of the Mechanical APDL Basic Analysis Guide categorizes boundary conditions in six physics regimes: structural, thermal, magnetic, electric, acoustic, and diffusion. The options in each physics environment include:

Structural—displacements, velocities, accelerations, forces, pressures, temperatures (for thermal strain), and gravity.

Thermal—temperatures, heat flow rates, convections, internal heat generation, and infinite surfaces.

Magnetic—magnetic potentials, magnetic fluxes, magnetic current segments, source current densities, and infinite surfaces.

Electric—electric potentials (voltage), electric current, electric charges, charge densities, and infinite surfaces.

Acoustic—pressures and displacements.

Diffusion—concentrations and diffusion flow rates.

8.3.2. Boundary Conditions in ANSYS—Organized by Entity

Boundary conditions in ANSYS can be applied to either solid model or finite element model entities. Solid model boundary conditions have two main advantages. First, they are easier to apply because solid model entities are easier to select. Second, because solid model entities are independent of the finite element mesh, you can refine or remesh your model after applying them. Solid model boundary conditions also have two main disadvantages. Solid model and finite element model loads can have different coordinate systems and loading directions. Thus, it is possible to accidentally apply solid model loads in incorrect or conflicting directions. And, some solid model loads cannot be displayed in the Graphics Window.

Ultimately, all BCs must be applied to the finite element model in order for a solution to be calculated. Thus, finite element boundary conditions provide more control over the model and more insight into its final state. However, finite element BCs can be more difficult to apply because finite element model entities are more difficult to select. In addition, finite element BCs must be deleted before the mesh can be cleared and must be reapplied once the mesh has been altered.

8.3.3. Boundary Conditions in ANSYS—Organized by Interaction

Boundary conditions in ANSYS are grouped into seven categories based on their interaction with the model. These categories are DOF constraints, forces (i.e., concentrated loads), surface loads, body loads, inertia loads, ocean loads, and coupled field loads.

8.3.3.1. Degree of Freedom Constraints

Constraints in ANSYS set one of the element degrees of freedoms for a specific node to a known value. The value can be of any magnitude and either sign (+ or −). Examples of constraints include specified (or zero) displacements, specified temperatures, and flux-parallel boundary conditions. Symmetry boundary conditions in structural analyses are also considered to be DOF constraints. The full list of available DOF constraints is shown in Section 2.5.3 of the Mechanical APDL Basic Analysis Guide.

8.3.3.2. Forces (i.e., Concentrated Loads)

"Forces" in ANSYS are concentrated loads applied to a single location (node) in the model. Examples include forces and moments in a structural analysis, heat flow rates in a thermal analysis, current in an electric analysis, and magnetic flux in a magnetic field analysis. The full list of available concentrated loads is shown in Section 2.5.6 of the Mechanical APDL Basic Analysis Guide.

8.3.3.3. Surface Loads

Surface loads are distributed loads applied over multiple elements or multiple nodes on the surface of a model. Surface loads can be uniform or can vary over the surface. Examples of surface loads include pressures in a structural analysis, convections and heat fluxes in a thermal analysis, and surface charge density in an electric analysis. When surface loads are applied in a two dimensional analysis, there is an assumption that the 2D model has a unit depth. This permits a 2D surface load to be applied along a line with the same magnitude and units that would be used for a 3D analysis. The full list of available surface loads is shown in Section 2.5.7 of the Mechanical APDL Basic Analysis Guide.

8.3.3.4. Body Loads

Body loads are volumetric or field loads. Examples include temperature in a structural analysis, frequency in a harmonic structural analysis, heat generation rate in a thermal analysis, and current density in a magnetic field analysis. The full list of available body loads is shown in Section 2.5.8 of the Mechanical APDL Basic Analysis Guide.

8.3.3.5. Inertia Loads

Inertia loads result from the inertia (mass matrix) of a body and include gravitational acceleration, angular velocity, and angular acceleration. Mass must be included in the model either as a density material property or via mass elements, such as MASS21. Inertia loads are used mainly in structural analyses. A detailed discussion about inertial loads can be found in Section 2.5.9 of the Mechanical APDL Basic Analysis Guide.

8.3.3.6. Ocean Loads

Ocean loads are line or surface effect loads that represent the influence of waves, current, drag, and buoyancy in hydrostatic and hydrodynamic analyses. Ocean loads are considered to be specialty loads. A detailed discussion of ocean loads can be found in Section 2.5.10 of the Mechanical APDL Basic Analysis Guide.

8.3.3.7. Coupled Field Loads

Coupled field loads are loads whose values are derived from a previous analysis. For example, you can use the temperatures calculated in a thermal analysis or the magnetic forces calculated in a magnetic field analysis as the inputs to a structural analysis. Coupled field loads are an integral part of multiphysics analysis. For more information, see the Mechanical APDL Coupled-Field Analysis Guide.

8.4. Applying Boundary Conditions

Boundary conditions in ANSYS can be applied using the GUI path: **Main Menu > Solution > Define Loads > Apply.** Below this menu tree level, the options displayed in the GUI are filtered based on the elements and degrees of freedom that have been defined. The GUI will not display loads or constraints that are unavailable for the currently defined element types and will not allow you to apply them. Boundary conditions can also be applied directly by using commands.

Constraints are applied to the model using the D family of commands (**D, DK, DL**, and **DA**). The **D** command applies a constraint directly to a node. The **DK, DL**, and **DA** commands apply constraints to the selected keypoint(s), lines(s), and area(s). These constraints are transferred to the underlying nodes after the **SOLVE** command is issued.

Forces are applied to the model using the F family of commands (**F** and **FK**). No commands exist to apply nodal type loads to lines, areas, or volumes.

Surface loads are applied to the model using the SF family of commands (**SF, SFE, SFL, SFA**, and **SFBEAM**). The **SFGRAD** command specifies a linear gradient for all **SF, SFE, SFL**, and **SFA** commands issued after it is activated. Similarly, the **SFFUN** command can be used to specify any type of varying surface load for all **SF** and **SFE** commands that follow it. Since the **SFGRAD** and **SFFUN** commands affect all surface load commands that are issued until the specification is removed, they should be deactivated immediately after use.

Body loads are applied to the model using the BF family of commands (**BF, BFE, BFK, BFL, BFA, BFV**, and **BFUNIF**). The **BFUNIF** command applies a uniform body force load to all nodes in the model. The **TUNIF** command is a special version of the **BFUNIF** command that is used only to apply a uniform structural temperature.

Inertia loads are applied to the model using the ACEL family of commands (**ACEL** and **CMACEL**), the OMEGA family of commands (**OMEGA, CGOMEGA, CMOMEGA, CMDOMEGA, DCGOMEGA**, and **DOMEGA**), and **CORIOLIS.** Inertia loads have no effect for elements that do not have mass.

8.4.1. Deleting Boundary Conditions

Boundary conditions can be deleted using the GUI path: **Main Menu > Solution > Define Loads > Delete**. Below this menu tree level, you will be able to choose to delete all boundary conditions, all boundary conditions of a specific type (solid model BCs, finite element model BCs, all inertia loads, all constraints, all forces, all surface loads, and all body loads), or individual boundary conditions. Just as the GUI will not display options to apply boundary conditions that are not available for the current model, it will not display options to delete boundary conditions that have not been defined.

The **LSCLEAR** command can be used to delete all boundary conditions in the model, all solid model BCs, all finite element BCs, or all inertia loads depending on the argument supplied. Separate commands also exist to delete each type of boundary condition for each entity type. For example, the **DKDELE, DLDELE**, and **DADELE** commands delete constraints on keypoints, lines, and areas, respectively. The full list of commands to delete boundary conditions can be found in Section 2.6 of the Mechanical APDL Command Reference.

8.4.2. Confirming Boundary Conditions in ANSYS

Applying the correct boundary conditions is critical for obtaining the correct solution to a given problem. If the assumptions made about the loads and constraints are incorrect, or if the boundary conditions are applied incorrectly, the program will solve the wrong problem. There are four ways to ensure that your boundary conditions have been applied correctly: you can review the messages provided in the Output Window, you can evaluate the immediate graphical feedback supplied by the program, you can list the boundary conditions that have been applied to the model, and you can plot the boundary conditions that have been applied to the model.

8.4.2.1. Immediate Feedback

As you have seen in the exercises for the previous chapters, ANSYS displays symbols on the model geometry to show that new boundary conditions have been applied. Constraints are shown with triangles. Forces are shown with arrows. Surface loads are shown with outlines (by default) or arrows (if selected in the Symbols menu). Body loads are shown either with contours (by default) or arrows (if selected in the Symbols menu). Inertia loads are shown at global origin: linear accelerations are shown with single-headed arrows, rotational velocities are shown with double-headed arrows, and rotational accelerations are shown with triple-headed arrows. Coriolis accelerations are not shown graphically.

The orientation of the triangle and arrow symbols indicates the direction of the applied boundary condition. The sign of the boundary condition value (positive or negative) controls the orientation. These symbols disappear when another graphics command is issued, so replotting, resizing, or rotating the model will remove these symbols.

8.4.2.2. Plotting Boundary Conditions

You can redisplay the boundary conditions in the Graphics Window by adjusting the symbols options in the Plot Controls menu (**Utility Menu > PlotCtrls > Symbols ...**). The model should replot by default. The outlines and contours for the surface and body loads are difficult to see, so you should always change their symbols to arrows.

While symbols provide helpful information, they can obscure important features of your model. For more complex models, it may be necessary to select and plot specific entities instead of viewing all of the entities in the model. It may also be necessary to turn off symbols after you have confirmed that the boundary conditions have been applied correctly.

8.4.2.3. Listing Boundary Conditions

You can list the boundary conditions that have been applied to the active set of entities in the model by using the GUI path: **Utility Menu > List > Loads**. In complex models with many boundary conditions, you may find it useful to select specific entities in the model and then list their BCs. For example, you could select the line at $x=0$ and then list the surface loads that have been applied to that line. You cannot list the BCs for entities that are not in the currently active set. For example, if the line at $x=0$ is not currently selected, the surface loads on that line will not be listed even though they were applied. If you wish to list the BCs for entities that are not in the active set, you will need to select those entities or to issue a Select All command (`ALLSEL,ALL`).

8.5. Potential Pitfalls Associated with Applying Boundary Conditions in ANSYS

8.5.1. No DOF to Constrain

You cannot supply a constraint for a degree of freedom that does not exist. For example, it may be necessary to prohibit in-plane rotation for a body that has DOFs in only x and y (UX and UY). However, if there is no rotational degree of freedom (ROTZ), a rotational constraint cannot be applied. Instead, the rotation must be constrained through a combination of x and y constraints (see section 8.5.2 for more details).

Similarly, you cannot apply a load for a degree of freedom that does not exist. For example, you cannot apply a moment (MZ) to a node from a continuum element that has only x and y degrees of freedom. Instead, the moment must be created by applying a force with an appropriate magnitude at an appropriate location.

8.5.2. Too Few Constraints

In the real world, bodies at rest tend to remain at rest unless acted upon by an external force. When exposed to small perturbations, bodies still tend to stay at rest because forces like friction and gravity counteract these perturbations.

In a finite element model, round-off errors generated by the solution can create small nodal forces that act as perturbations to the model. However, countering forces such as friction and gravity do not exist in finite element models (by default). Therefore, it is almost always necessary to supply constraints to preclude any type of nonpermissible motion or response by the model.

For 2D structural analyses that use elements with two degrees of freedom (UX and UY), at least three constraints are needed: one for displacement in x, one for displacement in y, and a third either in x or y to prevent rotation about the z-axis. For example, Figure 8.1 (left) shows a simple model with one constraint in x and another constraint in y placed on the node located at the origin. This satisfies the requirement to provide one constraint per degree of freedom in the degree of freedom set. But in this model, rotation about the z-axis is possible. Figure 8.1 (right) shows the same model with an additional constraint to prevent rotation.

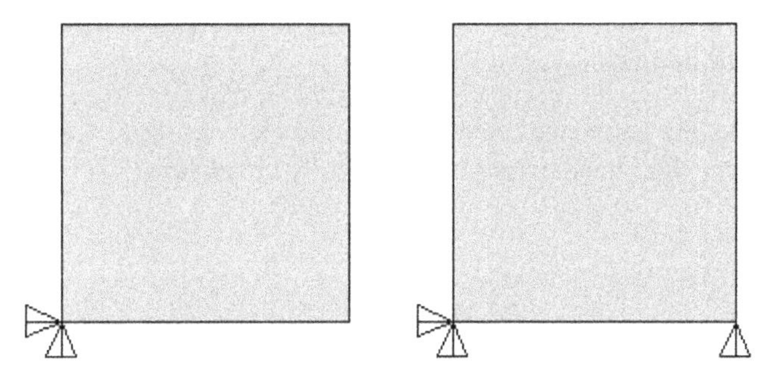

Figure 8.1 Under-Constrained (left) and Properly Constrained (right) Areas.

For 3D structural analyses using elements that have three degrees of freedom (UX, UY, and UZ), a minimum of six constraints must be supplied: three to constrain the model in x, y, and z and another three to constrain the rotations about x, y, and z. For thermal problems, at least one node must have a defined temperature (although this can be defined either as a constraint or as part of a convection boundary condition). Similar requirements exist for problems in other physics regimes.

If a model is obviously under constrained, ANSYS will issue an error after the solution has been initiated. The solution will then terminate, so you can correct the problem.

8.5.3. Too Many Constraints

It is also possible to over constrain the model. For example, you can over constrain a 2D structural analysis of a simple cantilever beam by constraining all of the nodes at the fixed end of the beam in x and y (Figure 8.2 left). Fully constraining these nodes is usually undesirable because it prevents contraction of the beam due to the Poisson effect and may generate a large artificial stress at the constraint location. Normally, it is better to constrain all nodes at the fixed end of the beam in the direction parallel to the long axis of the beam and to constrain a single node perpendicular to the axis of the beam (Figure 8.2 right).

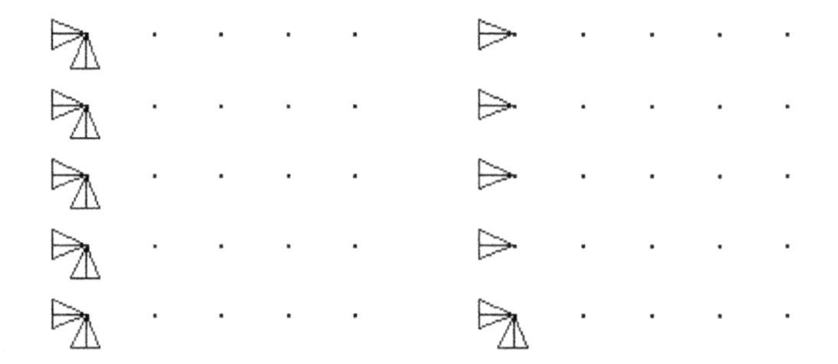

Figure 8.2 Over-Constrained (left) and Properly Constrained (right) Nodes.

8.5.4. Mixing Loads and Constraints

If you apply a load and a constraint to the same entity at the same location, the constraint will be used and the load will be 'ignored.' Constraints supersede loads after the **SOLVE** command is issued when the simultaneous equations are assembled. This occurs because the values of the DOFs at those location are specified by the constraints and thus do not need to be calculated from the applied loads. However, the loads continue to be defined and are included in the nodal force vectors. As a result, these applied loads will contribute to the reaction forces in the final solution. There is no way to prevent constraints from superseding loads and ANSYS will not warn you when it occurs.

Example 8.1 lists the commands to create a finite element model of a 1×1 aluminum 6061-T6 plate whose lower right-hand corner has been fully constrained.

Example 8.1

```
/PREP7                      ! Enter the Preprocessor
RECTNG,0,1,0,1              ! Create a 1x1 rectangle
ET,1,182                    ! Use SOLID182 elements
MP,EX,1,7.3e10              ! Define Young's modulus for aluminum 6061-T6
MP,PRXY,1,0.33              ! Define Poisson's ratio for aluminum 6061-T6
ESIZE,0.1                   ! Mesh with an element size of 0.1
AMESH,ALL                   ! Mesh the area
/SOL                        ! Enter the Solution processor
NSEL,S,LOC,X,1              ! Select the nodes at x=1
NSEL,R,LOC,Y,0              ! Reselect the nodes at y=0
D,ALL,ALL                   ! Constrain all DOFs for the selected node
```

Example 8.2 uses the model created in Example 8.1 to partially demonstrate the effect of mixing loads and constraints. The first two commands select the node at (0,0). Next, a displacement of 0 m in the y direction and a force of 40 N in the y direction are applied to that node. The **DLIST** command confirms that the constraint has been successfully applied and the **FLIST** command confirms that the load has been successfully applied. Then, all of the entities in the model are selected and the model is solved. Finally, we enter the General Postprocessor and plot the deformation in y for the entire model.

If you run this model, you will see that the displacement contour plot is a solid red color, indicating no deformation in the model. This is because the constraint superseded the load after the **SOLVE** command was issued. This resulted in a zero displacement at (0,0) in the final analysis instead of an applied load of 40 N. Since no external load was applied to this model, no deformation could occur.

Example 8.2

```
NSEL,S,LOC,X,0              ! Select the nodes at x=0
NSEL,R,LOC,Y,0              ! Reselect the nodes at y=0
D,ALL,UY,0                  ! Constrain all selected nodes in y
F,ALL,FY,40                 ! Apply a force of 40 N in y on selected nodes
DLIST,ALL                   ! List DOF constraints for all selected nodes
FLIST,ALL                   ! List the applied forces for all selected nodes
ALLSEL,ALL                  ! Select everything
SOLVE                       ! Solve the model
/POST1                      ! Enter the General Postprocessor
PLNSOL,U,Y                  ! Plot the displacement in Y
```

Example 8.3 uses the model created in Example 8.1 to fully demonstrate the effect of mixing loads and zero constraints. The first two commands select the node at (0,0). Next, a displacement of 0 m in the *y* direction and an upward force of 40 N in the *y* direction are applied to that node. Then, the node at (0.5,1) is selected and a downward force of 100 N in the *y* direction is applied to that node. All of the entities in the model are selected and the model is solved. Finally, we enter the General Postprocessor, plot the deformation in *y* for the entire model, and list the reaction forces in the model.

If you run this model, you will see that the deformation in the model is symmetric, indicating that the constraint in the lower left corner has overridden the first applied load. However, if the constraint had truly replaced the first applied load, the reaction forces would be symmetric with +50 N at each bottom corner. Instead, the reaction force in the lower left corner is +10 N. This indicates that the applied force of +40 N was added to the reaction force at that location.

Example 8.3

```
NSEL,S,LOC,X,0              ! Select the nodes at x=0
NSEL,R,LOC,Y,0              ! Reselect the nodes at y=0
D,ALL,UY,0                  ! Constrain all selected nodes in y
F,ALL,FY,40                 ! Apply an upward force of 40 N in y
NSEL,S,LOC,X,0.5            ! Select the nodes at x=0.5
NSEL,R,LOC,Y,1              ! Reselect the nodes at y=1
F,ALL,FY,-100               ! Apply a downward force of 80 N in y
ALLSEL,ALL                  ! Select everything
SOLVE                       ! Solve the model
/POST1                      ! Enter the General Postprocessor
PLNSOL,U,Y                  ! Plot the displacement in Y
PRRSOL                      ! List the reaction forces
```

8.5.5. Repeating Boundary Conditions

If you apply the same type of boundary condition to the same entity at the same location, the second value will replace the first (by default). The replacement occurs when the second boundary condition is applied and can be verified immediately.

Example 8.4 uses the model created in Example 8.1 to demonstrate the effect of repeated boundary conditions. First, the node at (0,0) is selected and a displacement of 2 cm in the *y* direction is specified. The **DLIST** command verifies that the displacement has been successfully applied. Next, a displacement of 10 cm in the *y* direction is specified for the same node. The second specification replaces the first, leading to a final displacement of 10 cm on the selected node. The second **DLIST** command verifies the replacement.

Example 8.4

```
NSEL,S,LOC,X,0          ! Select nodes at x=0
NSEL,R,LOC,Y,0          ! Reselect the nodes at y=0
D,ALL,UY,0.02           ! Apply a disp. of 2 cm in y on selected nodes
DLIST,ALL               ! List DOF constraints for all selected nodes
D,ALL,UY,0.10           ! Apply a disp. of 10 cm in y on selected nodes
DLIST,ALL               ! List DOF constraints for all selected nodes
```

You can set an option to permit repeated finite element BCs to be added (allowing them to accumulate) or to be ignored (so the original specification remains), but use of this option is not recommended. If there are multiple sources of boundary conditions for a single location, these BCs should be combined and then applied to the model using a single command with a single value. Loads that are not needed should not be applied.

There are no accumulation commands for solid model BCs. A repeated solid model boundary condition will always replace the first.

8.5.6. Mixing Solid Model and Finite Element Model Boundary Conditions

If you apply a solid model boundary condition and a finite element model boundary condition to the same entity at the same location, the solid model BC will replace the finite element model BC regardless of the order in which they were defined. This occurs because solid model BCs are transferred to the finite element model when the **SOLVE** command is issued. Thus, the solid model BCs are always applied to the model after the finite element BCs.

Example 8.5 uses the model created in Example 8.1 to demonstrate the effect of mixing solid model and finite element model BCs. First, the keypoint at (0,0) is selected and a displacement of 4 cm in the *y* direction is applied. Because nodal constraints were applied in Example 8.1, ANSYS will now issue a warning stating “Both solid model and finite element model boundary conditions have been applied to this model. As solid loads are transferred to the nodes or elements, they can overwrite directly applied loads.” Next, the node at (0,0) is selected and a displacement of 2 cm in the *y* direction is applied. The **DKLIST** command confirms that the keypoint displacement has been successfully applied, and the **DLIST** command confirms that the nodal displacement has been successful applied. Then, all of the entities in the model are selected and the model is solved. Finally, we enter the General Postprocessor and plot the deformation in *y* for the entire model.

Example 8.5

```
KSEL,S,LOC,X,0          ! Select keypoints at x=0
KSEL,R,LOC,Y,0          ! Reselect keypoints at y=0
DK,ALL,UY,0.04          ! Apply a disp. of 4 cm in y on selected KPs
NSEL,S,LOC,X,0          ! Select nodes at x=0
NSEL,R,LOC,Y,0          ! Reselect nodes at y=0
D,ALL,UY,0.02           ! Apply a disp. of 2 cm in y on selected nodes
DKLIST,ALL              ! List DOF constraints for all selected keypoints
DLIST,ALL               ! List DOF constraints for all selected nodes
ALLSEL,ALL              ! Select everything
SOLVE                   ! Solve the model
/POST1                  ! Enter the General Postprocessor
PLNSOL,U,Y              ! Plot the displacement in Y
```

If you run this model, you will see that the displacement contour plot indicates a maximum deformation of 4 cm in the *y* direction. This is because the solid model displacement replaced the finite element displacement after the **SOLVE** command was issued.

The transfer of solid model boundary conditions does not delete all finite element BCs that have been applied to the model. It only replaces loads and constraints that have been applied to the same entities in the same coordinate direction. For example, if pressure loads are applied to one part of the finite element model and to a different (nonoverlapping) part of the solid model, the model will include both pressures for the solution. Overlapping sections will have only the solid model pressure value applied.

You cannot prevent the solid model loads from transferring to the finite element model or prevent the solid model loads from replacing their finite element counterparts. (Accumulation commands have no effect on this procedure.) However, you can use the GUI path: **Main Menu > Solution > Define Loads > Operate > Transfer to FE > All Solid Lds** or the **SBCTRAN** command to transfer the solid model loads to the finite element model. You can then list the finite element loads for your model and verify that the final values are correct before issuing the **SOLVE** command.

8.6. Initial Conditions in ANSYS

Initial conditions refer to the state of a model at time $t=0$ in a time-dependent analysis. If no initial conditions are specified, ANSYS will assume that the initial values for the DOFs (displacement, temperature, etc.) and for the first derivatives of the DOFs (velocity, heat flux, etc.) are zero. These will be the correct initial conditions for some but not all problems. For example, objects don't usually have an initial temperature of zero.

If the default initial conditions do not reflect the system to be analyzed, there are two ways to define them. First, you can set the initial conditions by using the GUI path: **Main Menu > Solution > Define Loads > Apply > Initial Condit'n > Define** or by using the **IC** command. Second, you can solve the initial state of the model as a steady-state problem at an initial time that is approximately 0 with time integration effects turned off (`TIMINT,OFF`). (ANSYS does not permit solutions at time $t=0$, so you must use a very short time like t = 1e-6 instead.) Then, you can begin the transient analysis from the steady-state solution. This requires you to increment the time to something like $t=1$, change the boundary conditions to the unsteady state, and turn time integration effects back on (`TIMINT,ON`). This second approach will be used in Exercise 8-1.

8.7. Solution Options

The default solution options used by ANSYS are based on decades of experience and testing and represent the best judgment of the user community. Thus, the models in most basic analyses can be solved once the boundary conditions have been applied. However, for more advanced analyses, you may need to set some solution options before issuing the **SOLVE** command. In addition, there are no default options for time stepping and some transient and/or nonlinear options, so you must supply these options when needed.

This section discusses some of the solution options available in ANSYS, including the type of analysis to be performed; small vs large displacement options; load step, substep, and time step options; output options; and the choice of solver.

8.7.1. Analysis Type

ANSYS can perform seven types of analysis: static (default), transient, modal, spectrum, harmonic, eigen buckling, and substructuring/component mode synthesis (CMS).

Static analyses are either time independent or involve time without mass or inertia.

Transient analyses are time-dependent analyses that include mass or inertial effects.

Modal analyses identify the natural frequencies of a given system.

Spectrum analyses are extensions of modal analyses where a load spectrum is applied to each of the natural frequencies of the system to determine the response of the system at each of those frequencies.

Harmonic analyses identify the steady-state solution for systems with harmonically varying applied loads. If damping is included in a harmonic analysis, both the real and the complex solutions are calculated.

Eigen buckling is linear buckling that predicts the theoretical buckling load of an ideal structure. It is faster but less accurate than nonlinear buckling where an increasing load is applied during a nonlinear static analysis until the structure can no longer support the load.

Substructuring involves the creation of a matrix that is formed from other ANSYS elements where the active degree of freedom set is reduced to only those needed to interact with other parts of the model. This matrix is usually referred to as "superelement" in the documentation and literature.

To set the analysis type, use the GUI path: **Main Menu > Solution > Analysis Type > New Analysis** or use the **ANTYPE** command.

8.7.2. Small vs Large Displacement

By default, ANSYS assumes that the displacements and rotations in your model will be "small." From a mathematical perspective, this means that the second order (squared) terms in the constitutive equations for your model have a negligible impact on the calculation of your solution and are ignored. From a finite element perspective, this means that any changes in the stiffness of the individual elements in your model due to changes in the elements' shape and orientation will be negligible. Thus, the element stiffness matrices are not updated to account for these changes. These assumptions result in a linear analysis that is fast and efficient.

If large displacements, large strains, large rotations, or major distortions are expected or present, you must use the large displacement solution option and perform a nonlinear analysis. This option is slower because multiple iterations are usually required to find the solution. However, it produces better results when the conditions for a linear analysis are not met.

You can specify whether to assume small or large displacements for static and transient analyses using the "Analysis Options" drop down menu in the Basic tab of the Solution Controls dialog box (**Main Menu > Solution > Analysis Type > Sol'n Controls**) (Figure 8.3) or by issuing the **NLGEOM** command. Some analysis types, including modal, harmonic, eigen buckling, and substructuring analyses, are linear analyses that only support small displacements.

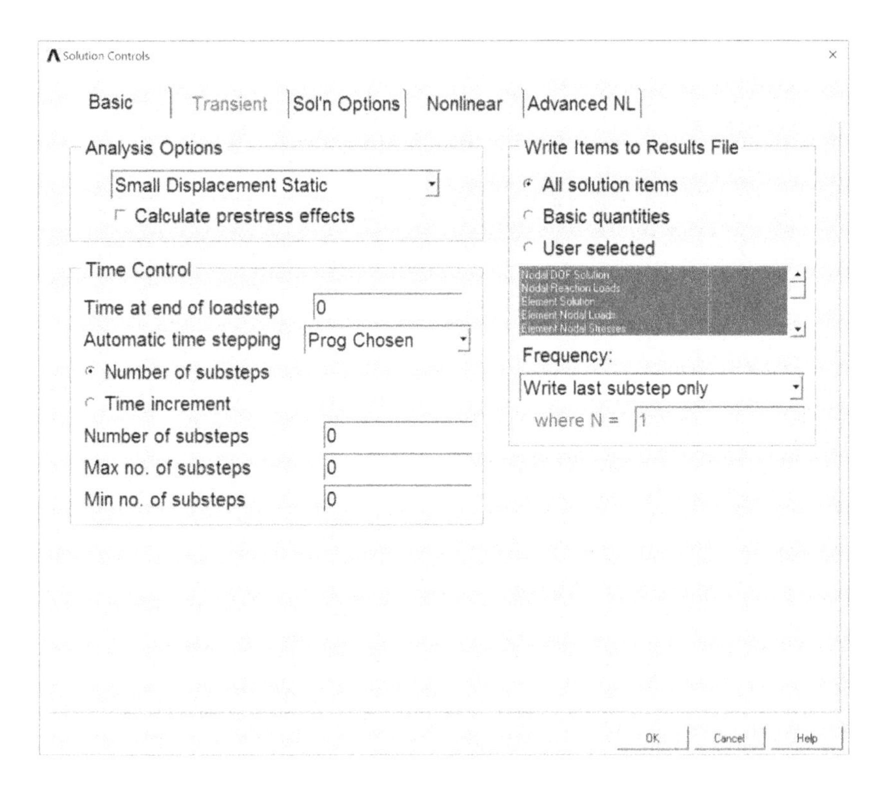

Figure 8.3 Basic Tab of the Solution Controls Dialog Box.

8.7.3. Load Steps

A load step in ANSYS is a set of boundary conditions that is solved to create a set of results. In static analyses, multiple load steps can be used to create multiple sets of results for a single finite element model. In transient analyses, load steps allow you to create a transient load history curve and apply the load in a series of discrete states. The number of load steps to use and the time associated with those load steps depend on the nature of the input transient values and the rate of change of these input values. The incremental time between load steps does not need to be constant.

You can include multiple load steps in a model by solving the model, modifying the model boundary conditions, and then solving the model again. You can continue to add load steps until you leave the Solution processor. This is referred to as the "multiple **SOLVE** method" in Section 4.6.1 of the Mechanical APDL Basic Analysis Guide.

8.7.4. Substeps

Substeps are points (or times) within a load step at which solutions are calculated. A solution must be calculated for each load step, so one substep is automatically included for each load step. Load steps can be divided into multiple (equally spaced) substeps to calculate additional solutions per load step.

In nonlinear static analyses, substeps are used to apply loads gradually. In transient analyses, they are used to divide the load steps into incremental time steps, so inertial effects can be calculated accurately. In both cases, additional substeps can increase the accuracy of the solution and facilitate solution convergence. Substeps provide no benefit in linear static analyses and should not be used.

If multiple substeps are specified, you must choose whether the load should be ramped or stepped. Ramped loads divide the full load step load into a number of equal load increments. One increment is added to the model at each substep, so the full value of the load will be applied at the end of the load step. Stepped loads apply the full value of the load to the model at the first substep of the load step. The load remains constant for all subsequent substeps of the load step. (See Section 2.4 in the Mechanical APDL Basic Analysis Guide for an illustration of these two options.) Ramped loads are the default for static and transient analyses when time integration is turned off. They should always be used in nonlinear static analyses. Stepped loads are the default for transient analyses when time integration is turned on. You can set this option in the Transient tab of the Solution Controls dialog box or by using the **KBC** command.

The number of substeps required for a given analysis depends on the application. As you gain experience with ANSYS you will develop a feel for the number of substeps required. Until then, the following rules of thumb may be helpful. For transient analyses, you need at least 30 substeps over the fundamental natural frequency of the system. For viscoelastic and viscoplastic analyses, you should use at least 10 substeps over the currently active time constant. Otherwise, the number of substeps can be determined using a convergence analysis. This procedure involves estimating the required number of substeps, solving the model, doubling the number of substeps, and solving the model again. If the two solutions are significantly different, the number of substeps should be increased again until the differences between the solutions are negligible. You can define the number of substeps to use in the Basic tab of the Solution Controls dialog box or by using the **NSUB(ST)** command. You can also activate auto time stepping to adjust the number of substeps in nonlinear static and transient analyses. See section 8.7.6 for details.

There is no practical limit to the number of load steps or substeps that can be used in a single analysis. However, the default number of solutions that can be placed in a results file is 10,000. This can be increased using the **/CONFIG** command. For example, `/CONFIG,NRES,20000` would set the maximum number of solutions to 20,000.

8.7.5. Time Steps

When used in transient analyses, substeps are often referred to as time steps. Since one solution is calculated per time step, specifying more time steps will make a transient solution more detailed and continuous in time. It will also increase the cost (solution time) of the analysis.

Time is actually present in all analyses performed in ANSYS since time is used as the Newton-Raphson load parameter. By default, the first load step of a static analysis ends at time $t=1$ and time automatically increments by 1 unit per load step. This conveniently assigns integer time values to the load steps in a static analysis.

For transient analyses, you must specify a time at the end of each load step based on the desired or required input load history. This time must be greater than the time at the end of the previous load step. You can specify the time at the end of the current load step in the Basic tab of the Solution Controls dialog box or by using the **TIME** command.

Once the time at the end of the load step has been defined, you can specify either the number of substeps (time steps) for the load step or the time increment (time step size) for the analysis. The time increment is simply the length of time used by the load step divided by the number of substeps. For this reason, the two options are equivalent and thus mutually exclusive. Either can be specified in the Basic tab of the Solution Controls dialog box. ANSYS will use whichever is specified last. You can also specify the time step size using the **DELTIM** command.

8.7.6. Auto Time Stepping

Automatic time stepping is an option that allows ANSYS to change the size of the time step based on the convergence rate of the analysis. If the program has difficulty achieving solution convergence, the number of substeps may be increased (i.e., the incremental time may be decreased). Similarly, if convergence is easily achieved, the number of substeps may be decreased (i.e., the incremental time may be increased). This option is used mainly for transient and nonlinear analyses. By default, the program decides whether or not to use auto time stepping.

You can control auto time stepping in the Basic tab of the Solution Controls dialog box or by using the **AUTOTS** command.

8.7.7. Output Options

By default, ANSYS calculates and stores all results that are available for the defined element types except for the UserMat state variables and the integration point locations (`OUTRES,ALL`). This includes both the nodal results (e.g., nodal DOF values, nodal reaction loads, and nodal velocities and accelerations for transient analyses) and the element results (e.g., stresses, strains, fluxes, gradients, element miscellaneous data, and the element loads needed to calculate reaction loads). It is possible (but not recommended) to output fewer results items. This can be done in the Basic tab of the Solution Controls dialog box or by using the **OUTRES** command. You can output additional results items, such as the state variables and the integration point locations, by issuing additional **OUTRES** commands.

For static and transient analyses, the solution data is written to the results file for the last sub-step of each load step. Thus, you will have one set of results per load step. For harmonic analyses, solution results are written for every substep. You can change the frequency at which data is saved in the Basic tab of the Solution Controls dialog box or by using the **OUTRES** command. In general, we recommend outputting all results for all solution times (`OUTRES,ALL,ALL`).

8.7.8. Solvers

ANSYS has five shared memory solvers and three distributed memory solvers available for use. The five shared memory solvers are the sparse direct solver, the Preconditioned Conjugate Gradient (PCG) solver, the Jacobi Conjugate Gradient (JCG) solver, the Incomplete Cholesky Conjugate Gradient (ICCG) solver, and the Quasi-Minimal Residual (QMR) solver. The three distributed memory solvers are the distributed memory sparse direct solver, the distributed memory PCG solver, and the distributed memory JCG solver. Both the shared memory and distributed memory solvers require a high-performance computing (HPC) license to use more than two processors.

The sparse solver is the default solver for all types of analyses except for modal and buckling analyses. (Modal and buckling analyses have no default solvers.) The sparse solver is a frontal type solver that has been optimized for sparsely populated matrices like those found in finite element analysis. It is recommended for nonlinear problems and is well suited for ill-conditioned problems.

The PCG solver (also known as the Power Solver) is an iterative conjugate gradient solver. It is the most robust iterative solver and is well suited for large, linear problems built from continuum elements. It requires large amounts of memory because it requires the entire stiffness matrix and its preconditioning matrix to be stored in memory at the same time.

The JCG solver is well suited for large single-field problems such as thermal, magnetic, and acoustic analyses. It does not include preconditioning, so the JCG memory requirement is approximately half of the PCG solver requirement. The JCG solver can be used as an alternative to the PCG solver but is generally not as robust.

The ICCG solver is a conjugate gradient solver that has been optimized for problems that involve more than one physics. It is well suited for problems involving thermal stresses, thermal electric applications, and fluid structure interaction. It is not recommended for novice users.

The QMR solver is an iterative solver for use with high-frequency electromagnetic analyses. It should not be used for any other application.

Table 4.1 of the Mechanical APDL Basic Analysis Guide provides some general guidelines to help you select the best solver to use for a given application. Most beginning ANSYS users should use either the sparse solver or the PCG solver.

8.8. Initiating a Solution

Once the solution options are set, you can initiate a solution by using the GUI path: **Main Menu > Solution > Solve > Current LS** (load step) or by issuing the **SOLVE** command.

8.8.1. Confirming the Solution Options

If the solution is initiated using the GUI (**Main Menu > Solution > Solve > Current LS**), two pop-up dialog boxes will appear: the Solution Options window (Figure 8.4) and the Solve Current Load Step dialog box (Figure 8.5).

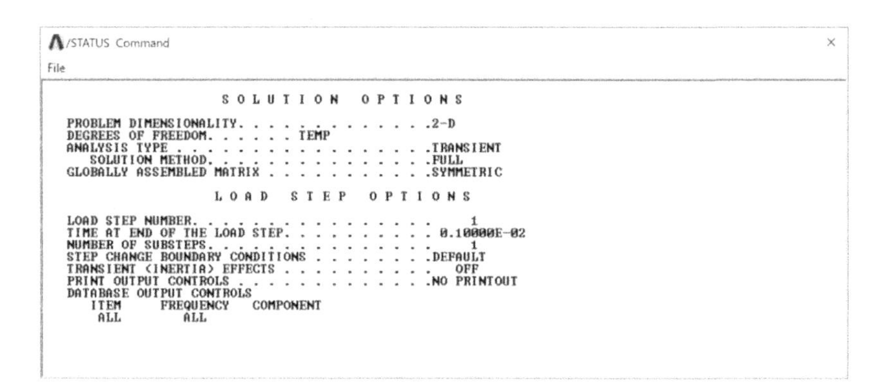

Figure 8.4 Solution Options Window.

The Solution Options window and the Output Window list some of the solution options and load step options that have been defined or will be used in the analysis. These include:

- The problem dimensionality—1D, 2D or 3D.
- The degrees of freedom in the model—UX, UY, TEMP, etc.
- The analysis type—Static, transient, etc.
- The load step number—Defaults to 1 plus the previous load step number.
- The number of substeps—Defaults to 1.
- The time at the end of the load step—Defaults to 1 plus the previous load step time.
- Transient effects—On or off (the default depends on the analysis type).

- Print output control—Specifies whether to print results in the Output Window.
- Database output controls—Specifies which results to output for which substeps.

More or less information may be displayed depending on the analysis and the solution options selected.

For transient analyses, it is important to review the solution options before clicking OK in the Solve Current Load Step dialog box. If you want to solve the initial case for a transient analysis using a short time step, transient (inertia) effects should be turned off. If they are not turned off, ANSYS will begin the transient analysis from the zero DOF, zero DOF gradient state instead of creating the initial condition state. If the initial case is solved using a short time step, transient effects need to be turned back on to solve the model after time begins.

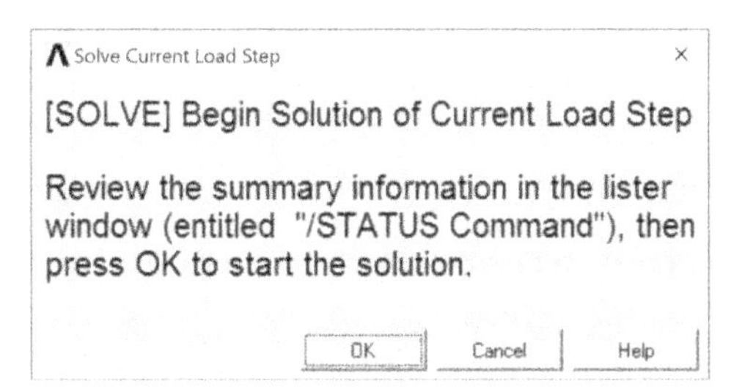

Figure 8.5 Solve Current Load Step Dialog Box.

If you issue the **SOLVE** command via the command prompt, the Solutions Options window will not be displayed. To see the solution options, you will need to issue the **/STATUS** command (`/STATUS,SOLU`).

8.8.2. Solution Checking

When the **SOLVE** command is issued via the command prompt or when you click OK in the Solve Current Load Step dialog box, control of the program is transferred to the solver. Before assembling the system of equations, the solver performs a series of checks to ensure that the submitted model is complete. For example, the solver checks the finite element model and will issue a warning if any nodes or elements in the model are unselected. It checks the material property library against the element library to ensure that all required material properties are defined and will issue a warning if any are missing. It checks the applied boundary conditions and will issue a warning if no constraints have been applied to the model, etc.

If any warning or error messages are generated, a pop-up box will appear to notify you of the total number of warnings and errors (Figure 8.6). In addition, up to 5 warning and error messages will be printed in the Output Window (or sent to the output file). After 5 messages, printed output will be suppressed and additional messages will appear only in the error file. You should check all warning and error messages carefully before continuing. Once you confirm that the **SOLVE** command should be executed, ANSYS will proceed with the assembly and solution of the problem's system of equations.

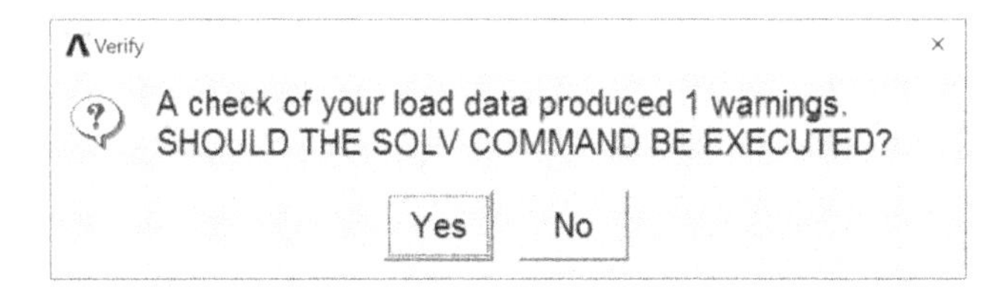

Figure 8.6 Verify Solve Command Dialog Box.

8.9. During Solution

The time required to calculate a solution can range from seconds or minutes for small linear models to hours or days for complex nonlinear analyses. Once ANSYS begins to calculate a solution, you cannot issue new commands or interact with the program. However, you can track the progress of the solution based on the feedback provided by the program.

8.9.1. Feedback During Solution

For all analyses, solution feedback is provided in the Output Window. For nonlinear analyses, additional information is provided by the Graphical Solution Tracker. Most of the information presented will be of little use to new users. But you can, and should, interpret this information to determine if the program is functioning normally or if the solution should be terminated.

8.9.1.1. General Feedback for Linear Analyses

The solution feedback for linear analyses includes:

- Information related to the mathematics of the problem such as the maximum and minimum matrix coefficients for the system of equations.
- Information related to the solver such as the type of solver being used, the number of equations to be solved, and the memory requirements of the solver.
- A notification of the completion of each load step and substep and the time used for each calculation.
- A summary of the results information written to file such as the file name(s) and size(s) if the solution was successful.
- An explanation if the solution was terminated by the program.

Solution feedback is written to the Output Window as soon as it is available. For linear models, this process can be almost instantaneous. If the Output Window indicates that the solution has been initiated but does not seem to be progressing after a few minutes, ANSYS may not have enough memory to assemble the equations. In this case, it is best to terminate the analysis and either increase the memory available to the computer or reduce the size of the model.

8.9.1.2. Additional Feedback for Nonlinear Analyses

For nonlinear models, ANSYS provides additional feedback, so you can monitor the convergence of the solution. To understand this additional feedback, you must first understand how a nonlinear solution is calculated in ANSYS.

Because the system of equations assembled for nonlinear analyses include combinations of first order and higher order terms associated with the degrees of freedom, a solution cannot be calculated directly. Instead, ANSYS uses an iterative procedure that begins with an estimate of the solution. The estimate is used to calculate a new solution to the problem and the new solution is compared to the original estimate. If the two values are within a specified tolerance, the solution is accepted. Otherwise, the new solution is used as the new estimate and the solver iterates again.

ANSYS uses the Newton-Raphson procedure to calculate the iterative solution. This method works well as long as the original estimate is close to the true solution. To ensure a good initial approximation, the model boundary conditions are applied to the model slowly (as substeps) and the model is allowed to reach equilibrium before a new incremental load is applied. If the substeps are small, the changes in the behavior of the model will be small and the original estimates should be adequate. More substeps require more total calculations but fewer iterations (called equilibrium iterations) per substep. This increases the likelihood that the final solution will be found.

At the end of each equilibrium iteration, ANSYS calculates at least one "force" convergence value and compares it to some criterion. If the force convergence value is greater than the criterion, at least one additional equilibrium iteration is required. When the convergence norm drops below the convergence criterion, the substep has converged and the applied load may be incremented for the next substep. This process continues until convergence is obtained for the full load. (See Section 14.11 of the Mechanical APDL Theory Reference for more information.)

The additional solution feedback reported in the Output Window is related to the force convergence value(s) and the associated criteria. The Graphical Solution Tracker (Figure 8.7) plots the force convergence values and their criteria at the end of each equilibrium iteration.

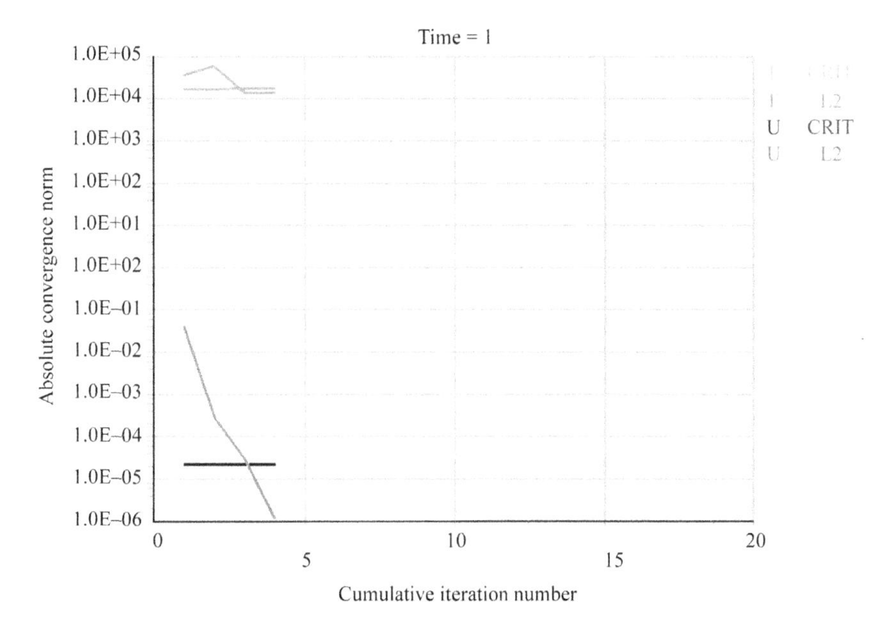

Figure 8.7 Graphical Solution Tracker for Nonlinear Analyses.

New messages should appear in the Output Window approximately every 20 seconds. You can use this information to verify that the program is still attempting to generate a solution. You can also use the information in the Output Window and in the Graphical Solution Tracker to identify when the solution is oscillating and needs to be restarted.

8.9.2. Solution Failure Modes

The solvers in ANSYS are very robust and usually succeed in generating solutions without complication. However, solver failures do occur. This section discusses the four most common solution failure modes that you may encounter when using ANSYS: solution nonconvergence, solution oscillation, a frozen program, and a crashed program.

Solution nonconvergence refers to a failure to achieve equilibrium during a nonlinear analysis. ANSYS will perform equilibrium iterations until equilibrium is reached or until the maximum number of permissible iterations has been performed. (The default maximum number of iterations depends on the type of analysis and can be changed by using the **NEQIT** command.) If equilibrium has not been achieved for a particular substep when that limit is reached and if auto time stepping is on, the solver will attempt to reduce the applied load by reducing the time step. When the minimum time step size has been reached and the maximum number of equilibrium iterations has been used, the solution will terminate. This usually means that the boundary conditions need to be applied in smaller increments over more load steps and substeps. If this occurs, reset your solution options and try rerunning the analysis.

Solution oscillation occurs when the solver is unable to produce a good estimate of the true solution. This forces the program to search for a good estimate rather than allowing it to make steady progress towards the correct the final value. This behavior often results in a visible oscillation in the Graphical Solution Tracker plots. A model with solution oscillation may never converge. To address solution oscillation, you should terminate the current solution, apply your boundary condition in smaller increments over more load steps and substeps, and rerun the analysis.

As noted in section 8.9.1.1, ANSYS sometimes stops supplying information in the Output Window. If the output stops, the solver has probably stopped. Sometimes this is caused by a problem with the computer and sometimes it is a problem with the model. If the problem is related to the computer, rebooting may resolve the issue. If it is a problem with the model, there is probably insufficient memory to assemble the system of equations. If this occurs, you must manually terminate the solution. You should then either reduce the number of elements in your model and rerun the analysis or solve the model on a computer with more memory.

Finally, you may encounter a fatal error that crashes the program. For example, a fatal error will occur if an analysis completely fills the hard disk. If a fatal error occurs, the program will either terminate without warning or it will display a dialog box to inform you that a fatal error has occurred. Clicking OK in the dialog box will terminate the program immediately. The program will not give you an opportunity to save your work and will not allow any other course of action. If a fatal error occurs, you should review the information in the Output Window (before clicking OK in the dialog box) and/or the error file (especially if the program terminates without warning), so you can prevent the error from occurring again.

Although solution failures are rare, they are also often unrecoverable. For this reason, we strongly recommend saving your model before initiating a solution.

8.9.3. Terminating a Running Job

If you encounter a solution failure or need to change something in your model, it may be necessary to terminate a solution while the solver is still running. You cannot cleanly terminate a linear analysis in ANSYS. You can only crash the program. This can be done by closing the Output Window, by closing ANSYS in the Windows Task Manager, or by rebooting your computer. All of these options will leave a lock file in your working directory.

Nonlinear analyses can be terminated by clicking the "Stop" button in the dialog box that appears above the Graphical Solution Tracker (Figure 8.8). This creates a file named jobname.abt in the working directory that contains only the word "nonlinear." ANSYS searches the working directory at the beginning of each equilibrium iteration for this file. If ANSYS finds the file, the solution is aborted. You can also terminate a nonlinear analysis by manually creating this file in the working directory.

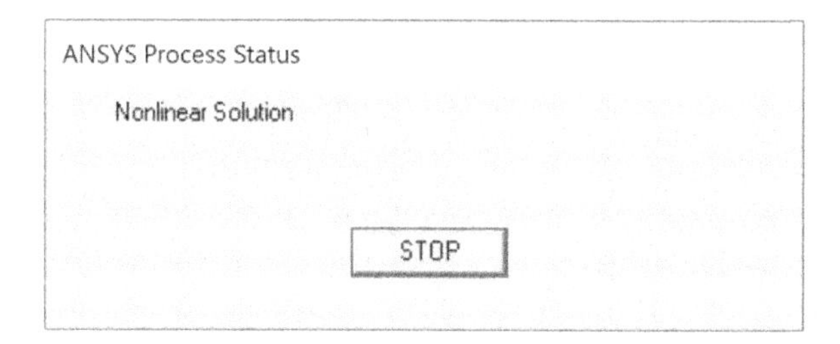

Figure 8.8 "Stop" Button to Terminate a Nonlinear Solution.

8.10. After Solution

It is good practice to verify that a new model is behaving as expected before running a series of load cases or an extensive transient analysis. So it is logical to proceed directly to postprocessing after generating a solution and to return to Solution at a later time to run the model again under different conditions. Unfortunately, ANSYS may not behave as expected when you return.

You learned in chapter 2 that ANSYS was originally a command-driven batch-processing program. The software ran in a data center environment where input was provided via a batch file (originally a box of punch cards), and the results were printed on paper. Today, ANSYS is a fully interactive program, but it still expects you to do all of the postprocessing after all of the variations have been solved. So each time you enter the Solution processor, ANSYS thinks that you have a new model to solve.

If you have a new model to solve or if something about the previous finite element model has changed (e.g., it has more nodes and elements, uses different materials, has different real constants, etc.), then the current model in the database will be inconsistent with the model previously written to the results file. This means that the new results file will be incompatible with the old results file and that a new results file will be needed. Based on this logic, ANSYS creates a new results file when you issue the **SOLVE** command for the first time after (re)entering Solution. It will write additional solution results to this results file until you leave Solution. If you reenter Solution and **SOLVE** again, by default the previous results file will be overwritten and all of your previous results will be lost.

There are two ways to prevent this from happening: you can change the jobname or you can restart your analysis.

8.10.1. Change the Jobname

It is often desirable to have all of your results on the same results file. But if this is not necessary, then you can change the jobname for your analysis after reentering Solution and before issuing the next **SOLVE** command. This will save your new results to a new results file and protect your old data. This approach was used in exercise 6-1.

8.10.2. Restart the Analysis

Finally, you may be able to restart the analysis. Restarting instructs ANSYS to continue the analysis from a specific load step and substep number. By default, a restart analysis is a continuation of the previous analysis and the new results will be appended to the old file after the last load step and substep. This is referred to as a single frame restart. You may also perform a restart at a specific time in the original analysis. This is referred to as a multiframe restart. All results on the results file after the restart point will be deleted to avoid conflicts with the new analysis. You can set the restart controls using the GUI path: **Main Menu > Solution > Analysis Type > Restart** or the **RESCONTROL** command.

Not all analyses can be restarted. In order to restart:

- The analysis type must either be static, transient, or harmonic.
- The physics must either be structural, thermal, or thermal-structural.
- At least one iteration must have been completed in the initial run.
- The initial run must not have terminated abnormally (no killed jobs, system crashes, etc.).
- The initial analysis and the restarted analysis must be done using the same version of ANSYS.

See Section 4.8 of the Mechanical APDL Basic Analysis Guide for more details.

Exercise 8-1

Time Varying Heat Conduction Through a Composite Wall

Overview

In this exercise, you will perform a transient thermal analysis of a hallway with an exterior composite wall. In the first part of the analysis, the air in the hallway will have a uniform temperature and you will solve for the temperature distribution in the wall. Then, the furnace in the building will break and you will determine how quickly the hallway and wall cool.

The wall is composed of a layer of brick that is 0.075 m thick, a layer of rock wool that is 0.025 m thick, and a layer of fiberboard insulation that is 0.025 m thick. The hallway is 1.5 m wide and 3.25 m high. Because this is a 1D problem, only 0.25 m of the height will be included (Figure 8-1-1). This reduces the number of elements in the problem and thus the solution time with no loss of accuracy.

The outside air is a constant −20°C. There is a light wind outside the house that creates an external convective heat transfer coefficient of 10 W/m^2K. The air inside the hallway has an initial constant temperature of 20°C. The natural convection inside the hallway is assumed to be negligible.

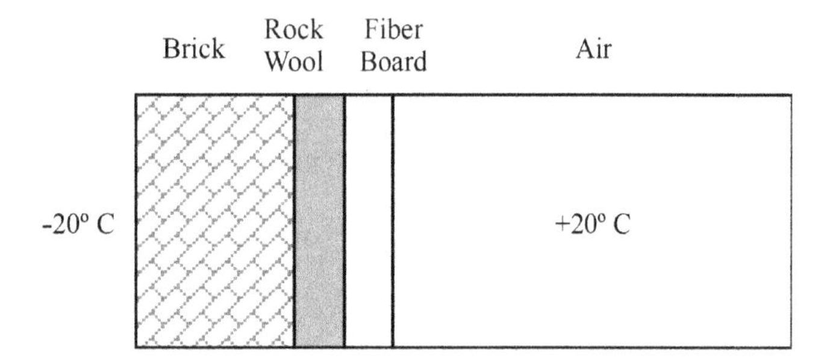

Figure 8-1-1 Schematic of the Wall and Hallway Cross Section.

ANSYS Mechanical APDL for Finite Element Analysis.
DOI: http://dx.doi.org/10.1016/B978-0-12-812981-4.00025-3

Model Attributes

Material Properties

- Brick—Thermal conductivity—0.720 W/mK
- Brick—Specific heat—835 J/kgK
- Brick—Density—1920 kg/m^3
- Rockwool—Thermal conductivity—0.038 W/mK
- Rockwool—Specific heat—418 J/kgK
- Rockwool—Density—130 kg/m^3
- Fiberboard—Thermal conductivity—0.055 W/mK
- Fiberboard—Specific heat—1300 J/kgK
- Fiberboard—Density—290 kg/m^3
- Air—Thermal conductivity—0.0263 W/mK
- Air—Specific heat—1007 J/kgK
- Air—Density—1.1614kg/m^3

Loads

- Forced convection at the outside (left side) of the wall h = 10 W/m^2K.

Constraints

- Outside temperature $T = -20°C$.

Initial Conditions

- Temperature of the air inside the hallway $T = +20°C$.

File Management

Create a new folder in your "Intro-to-ANSYS" folder for Exercise8-1

Change the Working Directory and the Jobname

Start ANSYS

Step 1: Define Geometry

1-1. Create a rectangle with corners at (0,0) and (0.075,0.25)

- Preprocessor > Modeling > Create > Areas > Rectangle > By 2 Corners

1-2. Create a rectangle with corners at (0.075,0) and (0.1,0.25)

1-3. Create a rectangle with corners at (0.1,0) and (0.125,0.25)

1-4. Create a rectangle with corners at (0.125,0) and (1.625,0.25)

1-5. Turn on keypoint numbering

- Utility Menu > PlotCtrls > Numbering ...

In this plot, some of the keypoints and lines overlap. This occurs because a separate set of keypoints and lines was created for each area. These duplicate entities must be merged to ensure that the solid model entities are properly connected, and that the loads (heat) will transfer properly across the boundaries.

1-6. Merge all solid model entities

- Preprocessor > Numbering Ctrls > Merge Items
- For "Label Type of item to be merged" scroll down and choose "All"
- Click OK

Recall from chapter 3, that merging "all" entities will merge all solid model entities, all finite element model entities, and all user-assigned element attributes. This argument should not be used after meshing unless you also intend to merge the element attributes.

1-7. Turn off keypoint numbering

1-8. Save your progress

Step 2: Define Element Types

2-1. Define PLANE55 as the element type for this model

Step 3: Define Material Properties

3-1. Define the material properties for the brick (material #1)

- Preprocessor > Material Props > Material Models (Figure 8-1-2)

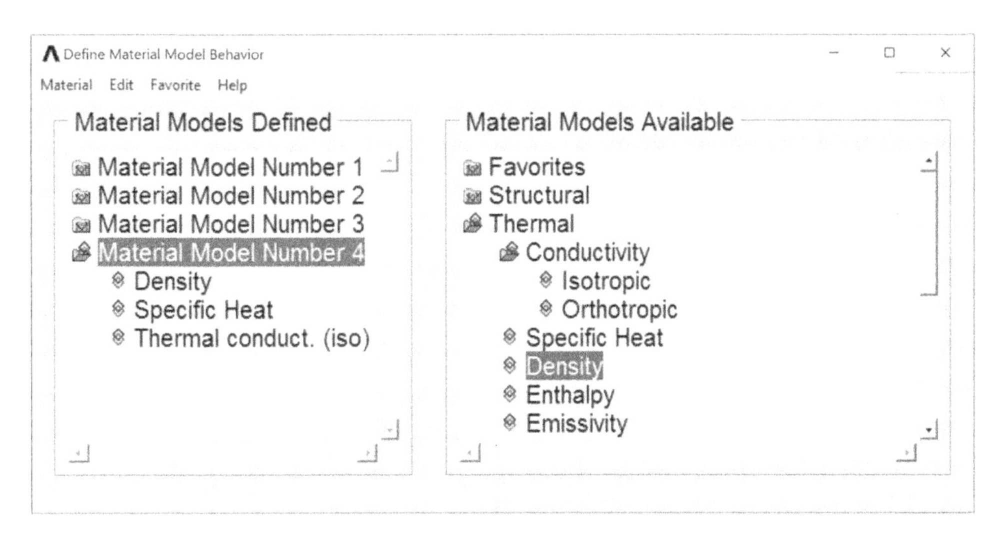

Figure 8-1-2 Four Material Models Defined.

3-2. Define the material properties for the rock wool (material #2)

- Preprocessor > Material Props > Material Models
- Material > New Model . . .

3-3. Define the material properties for the fiber board (material #3)

- Preprocessor > Material Props > Material Models
- Material > New Model . . .

3-4. Define the material properties for the air (material #4)

- Preprocessor > Material Props > Material Models
- Material > New Model . . .

3-5. Save your progress.

Step 4: Mesh

Because this model has four different materials, you will need to set the element (mesh) attributes for each of the four areas. As noted in chapter 6, there are many ways to set the mesh attributes for a model. You will use four different methods in this exercise: specifying the global element (mesh) attributes via the GUI, specifying the global element (mesh) attributes via the command line, specifying the element attributes for all areas in the model, and specifying the element (mesh) attributes for the picked areas.

4-1. Turn on area, node, and element numbering

- Utility Menu > PlotCtrls > Numbering. . .
- Check the box for "AREA Area numbers"
- Check the box for "NODE Node numbers"
- Select "Material numbers" from the "Elem/Attrib numbering" drop down menu
- Change "[/NUM] Numbering shown with" to "Colors only"
- Click OK

4-2. Replot areas if necessary

4-3. Set the mesh attributes for the brick

- Preprocessor > Meshing > Mesh Attributes > Default Attribs
- Confirm that element type #1 is selected
- Confirm that material #1 is selected
- Click OK

This sets the default global element attributes for the model. ANSYS will continue to use these as the default element attributes for all meshing operations until changed.

4-4. Set the element sizing for the model

- Preprocessor > Meshing > MeshTool
- In the Size Controls section of the Mesh Tool, click on the Global "Set" button
- Set the "SIZE Element edge length" to 0.0125
- Click OK

This will specify the element size for all meshing operations until changed. The element size value was chosen to ensure that at least two elements will be generated through the thickness of each area.

4-5. Map mesh the brick wall

- Confirm that the "Mapped" radio button is checked to produce a mapped mesh
- Click the Mesh button
- Choose the area that represents the brick wall (Area 1 on the left)
- Click OK

There should be six elements through the thickness of the brick wall.

4-6. Set the mesh attributes for the rock wool

- In the command line, type:
  `MAT,2`

This sets the default global material reference number to 2. You can confirm this by reopening the Meshing Attributes dialog box using the GUI path: ***Preprocessor > Meshing > Mesh Attributes > Default Attribs****. Material 2 will remain the default material until changed.*

4-7. Map mesh the rock wool wall

- In the command line, type:
  `AMESH,2`

This meshes Area 2 (the rock wool) using the default element sizing and the new default material attribute. There should be two elements through the thickness of the rock wool.

4-8. Set the mesh attributes for the fiberboard wall

- Preprocessor > Meshing > Mesh Attributes > All Areas
- Set the "MAT Material number" to 3
- Click OK

This sets the default area element attributes in the model. ANSYS will continue to use these as the defaults for all area meshing operations until they are changed. These area defaults will override (but not change) the global defaults.

4-9. Map mesh the fiberboard wall

There should be two elements through the thickness of the fiberboard.

4-10. Replot areas if necessary

4-11. Set the mesh attributes for the air

- Preprocessor > Meshing > Mesh Attributes > Picked Areas
- Click on the air (Area 4 on the right)

Remember that you need to click near the centroid of the area. Area 4 is much larger than the others, so you may have to zoom out to select Area 4. You can also enter "4" in the List of Items text box.

- Click OK
- Set the "MAT Material number" to 4
- Click OK

*This operation uses the **AATT** command to set the element attributes for the picked area only. The default global and area element attributes are unchanged.*

4-12. Map mesh the air

- Use the Mesh Tool to mesh the air

The completed finite element mesh is shown in Figure 8-1-3.

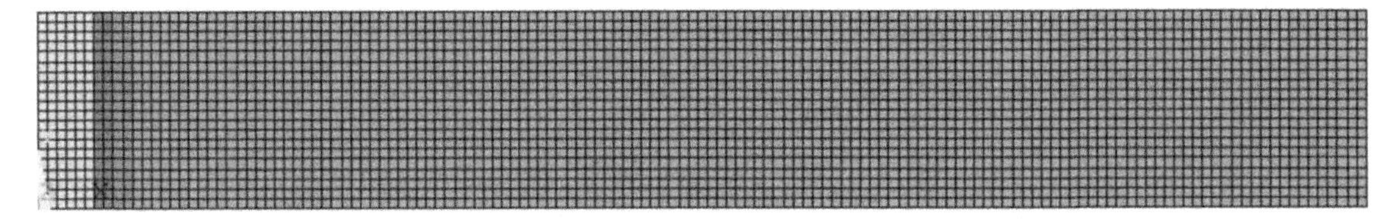

Figure 8-1-3 Finite Element Mesh with Material Numbering On.

4-13. Save your progress.

Step 5: Apply Constraint BCs

This series of operations will select the nodes associated with the air and set their initial temperatures to 20°C. Because the initial temperature of the fiberboard is unknown, the nodes shared with the fiberboard will not be included in the selection set.

There are several ways to do this selection. The steps below will select the nodes attached to the wall and then invert the set to select the nodes attached only to the air. You could select the same set of nodes using two other selection procedures. You could either select the nodes attached to Area 4 and then unselect the nodes on the line that is shared by the fiberboard or you could select the nodes in the air by location (x_{min}, x_{max}). Because the areas were map meshed with a known element size of 0.0125 m, you know that there are nodes located at $x = 0.125$, $x = 0.1375$, $x = 0.1500$, etc. The selection range minimum x_{min} must be greater than 0.125, so the nodes on the inner wall are not included, but it must also be less than 0.1375, so none of the air nodes are missed. Similarly, the selection range maximum x_{max} must be greater than 1.625 to ensure that no air nodes are missed.

5-1. Select all three areas in the wall by location

There are five ways to select the areas that represent the wall. You can select by num/pick (click on the three areas), by num/pick (leave the "List of Items" radio box checked and enter 1,2,3 in the text box), by num/pick (check the "Min, Max, Inc" radio button and enter "1,3" or "1,3,1" in the text box), by location using the x coordinates (enter "0,0.125" in the text box), or by attributes using the material number (enter "1,3" or "1,3,1" in the text box). We encourage you to select by location or by attributes because these are the most reliable and robust selection methods. They are also the best selection methods to use with input files.

- Utility Menu > Select > Entities
- Choose Areas, By Location, X coordinates
- Enter "0,0.125" in the text box
- Click OK

5-2. Plot the selected areas to confirm that the selection operation was successful

- Utility Menu > Plot > Areas (Figure 8-1-4, left)

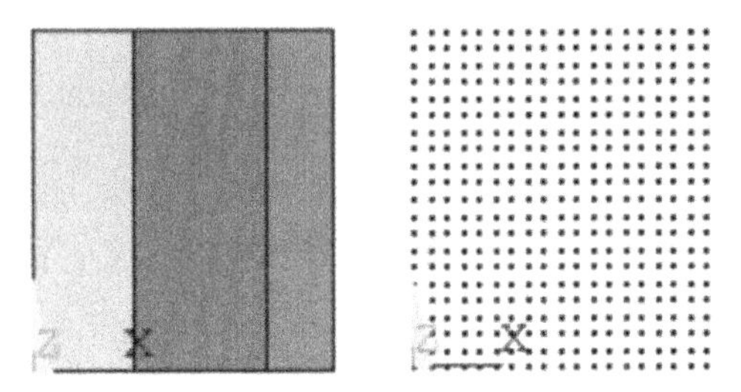

Figure 8-1-4 Area Plot of the Three Areas in the Wall (left) and Plot of the Nodes Attached to Those Areas (right).

5-3. Select the nodes attached to those areas

- Utility Menu > Select > Entities
- Choose Nodes, Attached to, Areas, all
- Click OK

5-4. Plot the selected nodes to confirm that the selection operation was successful

- Utility Menu > Plot > Nodes (Figure 8-1-4, right).

5-5. Invert the selected node set

- Utility Menu > Select > Entities
- Ensure that Nodes is still selected
- Click the Invert button
- Do not click OK

The inversion operation does not require any confirmation. ANSYS will invert the selected set as soon as you click the "Invert" button.

- Click Cancel to close the dialog box.

5-6. Plot the selected nodes to confirm that the selection operation was successful

A plot of the selected set of nodes is shown in Figure 8-1-5.

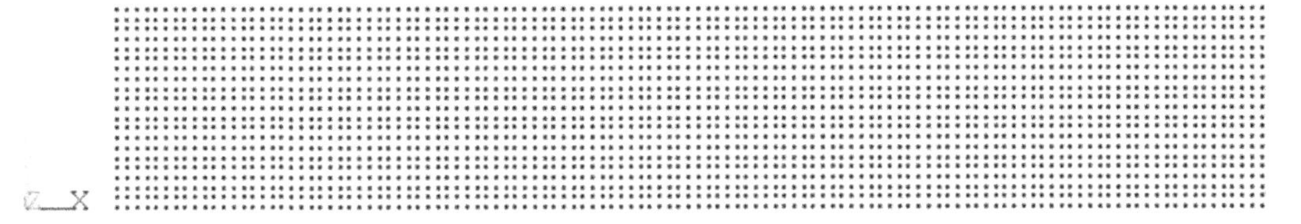

Figure 8-1-5 Plot of the Inverted Set of Nodes Representing the Air.

5-7. Create a component from the selected nodes

- Utility Menu > Select > Comp/Assembly > Create Component
- Type "AIR" for the component name
- Confirm that "Entity Component is made of" is set to "Nodes"
- Click OK

As noted in chapter 7, *a component is a group of model entities. Components are used to facilitate future selection operations.*

5-8. Apply an initial temperature of 20°C to the nodes in the air

- Solution > Define Loads > Apply > Thermal > Temperature > On Nodes
- Pick All
- Choose TEMP
- Enter 20 for "VALUE Load TEMP value"
- Click OK

All visible nodes should now have yellow arrows pointing to the left.

5-9. Save your progress

Step 6: Apply Load BCs

6-1. Select the nodes on the line at $x = 0$

- Utility Menu > Select > Entities
- Choose Nodes, By Location, X coordinates
- Enter 0 in the text box
- Ensure that From Full is selected
- Click OK

6-2. Plot the selected nodes to confirm that the selection operation was successful

The selected nodes should form a vertical line above the triad (i.e., the origin).

6-3. Apply a convection BC to the selected nodes

- Solution > Define Loads > Apply > Thermal > Convection > On Nodes
- Pick All
- Enter 10 for "VALI Film coefficient"
- Enter −20 for "VAL2 Bulk temperature"
- Click OK

Red lines should appear between all of the visible nodes.

*If the convection load had been applied to the line at $x = 0$, the load would have transferred to the nodes when the **SOLVE** command was issued and the result would have been the same. The convection was applied to the nodes to avoid mixing solid model and finite element model BCs.*

6-4. Select everything

- Utility Menu > Select > Everything

6-5. Plot all nodes in the model to confirm that all boundary conditions were correctly applied

- Utility Menu > Plot > Nodes

Figure 8-1-6 shows the full set of applied boundary conditions (i.e. the initial conditions) in the model.

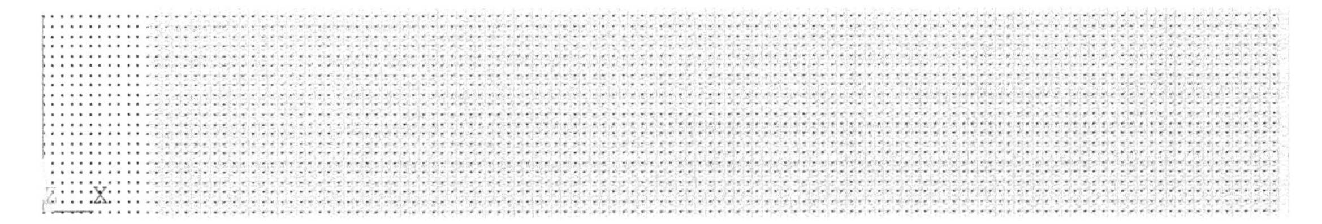

Figure 8-1-6 All Nodes in Model with Applied BCs.

6-6. Save your progress.

Step 7: Set the Solution Options

7-1. Set the solution type to a transient analysis

- Solution > Analysis Type > New Analysis
- Click the Transient radio button
- Click OK to close the New Analysis dialog box
- Click OK to accept the default solution method and close the Transient Analysis dialog box

7-2. Set the solution controls

- Solution > Analysis Type > Sol'n Controls
- Click on the Basic tab
- In the "Time Control" section, enter "0.001" for "Time at end of loadstep"

You cannot solve the model for time t = 0. However, you can solve the model for a very short time that is effectively zero.

- In the "Write Items to Results File" section, confirm that "All solution items" is selected
- In the "Frequency" drop down menu, scroll up and choose "Write every substep"
- Click OK

7-3. Turn off Time Integration

- Solution > Analysis Type > Sol'n Controls
- Click on the Transient tab
- In the "Full Transient Options" section, uncheck "Transient effects"
- Click OK

Or

- In the command line, type:

```
TIMINT,OFF
```

By default, time integration is turned on for transient analyses. You must turn off time integration so the steady-state behavior of the system can be calculated.

Step 8: Solve

8-1. Select everything

- Utility Menu > Select > Everything

8-2. Plot the selected nodes to confirm that the selection operation was successful

8-3. Save your progress

8-4. Solve

- Solution > Solve > Current LS
- Do not click OK

8-5. Review the Solution Options and Load Step Options (Figure 8-1-7)

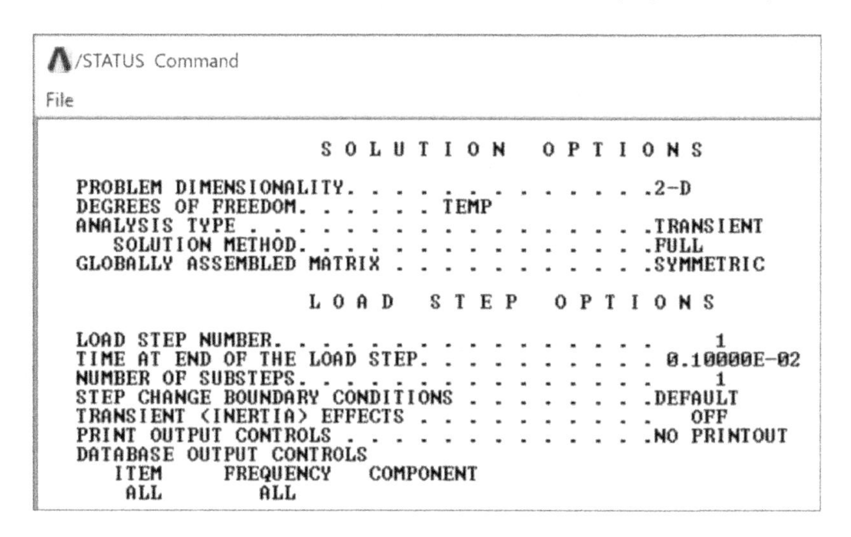

Figure 8-1-7 Solution Options Window for the First Load Step.

For this analysis:

- *The analysis type should be transient.*
- *The solution method should be full.*
- *The load step number should be 1.*
- *The time at the end of the load step should be 0.001 (i.e., 0.1e-2).*
- *Transient (inertia) effects should be off (not blank).*
- *No time integration parameters should be listed.*
- *For the database output controls, both the item and the frequency should be listed as "all" (write all results for all substeps).*

If the information in the Solution Options window is incorrect, click "Cancel" in the Solve Current Load Step dialog box to return to the Solution processor and correct the solution options.

- Click OK to start the analysis

Step 9: Adjust the BCs

9-1. Select the AIR component

- Utility Menu > Select > Comp/Assembly > Select Comp/Assembly . . .
- Select component/assembly by component name
- Click OK
- Select Component AIR
- Click OK

9-2. Plot the selected nodes to confirm that the selection operation was successful

9-3. Delete all temperature constraints on the selected nodes

- Solution > Define Loads > Delete > Thermal > Temperature > On Nodes
- Click Pick All
- Choose TEMP as the "Lab DOFs to be deleted"
- Click OK

Step 10: Set the Solution Options for the Transient Case

10-1. Set the time solution controls

- Solution > Analysis Type > Sol'n Controls
- Click on the Basic tab
- In the "Time Control" section, enter 18000 for "Time at end of loadstep"
- Enter 600 for the "Number of substeps"

This directs the program to solve the model for 18,000 seconds (5 hours) and to calculate 600 solutions (every 30 seconds).

- In the "Automatic time stepping" drop down menu, choose "Off."

This ensures a constant time step of 30 seconds.

- Click on the "Transient" tab
- In the "Full Transient Options" section check the "Transient effects" box
- Click OK

Now the transient model can be solved.

Step 11: Solve the Transient Case

11-1. Select everything

11-2. Save your progress

11-3. Solve

- Solution > Solve > Current LS
- Do not hit OK

11-4. Review the solution options and load step options (Figure 8-1-8)

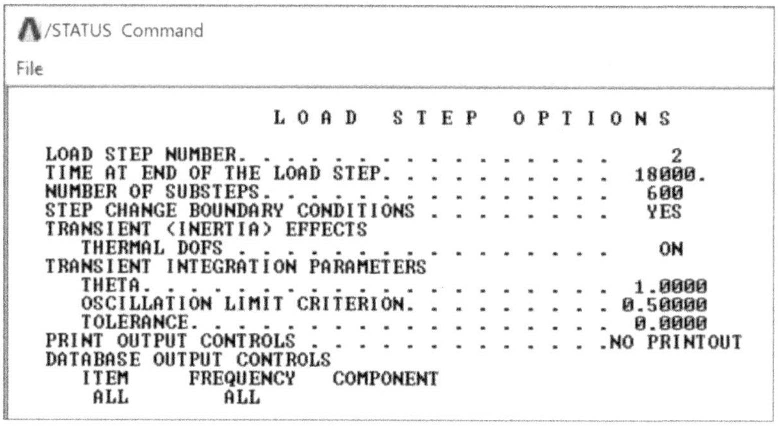

```
/STATUS Command
File

                     L O A D   S T E P   O P T I O N S

LOAD STEP NUMBER. . . . . . . . . . . . . . . . . .        2
TIME AT END OF THE LOAD STEP. . . . . . . . . . .     18000.
NUMBER OF SUBSTEPS. . . . . . . . . . . . . . . .       600
STEP CHANGE BOUNDARY CONDITIONS . . . . . . . . .       YES
TRANSIENT (INERTIA) EFFECTS
   THERMAL DOFS . . . . . . . . . . . . . . . . .        ON
TRANSIENT INTEGRATION PARAMETERS
   THETA. . . . . . . . . . . . . . . . . . . . .    1.0000
   OSCILLATION LIMIT CRITERION. . . . . . . . . .  0.50000
   TOLERANCE. . . . . . . . . . . . . . . . . . .   0.0000
PRINT OUTPUT CONTROLS . . . . . . . . . . . . . .NO PRINTOUT
DATABASE OUTPUT CONTROLS
   ITEM      FREQUENCY    COMPONENT
    ALL         ALL
```

Figure 8-1-8 Solution Options Window for the Second Load Step.

For this analysis:

- *The load step number should be 2.*
- *The time at the end of the load step should be 18,000.*
- *The number of subsets should be 600.*
- *Transient (inertia) effects should be on.*
- *Three transient integration parameters should be listed (theta, oscillation limit criterion, and tolerance).*
- *For the database output controls, both the item and the frequency should be listed as "all" (write all results for all substeps).*

If the information in the Solution Options window is incorrect, click "Cancel" in the Solve Current Load Step dialog box to return to the Solution processor and correct the solution options. If the Solution Options window shows a load step number of 1, return to step 5 and rerun the model for the initial steady-state condition before proceeding with the transient model.

- Click OK to start the analysis

After you click OK, look at the Output Window to see the feedback for each load step scroll by.

11-5. Save your results

Step 12: Postprocess the Results

The general postprocessor (POST1) can only list or plot results that are currently in the database. Only one set of results can be in the database at any given time. At the end of solution, the results in the database are always from the last substep. Before you can examine the results from another substep, the results from that substep must be loaded into the database.

12-1. Load the steady state results into the database

- General Postproc > Read Results > First Set

The Output Window will confirm that this operation has been performed.

12-2. Plot the steady-state temperature distribution in the wall and the hallway

- General Postproc > Plot Results > Contour Plot > Nodal Solu
- DOF Solution > Nodal Temperature.

Figure 8-1-9 shows the initial steady-state temperature distribution. Note that the temperature gradient is mainly within the wall and the outside temperature is a little warmer than ambient because of the heat transfer from the house.

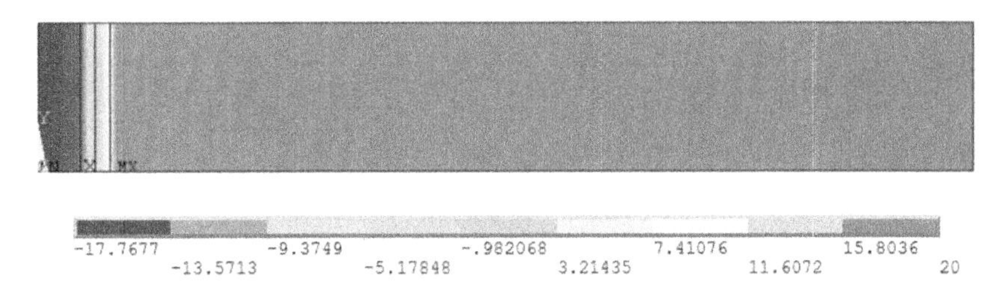

Figure 8-1-9 Steady-State Temperature Distribution Before the Furnace Breaks (3D Graphics).

12-3. Load the results for 1 hour after the furnace breaks

- General Postproc > Read Results > By Time/Freq
- Enter 3600 in the textbox for "TIME Value of time or freq"
- Click OK

The time at 1 h = (60 s/min) x (60 min/h) = 3600 s.

12-4. Plot the temperature in the hallway 1 hour after the furnace breaks

- General Postproc > Plot Results > Contour Plot > Nodal Solu
- DOF Solution > Nodal Temperature

This shows that the temperature in the wall is already below zero. Any pipes that are embedded in the wall are at risk of freezing.

12-5. Load the results for 5 hours after the furnace breaks

- General Postproc > Read Results > Last Set

12-6. Plot the temperature in the hallway 5 hours after the furnace breaks

- General Postproc > Plot Results > Contour Plot > Nodal Solu
- DOF Solution > Nodal Temperature

Figure 8-1-10 shows that the temperature at the far side of the hallway has dropped to almost 15°C. In the area near the wall, the temperature is below −10°C. Notice also that the temperature of the outside wall has dropped from −17.8°C to −18.5°C because of the reduced temperature in the hallway.

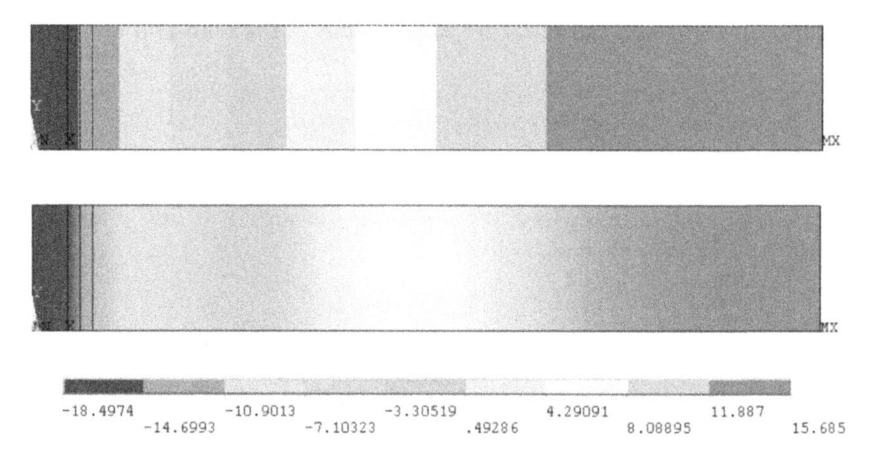

Figure 8-1-10 Temperature Distribution 5 Hours After the Furnace Breaks: Win32 Graphics (top), 3D Graphics (bottom).

These results also show that the model is unrealistic. The warmer air closer to the interior of the house will rise, whereas the cooler air near the exterior wall will sink. This will create natural convection currents that will mix the air in the room and cool it in a more uniform manner.

12-7. Animate the temperature drop over time

- Utility Menu > PlotCtrls > Animate > Over Time...
- Enter 20 in the text box for "Number of animation frames"
- Enter 1 in the first text box for "Range Minimum, Maximum"
- Enter 18000 in the second text box for "Range Minimum, Maximum"
- For the "[PLDI, PLNS, PLVE, PLES] Display Type" verify that "DOF solution" and "Temperature TEMP" are chosen
- Click OK
- In the Animation Controller, set the animation to "Forward Only"
- Close the Animation Controller when you are finished looking at the animation

Step 13: Time History Postprocessing

The General Postprocessor (POST1) displays the results for the entire model at one time. To look at the results for a particular part of the model for all time, you must use the Time History Postprocessor (POST26).

In this section you will examine the temperature drop in six locations throughout the model: the outside of the wall (x = 0), the inside of the brick/outside of the rock wool (x = 0.075), the inside of the rock wool/outside of the fiber board (x = 0.1), the inside of the wall (x = 0.125), the mid-point of the hallway (x = 0.875), and the far side of the hallway (x = 1.625). Since this is a 1D problem, the location of the results in y is unimportant. You will use the vertical mid-point of the model (y = 0.125) for convenience.

13-1. Plot the nodes in the model

13-2. Pan and zoom the model until all 21 nodes at the left edge of the model are clearly visible

- Utility Menu > PlotCtrls > Pan Zoom Rotate...
- Choose the "Box Zoom" option

13-3. Start the Variable Viewer

- Main Menu > TimeHist Postpro

This should open the Variable Viewer (Figure 8-1-11) automatically. If it does not, use the GUI Path: ***TimeHist Postpro > Variable Viewer*** *to open it.*

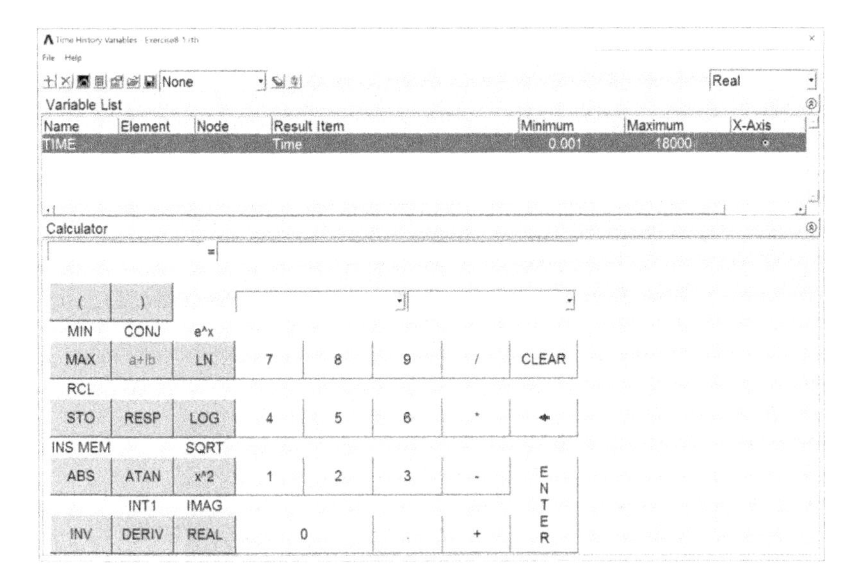

Figure 8-1-11 Time History Variable Viewer.

13-4. Close the Variable Viewer calculator

- Click on the double upward arrow ⓐ to collapse the calculator

Figure 8-1-12 shows the Variable Viewer after the calculator has been collapsed.

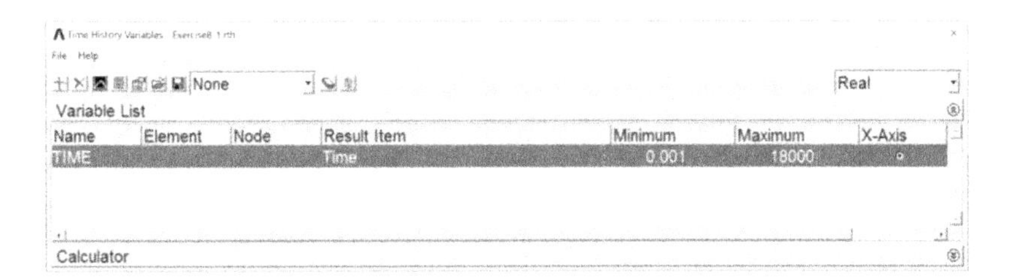

Figure 8-1-12 Time History Variable Viewer After Collapsing the Calculator.

POST26 requires you to identify the numbers of the nodes or elements whose results will be retrieved and plotted. The following operations will identify the nodes in the area of interest and define postprocessing variables for those nodes.

13-5. Add Variable 2 to the Variable Viewer

- Click on the Add Data button in the Variable Viewer menu bar
- Choose Nodal Solution > DOF Solution > Nodal Temperature in the dialog box
- Change the "Variable Name" from "TEMP_2" to "OutsideBrick_2"
- Leave the "Sector Number" blank
- Click OK

This will open the Picker, so you can select the node to associate with this variable.

- Click on the node at the vertical mid-point of the leftmost line in the model

This node is 11th from the top and from the bottom of the model and is located at (0,0.125,0). It is Node 43 in Revision 17.2.

- Click OK

The variable "OutsideBrick_2" is officially Variable 2. You can confirm this in the Output Window (Figure 8-1-13).

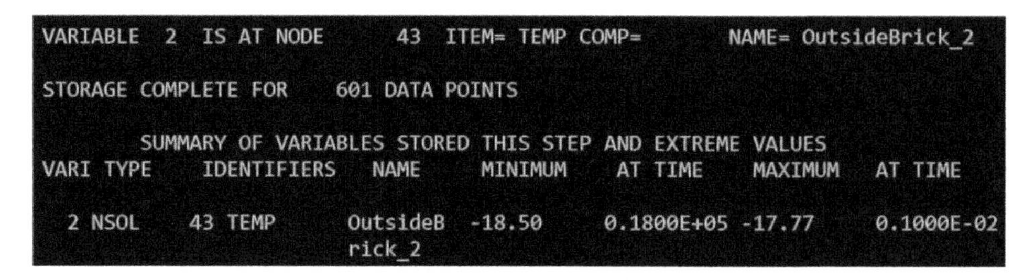

Figure 8-1-13 Output Window Verification of Variable Definition.

13-6. Add the remaining variables to the Variable Viewer

Repeat the procedure above to add the other variables to the Variable Viewer or enter the following commands in the command prompt to add the remaining variables:

```
NSOL,3,NODE(0.075,0.125,0),TEMP,,InsideBrick_3
NSOL,4,NODE(0.100,0.125,0),TEMP,,InsideWool_4
NSOL,5,NODE(0.125,0.125,0),TEMP,,InsideWall_5
NSOL,6,NODE(0.875,0.125,0),TEMP,,HallCenter_6
NSOL,7,NODE(1.625,0.125,0),TEMP,,HallEnd_7
```

The second argument for the ***NSOL*** *command is the node number. We don't know which node numbers to specify. However, we know their locations, therefore we can specify them using the node GET function NODE(x,y,z). This function returns the node number of the node closest to the specified (x,y,z) coordinates. This is advanced APDL command syntax and beyond the scope of this book. It has been included in this exercise for convenience.*

If you add the other variables using the command line, you will need to click the Refresh Time-History Data (F5) button to display all defined variables in the Time History Variable Viewer (Figure 8-1-14).

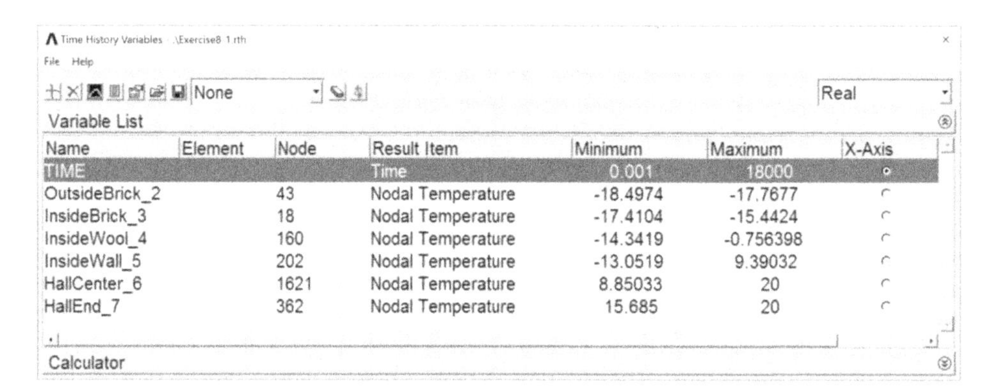

Name	Element	Node	Result Item	Minimum	Maximum	X-Axis
TIME			Time	0.001	18000	
OutsideBrick_2		43	Nodal Temperature	-18.4974	-17.7677	
InsideBrick_3		18	Nodal Temperature	-17.4104	-15.4424	
InsideWool_4		160	Nodal Temperature	-14.3419	-0.756398	
InsideWall_5		202	Nodal Temperature	-13.0519	9.39032	
HallCenter_6		1621	Nodal Temperature	8.85033	20	
HallEnd_7		362	Nodal Temperature	15.685	20	

Figure 8-1-14 All Time History Variables for Plotting.

13-7. Plot the variables

- Click on variable 2 in the Variable Viewer to select it
- Press the Ctrl button and click with the mouse to select variables 3−7 listed in the Variable Viewer
- (Do not click on the TIME variable)
- Click on the Graph Data button in the Variable Viewer menu bar

Or

- Enter the following command in the command prompt:
  `PLVAR,2,3,4,5,6,7`

Time will be used as the x-axis automatically.

The graph of temperature vs. time for the six specified locations in the model is shown in Figure 8-1-15.

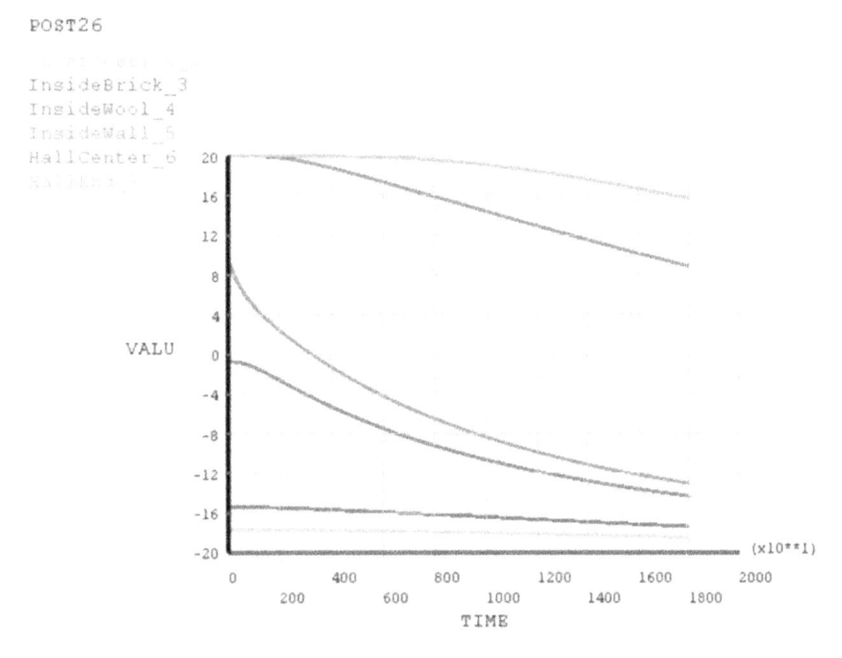

Figure 8-1-15 Temperature vs Time for Six Locations in the Model.

Close the Program

- Utility Menu > File > Exit ...
- Choose "Quit—No save!"
- Click OK

Sample Input File

```
/PREP7                          ! Enter the Preprocessor
RECTNG,0,0.075,0,0.25           ! Create a rectangle to represent the brick
RECTNG,0.075,0.1,0,0.25         ! Create a rectangle to represent the rock wool
RECTNG,0.1,0.125,0,0.25         ! Create a rectangle to represent the fiberboard
RECTNG,0.125,1.625,0,0.25       ! Create a rectangle to represent the air
NUMMRG,ALL                      ! Merge all solid model entities
ET,1,PLANE55                    ! Use PLANE55 elements
MP,KXX,1,0.72                   ! Define thermal conductivity for material #1
MP,C,1,835                      ! Define specific heat for material #1
MP,DENS,1,1920                  ! Define density for material #1
MP,KXX,2,0.038                  ! Define thermal conductivity for material #2
MP,C,2,418                      ! Define specific heat for material #2
MP,DENS,2,130                   ! Define density for material #2
MP,KXX,3,0.055                  ! Define thermal conductivity for material #3
MP,C,3,1300                     ! Define specific heat for material #3
MP,DENS,3,290                   ! Define density for material #3
MP,KXX,4,0.0263                 ! Define thermal conductivity for material #4
MP,C,4,1007                     ! Define specific heat for material #4
MP,DENS,4,1.1614                ! Define density for material #4
ESIZE,0.0125                    ! Set element edge length to 0.0125 m
MAT,1                           ! Use material #1
AMESH,1                         ! Mesh Area 1
MAT,2                           ! Use material #2
AMESH,2                         ! Mesh Area 2
MAT,3                           ! Use material #3
AMESH,3                         ! Mesh Area 3
MAT,4                           ! Use material #4
AMESH,4                         ! Mesh Area 4

/SOLU                           ! Enter the Solution processor
ASEL,S,LOC,X,0,0.125            ! Select all areas associated with the wall
NSLA,S,1                        ! Select all nodes attached to selected areas
NSEL,INVERT                     ! Invert selection to get air only
D,ALL,,20,,,,TEMP               ! Set temperature for all selected nodes to 20
CM,AIR,NODE                     ! Create a component for selected nodes "AIR"
NSEL,S,LOC,X,,0                 ! Select all of the nodes at x=0
SF,ALL,CONV,10,-20              ! Apply convection h=10, T=-20 to selected nodes

ANTYPE,4                        ! Analysis type=4 (Transient analysis)
OUTRES,ALL,ALL                  ! Output all results for all substeps
TIME,0.001                      ! Set time at the end of loadstep to 0.001 s
TIMINT,OFF                      ! Turn time integration off
ALLSEL                          ! Select everything
SOLVE                           ! Solve the steady state model

CMSEM,S,AIR                     ! Select the node component AIR
DDELE,ALL,ALL                   ! Delete all constraints from selected nodes
TIMINT,ON                       ! Turn time integration off
TIME,18000                      ! Time at end of last loadstep=18000 s
NSUB,600                        ! Solve 600 substeps (every 30 seconds)
AUTOTS,OFF                      ! Turn auto time stepping off
ALLSEL                          ! Select everything
SOLVE                           ! Solve the transient model
```

```
FINISH                          ! Finish and Exit Solution

/POST1                          ! Enter the General Postprocessor
SET,1                           ! Read first set of results
PLNSOL,TEMP,,0                  ! Plot the nodal temperature
SET,,,1,,3600                   ! Read results at t=3600 seconds
PLNSOL,TEMP,,0                  ! Plot the nodal temperature
SET,LAST                        ! Read the last set of results
PLNSOL,TEMP,,0                  ! Plot the nodal temperature
ANTIME,20,0.5,,1,0,1,18000      ! Animate the results from the last contour plot
FINISH                          ! Finish and Exit General Postprocessor

/POST26                         ! Enter the Time History Postprocessor
NSOL,2,NODE(0.000,0.125,0),TEMP,,OutsideBrick_2  ! Define variable 2
NSOL,3,NODE(0.075,0.125,0),TEMP,,InsideBrick_3   ! Define variable 3
NSOL,4,NODE(0.100,0.125,0),TEMP,,InsideWool_4    ! Define variable 4
NSOL,5,NODE(0.125,0.125,0),TEMP,,InsideWall_5    ! Define variable 5
NSOL,6,NODE(0.875,0.125,0),TEMP,,HallCenter_6    ! Define variable 6
NSOL,7,NODE(1.625,0.125,0),TEMP,,HallEnd_7       ! Define variable 7
PLVAR,2,3,4,5,6,7               ! Plot variables 2 through 7 vs. TIME
FINISH                          ! Finish and Exit the Time History Postprocessor

SAVE                            ! Save the database
!/EXIT                          ! Exit ANSYS
```

Chapter

9

Postprocessing

Suggested Reading Assignments:
Mechanical APDL Basic Analysis Guide: Chapters 5–7
Mechanical APDL Basic Analysis Guide: Chapter 8 Section 8.2
Mechanical APDL Basic Analysis Guide: Chapter 11
Mechanical APDL Modeling and Meshing Guide: Chapter 3 Sections 3.3–3.5
Mechanical APDL Element Reference: Chapter 3 Sections 3.2–3.3

CHAPTER OUTLINE

This chapter provides an overview of postprocessing in ANSYS. You will learn about the types of results that are and can be calculated in ANSYS, where those results are stored, and how to access them. Next, you will learn about various results display options. You will learn how to list, plot, graph, animate, operate on, and combine results, and how to save postprocessing graphics and information. Finally, you will learn about model verification and validation.

9.1. Postprocessing Overview

Postprocessing refers to all activities associated with the retrieval, manipulation, and display of results from a finite element model. Postprocessing has six basic steps: choosing which results to retrieve; retrieving those results; storing the retrieved results for additional processing (if necessary or desired); displaying and viewing the retrieved results; verifying the model so its results can be used; and interpreting the results to support engineering practice. The first and last steps

ANSYS Mechanical APDL for Finite Element Analysis.
DOI: http://dx.doi.org/10.1016/B978-0-12-812981-4.00009-5

depend on the model's purpose. For example, the results can be used to demonstrate how a proposed design will function, to predict whether a proposed design will be able to meet its requirements, to optimize a design or a set of design parameters, to analyze the performance of an existing design, and to determine the cause(s) of failure in an existing device. Choosing the results to retrieve, model verification, and results interpretation require good engineering judgment and are independent of the finite element program used. This chapter focuses on results retrieval, storage, manipulation, and display, which are program specific.

In ANSYS, you can retrieve a single result such as the minimum temperature or maximum stress in a model. You can operate on results to calculate information such as the average temperature of a body or the total contact area between two surfaces. You can list or plot the results for a single location, for a subset of locations in the model, along a predefined path, or for the entire model for a given load step or at a given instant in time. You can list and graph results for a single location or a set of locations over multiple load steps or time. And, you can animate results plots over multiple load steps or time. Most postprocessing tasks associated with results retrieval, storage, manipulation, and visualization can be performed using one of the ANSYS postprocessors (POST1 or POST26). However, some postprocessing tasks require APDL. Results data can also be exported and processed using another program. APDL operations and data export are advanced postprocessing topics and beyond the scope of this book.

9.2. Types of Results

In ANSYS, the data produced during solution is referred to as either primary or derived data. It is also referred to as either nodal solution data or element solution data. The documentation uses both sets of terminology, sometimes interchangeably.

The primary data are the degree of freedom solutions. This information is calculated and reported at the nodes. The derived data are then calculated for each element (usually at the integration points) using the degree of freedom solution. The derived data may be reported at each node of each element, at each integration point of each element, or at the centroid of each element. Because of this, the derived data is sometimes referred to as the "integration point solution." Examples of the primary and derived data from Table 5.1 in the Mechanical APDL Basic Analysis Guide are shown in Figure 9.1.

Table 5.1: Primary and Derived Data for Different Disciplines

Discipline	Primary Data	Derived Data
Structural	Displacement	Stress, strain, reaction, etc.
Thermal	Temperature	Thermal flux, thermal gradient, etc.
Magnetic	Magnetic Potential	Magnetic flux, current density, etc.
Electric	Electric Scalar Potential	Electric field, flux density, etc.
Fluid	Velocity, Pressure	Pressure gradient, heat flux, etc.
Diffusion	Concentration	Concentration gradient, diffusion flux, etc.

Figure 9.1 Examples of Primary and Derived Results Data by Physics.
Source: Mechanical APDL Basic Analysis Guide.

The nodal solution data for the degrees of freedom is simply the primary data. Similarly, the element solution data is the derived data. The averaged nodal solution data for the derived results (stresses, strains, fluxes, gradients, etc.) is calculated by extrapolating the element solution data back to the nodes and then averaging the element solution data from all elements attached to each node. This provides a better estimate of the true values of the derived results at each location and allows the derived results to be reported and retrieved for the nodes. An averaged nodal solution and its associated element solution from exercise 6-1 are shown in Figure 9.2.

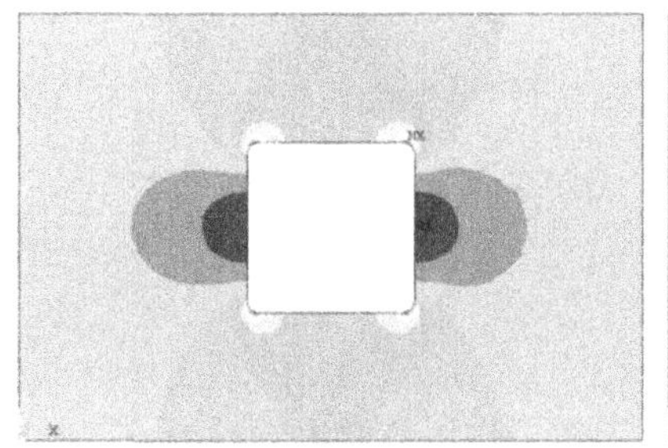
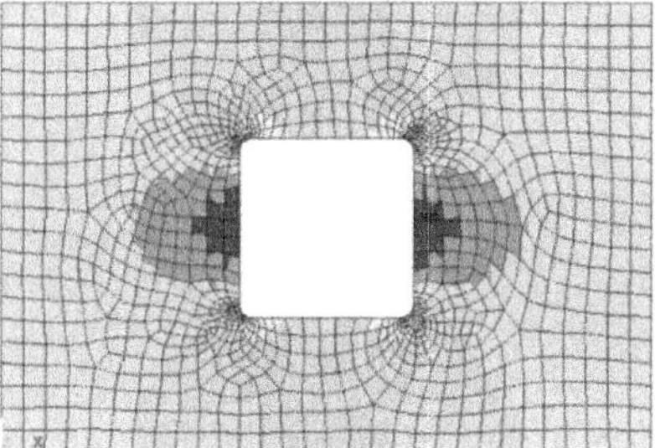

Figure 9.2 Thermal Flux Contour Plots from Exercise 6-1: Averaged Nodal Solution (left) and Element Solution (right).

The General Postprocessor rarely requires you to differentiate between the nodal DOF solution and the averaged nodal solution for the derived results. However, sometimes the Time History Postprocessor does. For example, **NSOL** is the POST26 command to define a variable that will contain a series of nodal DOF data, **ESOL** is the POST26 command to define a variable that will contain a series of element data, and **ANSOL** command is the POST26 command to define a variable that will contain the averaged nodal solution data for a series of derived results. Therefore, it is important to be aware that the distinction exists.

Because the (averaged) nodal solution represents the best estimate of the 'true' results, it is most commonly used for model validation and engineering decision-making. The element solution data provides a more accurate representation of the behavior of the model and is therefore most commonly used for model verification.

9.3. Available Results

The results that can be output for an analysis and where that information is stored depend on the element(s) used, the loads defined, the key options selected, the solution options specified, and whether the program was run in interactive or batch mode.

The solution output that can be calculated and stored for each element is described in the Output Data section of its Element Library entry. For most elements, this includes the nodal degree of freedom results and the element output described in the Element Output Definitions table (Figure 9.3). A few elements can have additional output. This information is usually restricted to the output file and is listed in the element's Miscellaneous Element Output table (Figure 9.4). As noted in chapter 8, the solution output that is actually calculated and stored depends on the options specified by the **OUTRES** command.

Table 183.1: PLANE183 Element Output Definitions

Name	Definition	O	R
EL	Element number	-	Y
NODES	Nodes - I, J, K, L (for KEYOPT(1) = 0 and I, J, K (for KEYOPT(1) = 1)	-	Y
MAT	Material number	-	Y
THICK	Thickness	-	Y
VOLU:	Volume	-	Y
XC, YC	Location where results are reported	Y	4
PRES	Pressures P1 at nodes J, I; P2 at K, J; P3 at L, K; P4 at I, L (P4 only for KEYOPT(1) = 0	-	Y
TEMP	Temperatures T(I), T(J), T(K), T(L) (T(L) only when KEYOPT(1) = 0)	-	Y
S:X, Y, Z, XY	Stresses (SZ = 0.0 for plane stress elements)	Y	Y
S:1, 2, 3	Principal stresses	-	Y
S: INT	Stress intensity	-	Y
S:EQV	Equivalent stress	-	Y
EPEL:X, Y, Z, XY	Elastic strains	Y	Y
EPEL:EQV	Equivalent elastic strain [7]	-	Y
EPTH:X, Y, Z, XY	Thermal strains	3	3
EPTH:EQV	Equivalent thermal strain [7]	-	3
EPPL:X, Y, Z, XY	Plastic strains[8]	1	1
EPPL:EQV	Equivalent plastic strain [7]	-	1
EPCR:X, Y, Z, XY	Creep strains	2	2
EPCR:EQV	Equivalent creep strains [7]	2	2
EPTO:X, Y, Z, XY	Total mechanical strains (EPEL + EPPL + EPCR)	Y	-
EPTO:EQV	Total equivalent mechanical strains (EPEL + EPPL + EPCR)	Y	-

Figure 9.3 Partial Element Output Definitions Table for PLANE183.

Table 13.2: PLANE13 Miscellaneous Element Output

Description	Names of Items Output	O	R
Integration Pt. Solution	SINT, SEQV, EPEL, S, MUX, MUY, H, HSUM, B, BSUM	1	-
Nodal Solution	SINT, SEQV, S, H, HSUM, B, BSUM	2	-

Figure 9.4 Miscellaneous Element Output Table for PLANE13.

Where the results are stored depends on several factors. In both interactive and batch mode, the results are written to the results file. New results are appended to the end of the results file for each new load step, substep (if specified), time step, and/or **SOLVE**. Thus, the results file will contain all results sets for the entire analysis. (Recall that exiting and reentering Solution and then issuing a new **SOLVE** command starts a new analysis and will overwrite the entire results file unless you take steps to prevent this.) In interactive mode, results are also written to the database as they are calculated. Since the database can only contain one set of results at a time, each new solution will overwrite any results that were previously stored in the database. Therefore, by default the database will contain (only) the most recent set of results. Finally, results can be written to the output file if desired and specified (see section 9.4 for more information).

Not all results can be written to both the results file and the output file. The two rightmost columns in the Element Output Definitions and Miscellaneous Element Output tables indicate which results are available in which file. A "Y" indicates that the results are available if they were calculated. A "-" indicates that the results are not available in the indicated file even if they were calculated. A number indicates that the results are only calculated, and thus only available, under certain conditions. The number is a hyperlink to a footnote at the bottom of the table that explains the limitation for that specific element. For example, the "1" in the right column of Figure 9.3 links to a footnote that indicates that plastic strains (EPPL: X, Y, Z, XY) are only calculated for PLANE183 elements if nonlinear material properties are used or if large deflections are enabled. The "2" indicates that creep strains are only calculated if the element "has a creep load" (i.e., if creep capabilities have been included in the model). The "3" indicates that thermal strains are only calculated if the element "has a thermal load" (i.e., if thermal expansion capabilities have been included in the model).

In general, all nodal and element results are 'available' from the results file and only a limited number of results (usually the primary results and some derived results) are available in the output file. However, some of the information that is 'available' from the results file (e.g., the averaged nodal results for the derived data) is calculated and stored in the database when requested instead of being stored on the results file.

9.4. Accessing Results From the Output File

Writing results to and reading results from the output file can be a complicated undertaking. Results can only be sent to the output file in batch mode. In addition, because the output file can become very large, results output is automatically suppressed for models with more than 10 elements. Therefore, under normal circumstances the output file will not contain any model results.

To write results to the output file, the instructions to create and solve the model must be contained in an input file, that input file must be run in batch mode, and the input file must specify that all results should be written to the output file (i.e., an `OUTPR,ALL,ALL` must be included before the **SOLVE** command). An input file with these features can be found at the end of exercise 2-1. The examples in this section are from an output file generated using that input file.

After these conditions are met, the output file can be opened from the working directory and viewed with any plain text editor. By default, the output file has the same name as the jobname and has a .OUT extension (i.e., jobname.out). Depending on your operating system settings, it may also appear as the jobname without an extension. If you did not change the output file to have a unique or obvious name and do not see a file with a .OUT extension, the output file can be identified in Windows by viewing the folder Details and looking for the file Type listed as "OUT File."

The output file contains information and echoes from the program, which are similar to those in the Output window. The output file begins with a series of legal notices, followed by information about the installed software and the command line arguments, a release and time stamp, a listing of the input file used for the analysis, and another release and time stamp. Next, the output file lists the preprocessing and solution output, including solution options, load step options, and the solution echo. Another release and time stamp is added after the solution is finished. Then, the model results are listed.

During solution, the primary (degree of freedom) results are calculated first. Therefore, the primary results are the first results listed in the output file. All degree of freedom values are calculated before the output is generated, so the nodal results are usually in ascending order. Figure 9.5 shows the beginning of the printed nodal solution—the displacements in x (UX) and in y (UY) for Nodes 1 through 10 for load step 1, substep 1, and cumulative iteration 1—from exercise 2-1. Similar results are available for every node in the model. Another release and time stamp follows the nodal results.

```
***** DEGREE OF FREEDOM SOLUTION *****     TIME =    1.0000
LOAD STEP=      1   SUBSTEP =      1   CUM. ITER.=      1

NOTE - ALL VECTOR DOFS ARE IN NODAL COORDINATE SYSTEMS.

  NODE       UX              UY
     1    0.209181E-05    0.329793E-08
     2    0.209181E-05   -0.466477E-06
     3     0.00000       -0.463178E-06
     4     0.00000         0.00000
     5    0.127199E-05   -0.545582E-07
     6    0.104306E-05   -0.880386E-07
     7    0.814123E-06   -0.544033E-07
     8    0.814125E-06   -0.408776E-06
     9    0.104295E-05   -0.375083E-06
    10    0.127199E-05   -0.408616E-06
```

Figure 9.5 Printout of Selected Nodal Results for Exercise 2-1.

The derived results are calculated after the degree of freedom solution. Thus, the element solution is listed in the output file after the DOF solution. Figure 9.6 shows the beginning of the printed element solution for exercise 2-1. The element coordinate stresses (S), the total mechanical strains (EPTO), and the elastic strains (EPEL) are shown in the element coordinate system (11 = X, 22 = Y, 12 = XY) for each of the four integration points of element #20159 for load step 1 and substep 1. Similar results are available for every element in the model. Note that the element results are printed in solution (calculation) order and thus do not necessarily follow the element numbering order. This minimizes the amount of time needed to calculate the element results and the amount of memory needed to store the information before writing the output to file.

```
 ***** ELEMENT SOLUTION *****             TIME =    1.0000
 LOAD STEP=      1   SUBSTEP =      1   CUM. ITER.=      1

Material Point output for element   20159   TYPE - PLANE183

   Intg.Pt. "S"      Stresses (11,22,33,12)
       1      0.99143E+06 -0.10050E+02  0.00000E+00  0.53816E+01
       2      0.99163E+06 -0.72761E+00  0.00000E+00  0.14460E+01
       3      0.99162E+06 -0.73342E+00  0.00000E+00  0.54148E+01
       4      0.99142E+06 -0.10092E+02  0.00000E+00  0.20105E+02

   Intg.Pt. "EPTO"  Strains  (11,22,33,12)
       1      0.13563E-04 -0.44758E-05 -0.44756E-05  0.19583E-09
       2      0.13565E-04 -0.44766E-05 -0.44766E-05  0.52618E-10
       3      0.13565E-04 -0.44766E-05 -0.44766E-05  0.19704E-09
       4      0.13563E-04 -0.44758E-05 -0.44756E-05  0.73161E-09

   Intg.Pt. "EPEL"  StrainsEL(11,22,33,12)

       1      0.13563E-04 -0.44758E-05 -0.44756E-05  0.19583E-09
       2      0.13565E-04 -0.44766E-05 -0.44766E-05  0.52618E-10
       3      0.13565E-04 -0.44766E-05 -0.44766E-05  0.19704E-09
       4      0.13563E-04 -0.44758E-05 -0.44756E-05  0.73161E-09
```

Figure 9.6 Printout of Selected Element Results for Exercise 2-1.

The results for new solutions are appended to the end of the output file. This means that the full set of results for a new load step, substep, and/or time step will follow the full set of results for the previous load step, substep, and/or time step. This substantially increases the difficulty of locating specific results information for an analysis with multiple load steps and/or time steps.

There are three main conclusions to draw from this discussion. First, results are nothing more than numbers associated with one or more degrees of freedom and one or more finite element entities. Second, the numbers themselves usually have relatively little meaning unless they are combined and/or queried and then displayed in a manner that allows them to be understood and interpreted. And third, manually locating, reviewing, and/or visualizing the printed output to obtain specific information about a given load step or substep is tedious, error-prone, and impractical, even for the smallest finite element models. For these reasons, results are almost always accessed from the results file via the database and then listed, plotted, graphed, or animated using one of the postprocessors.

9.5. Accessing Results From the Results File

As noted in chapter 2, the results file is a binary file that is intended to be accessed through the ANSYS database. (Information about how to access the results file directly can be found in the Mechanical APDL Programmer's Reference.) The results file has the same file name as the jobname. However, the extension of the results file depends on the physics of your analysis. For structural and coupled field analyses, the results file has a .RST extension (i.e., jobname.rst). For thermal and diffusion analyses, the results file has a .RTH extension (i.e., jobname.rth). For magnetic field analyses, the results file has a .RMG extension (i.e., jobname.rmg). For fluid analyses, the extension is either .RST (if structural degrees of freedom have been included in the model) or .RTH (if they have not).

Data is stored on the results file by data set number. Data set numbers are assigned based on the order in which the results were calculated. (The first data set calculated is set #1, the second data calculated set is set #2, and so on.) ANSYS also stores the time, load step number, and substep number for each data set in a table that refers back to the data set numbers. This allows you to look up and load results sets by any of these characteristics.

Results sets in ANSYS are also associated with a cumulative iteration number. This represents the number of times that the model has passed through the equation solver. For linear analyses, only one iteration is required to obtain a solution. Therefore, the cumulative iteration number and the data set number will be the same. However, nonlinear analyses usually require multiple equilibrium iterations to reach a solution. Therefore, the cumulative number will be the total number of iterations used including equilibrium iterations that did not converge. You cannot choose a results set based on cumulative iteration number, but you will see references to it in the output file and when viewing the available results sets in the GUI.

You can list the data sets that are available on the results file by following the GUI path: **Main Menu > General Postproc > Results Summary** or by issuing the command `SET,LIST`. Figure 9.7 shows a typical Results Summary listing that includes the data set number, the time associated with that data set, and the load step, substep and cumulative iteration numbers for the data set.

SET,LIST Command

File

***** INDEX OF DATA SETS ON RESULTS FILE *****

SET	TIME/FREQ	LOAD STEP	SUBSTEP	CUMULATIVE
1	200.00	1	1	1
2	400.00	1	2	2
3	600.00	1	3	3
4	800.00	1	4	4
5	1000.0	1	5	5
6	1200.0	1	6	6
7	1400.0	1	7	7
8	1600.0	1	8	8
9	1800.0	1	9	9
10	2000.0	1	10	10
11	2200.0	1	11	11

Figure 9.7 Partial List of Data Sets from Exercise 8-1.

9.5.1. Accessing the Results File Through the General Postprocessor

The General Postprocessor operates on the data stored in the database. Therefore, it can only access and process one results set at a time. You can load a set of results into the database by following the GUI path: **Main Menu > General Postproc > Read Results** (Figure 9.8). Below this level, there are options to select the first set, the next set, the previous set, or last set on the results file. The next and the previous results set are based on the set that is currently in the database. (The data set after the last set is the first set and the data set before the first set is the last set). You can also choose a data set by one of its characteristics (set number, load step number, or time/frequency) or "By Pick." The "By Pick" option lists the available data sets on the results file, so you can choose based on all of the data set characteristics (Figure 9.9). Finally, you can load a set of results by issuing the **SET** command.

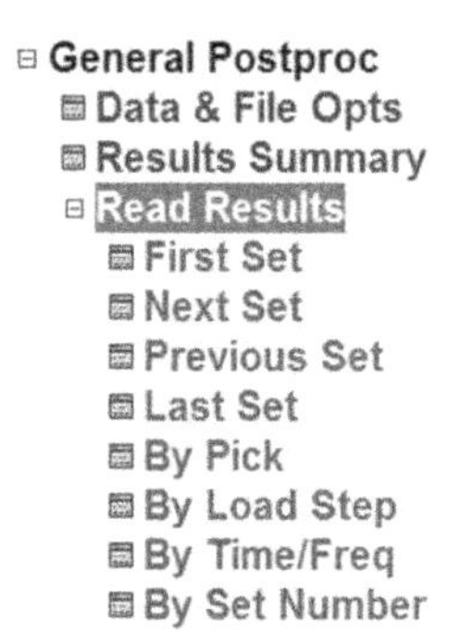

Figure 9.8 Choosing a Results Set to Load Into the Database for Postprocessing.

If the load step number is used to identify a data set and no substep number is specified, you will get the last substep for that load step. If time is used to identify a data set and the specified time is between two solution times, ANSYS will linearly interpolate the results using the results from the time before and after the specified time. If the specified time is after the last time, the last time will be used.

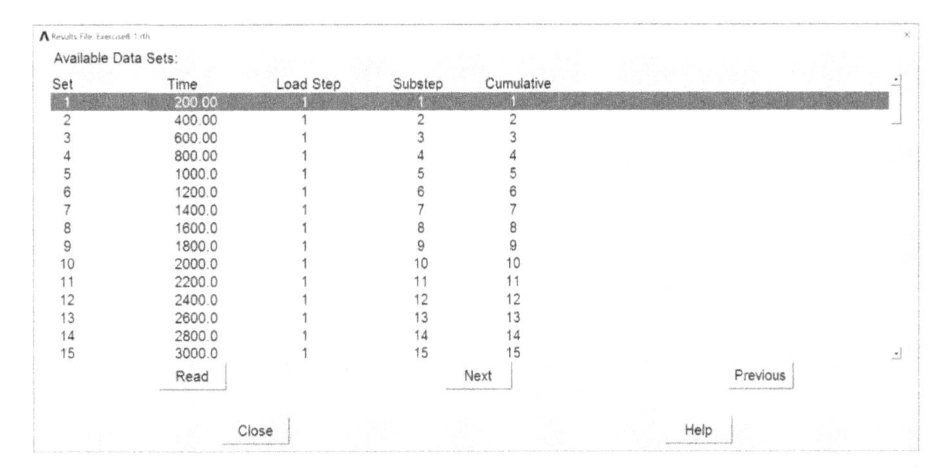

Figure 9.9 Picking a Data Set to Load from Exercise 8-1.

Once a data set from the results file has been read into the database, you can begin performing postprocessing operations (e.g., listing, plotting, etc.) on that data.

9.5.2. Accessing Results from the Results File Through the Time History Postprocessor

The Time History Postprocessor does not operate on a single set of results stored in the database. Instead, it operates on variables that store the results for a single node or a single element integration point (i.e., for a single location) for all load steps, substeps (if applicable), time steps, and/or frequencies in the analysis (by default), or for a subset of these data sets (if specified). The Time History Postprocessor queries the results file to obtain the necessary information when POST26 variables are defined and stored.

9.5.3. Accessing Results via Component Name or Sequence Number

Almost all results can be listed, plotted, graphed, and animated. However, the commands and GUI paths in both postprocessors depend on whether the results are identified by an item and a component name or by an item and a sequence number.

All nodal results and many element results are identified by an item and component name. These can be accessed through the Result Item dialog boxes in both postprocessors (Figure 9.10). The first column of each element's Element Output Definitions table lists the results by Item: Component. For example, the Element Output Definition for PLANE183 shown in Figure 9.3 indicates that the coordinate stresses, principal stresses, and equivalent stress have an item of "S." The component names for the coordinate stresses are "X," "Y," "Z," and "XY." The component names for the principal stresses are "1," "2," and "3." The component name for the equivalent stress is "EQV." Therefore, the command to create a contour plot of the nodal solution for the equivalent stress would be `PLNSOL,S,EQV`. Similarly, the command to create a contour plot of the element solution for the equivalent stress would be `PLESOL,S,EQV`. Most of the results for continuum elements (PLANEs, SOLIDs, etc.) are identified using the component method.

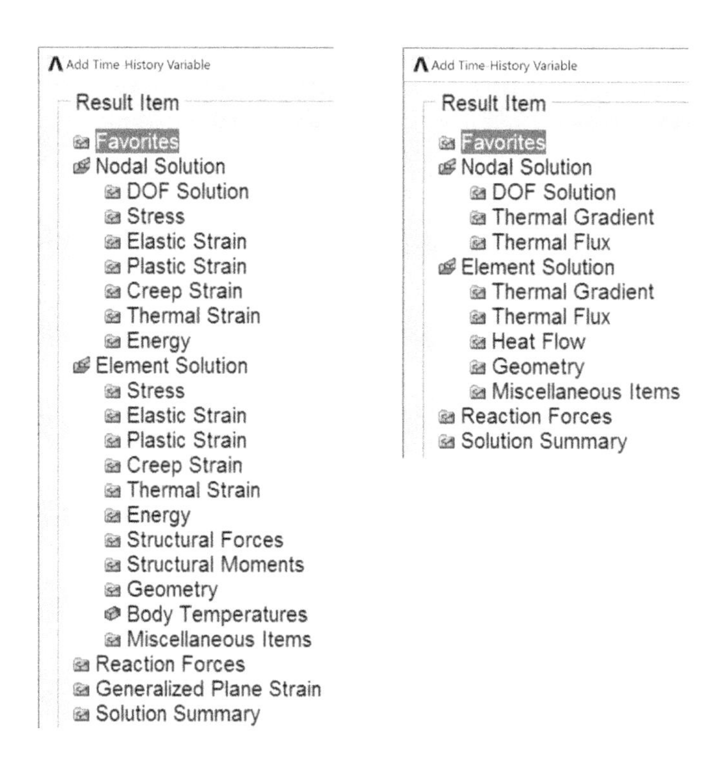

Figure 9.10 Component Name Results Items Available for a Structural Analysis (Exercise 2-1) (left) and for a Thermal Analysis (Exercise 8-1) (right) as Shown in the Time History Postprocessor.

Results that are identified by an item and a sequence number tend to be unaveraged element results or element results that are not naturally single valued. This includes all integration point data; all derived data for structural line elements, thermal line elements, and contact elements; and layer data for layered elements. Shell element results are also often retrieved using the sequence number method. Common "items" include SMISC (summable miscellaneous data), NMISC (non-summable miscellaneous data), LS (line element elastic stresses), LEPEL (line element elastic strains), LEPTH (line element thermal strains), LEPPL (line element plastic strains), LEPCR (line element creep strains), and LBFE (line element temperatures). The sequence numbers always have numerical values. Sequence number method data is usually accessed via element tables. When these results can be accessed through the Result Item menus, they are shown as "Miscellaneous Items."

The Element Output Definitions for results accessed using the sequence number method do not have colons. For example, the Element Output Definition for PLANE183 shown in Figure 9.3 indicates that pressures (P1–P4) and thickness must be retrieved through item and sequence number. The item and sequence numbers for each element are given in its Item and Sequence Numbers Table (Figure 9.11). Using the Item and Sequence number table for PLANE183, we can determine that the command to create an element table named "P1NodeI" for pressure P1 at Node I would be `ETABLE,P1NodeI,SMISC,2`. The command to create an element table named "P1NodeJ" for Pressure P1 and Node J would be `ETABLE,P1NodeJ,SMISC,1`. Element tables are discussed in more detail in Section 9.8.2.4.

Table 183.2: PLANE183 Item and Sequence Numbers

Output Quantity Name	ETABLE and ESOL Command Input									
	Item	E	I	J	K	L	M	N	O	P
P1	SMISC	-	2	1	-	-	-	-	-	-
P2	SMISC	-	-	4	3	-	-	-	-	-
P3	SMISC	-	-	-	6	5	-	-	-	-
P4[*1*]	SMISC	-	7	-	-	8	-	-	-	-
THICK	NMISC	1	-	-	-	-	-	-	-	-

Figure 9.11 Item and Sequence Number Definitions for PLANE183.

9.6. Results Coordinate Systems

All results in ANSYS are calculated, stored, and displayed in a coordinate system. The three most important coordinate systems for postprocessing are the nodal coordinate system, the element coordinate systems, and the results (display) coordinate system. It is important to understand these coordinate systems and to choose an appropriate results display coordinate system before postprocessing your results.

9.6.1. The Nodal Coordinate System

The nodal coordinate system determines the orientation of the degrees of freedom at each node. It is used to calculate and store all nodal results, including the primary degree of freedom solutions, the nodal loads, and the reaction loads. The nodal coordinate system has no impact on the location of the nodes (which is still determined by the global coordinate system) or on the display of the nodal results (which is determined by the results coordinate system).

By default, the nodal coordinate system is parallel to the global Cartesian coordinate system. It is possible to modify the nodal coordinate system using the methods described in Section 3.3 of the Mechanical APDL Modeling and Meshing Guide. If you modify the nodal coordinate system, you can list the nodal coordinate rotation angles with respect to the global Cartesian system using the **NLIST** command or the GUI path: **Utility Menu > List > Nodes**.

9.6.2. Element Coordinate Systems

Each element has its own element coordinate system. The element coordinate system is used to determine the orientation of nonisotropic material properties, to apply pressures, and to calculate and store the element results. All element coordinate systems are right-handed orthogonal systems. In general, line elements are oriented with the element x-axis running from Node I to Node J. In general, shell elements are oriented with the element x-axis running from Node I to Node J, with the z-axis normal to the shell surface, and with the positive direction determined by the right-hand rule. And, in general, 2D and 3D solid elements have element coordinate systems parallel to the global Cartesian coordinate system. However, there are many exceptions. The element coordinate system for each element type can be found in the first figure of its Element Library entry in the Mechanical APDL Element Reference.

Many element coordinate systems can be modified by setting a key option when the element is defined. In addition, the element coordinate systems for area and volume elements can be aligned with a previously defined local coordinate system using the **ESYS** command or the GUI path: **Main Menu > Preprocessor > Modeling > Create > Element > Elem Attributes**. If both a key option and the **ESYS** command are used, the **ESYS** definition will determine the element coordinate system.

9.6.3. The Results (Display) Coordinate System

The results coordinate system determines how the nodal and element results are displayed during postprocessing. The results coordinate system can be changed by using the **RSYS** command or by following the GUI path: **Main Menu > General Postproc > Options for Outp**. After the **RSYS** command is issued, all postprocessing operations (listing, plotting, etc.) will be performed in the new results coordinate system.

The predefined results coordinate systems in ANSYS are:

RSYS 0—Global Cartesian (X, Y, Z) (the default for most analyses)

RSYS 1—Global cylindrical (R, θ, Z)

RSYS 2—Global spherical (R, θ, ϕ)

SOLU—The Solution coordinate system (the default for spectrum analyses)

LSYS—The layer coordinate system (for layered shell and layered solid elements).

The Solution coordinate system for nodal results is the nodal coordinate system. The Solution coordinate system for element results is the element coordinate system for each element. If the nodal or element coordinate system is not defined, the Global Cartesian coordinate system is used. The layer coordinate system is the coordinate system for the current layer of interest in layered shell and solid elements. The layer of interest is defined using the **LAYER** command.

You can use a previously defined local coordinate system as the results coordinate system or define a new local coordinate system to simplify postprocessing activities. As seen in chapter 3, local coordinate systems must be assigned a number greater than or equal to 11 and can be created using the GUI path: **Utility Menu > WorkPlane > Local Coordinate Systems > Create Local CS** or the **LOCAL** command.

Changing the results coordinate system can be extremely useful, especially when postprocessing models where the geometry is fundamentally cylindrical or spherical. For example, Figure 9.12 shows the results for a circular 6061-T6 aluminum plate (radius 1 m) fixed at the center and constrained to prevent rotation with a uniform pressure of 100 N applied to the edge. Plotting the displacements in the x and y directions in the default global Cartesian coordinate system (`RSYS,0`) provides an accurate but unintuitive view of the system (Figure 9.12, left). Plotting the displacements in the 'x' (radial) and 'y' (angular) in global cylindrical coordinates (`RSYS,1`) shows a uniform inward radial displacement and no angular displacement as expected (Figure 9.12, right).

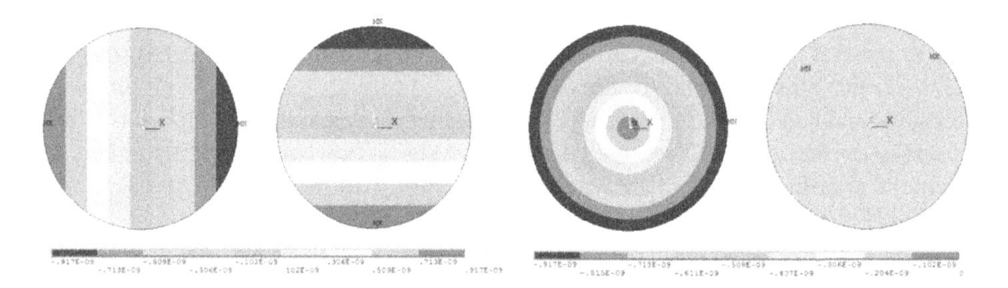

Figure 9.12 X Component of Displacement (RSYS0, left), Y Component of Displacement (RSYS0, left center), Radial Displacement ("X Component of Displacement," RSYS1, right center), and Angular Displacement ("Y Component of Displacement," RSYS1, right) for a Circular Plate Under a Uniform External Pressure (Win32 Graphics).

The **RSYS** command has more exceptions and therefore requires more careful application than the **CSYS** command. For example, the **RSYS** command has no effect on beam or pipe stresses. These are always displayed in the element coordinate system. In addition, `RSYS,SOLU` must be used with care. The element coordinate systems for shell elements and certain solid elements can differ from element to element. This can make nodal averaging of component element results (e.g., the coordinate stresses) invalid. Finally, rotation of nodal results is only valid if the resulting component set is consistent with the degree of freedom set at the node. For example, nodal results with 2 degrees of freedom (x and y) should not be rotated out of the xy plane. Similarly, if the results from nodes without a degree of freedom in z are rotated, any z component that results from that rotation will not be available for listing or plotting.

The results coordinate system will be changed to facilitate postprocessing in exercise 9-3.

9.7. Full Graphics vs PowerGraphics

ANSYS has two modes for calculating and displaying results: Full Graphics and PowerGraphics. Full Graphics is the default in batch mode. PowerGraphics is the default in interactive mode. There are three ways to toggle the graphics: (1) by using the GUI path: **Utility Menu > PlotCtrls > Style > Hidden Line Options** . . ., (2) by using the POWRGRPH button in the ANSYS toolbar, or (3) by using the **/GRAPHICS** command (`/GRAPHICS,FULL` or `/GRAPHICS,POWER`). All postprocessing operations (listing, plotting, etc.) will be performed using the default graphics option or the graphics option specified by the **/GRAPHICS** command. Therefore, it is important to understand the differences between the two and to choose the appropriate graphics option before postprocessing your results.

Full Graphics displays the results for the entire active set of elements. In contrast, PowerGraphics (usually) only displays results for element surfaces that are on the exterior of the active set of elements. This means that PowerGraphics is often much faster and more efficient for 3D analyses. However, because PowerGraphics displays maximum and minimum values that are valid only for the exterior surfaces of the model, they may not be true representations of the maxima and minima for the selected set of entities. These are the most important features of Full Graphics and PowerGraphics for new users.

Both Full Graphics and PowerGraphics print (list) the degree of freedom results for all nodes in the selected set. Thus, listed DOF results from Full Graphics and PowerGraphics will always agree. This behavior is undocumented, but it can be verified easily and is demonstrated in exercise 9-2.

Full Graphics uses the same averaging schemes for plotting and printing the nodal average solution; therefore, the printed and plotted results should always be self-consistent. In contrast, PowerGraphics uses different averaging schemes for plotting and printing; therefore, the maxima and minima for the same set of nodal averaged results may be different when plotting and listing with PowerGraphics.

Because PowerGraphics only considers the model surface, Full Graphics and PowerGraphics also sometimes average results differently. For example, PowerGraphics selects the external surfaces of the set of elements and then selects the nodes associated with those surfaces. Therefore, results for an element that has a single node on the exterior of the model but no surface will be included in a Full Graphics plot but not in a PowerGraphics plot.

In addition, Full Graphics and PowerGraphics behave differently at geometric, element layer, material, and real constant discontinuities. Full Graphics assumes that the solution is continuous and therefore always averages results according to the program defaults. This can lead to inaccurate results near discontinuities (where the solution is not continuous and should not be treated as such). PowerGraphics never averages nodal results (e.g., stresses, fluxes, etc.) across discontinuous surfaces. This fact is poorly documented, but it can be verified easily and is demonstrated in exercise 9-2. By default, PowerGraphics also does not average results at material discontinuities. You can control if it averages results at material and real constant discontinuities using the **AVRES** command. For these reasons, the documentation states that

PowerGraphics provides 'better' results than Full Graphics near discontinuities. Nevertheless, Section 8.2 of the Mechanical APDL Basic Analysis Guide recommends always postprocessing each side of a discontinuity separately, regardless of the graphics option chosen. If the results in the region near a discontinuity are important to the outcome of the analysis, the authors recommend modeling the discontinuity in detail (i.e., using a very fine mesh), postprocessing the results in great detail (e.g., both sides separately and together), and using multiple methods to validate the results.

Full Graphics is valid for all element types. In contrast, PowerGraphics cannot be used with circuit elements, "membrane stiffness only" options, and some diffusion analysis results (e.g., **CONC, CG, DF**, and **EPDI**). Some results data are not supported by PowerGraphics (e.g., failure criteria, energy errors, etc.). See the **PLNSOL, PLESOL, PRNSOL**, and **PRESOL** command documentation for more details. PowerGraphics does not support safety factor calculations. PowerGraphics is not supported by some commonly used commands (e.g., **ESYS, NSEL, PRETAB, PLETAB**, etc.). And, it does not support all options of other commonly used commands (e.g., **/ESHAPE**). In some of these cases, the program will provide the requested results using Full Graphics. In other cases, you will have to manually change to Full Graphics to access the requested results or features.

Finally, Full Graphics allows you to exclude one or more individual nodes from postprocessing calculations performed on the elements that contain those nodes. This procedure is not possible with PowerGraphics. This topic is otherwise beyond the scope of this book.

Because the behavior of the program with PowerGraphics is complicated and poorly documented, we recommend that new users postprocess full 3D models using Full Graphics and use both Full Graphics and PowerGraphics to investigate the behavior of their models near discontinuities.

Exercises 9-2 and 9-3 will explore the use of PowerGraphics and Full Graphics in detail.

9.8. Displaying and Viewing Results

In general, results in ANSYS can be listed, plotted, animated, and graphed. These operations display the results for the active set of nodes or elements in the model (depending on the operation) using the default or specified results display coordinate system and the default or specified graphics mode.

9.8.1. Listing Results

Listing or printing results is similar to looking at results in the output file (but more convenient). Most results for a given time or load step can be listed through the Utility Menu (**Utility Menu > List > Results**) or through the General Postprocessor (**Main Menu > General Postproc > List Results**). Both GUI paths issue the same commands and function in the same way.

Nodal results can be listed directly using the **PRNSOL** command. Nodal results are listed in ascending order by node number. A summary is also included at the bottom of the window to indicate the maximum and/or minimum values in the results set and the nodes where those values exist (Figure 9.13). Often, the program will return more results than requested. For example, requesting the nodal displacements for one component (*x*, *y*, or *z*) will return only those results. However, requesting vector summed results (e.g., displacement vector sum, rotation vector sum, thermal flux vector sum, etc.) will return all of the component displacement results (*x*, *y*, and *z*) in addition to the vector summed results. Similarly, requesting a nodal stress or strain will return a list of all 6 coordinate values (*x*, *y*, *z*, *xy*, *yz*, and *xz*). And, requesting a nodal principle stress or strain, stress or strain intensity, or the equivalent stress or strain will return all 5 stresses or strains (1, 2, 3, INT, and EQV).

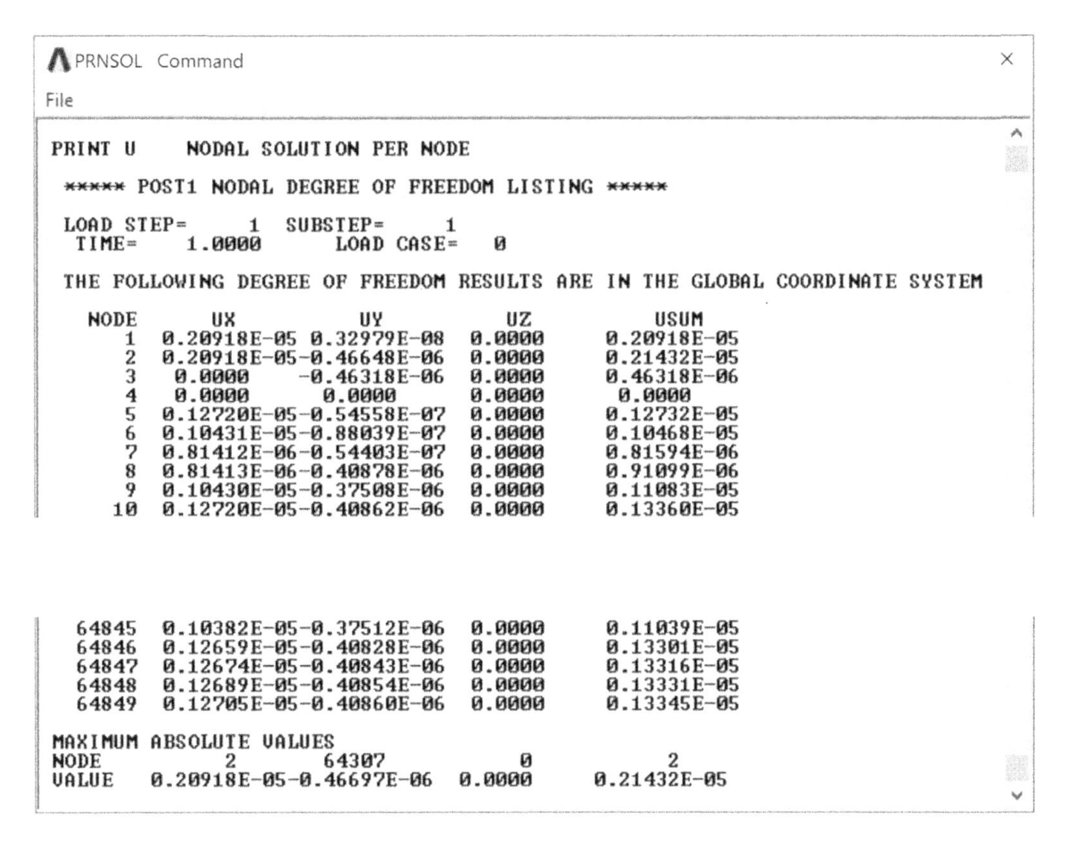

```
PRNSOL Command
File

PRINT U     NODAL SOLUTION PER NODE

 ***** POST1 NODAL DEGREE OF FREEDOM LISTING *****

 LOAD STEP=      1  SUBSTEP=      1
  TIME=    1.0000      LOAD CASE=   0

 THE FOLLOWING DEGREE OF FREEDOM RESULTS ARE IN THE GLOBAL COORDINATE SYSTEM

   NODE       UX          UY          UZ          USUM
      1  0.20918E-05 0.32979E-08  0.0000      0.20918E-05
      2  0.20918E-05-0.46648E-06  0.0000      0.21432E-05
      3   0.0000     -0.46318E-06  0.0000      0.46318E-06
      4   0.0000      0.0000      0.0000       0.0000
      5  0.12720E-05-0.54558E-07  0.0000      0.12732E-05
      6  0.10431E-05-0.88039E-07  0.0000      0.10468E-05
      7  0.81412E-06-0.54403E-07  0.0000      0.81594E-06
      8  0.81413E-06-0.40878E-06  0.0000      0.91099E-06
      9  0.10430E-05-0.37508E-06  0.0000      0.11083E-05
     10  0.12720E-05-0.40862E-06  0.0000      0.13360E-05

  64845  0.10382E-05-0.37512E-06  0.0000      0.11039E-05
  64846  0.12659E-05-0.40828E-06  0.0000      0.13301E-05
  64847  0.12674E-05-0.40843E-06  0.0000      0.13316E-05
  64848  0.12689E-05-0.40854E-06  0.0000      0.13331E-05
  64849  0.12705E-05-0.40860E-06  0.0000      0.13345E-05

MAXIMUM ABSOLUTE VALUES
NODE            2       64307        0            2
VALUE    0.20918E-05-0.46697E-06  0.0000     0.21432E-05
```

Figure 9.13 Beginning (top) and End (bottom) of Nodal Displacement Listing from Exercise 2-1.

Element results can be listed directly using the **PRESOL** command. Element results are listed in ascending order by element number. Results are shown for each node (or integration point) of each element (Figure 9.14). Requesting an element solution component stress or strain will return a list of all six coordinate values (*x*, *y*, *z*, *xy*, *yz*, and *xz*) per node per element. Requesting an element solution principle stress or strain, stress or strain intensity, or the equivalent stress or strain will return all five stresses or strains (1, 2, 3, INT, and EQV) per node per element. There is no summary at the bottom of the element solution lists and the maximum and minimum are not automatically reported.

```
PRESOL Command
File

PRINT S     ELEMENT SOLUTION PER ELEMENT

***** POST1 ELEMENT NODAL STRESS LISTING *****

 LOAD STEP=     1  SUBSTEP=     1
  TIME=    1.0000      LOAD CASE=   0

 THE FOLLOWING X,Y,Z VALUES ARE IN GLOBAL COORDINATES

 ELEMENT=      1          PLANE183
   NODE     SX          SY          SZ          SXY         SYZ         SXZ
   4553  0.12482E+07 0.26567E+06  0.0000      67422.      0.0000      0.0000
  20382  0.12732E+07 0.24508E+06  0.0000      70802.      0.0000      0.0000
    363  0.12218E+07 0.21280E+06  0.0000      76398.      0.0000      0.0000
  20398  0.11966E+07 0.23167E+06  0.0000      71551.      0.0000      0.0000

 ELEMENT=      2          PLANE183
   NODE     SX          SY          SZ          SXY         SYZ         SXZ
   4557  0.11233E+07 0.14448E+06  0.0000      70903.      0.0000      0.0000
  20391  0.11396E+07 0.12970E+06  0.0000      74518.      0.0000      0.0000
    362  0.11270E+07 0.10590E+06  0.0000      71454.      0.0000      0.0000
  20393  0.11124E+07 0.11959E+06  0.0000      68997.      0.0000      0.0000
```

Figure 9.14 Beginning of Element Solution Coordinate Stresses from Exercise 2-1.

The element results listings for some section-based elements also include the results for the "section nodes." For example, Figure 9.15 shows the beginning of the element solution principle stresses from exercise 4.1. The results list identifies the element number (1) and the element node numbers (1 and 4) for that element before listing results for each of the section nodes (1, 3, 5, 11, 13, 15, 21, 23, and 25) per node. Note that the section node numbers are the same for every element node. You can find a schematic of the BEAM189 element with the three element nodes in Figure 189.1 of the ANSYS Mechanical APDL Element Library. You can also find schematics for some of the standard library sections for BEAM189 in Figure 189.3 of the Element Library (Figure 9.16, left). Unfortunately, these schematics do not show the section node numbering. The best way to understand these results is to plot the section using the GUI path: **Main Menu > Preprocessor > Sections > Beam > Plot Section** or using the **SECPLOT** command (Figure 9.16, right).

PRESOL Command

File

```
PRINT S     PRIN ELEMENT SOLUTION PER ELEMENT

 STRESSES AT BEAM SECTION NODAL POINTS

ELEMENT =       1  SECTION ID =       1

ELEMENT NODE = 1

    SEC NODE        S1           S2             S3            SINT          SEQU
       1         0.0000       0.0000       -0.30000E+08   0.30000E+08   0.30000E+08
       3         0.0000       0.0000        0.0000         0.0000        0.0000
      13       0.63611E-09    0.0000        0.0000       0.63611E-09   0.63611E-09
      11         0.0000       0.0000       -0.30000E+08   0.30000E+08   0.30000E+08
       5       0.30000E+08    0.0000        0.0000       0.30000E+08   0.30000E+08
      15       0.30000E+08    0.0000        0.0000       0.30000E+08   0.30000E+08
      23         0.0000       0.0000        0.0000         0.0000        0.0000
      21         0.0000       0.0000       -0.30000E+08   0.30000E+08   0.30000E+08
      25       0.30000E+08    0.0000        0.0000       0.30000E+08   0.30000E+08

    Max=       0.30000E+08    0.0000        0.0000       0.30000E+08   0.30000E+08

    Min=         0.0000       0.0000       -0.30000E+08    0.0000        0.0000

ELEMENT NODE = 4

    SEC NODE        S1           S2             S3            SINT          SEQU
       1         0.0000       0.0000       -0.15000E+08   0.15000E+08   0.15000E+08
       3         0.0000       0.0000        0.0000         0.0000        0.0000
      13         0.0000       0.0000       -0.17044E-09   0.17044E-09   0.17044E-09
      11         0.0000       0.0000       -0.15000E+08   0.15000E+08   0.15000E+08
       5       0.15000E+08    0.0000        0.0000       0.15000E+08   0.15000E+08
      15       0.15000E+08    0.0000        0.0000       0.15000E+08   0.15000E+08
      23         0.0000       0.0000        0.0000         0.0000        0.0000
      21         0.0000       0.0000       -0.15000E+08   0.15000E+08   0.15000E+08
      25       0.15000E+08    0.0000        0.0000       0.15000E+08   0.15000E+08

    Max=       0.15000E+08    0.0000        0.0000       0.15000E+08   0.15000E+08

    Min=         0.0000       0.0000       -0.15000E+08    0.0000        0.0000
```

Figure 9.15 Beginning of Element Solution Principle Stresses from Exercise 4-1.

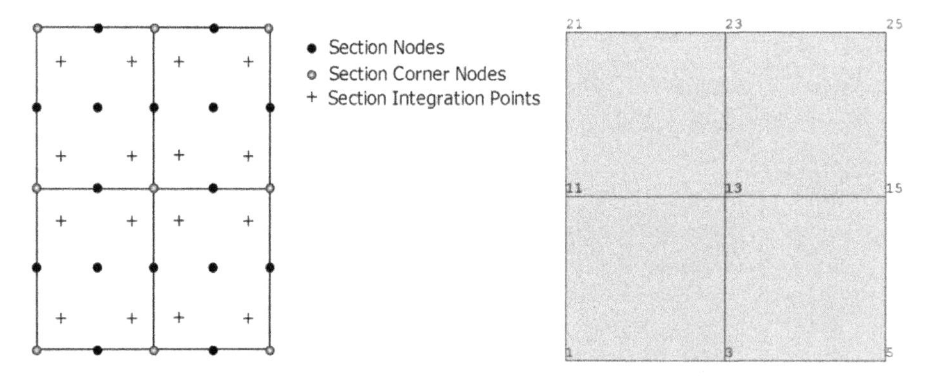

Figure 9.16 Schematic of Section Nodes for a Rectangular Section from the ANSYS Mechanical APDL Element Library (left) and Section Plot with Section Node Numbering (right) for BEAM189.

Element table results can be listed through the Utility Menu (**Utility Menu > List > Results > Element Table Data** . . .) or through the General Postprocessor (**Main Menu > General Postproc > Element Table > List Elem Table**) or using the **PRETAB** command. A summary is included at the bottom of the window to indicate the maximum and/or minimum values in the results set and the elements where those values exist. Element table results can only be listed if an element table has been defined (see Section 9.8.2.4).

Results along a specific path can be listed through the General Postprocessor (**Main Menu > General Postproc > Path Operations > Plot Path Item > List Path Items**) or using the **PRPATH** command. Path items can only be listed if a path has been defined and results have been mapped to it (see Section 9.8.2.5).

The results for Time History Postprocessing variables can be listed using the GUI path: **Main Menu > TimeHist Postpro > List Variables** or using the **PRVAR** command (Figure 9.17). The model results must be defined as variables before they can be listed. Up to six variables can be listed at one time. Time is automatically listed and does not count as one of the six (unless it is requested again). There is no summary at the bottom of variable results lists, and the maximum and minimum are not automatically reported.

PRVAR Command

File

***** ANSYS POST26 VARIABLE LISTING *****

TIME	43 TEMP OutsideBrick_2	18 TEMP InsideBrick_3	160 TEMP InsideWool_4	202 TEMP InsideWall_5	1621 TEMP HallCenter_6	362 TEMP HallEnd_7
200.00	-1.98014	-0.149808E-02	0.734212E-01	2.37672	20.0000	20.0000
400.00	-3.14255	-0.787622E-02	0.231752	4.02703	20.0000	20.0000
600.00	-3.93324	-0.231998E-01	0.462390	5.22350	20.0000	20.0000
800.00	-4.53160	-0.509747E-01	0.745789	6.13021	20.0000	20.0000
1000.0	-5.01688	-0.934936E-01	1.06277	6.84707	20.0000	20.0000
1200.0	-5.42813	-0.151737	1.39720	7.43572	20.0000	20.0000
1400.0	-5.78681	-0.225591	1.73659	7.93471	20.0000	20.0000
1600.0	-6.10595	-0.314163	2.07183	8.36856	20.0000	20.0000
1800.0	-6.39412	-0.416094	2.39656	8.75311	20.0000	20.0000
2000.0	-6.65732	-0.529800	2.70659	9.09880	20.0000	20.0000

Figure 9.17 Beginning of Variable Results List from Exercise 8-1.

Specialty listing commands exist for other types of results. These include results for energies (**PRENERGY**), percent errors (**PRERR**), far fields (**PRFAR**), solution summary data (**PRITER**), joint element output (**PRJSOL**), near-zone pressures (**PRNEAR**), summed element nodal loads (**PRNLD**), orbital motion characteristics of rotating structures (**PRORB**), constrained node reaction solutions with the **FORCE** command (**PRRFOR**), constrained node reaction solutions (without the **FORCE** command) (**PRRSOL**), linearized stresses along a section path (**PRSECT**), vector data (**PRVECT**), spot weld solutions (**SWLIST**), and Superelement DOF results (**SEDLIST**).

9.8.2. Plotting Results

Plotting is one of the most common and convenient methods for reviewing the results of a finite element model. Plotting creates a visual display of the results, which provides an overview of the model behavior and allows you to identify the critical regions of the model. The most common plots that can be created are contour plots, displacement plots, vector plots, element table plots, and path plots. These are all available through the General Postprocessor.

9.8.2.1. Contour Plots

Contour plots display the results of a single data set (time step, load step, or substep) over the model geometry. The range of values in the results set is divided into a number of subranges (nine by default), and each subrange is assigned a color. The colors are then mapped over the geometry to indicate the result values at each location in the model. Most results for continuum (e.g., PLANE and SOLID) elements can be displayed as contour plots. For beam and shell elements, it may be necessary to issue the **/ESHAPE** command before creating a contour plot or to plot the results using element tables instead.

Contour plots of the nodal solution can be created using the GUI path: **Main Menu > General Postproc > Plot Results > Contour Plot > Nodal Solu** or using the **PLNSOL** command. Contour plots of the element solution can be created using the GUI path: **Main Menu > General Postproc > Plot Results > Contour Plot > Element Solu** or using the **PLESOL** command.

There are three important display options for contour plots: the undisplaced shape key, the scale factor, and the contour scale. The displaced shape option controls if and how the deformed model is shown in comparison to the undeformed (original) geometry. The "Deformed shape only" is the default and provides no basis for comparison (Figure 9.18, left). The "Deformed shape with undeformed edge" option overlays the contour plot on an outline of the original model (Figure 9.18, center). The "Deformed shape with undeformed model" option overlays the contour plot on the original finite element model (Figure 9.18, right).

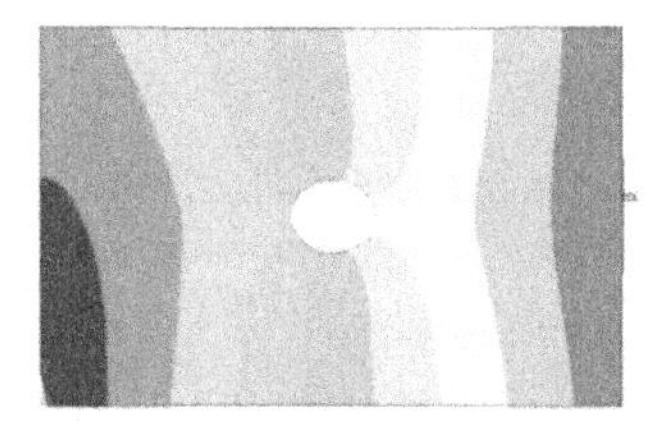
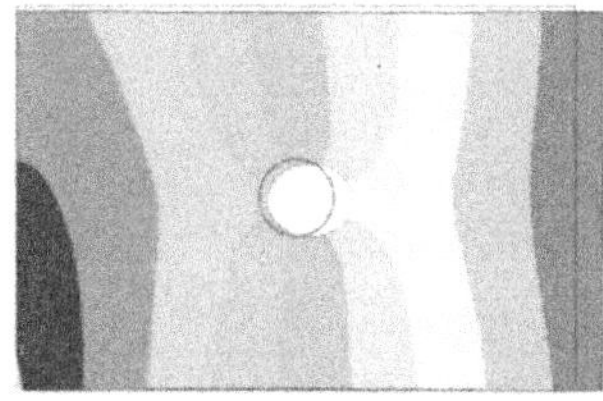
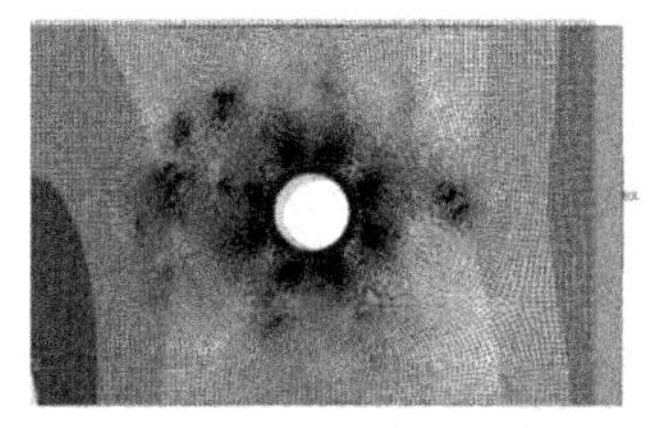

Figure 9.18 Displacement Vector Sum from Exercise 5-1: Deformed Shape Only (left), Deformed shape with Undeformed Edge (center), and Deformed Shape with Undeformed Model (right) (Win32 Graphics).

The scale factor determines if and how the deformed model is displayed in comparison to its as-calculated state. The deformations in most finite element models are relatively small and would be difficult or impossible to view if they were plotted as calculated (at their true scale). Therefore, ANSYS automatically scales the maximum displacement to 5% of the model dimension. The resulting scale factor is shown in grey to the right of the Scale Factor drop down menu in the Contour Solution Data dialog boxes. For example, Figure 9.18 uses an automatically calculated scale factor of 16.89 (Figure 9.19). Three addition scale options are available: True Scale (1:1), User Specified, and Off. If the User Specified option is chosen, you can specify the scaling factor to use in the text box. Figure 9.20 shows the geometry from Figure 9.18 at its true scale (left), with its auto calculated scale of 16.89 (center), and with a user specified scale of 75 times (right). Note that the slight necking in the model is visible at 75x.

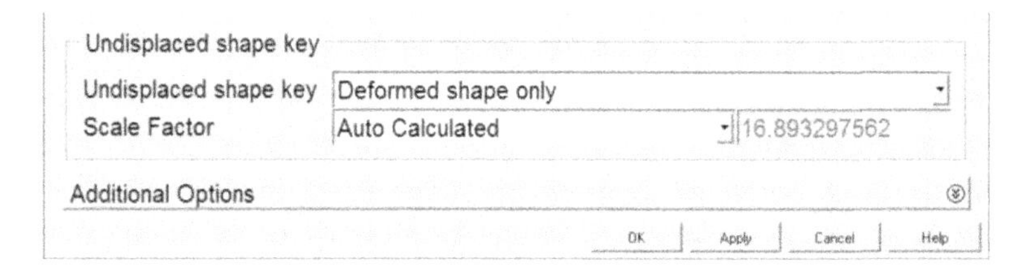

Figure 9.19 Contour Plot Controls Showing the Auto Calculated Scale Factor for Figure 9.18.

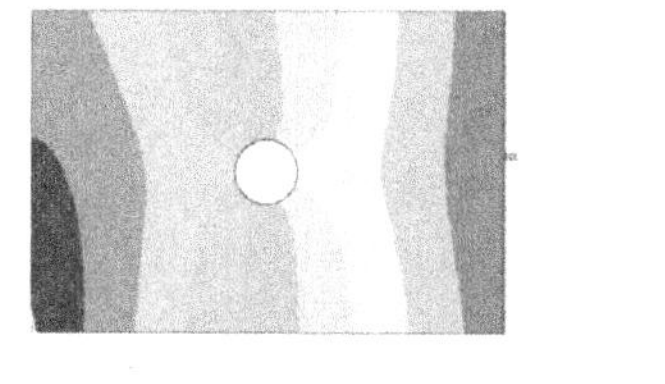

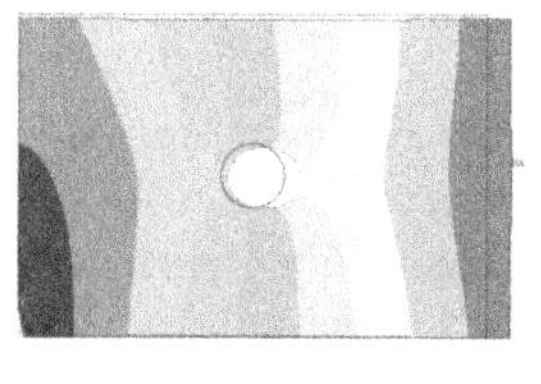

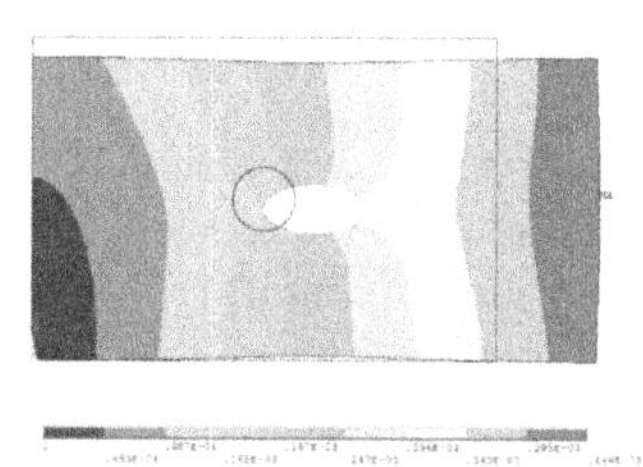

Figure 9.20 Contour Plot Scaling for Figure 9.18: True Scale (left), Auto Calculated (center), and User Specified at 75x (right) (Win32 Graphics).

Finally, you can control the color contour display using the menus accessible via the GUI path: **Utility Menu > PlotCtrls > Style > Contours**. Here, it is possible to specify the number of contour intervals to use and to set the minimum and maximum contour values. Using the same number of contour intervals and the same minimum and maximum values on multiple plots allows meaningful side-by-side comparison of results from the same or similar models. This is a very powerful technique and will be used in exercises 9-1 and 9-2.

9.8.2.2. Displacement Plots

Displacement plots are similar to contour plots but simpler. Displacement plots display the deformed model geometry either alone, with the undeformed edge, or overlaid on an outline of the undeformed geometry. The deformed finite element model is shown in a solid blue color (no contours are included). Displacement plots are particularly useful for reviewing the mode shapes from a modal analysis. Displacement plots can be created using the GUI path: **Main Menu > General Postproc > Plot Results > Deformed Shape** or using the **PLDISP** command. Examples of displacement plots can be found in exercise 9-3.

9.8.2.3. Vector Plots

Vector plots combine the x, y, and z components of a result to calculate a result vector at each location in the active set. The result at each location is then displayed as an arrow whose length and color represent the magnitude of the requested result and whose direction is determined from the vector direction cosines. For example, Figure 9.21 shows a thermal flux vector plot from exercise 3-2 (right) and a close-up showing the details of the individual vectors (left). Note that the largest vectors have a red color and are located at the interface between the pipe and the pipe flange where the heat flux is the highest. Vector plots are very useful for viewing results that represent energy flows (e.g., thermal and magnetic gradients and fluxes). They can be created using the GUI path: **Main Menu > General Postproc > Plot Results > Vector Plot** or using the **PLVECT** command.

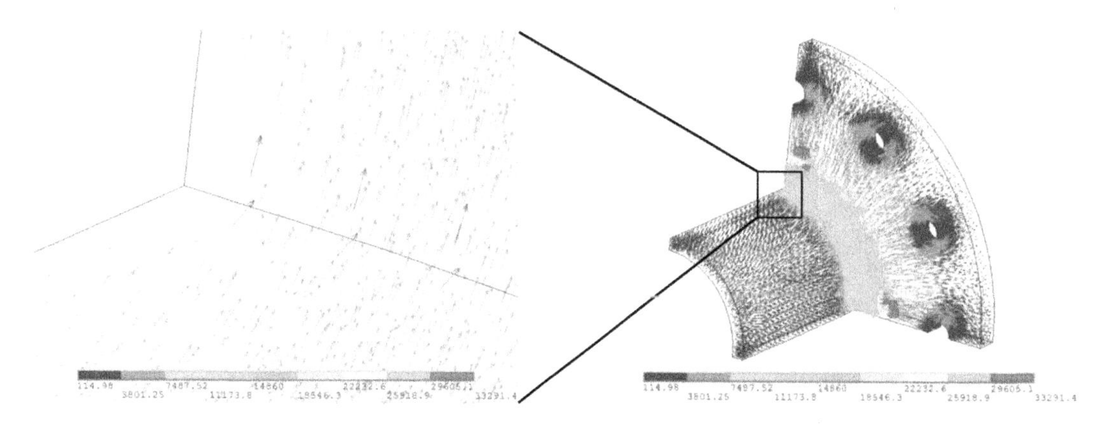

Figure 9.21 Thermal Flux Vector Plots from Exercise 3-2 (Win32 Graphics).

9.8.2.4. Element Table Plots

Although most results data can be displayed and viewed using the methods described above, some results (especially those that are not naturally single-valued or that are not associated with a geometric entity) must be displayed using element tables. This generally applies to the integration point data; the derived data for structural line elements, thermal line elements, and contact elements; and the layer data for layered elements.

Element tables can be created and filled using the GUI path: **Main Menu > General Postproc > Element Table > Define Table** or using the **ETABLE** command. Once an element table has been created, it can be plotted using the GUI path: **Main Menu > General Postproc > Plot Results > Contour Plot > Elem Table** or using the **PLETAB** command. This will produce a contour plot over the model geometry for a single element table (e.g., the results for Node I of each element) using linear interpolation to create a single results value for each element. The resulting plot may not be a good representation of the model behavior. In addition, if the geometry is a line, the results may be very difficult to see.

Instead, it is often better and more useful to plot element table items as contoured areas along the element using the GUI path: **Main Menu > General Postproc > Plot Results > Contour Plot > Line Elem Res** or using the **PLLS** command. This allows you to specify the values from two element tables (e.g., the results at Node I and Node J of each element). The results will then be plotted over the geometry as trapezoids to reflect the two different values.

Additional details, examples, and a comparison of element table plots can be found in exercise 9.1.

9.8.2.5. Path Plots

Path plots allow you to display the results along a two- or three-dimensional predefined path through the model. Path points must be within an element or on the surface of an element, but their locations are otherwise independent. Thus, paths can be used to retrieve data at any location in the model, regardless of the locations of the individual nodes or integration points. The results along the path are interpolated from the calculated results using the element shape functions.

Creating a path plot involves four steps. First, the path segment(s) must be defined. Then, the points on the path segment(s) must be specified. These operations can be performed via the GUI path: **Main Menu > General Postproc > Path Operations > Define Path** or by using the **PATH** and **PPATH** commands. To define the path, you will need to supply a path name, the number of points that define the path segments, the number of intermediate locations along the path segments, and the number of result items to be posted to the path. Third, the data must be posted to the path points. This can be done using the GUI path: **Main Menu > General Postproc > Path Operations > Map onto Path**. It can also be done using the **PDEF** command for individual results or the **PVECT** command for vector items. Finally, the path data can be listed, plotted, or graphed.

You can create a path plot on the geometry using the GUI path: **Main Menu > General Postproc > Path Operations > Plot Path Item > On Geometry** or by using the **PLPAGM** command. And, you can create a path plot on a graph using the GUI path: **Main Menu > General Postproc > Path Operations > Plot Path Item > On Graph** or by using the **PLPATH** command.

Example 9.1 lists the commands to create a path through the center line of the model from exercise 5-1, post the results for the equivalent (von Mises) stress onto the path, and then plot (Figure 9.22, left) and graph (Figure 9.22, right) the equivalent stress for the path. In general, path operations are easier to perform via the GUI than with commands. Therefore, we recommend that novices use the Sequential Method for transferring path operations to input files (see chapter 10 for more information).

Example 9.1

```
PATH,Example,2,30,20,       ! Create new path from 2 points w/ 30 points & 20 sets
PPATH,1,0,0.075,0,0,0,      ! 1st path point at (0.075,0,0)
PPATH,2,0,0.075,0.1,0,0,    ! 2nd path point at (0.075,0.1,0)
PDEF,,S,EQV,AVG             ! Interpolate equivalent stress onto path
PLPAGM,SEQV,1,'NODE'        ! Plot equivalent stress on the path overlaid on nodes
PLPATH,SEQV                 ! Graph the equivalent stress on the path
```

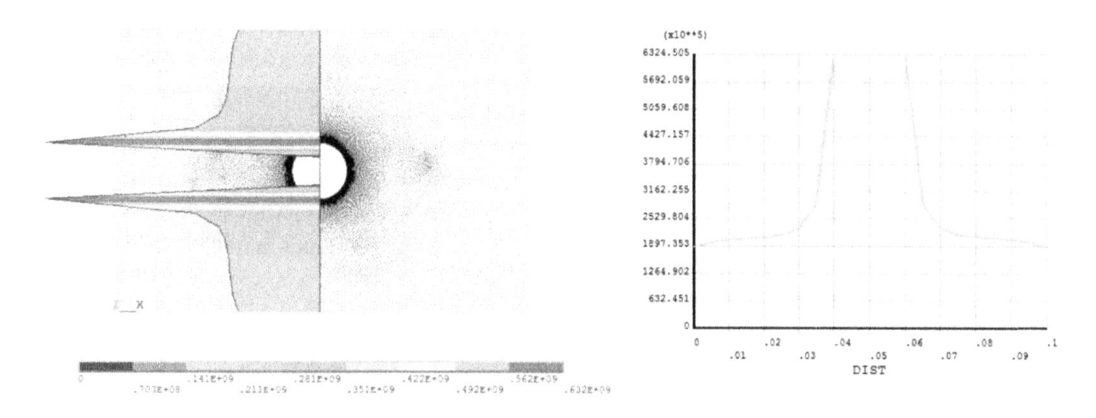

Figure 9.22 Equivalent Stress Plotted on a Path (left) and Graphed (right) using the Model from Exercise 5-1.

9.8.3. Animating Results

Animations display sequences of results from the General Postprocessor with a specified time delay between each plot. Although a number of animation options are available, two of the more useful options are animating Over Time and animating Over Results. Animating over time allows you to display a specified number of results plots over a specified range of load steps or time ranges. The images will be evenly spaced over the specified period. ANSYS will interpolate the data to fulfill this requirement if necessary. Animating over results will create one animation frame for each data set specified; no interpolation is necessary and none will be performed. An example animation can be found in exercise 8.1.

Animations can be created through the GUI path: **Utility Menu > PlotCtrls > Animate**. You can save an animation in interactive mode as an .AVI file by following the GUI path: **Utility Menu > PlotCtrls > Animate > Save Animation** ... or using the **/ANFILE** command.

In general, animations are easy to create in the GUI and difficult to create through commands. If you need to save the commands to re-create an animation in the future, we recommend using the Sequential Method to create the animation in interactive mode and then copy out and edit the commands for future use (see chapter 10 for more details).

Animations cannot be created or saved in batch mode.

9.8.4. Graphing Results

Finally, results values at specific locations in the model can be graphed against time (or load step) in the Time History Postprocessor (POST26). Although this functionality was intended for postprocessing transient analyses, it can be used for any type of result that can be calculated by ANSYS.

In order to graph results, they must first be stored as variables. The number of variables that can be stored for a given session must be defined before the results are read in from the results file.

By default, the number of variables that can be created is 10. However, this can be set to a maximum of 200 using the **NUMVAR** command.

Time is always variable number 1 and is automatically defined for you. You can store results data as additional variables using the GUI path: **Main Menu > TimeHist Postpro > Store Data**. For the degree of freedom solution, you can also use the **NSOL** command. For unaveraged element solution data, you can also use the **ESOL** command. For the averaged nodal solution data, you can also use the **ANSOL** command. Less common results items include the reaction forces (**RFORCE**), generalized plain strain results (**GSSOL**), and solution summary results (**SOLU**). Most of these commands require you to specify the node or element number associated with the desired data; however, the **GSSOL** and **SOLU** commands do not require a node or element to be specified.

By far, the easiest way to define variables is to use the Variable Viewer. The Variable Viewer opens by default as soon as you enter the Time History Postprocessor. You can always reopen it using the GUI path: **Main Menu > TimeHist Postpro > Variable Viewer**. The variable viewer consists of three parts: a command menu to create, delete, plot, and list the POST26 variables; a Variable List to identify the currently defined variable(s); and a calculator for creating combinations of variables. The Variable Viewer was used in exercise 8-1.

Each variable can only contain one type of results information. Additional types of results must be stored as additional variables. For example, if you want to be able to graph the displacement at a given location in x and y, you must define one variable for the x component of displacement for that node and another variable for the y component.

Once your results variables have been defined, they can be graphed using the GUI path: **Main Menu > TimeHist Postpro > Graph Variables** or using the **PLVAR** command. Example graphing operations can be found in exercises 6-1 and 8-1.

9.9. Postprocessing With Load Case Combinations

There are many ways to operate on results in ANSYS. Most are beyond the scope of this book. However, load case combinations are an important and relatively simple technique for operating on results data. For example, load case combinations allow you to change the sign of a set of results, take the absolute value of a set of results, scale a set of results, add two or more results sets together, subtract one set of results from another, and so on. This capability was originally developed as a means for combining the results of mode-spectrum analyses and for use with harmonic elements (both of which are also beyond the scope of this book).

Before you can perform any load case operations, one or more load cases must be defined. This can be done using the GUI path **Main Menu > General Postproc > Load Case > Create Load Case** or using the **LCDEF** command. Most load case operations act on the results portion of the database. Therefore, you should always zero (or clear) the results portion of the database before beginning a new load case combination or operation. This can be done using the GUI path **Main Menu > General Postproc > Load Case > Zero Load Case** or with the **LCZERO** command. At this point, you can read the first load case into the database via the GUI path **Main Menu > General Postproc > Load Case > Read Load Case** or using the **LCASE** command, and then operate on that set. Or you can begin to operate on the empty set of results in the database using one of the options in the load case menu tree or using the **LCOPER** command. Once the load case operations or combinations are complete, the results in the database can be processed like another other set of results. Finally, you can save a load case combination for future use using the GUI path: **Main Menu > General Postproc > Load Case > Write Load Case** or using the **LCWRITE** command.

It is important to note that not all results should be combined using load case operations. For example, load case combinations can be used on the displacements, stresses, and strains from elastic analyses. However, they should not be used on the results from nonlinear analyses or analyses whose results are non-zero in the unloaded state (e.g., most thermal analyses). ANSYS will not prevent you from creating load cases for nonlinear analyses, thermal analyses, or any other type of analysis where load case combinations are invalid. Instead, it will perform the operations exactly as requested and return invalid results. Therefore, load case combinations should only be used if you understand the conditions under which results can and should be combined, and if you take care to verify and validate the results after postprocessing is complete.

Additional details and examples of load case operations can be found in exercise 9-3 of this book. For more information, see Section 6.3.3 of the ANSYS Mechanical APDL Basic Analysis Guide.

9.10. Saving Postprocessing Graphics and Information

The lists, plots, and graphs generated during postprocessing are often needed for reports or presentations. Therefore, it is often desirable to save or export postprocessing information.

All listed information in ANSYS is provided as plain text. Therefore, it can be copied and pasted into another plain text editor normally. In interactive mode, you can save listed information from within the program by using the File menu in the upper left corner of the listing window. The File menu gives you three choices: Save as, Print, and Copy to Output. In batch mode, listed information is written to the output file by default. Since searching through the output file to find the listed data is time consuming and error-prone, we recommend sending listed output to a separate file. This can be done by issuing the **/OUTPUT** command with a new file name before issuing the listing command. After the listing operation is complete, output can be directed back to the output file by issuing the **/OUTPUT** command without a filename.

In interactive mode, the easiest way to export graphics from ANSYS is to take a screenshot and then process that image in another program. It is also the only way to guarantee that what you see is what you will get. Most of the images in this book were produced from screenshots. You can also send images from the program to a printer or to a file using the GUI path: **Utility Menu > PlotCtrls > Hard Copy**. If you send the hard copy to file, a dialog box will be displayed that will allow you to choose the file type (BMP, postscript, TIFF, JPEG, PNG, etc.), to choose whether or not to reverse the video of the plots (i.e., to toggle the background from black to white or vice versa), and to specify the file name. The image will be saved with that file name in your working directory. This is a GUI-only procedure; there is no command to perform this operation. The more complicated alternative is to use the **/SHOW** command to redirect plots from the program to a file. We do not recommend using this method in interactive mode. However, it is both a robust and necessary method for exporting (and therefore viewing) graphics in batch mode. The use of the **/SHOW** command for saving graphics in batch mode is demonstrated in exercise 10-3.

9.11. Model Verification and Validation

As an analyst, it is your responsibility to ensure that the results generated for your model are consistent with the laws of physics and with your expectations. ANSYS, Inc. has an extensive set of verification procedures to ensure that the program produces accurate results based on the input provided. Like all computer programs, the ANSYS source code has errors. However, errors are rare and most are fixed as soon as they are identified. You can report a suspected error through the ANSYS Customer Portal. But you should assume that problems with your results are due to errors in your model and not errors in the ANSYS software.

There are a number of techniques that can be used to verify the quality of your finite element models. The best validation technique is to compare the results of your model to experimental data. The next best option is to compare the analytical solution for a simplified version of your problem to your results to provide a first-order estimation of the solution quality. If a simplified analytical solution is not available (or calculable), you can perform a Fermi type calculation to estimate the order of magnitude of the expected results. Otherwise, you should make sure that the model behavior is consistent with expectations. For example, bodies under compressive loads should be compressed and bodies under tensile loads should become elongated. Heat should flow from areas with higher temperatures to areas with lower temperatures. Inconsistencies in these types of results are usually due to a sign error somewhere in the model. Bodies under small structural loads should have stresses well below the yield strength. Otherwise there may be an incorrect exponent in an applied load. Small deformations under large applied loads may indicate an overconstrained model. For every model, it is important to ask yourself: Do the results make sense?

You should also verify that the assumptions contained within the model are valid. For example, the maximum stress in a model with linear material properties should be well below the material yield strength. Otherwise, nonlinear material properties should be defined and the model should be solved again. Nodal results plots should be continuous and the differences between the nodal and element plots should be small. Otherwise, the finite element mesh density may be insufficient. Finally, you should review the error file to ensure that the analysis ran without errors and that any warning messages are easily explained.

As you gain experience with finite element analysis, this process will become a second nature to you. Until then, it is a good idea to ask an experienced colleague to review your model and assumptions to ensure that you haven't overlooked something important.

The importance of verifying the results of an analysis cannot be overstated. Results from a poorly verified analysis or results from a well-verified model that do not match the physical system can be costly in time, money, and reputation.

Exercise 9-1

Postprocessing an Axisymmetric Cylindrical Pressure Vessel Using Element Tables

Overview

In this exercise, you will build an axisymmetric model of a thin-walled cylindrical pressure vessel. The pressure vessel is made of stainless steel with a Young's modulus of 200 GPa and a Poisson's ratio of 0.28. The cylinder that forms the main body is 80 cm long. It has a radius of 20 cm and a thickness of 2 cm. The cylinder is sealed with hemispherical end caps. The vessel is pressurized to 1 MPa (Figure 9-1-1). The goal is to find the radial displacement of the cylindrical walls and the hoop and meridional stresses in the walls that are generated in response to the applied radial stress.

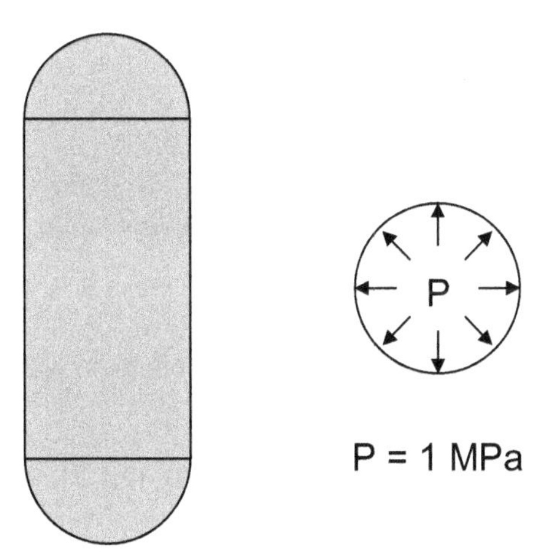

Figure 9-1-1 Schematic of the Pressure Vessel.

ANSYS Mechanical APDL for Finite Element Analysis.
DOI: http://dx.doi.org/10.1016/B978-0-12-812981-4.00026-5

Model Attributes

Material Properties for Stainless Steel

- Young's modulus—200 GPa
- Poisson's ratio—0.28

Loads

- 1 MPa pressure on all internal surfaces

Constraints

- No displacement at the bottom of the pressure vessel in any direction.
- No displacement at the top of the pressure vessel in the horizontal (x) direction.

File Management

Create a new folder in your "Intro-to-ANSYS" folder for Exercise9-1

Change the Working Directory and the Jobname

Start ANSYS

In this exercise, GUI paths have been replaced with commands for common operations that you should already be comfortable performing and for repetitive tasks. This makes the exercise quicker and easier so you can focus on postprocessing. It will also help prepare you to create input files in chapter 10.

Step 1: Define Geometry

1-1. Create a keypoint at (0,0)

- Preprocessor > Modeling > Create > Keypoints > In Active CS

1-2. Create a keypoint at (0.2,0.2)

- Enter the following command into the command prompt:

```
K,2,0.2,0.2
```

1-3. Create the remaining keypoints at (0.2,1), (0,1.2), (0,0.2), and (0,1)

- Enter the following commands into the command prompt:

```
K,3,0.2,1
K,4,0,1.2
K,5,0,0.2
K,6,0,1
```

The resulting keypoints are shown in Figure 9-1-2 (left).

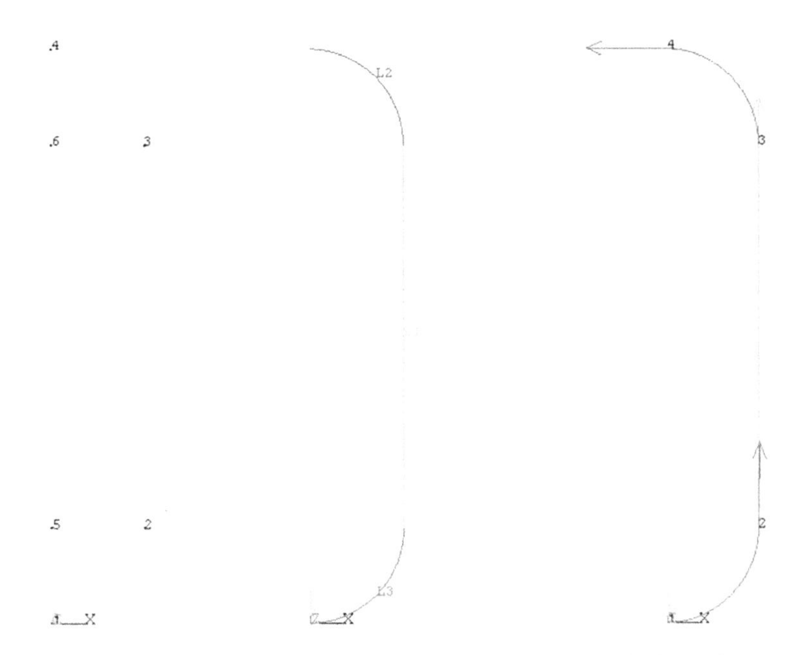

Figure 9-1-2 Keypoint Plot with Numbering On (left), Line Plot with Numbering On (center), and Line Plot with Line Numbering and Line Direction On (right).

1-4. Create a line between Keypoints 2 and 3 to form the outer edge of the cylinder

- Enter the following command into the command prompt:
  `L,2,3`

The order in which you specify the keypoints determines the orientation of the line that is created. When the line is meshed, the x axis of the shell elements will be aligned with the x axis of the line. By default, loads are applied to Surface 1 of each element (the "top" face) when more than one surface exists. Thus, a positive pressure applied to the line from Keypoint 2 to 3 or to the nodes attached to that line will be applied on the inner surface of the vessel and act outward.

1-5. Create an arc between the keypoints at (0.2,1) and (0,1.2)

- Preprocessor > Modeling > Create > Lines > Arcs > By End KPs & Rad
- Choose Keypoint 3 at (0.2,1) and then Keypoint 4 at (0,1.2)
- Click OK
- Choose Keypoint 6 at (0,1)
- Click OK
- Enter 0.2 as the value of the radius of the arc
- Click Apply

1-6. Create an arc between the keypoints at (0,0) and (0.2,0.2)

- Choose Keypoint 1 at (0,0) and then Keypoint 2 at (0.2,0.2)
- Use Keypoint 5 as the center of curvature

The resulting lines are shown in Figure 9-1-2 (center).

1-7. Turn line direction symbols on to confirm that the lines are correctly orientated

- Utility Menu > PlotCtrls > Symbols . . .
- In the fourth section, turn LDIR Line Direction on
- Click OK

1-8. Replot lines if necessary

- Enter the following command into the command prompt:
  `LPLOT`

The arrows for the bottom two lines should point up and the arrow for the top line should point to the left (Figure 9-1-2, right).

1-9. Turn line direction symbols off

- Utility Menu > PlotCtrls > Symbols ...
- In the fourth section, turn LDIR Line Direction off
- Click OK

Step 2: Define Element Types

2-1. Define SHELL61 as element type #1 for this model

- Enter the following command into the command prompt:
  `ET,1,61`

2-2. Set the key options for this element

- Preprocessor > Element Type > Add / Edit / Delete
- Click Options ...
- For Solution output at K6 choose "Ends & 1 int pt"

2-3. Define the real constants for your model

- Preprocessor > Real Constants > Add/Edit/Delete
- Specify 0.02 for "Shell thickness at Node I TK(I)"

SHELL61 elements can have a linearly varying thickness over the length of each element (Figure 9-1-3). In this exercise, we want a constant thickness for the entire model. Therefore, we only need to specify one value.

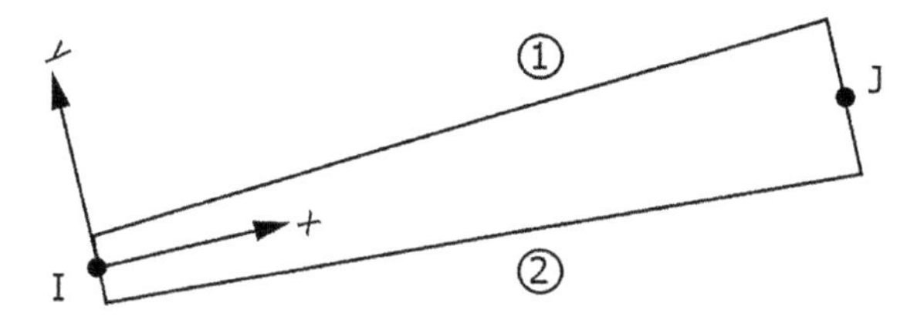

Figure 9-1-3 Schematic of SHELL61 Element Geometry (Shown with Optional Linearly Varying Thickness).

Step 3: Define Material Properties

3-1. Define the material properties for this model

- Enter the following commands into the command prompt:

```
MP,EX,1,200e9
MP,PRXY,1,0.28
```

The first command specifies the Young's modulus (EX) for material model #1 with a value of 200e9. The second command specifies the Poisson's ratio (PRXY) for material model #1 with a value of 0.28.

Step 4: Mesh

4-1. Mesh the model

- Enter the following commands into the command prompt:
```
ESIZE,0.01
LMESH,ALL
```

The first command sets the element edge length for the model to 0.01. The second command meshes all lines in the model.

4-2. Turn element numbering on

- Enter the following command into the command prompt:
```
/PNUM,ELEM,1
```

4-3. Plot elements

- Enter the following command into the command prompt:
```
EPLOT
```

4-4. Zoom in to examine mesh density

- Utility Menu > PlotCtrl > Pan Zoom Rotate
- Choose Box Zoom and draw a box that contains at least three elements

The resulting element plot should resemble the one shown in Figure 9-1-4 (left).

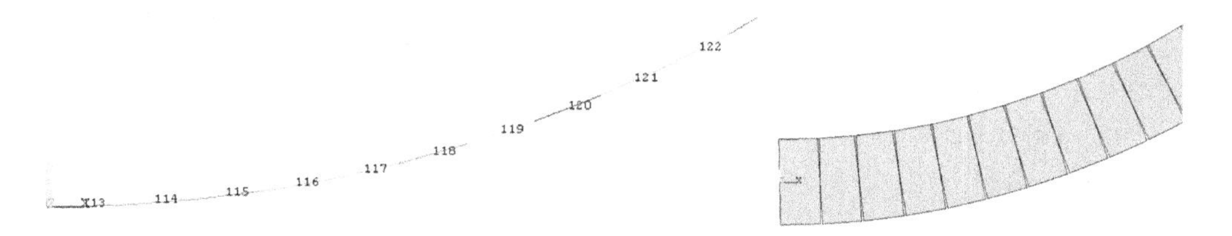

Figure 9-1-4 Element Plot with Element Numbering On (left) and with Element Shape Display On (right).

4-5. Turn element numbering off

- Enter the following command into the command prompt:
```
/PNUM,ELEM,0
```

4-6. Turn element shape display on

- Enter the following command into the command prompt:
```
/ESHAPE,1
```

4-7. Replot elements if necessary

- Enter the following command into the command prompt:
```
EPLOT
```

This shows that the element thickness was correctly specified (Figure 9-1-4, right).

4-8. Turn element shape display off

- Enter the following command into the command prompt:
```
/ESHAPE,0
```

Step 5: Apply Constraint Boundary Conditions

5-1. Enter the Solution processor

- Enter the following command into the command prompt:
```
/SOL
```

5-2. Change to isometric view and fit the image in the Graphics Window

- Enter the following commands into the command prompt:
```
/VIEW,1,1,1,1
/AUTO
```

These operations will take effect after the next plotting command.

5-3. Plot keypoints

- Enter the following command into the command prompt:
```
KPLOT
```

5-4. Fully constrain the keypoint at the bottom of the pressure vessel (Keypoint 1)

- Enter the following command into the command prompt:
```
DK,1,ALL,0
```

5-5. Constrain the keypoint at the top of the pressure vessel (Keypoint 4) in *x* and rotation about *z*

- Enter the following commands into the command prompt:
```
DK,4,UX,0
DK,4,ROTZ,0
```

These constrains will allow the pressure vessel to expand and contract along the y axis.

5-6. Constrain all nodes in *z*

- Enter the following command into the command prompt:
```
D,ALL,UZ,0
```

The UZ DOF is not needed for axisymmetric analyses like this one. Constraining UZ for all nodes in the model effectively removes this DOF from the solution.

Step 6: Apply Load Boundary Conditions

6-1. Plot nodes

- Enter the following command into the command prompt:
```
NPLOT
```

6-2. Set the boundary condition symbols

- Utility Menu > PlotCtrls > Symbols . . .
- Change "[/PSF] Surface Load Symbols" to "Pressures"
- Change "Show pres and convect" to "Arrows"
- Click OK

6-3. Apply a pressure (surface load) of 1 MPa to all nodes

- Enter the following command into the command prompt:
  `SF,ALL,PRES,1e6`

If you cannot see the arrow symbols clearly, zoom in. You may also need to replot the nodes. The arrows verify that the pressure has been applied to the correct (inner) face and in the correct (outward) orientation.

6-4. Save your progress

- Enter the following command into the command prompt:
  `SAVE`

Step 7: Set the Solution Options

The default solution options can be used for this analysis.

Step 8: Solve

8-1. Select everything, solve, and save

- Enter the following commands into the command prompt:

```
ALLSEL
SOLVE
SAVE
```

Step 9: Postprocess the Results

9-1. Change to front view and fit the image if necessary

- Enter the following commands into the command prompt:

```
/VIEW,1,,,1
/AUTO
```

9-2. Plot the displacement in x

- General Postproc > Plot Results > Contour Plot > Nodal Solu

This creates a contour line plot (Figure 9-1-5, left). The results are visible but difficult to see.

9-3. Turn element shape display on

- Enter the following command into the command prompt:
  `/ESHAPE,1`

9-4. Plot the displacement in x

- General Postproc > Plot Results > Contour Plot > Nodal Solu

Element shape display provides a thickness for the elements and makes the contour plot results easier to see (Figure 9-1-5, right).

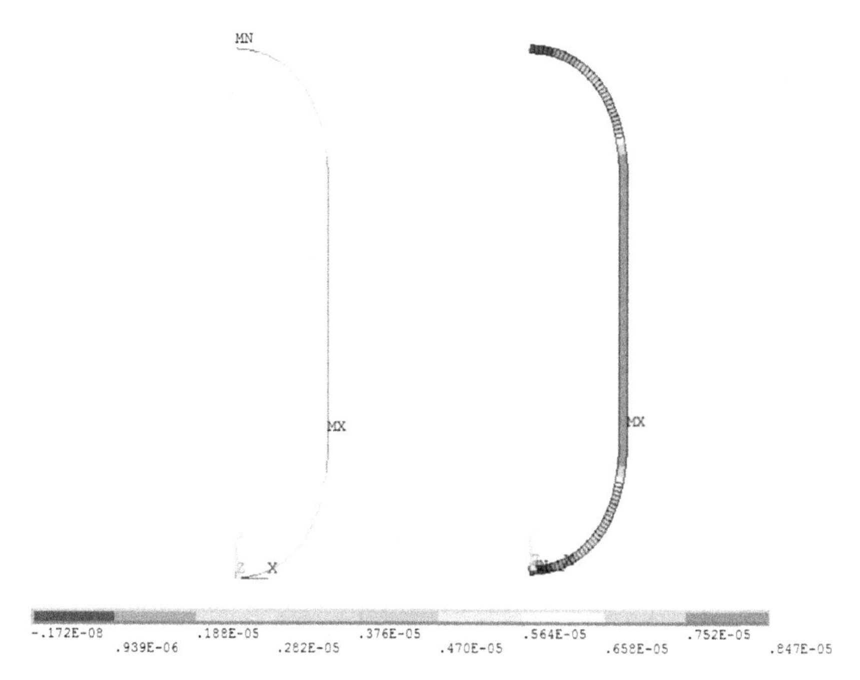

Figure 9-1-5 Plot of Displacement in X (Win32 Graphics): with Element Shapes Off (left) and with Element Shapes On (right).

9-5. Plot the nodal von Mises stress for the pressure vessel

- General Postproc > Plot Results > Contour Plot > Nodal Solu

The solid red plot indicates that these results do not exist or cannot be accessed using the PLNSOL command.

9-6. Plot the element von Mises stress for the pressure vessel

- General Postproc > Plot Results > Contour Plot > Element Solu

Similarly, these results do not exist or cannot be accessed using the PLESOL command.

9-7. Open the documentation for SHELL61 elements

- Utility Menu > Help > Help Topics
- In the search box, enter "SHELL61"
- Click on the link that says: SHELL61 Mechanical APDL > Element Reference > Element Library (Figure 9-1-6)

You may need to zoom in to view the information in the ANSYS Help Viewer. If necessary, right click and follow the path Zoom > Zoom In or use the shortcut "Ctrl ++".

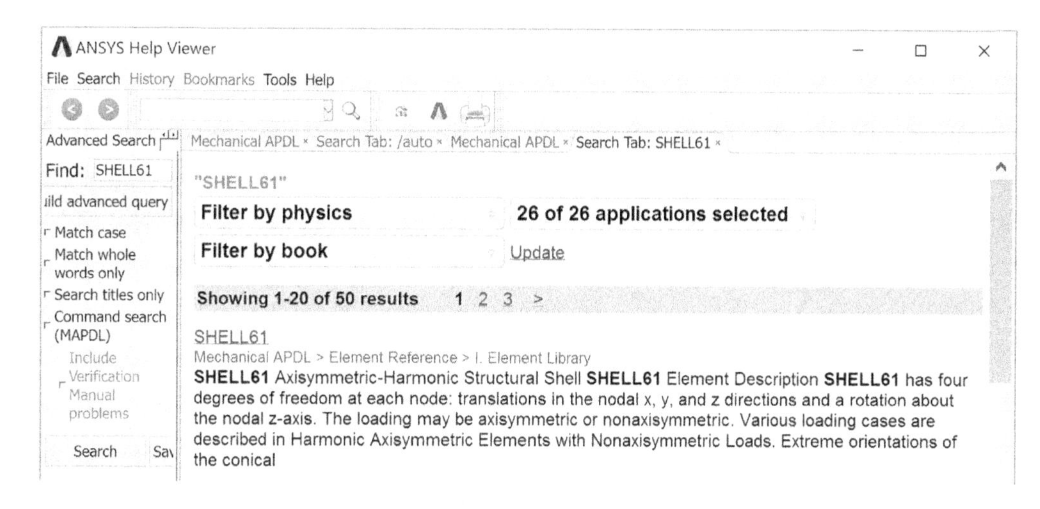

Figure 9-1-6 ANSYS Help Viewer Search for SHELL61.

Or

- In the Table of Contents, follow the path: Mechanical APDL > Element Reference > I. Element Library > Shell61

The beginning of the SHELL61 documentation is shown in Figure 9-1-7.

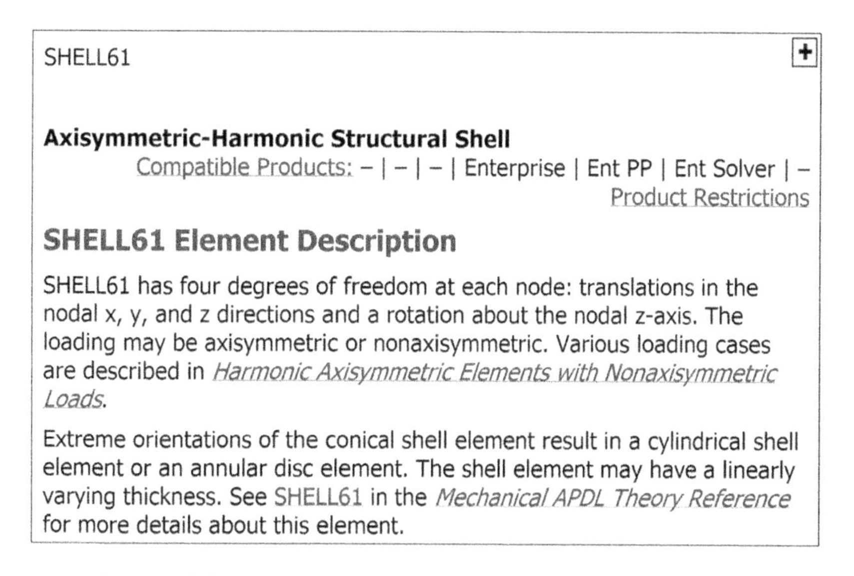

SHELL61

Axisymmetric-Harmonic Structural Shell

Compatible Products: – | – | – | Enterprise | Ent PP | Ent Solver | –

Product Restrictions

SHELL61 Element Description

SHELL61 has four degrees of freedom at each node: translations in the nodal x, y, and z directions and a rotation about the nodal z-axis. The loading may be axisymmetric or nonaxisymmetric. Various loading cases are described in *Harmonic Axisymmetric Elements with Nonaxisymmetric Loads*.

Extreme orientations of the conical shell element result in a cylindrical shell element or an annular disc element. The shell element may have a linearly varying thickness. See SHELL61 in the *Mechanical APDL Theory Reference* for more details about this element.

Figure 9-1-7 Mechanical APDL Element Reference Entry for SHELL61.

Most of the output data for shell elements is classified as "additional element output." Before we can plot additional output data, we must retrieve it and store it in an element table. We are interested in the top, bottom, and middle values for the hoop stress and the longitudinal (in the cylinder) or meridional (in the end caps) stress. Figure 9-1-8 shows the Stress Definitions and Figure 9-1-9 shows the Element Output Definitions table from the Mechanical APDL Element Reference for SHELL61. These show that the radial stress is referred to as the through-thickness stress STHK, the hoop stress is referred to as SH, and the longitudinal or meridional stress is referred to as SM.

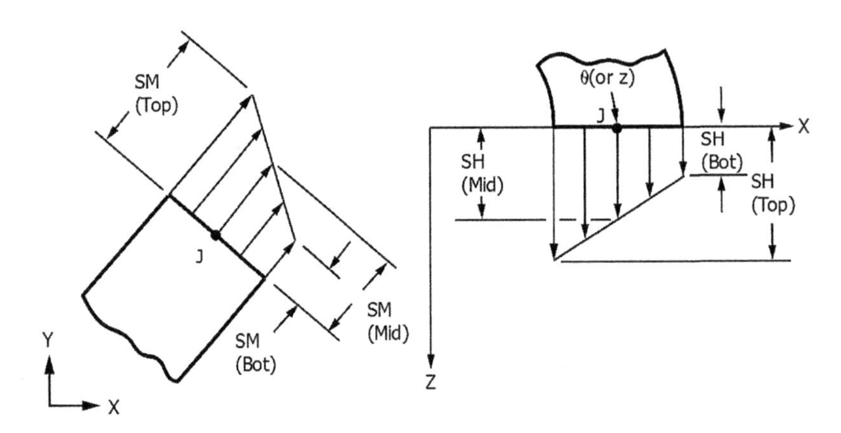

Figure 9-1-8 Stress Definitions for SHELL61 from the Mechanical APDL Element Reference.

Table 61.1: SHELL61 Element Output Definitions

Name	Definition	O	R
EL	Element Number	Y	Y
NODES	Nodes - I, J	Y	Y
MAT	Material number	Y	Y
LENGTH	Distance between node I and node J	Y	Y
XC, YC	Location where results are reported	Y	*2*
TEMP	Temperatures T1, T2, T3, T4	Y	Y
PRES	Pressures P1 (top) at nodes I,J; P2 (bottom) at nodes I,J	Y	Y
MODE	Number of waves in loading	Y	Y
ISYM	Loading key: 1 = symmetric, -1 = antisymmetric	Y	Y
T(X, Z, XZ)	In-plane element X, Z, and XZ forces at KEYOPT(6) location(s)	Y	Y
M(X, Z, XZ)	Out-of-plane element X, Z, and XZ moments at KEYOPT(6) location(s)	Y	Y
MFOR(X, Y, Z), MMOMZ	Member forces and member moment for each node in the element coordinate system	*1*	Y
PK ANG	Angle where stresses have peak values: 0 and 90/MODE°. Blank if *MODE* = 0.	Y	Y
S(M, THK, H, MH)	Stresses (meridional, through-thickness, hoop, meridional-hoop) at PK ANG locations, repeated for top, middle, and bottom of shell	Y	Y

Figure 9-1-9 Element Output Definitions for SHELL61 from the Mechanical APDL Element Reference.

These results are retrieved using the ETABLE command that has the syntax: ***ETABLE****,Lab,Item, Comp,Options. The label is defined by the user. The SHELL61 Item and Sequence Numbers table (Figure 9-1-10) shows that the item for all three stresses is "LS." The Comp depends on the node for which the result should be retrieved (I, J, or an intermediate location) and on the surface (top, middle, or bottom). An example command to retrieve the hoop stress for the top surface at node I would be* `ETABLE,THoopI,LS,3`. *An example command to retrieve the hoop stress for the top surface at Node J would be* `ETABLE,THoopJ,LS,27`.

Table 61.2: SHELL61 Item and Sequence Numbers (KEYOPT(6) = 0 or 1)

Output Quantity Name	ETABLE and ESOL Command Input			
	Item	I	IL1	J
Top				
SM	LS	1	13	25
STHK	LS	2	14	26
SH	LS	3	15	27

Figure 9-1-10 Select Element Table Item and Sequence Numbers for SHELL61 from the Mechanical APDL Element Reference.

9-8. Define and fill an element table for the top face hoop stress data for Node I

- General Postprocessor > Element Table > Define Table
- Click the “Add” button
- Enter “THoopI” for “Lab User label for item” (Figure 9-1-11)
- For Item,Comp Results data item, scroll to the bottom of the left column and choose “By sequence num”
- In the right column, choose “LS”
- In the text box under the right column, add a “3” so the full entry in the textbox reads “LS,3”
- Click Apply

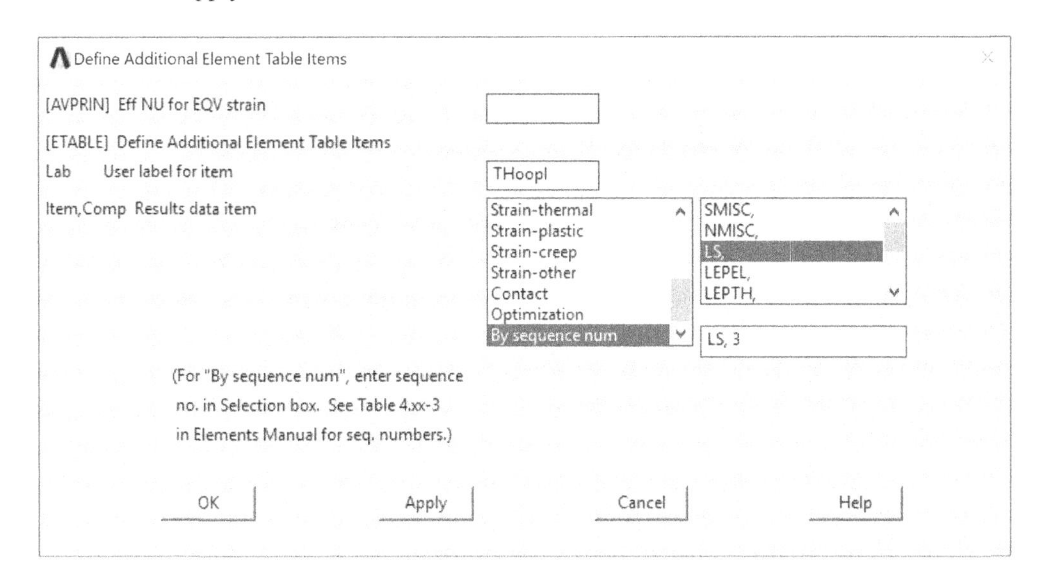

Figure 9-1-11 Define Element Table Items Dialog Box.

9-9. Define and fill an element table for the top face hoop stress data for Node J

- Enter “THoopJ” for Lab User label for item
- For Item,Comp Results data item, scroll to the bottom of the left column and choose “By sequence num”
- In the right column, leave “LS,” selected
- Change the entry in the textbox to read “LS,27”
- Click OK

The Element Table Data dialog box should indicate that both element tables have been created (Figure 9-1-12).

- Close the Element Table Data dialog box

9-10. Define element tables for the middle and bottom face hoop stresses

- Enter the following commands into the command prompt:

```
ETABLE,MHoopI,LS,7
ETABLE,MHoopJ,LS,31
ETABLE,BHoopI,LS,11
ETABLE,BHoopJ,LS,35
```

You can confirm these Comp numbers using Table 61.2 in the Mechanical APDL Element Reference.

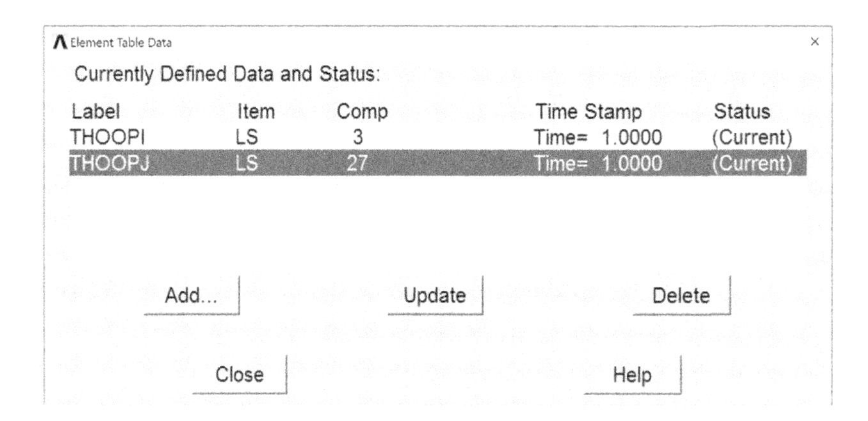

Figure 9-1-12 Currently Defined Element Tables.

9-11. Confirm that the element tables have been created correctly

- General Postprocessor > Element Table > Define Table
- Close the dialog box when you are finished reviewing the defined element tables

9-12. Specify the range of values to use for the result plot color contours

- Utility Menu > PlotCtrls > Style > Contours > Uniform Contours . . .
- Leave NCONT Number of contours set to 9
- Change the Contour intervals to "User specified"
- For VMIN Min contour value enter 4e6
- For VMAX Max contour value enter 1e7
- Click OK

This will ensure that all plots will have the same color contours for direct comparison. The values were chosen by inspection after creating the next three plots with automatic color contours. You may choose different values if you prefer.

9-13. Plot the top (inner) hope stress for Node I

- General Postprocessor > Plot Results > Contour Plot > Elem Table
- Choose "ThoopI" for Itlab Item to be plotted
- Click OK

The PLETAB command plots the results over the line elements. Therefore, the results can be difficult to see and interpret. The ***/ESHAPE*** *command has no effect on this plot.*

9-14. Plot the top (inner) hoop stress

- General Postprocessor > Plot Results > Contour Plot > Line Elem Res
- Choose "ThoopI" for LabI Elem table item at Node I
- Choose "ThoopJ" for LabJ the Elem table item at Node J
- Click OK

The PLLS command displays element table results as contoured areas along the elements.

9-15. Plot the middle hoop stress

- General Postprocessor > Plot Results > Contour Plot > Line Elem Res
- Choose "MHoopI" for the Elem table item at Node I
- Choose "MHoopJ" for the Elem table item at Node J
- Click OK

9-16. Plot the bottom (outer) hoop stress

- General Postprocessor > Plot Results > Contour Plot > Line Elem Res
- Choose "BHoopI" for the Elem table item at Node I
- Choose "BHoopJ" for the Elem table item at Node J
- Click OK

The top (inner), middle, and bottom (outer) hoop stresses are shown in Figure 9-1-13 (left, center, and right, respectively). These plots show that the hoop stress increases through the thickness of the pressure vessel wall and that the hoop stress is highest in the cylindrical section.

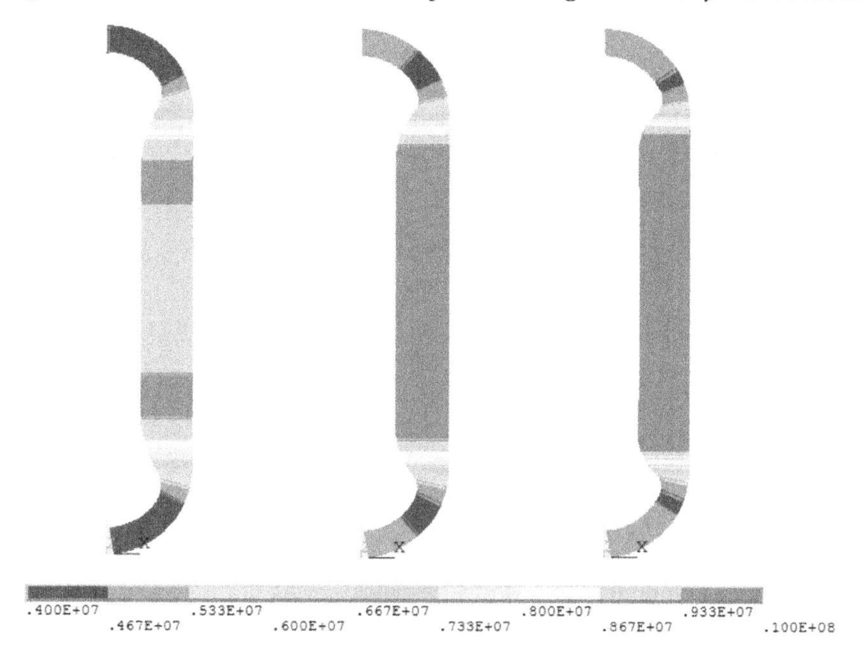

Figure 9-1-13 Hoop Stresses: Top (Inner) Surface (left), Middle of the Shell (center), and Bottom (Outer) Surface (right) (Win32 Graphics, Contours Fixed for Comparison).

9-17. Find the maximum hoop stress in the model

- Utility Menu > List > Results > Element Table Data ...
- Choose all six hoop stresses (THoopI, ThoopJ, MHoopI, MHoopJ, BHoopI, BHoopJ)
- Click OK
- Scroll to the bottom of the PRETAB Command window

This shows that the maximum hoop stress in the model is 0.997e7 on the bottom surface at Node I of Element 72 and Node J of Element 9.

- Close the window when you are finished

9-18. Define element tables for the top, middle, and bottom face meridional stresses (SM)

- Enter the following commands into the command prompt:

```
ETABLE,TMerI,LS,1
ETABLE,TMerJ,LS,25
ETABLE,MMerI,LS,5
ETABLE,MMerJ,LS,29
ETABLE,BMerI,LS,9
ETABLE,BMerJ,LS,33
```

You can confirm these Comp numbers using Table 61.2 in the Mechanical APDL Element Reference.

9-19. Confirm that the element tables have been created correctly

- General Postprocessor > Element Table > Define Table
- Close the dialog box when you are finished

9-20. Specify the range of values to use for the result plot color contours

- Utility Menu > PlotCtrls > Style > Contours > Uniform Contours . . .
- Change VMIN Min contour value enter 3e6
- Change VMAX Max contour value enter 6.5e6
- Click OK

Or

- Enter the following command into the command prompt:
  `/CONT,1,9,3e6,,6.5e6`

These values were chosen by inspection after creating the next three plots with automatic color contours. You may choose different values if you prefer.

9-21. Plot the top (inner) meridional stress

- Enter the following command into the command prompt:
  `PLLS,TMerI,TMerJ`

9-22. Plot the middle meridional stress

- Enter the following command into the command prompt:
  `PLLS,MMerI,MMerJ`

9-23. Plot the bottom (outer) meridional stress

- Enter the following command into the command prompt:
  `PLLS,BMerI,BMerJ`

Figure 9-1-14 shows the top (inner), middle, and bottom (outer) meridional stresses (left, center, and right, respectively).

9-24. Return to automatic color contours

- Utility Menu > PlotCtrls > Style > Contours > Uniform Contours . . .
- For Contour intervals, choose "Auto calculated"
- Click OK

Or

- Enter the following command into the command prompt:
  `/CONT,1,AUTO`

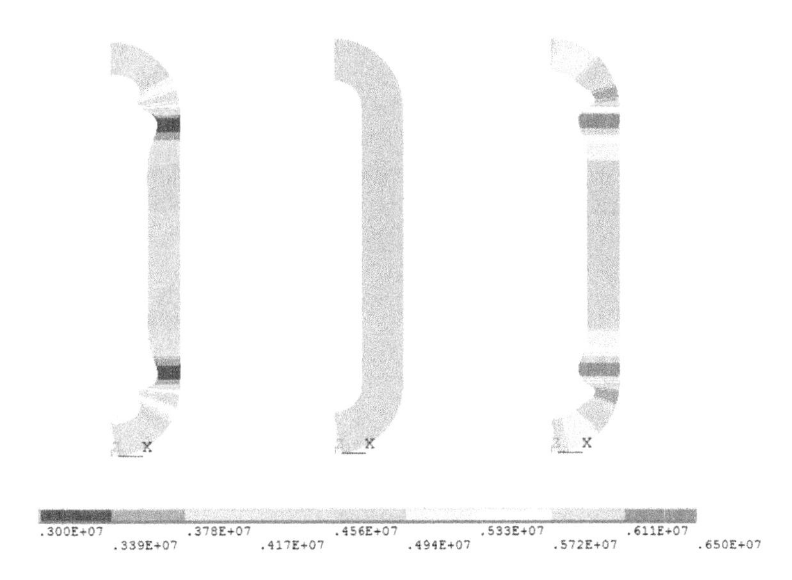

Figure 9-1-14 Meridional Stresses: Top (Inner) Surface (left), Middle of the Shell (center), and Bottom (Outer) Surface (right) (Win32 Graphics, Contours Fixed for Comparison).

9-25. Replot the middle meridional stress

- Enter the following command into the command prompt:

```
PLLS,MMerI,MMerJ
```

This shows that the meridional stresses in the middle of the pressure vessel wall are not uniform and confirms that the variations are small (Figure 9-1-15).

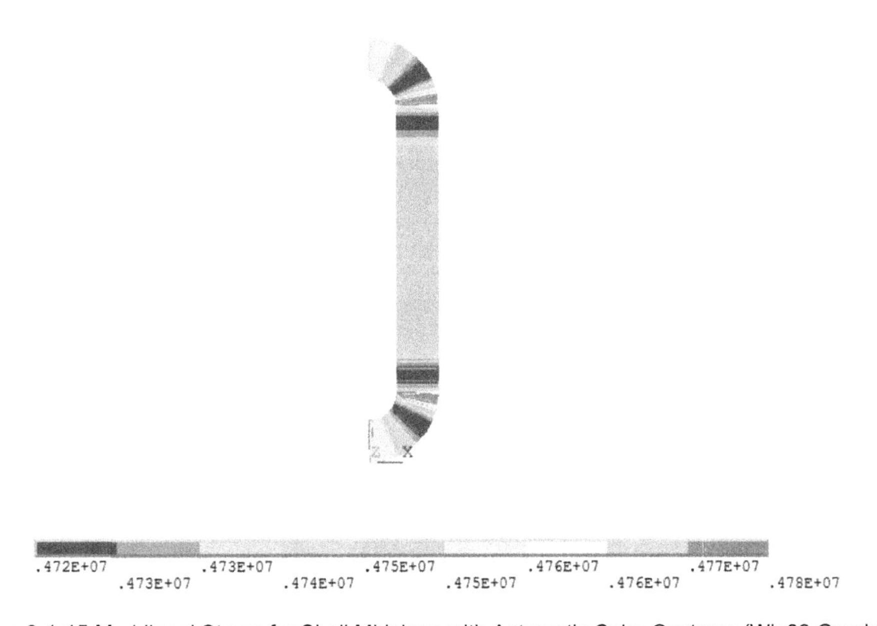

Figure 9-1-15 Meridional Stress for Shell Midplane with Automatic Color Contours (Win32 Graphics).

Step 10: Compare and Verify the Results

For a cylindrical pressure vessel, the hoop stress and the longitudinal stress away from the end caps are given by the following equations:

$$\sigma_{\text{hoop}} = (P)(r)/t \qquad \sigma_{\text{long}} = (P)(r)/2/t$$

where P is the internal pressure, r is the radius of the vessel, and t is the wall thickness. Therefore, it is expected that the hoop stress will have a value of 1e7 Pa and the longitudinal stress will have a value of 0.5e7 Pa. The model predicts a maximum hoop stress of 0.997e7 for a difference of 0.3%. The model predicts a longitudinal stress at the outer surface ~ 0.478e7 for a difference of 4.4%. Therefore, the model is in good agreement with the theory and can be used for engineering design and analysis.

Close the Program

- Utility Menu > File > Exit . . .
- Choose "Quit - No save!"
- Click OK

Sample Input File

```
/PREP7                  ! Enter the Preprocessor
K,1,0,0                 ! Create a keypoint at (0,0)
K,2,0.2,0.2             ! Create a keypoint at (0.2,0.2)
K,3,0.2,1               ! Create a keypoint at (0.2,1)
K,4,0,1.2               ! Create a keypoint at (0,1.2)
K,5,0,0.2               ! Create a keypoint at (0,0.2)
K,6,0,1                 ! Create a keypoint at (0,1)

L,2,3                   ! Create a line between KP 2 and 3
LARC,3,4,6,0.2,         ! Create arc between KP 3 and 4 centered on KP6
LARC,1,2,5,0.2,         ! Create arc between KP 1 and 2 centered on KP5
MP,EX,1,200e9           ! Define Young's modulus for material #1
MP,NUXY,1,0.28          ! Define Poisson's ratio for material #1
ET,1,SHELL61            ! Define element #1 to be a Shell51
KEYOPT,1,6,1            ! Output at nodes & 1 integration point
R,1,0.02                ! Define real constant shell thickness=0.02
ESIZE,0.01              ! Use an element edge length of 0.01
LMESH,ALL               ! Mesh all lines in the model
FINISH                  ! Finish and Exit the Preprocessor

/SOLU                   ! Enter the Solution processor
DK,1,ALL,0              ! Constrain keypoint in all DOFs
DK,4,UX,0               ! Constrain keypoint in x
DK,4,ROTZ,0             ! Constrain keypoint in rotation about z
D,ALL,UZ,0              ! Constrain all nodes in z
SF,ALL,PRES,1e6,        ! Apply a pressure of 1e6 Pa to all nodes
ALLSEL                  ! Select everything
SOLVE                   ! Solve the model
FINISH                  ! Finish and Exit Solution

/POST1                  ! Enter the General Postprocessor
PLNSOL,U,X,0,1          ! Plot nodal displacement in x
PLNSOL,S,EQV,0,1        ! Plot nodal von Mises stress
PRESOL,S,EQV,0,1        ! Plot element von Mises stress
ETABLE,THoopI,LS,3      ! Store the top hoop stress for node I
ETABLE,THoopJ,LS,27     ! Store the top hoop stress for node J
ETABLE,MHoopI,LS,7      ! Store the middle hoop stress for node I
```

```
ETABLE,MHoopJ,LS,31    ! Store the middle hoop stress for node J
ETABLE,BHoopI,LS,11    ! Store the bottom hoop stress for node I
ETABLE,BHoopJ,LS,35    ! Store the bottom hoop stress for node J
/CONT,1,9,4e6,,1e7     ! Set the values for the color contours
PLLS,THoopI,THoopJ     ! Plot hoop stress for top (inner) surface
PLLS,MHoopI,MHoopJ     ! Plot hoop stress for the middle of the shell
PLLS,BHoopI,BHoopJ     ! Plot hoop stress for bottom (outer) surface
ETABLE,TMerI,LS,1      ! Store the top meridional stress for node I
ETABLE,TMerJ,LS,25     ! Store the top meridional stress for node J
ETABLE,MMerI,LS,5      ! Store the middle meridional stress for node I
ETABLE,MMerJ,LS,29     ! Store the middle meridional stress for node J
ETABLE,BMerI,LS,9      ! Store the bottom meridional stress for node I
ETABLE,BMerJ,LS,33     ! Store the bottom meridional stress for node J
/CONT,1,9,3e6,,6.5e6   ! Set the values for the color contours
PLLS,TMerI,TMerJ       ! Plot the top meridional stress
PLLS,MMerI,MMerJ       ! Plot the middle meridional stres
PLLS,BMerI,BMerJ       ! Plot the bottom meridional stress
/CONT,1,AUTO           ! Use automatic color contours
PLLS,MMerI,MMerJ       ! Plot the middle meridional stress
FINISH                 ! Finish and Exit General Postprocessor

SAVE                   ! Save the database
!/EXIT                 ! Exit ANSYS
```

Exercise 9-2

Postprocessing a 3D Thermal Model With Geometric Discontinuities Using Power Graphics

Overview

In this exercise, you will perform a steady-state thermal analysis of a metal bar with sharp geometric discontinuities that has been modeled using continuum elements. The bar is made of high carbon steel with a thermal conductivity of 40 W/mK. The middle section of the bar is 40x40x40 cm. The outer sections of the bar are 20x20x60 cm. 500,000 W/m^3 of heat is generated over a volume of 5x5x5 cm in the center of the bar. The temperatures of the two ends of the bar are fixed at 100°C (Figure 9-2-1). The goal of this analysis is to visualize the temperature and heat flux distributions in the bar and to determine the maximum temperature in the system.

The postprocessing will explore the effect of PowerGraphics vs. Full Graphics on listing and plotting operations. It will also explore the impact of the selection set on the averaged nodal results. Finally, it will investigate various postprocessing operations at sharp geometric discontinuities.

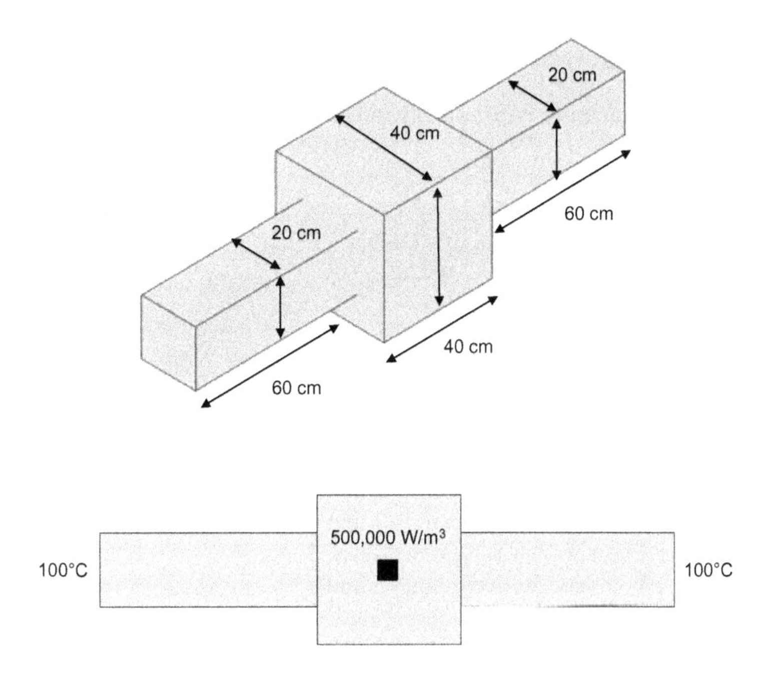

Figure 9-2-1 Schematic of the Steel Bar Geometry (Isometric View, top) and Thermal Boundary Conditions (Cross-Sectional View, bottom).

ANSYS Mechanical APDL for Finite Element Analysis.
DOI: http://dx.doi.org/10.1016/B978-0-12-812981-4.00027-7

Model Attributes

Material properties for High Carbon Steel

- Thermal conductivity—40 W/mK

Thermal Loads

- Heat generation of 500,000 W/m^3 over a volume of 5x5x5 cm in the center of the bar

Thermal Constraints

- Temperature of 100°C on the left and right sides of the bar

File Management

Create a new folder in your "Intro-to-ANSYS" folder for Exercise9-2

Change the Working Directory and the Jobname

Start ANSYS

Step 1: Define Geometry

Although the geometry for this model is very simple, the postprocessing will require a very coarse mapped mesh that is uniform throughout the bar. The best way to create this mesh is to generate and extrude a set of areas. In addition, we need to make sure that heat can cross all lines and areas in the model (i.e., that no internal lines or surfaces are adiabatic). There are many ways to do this, but the most robust option is to make sure that all boundaries are symmetric and then merge them so there is only one entity in each location to mesh.

1-1. Enter the Preprocessor

1-2. Create keypoints to represent the left side of the 40x40 cm cross section

- Create a keypoint at (−0.2, − 0.2)
- Create a keypoint at (−0.2, − 0.1)
- Create a keypoint at (−0.2, − 0.025)
- Create a keypoint at (−0.2,0.025)
- Create a keypoint at (−0.2,0.1)
- Create a keypoint at (−0.2,0.2)

1-3. Create lines to connect the keypoints on the left side of the cross section

- Create a line between Keypoints 1 and 2
- Create a line between Keypoints 2 and 3
- Create a line between Keypoints 3 and 4
- Create a line between Keypoints 4 and 5
- Create a line between Keypoints 5 and 6

1-4. Create keypoints to represent the top side of the 40x40 cm cross section

- Create a keypoint at (−0.1,0.2)
- Create a keypoint at (−0.025,0.2)
- Create a keypoint at (0.025,0.2)

- Create a keypoint at (0.1,0.2)
- Create a keypoint at (0.2,0.2)

1-5. Create lines to connect the keypoints on the top side of the cross section

These lines will be used to identify the extrusion path.

- Create a line between Keypoints 6 and 7
- Create a line between Keypoints 7 and 8
- Create a line between Keypoints 8 and 9
- Create a line between Keypoints 9 and 10
- Create a line between Keypoints 10 and 11

1-6. Plot lines

1-7. Drag the first five lines along the path defined by the second five lines

- Main Menu > Preprocessor > Modeling > Operate > Extrude > Lines > Along Lines
- Click the five lines on the left side of the screen
- Click OK
- Click the five lines on the top of the screen
- Click OK

The resulting area plot is shown in Figure 9-2-2.

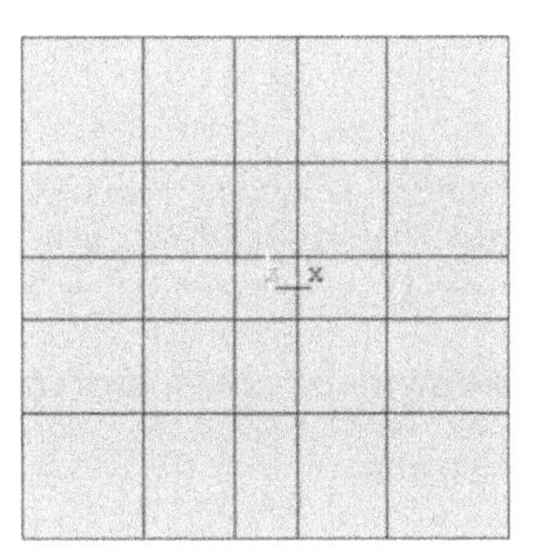

Figure 9-2-2 Area Plot.

1-8. Change to the isometric view

The isometric area plot is shown in Figure 9-2-3 (left).

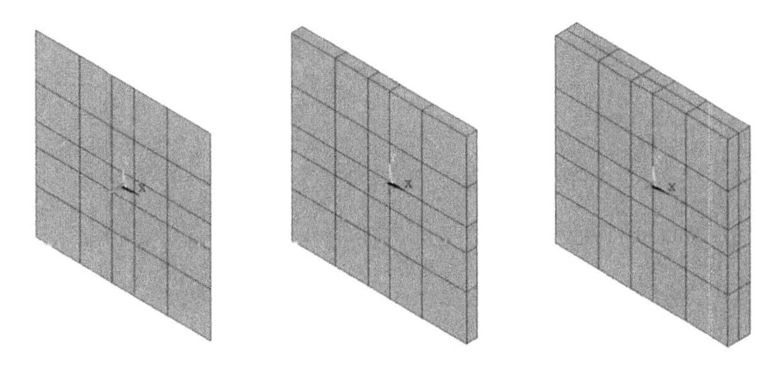

Figure 9-2-3 Isometric Views of the Base Areas (left), the First Extrusion (center), and Both Extrusions (right).

1-9. Extrude the areas by 2.5 cm (0.025 m) in the positive normal direction

- Main Menu > Preprocessor > Modeling > Operate > Extrude > Areas > By XYZ Offset
- Pick All
- Set the offset for DZ to 0.025
- Click OK

The resulting volume plot is shown in Figure 9-2-3 (center).

1-10. Select all areas at $z = 0$

- Enter the following command into the command prompt:

```
ASEL,S,LOC,Z,0
```

1-11. Extrude the set of areas at $z = 0$ by 2.5 cm (0.025 m) in the negative normal direction

- Main Menu > Preprocessor > Modeling > Operate > Extrude > Areas > By XYZ Offset
- Pick All
- Set the offset for DZ to −0.025
- Click OK

1-12. Select everything and plot volumes

- Enter the following commands into the command prompt:

```
ALLSEL
VPLOT
```

The resulting volume plot is shown in Figure 9-2-3 (right). The extruded block now has two 5x5x2.5 cm volumes at its center. We will apply the heat generation boundary condition to these two volumes.

1-13. Confirm that your geometry has been correctly created and save your work

1-14. Select all areas at $z = 0.025$

1-15. Extrude the active set of areas by 0.175 m in the positive normal direction

1-16. Select all areas at $z = -0.025$

1-17. Extrude the active set of areas by 0.175 m in the negative normal direction

1-18. Select everything and plot volumes

- Enter the following commands into the command prompt:

```
ALLSEL
VPLOT
```

1-19. Confirm that your geometry has been correctly created and save your work

1-20. Select all areas between (−0.1, − 0.1,0.2) and (0.1, 0.1,0.2)

- Enter the following commands into the command prompt:

```
ASEL,S,LOC,X,-0.1,0.1
ASEL,R,LOC,Y,-0.1,0.1
ASEL,R,LOC,Z,0.2
```

Note that the first command is ASEL,S (select a new set), but the following two commands are ASEL,R (reselect from the current set).

1-21. Extrude the active set of areas by 0.6 m in the positive normal direction

- Enter the following command into the command prompt:
```
VEXT,ALL,,,0,0,0.6
```

1-22. Select all areas between (–0.1, – 0.1, – 0.2) and (0.1, 0.1, – 0.2)

- Enter the following commands into the command prompt:
```
ASEL,S,LOC,X,-0.1,0.1
ASEL,R,LOC,Y,-0.1,0.1
ASEL,R,LOC,Z,-0.2
```

1-23. Extrude the active set of areas by 0.6 m in the negative normal direction

- Enter the following command into the command prompt:
```
VEXT,ALL,,,0,0,-0.6
```

1-24. Select everything and plot volumes

- Enter the following commands into the command prompt:
```
ALLSEL
VPLOT
```

1-25. Zoom out or zoom to fit if necessary

The volume plot with the completed geometry is shown Figure 9-2-4 (left).

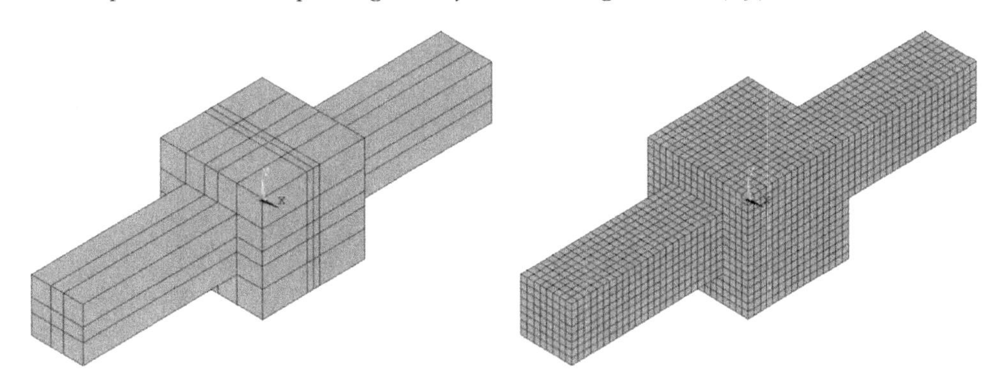

Figure 9-2-4 Completed Geometry (left) and Finite Element Mesh (right).

1-26. Merge all entities in the model

- Enter the following command into the command prompt:
```
NUMMRG,ALL
```

1-27. Save your progress

Step 2: Define Element Types

2-1. Define SOLID70 as element type #1 for this model

Step 3: Define Material Properties

3-1. Define an isotropic thermal conductivity for this model

Step 4: Mesh

4-1. Mesh the model

- Enter the following commands into the command prompt:

```
ESIZE,0.025
VMESH,ALL
```

The first command sets the element edge length for the model to 0.025. This is the largest element edge length that can be used to create a uniform and consistent mesh in the model. The second command meshes all volumes in the model. The resulting finite element mesh is shown in Figure 9-2-4 (right).

Step 5: Apply Constraint Boundary Conditions

5-1. Enter the Solution processor

- Enter the following command into the command prompt:

```
/SOL
```

5-2. Set the temperature of the left end of the bar to 100°C

- Enter the following commands into the command prompt:

```
NSEL,S,LOC,Z,0.8
D,ALL,TEMP,100
```

5-3. Set the temperature of the right end of the bar to 100°C

- Enter the following commands into the command prompt:

```
NSEL,S,LOC,Z,-0.8
D,ALL,TEMP,100
```

Step 6: Apply Load Boundary Conditions

6-1. Select the 5x5x5 cm volume at the center of the bar to apply the heat generation boundary condition

- Enter the following commands into the command prompt:

```
VSEL,S,LOC,X,-0.025,0.025
VSEL,R,LOC,Y,-0.025,0.025
VSEL,R,LOC,Z,-0.025,0.025
```

6-2. Plot volumes to ensure that only two small volumes have been selected

6-3. Apply a heat generation of 500,000 W/m^3 on the volume

- Main Menu > Solution > Define Loads > Apply > Thermal > Heat Generat > On Volumes
- Choose "Pick All"
- For "VALUE Load HGEN value" enter 500000
- Click OK

Step 7: Set the Solution Options

The default solution options can be used for this analysis.

Step 8: Solve

8-1. Select everything

8-2. Solve the model

8-3. Save your results

Step 9: Postprocess the Results to Explore the Impact of PowerGraphics vs. Full Graphics on Contour Plots of the Full Model

9-1. Enter the General Postprocessor

9-2. Ensure that PowerGraphics is on

- Utility Menu > PlotCtrls > Style > Hidden Line Options . . .
- For "[/GRAPHICS] Used to control the way a model is displayed Graphic display method is" ensure that "PowerGraphics" is selected
- Click OK

9-3. Plot the nodal temperature distribution in the model

Note that the maximum temperature in the legend is 112.878°C (Figure 9-2-5, left).

9-4. Turn PowerGraphics off (turn Full Graphics on)

- Click the POWRGRPH button in the ANSYS Toolbar
- Check the radio button for "OFF"
- Click OK

9-5. Plot the nodal temperature distribution in the model

Note that the maximum temperature in the legend is now 118.357°C (Figure 9-2-5, right).

As noted in chapter 9, Full Graphics considers all results associated with the selected set of nodes or elements, whereas PowerGraphics only considers the results for the surface of the model. Because the maximum temperature in this model is at the center of the bar, the contour plot with Full Graphics has a higher maximum temperature than the contour plot with Power Graphics. Therefore, the Full Graphics plot is a better representation of the true behavior of the model.

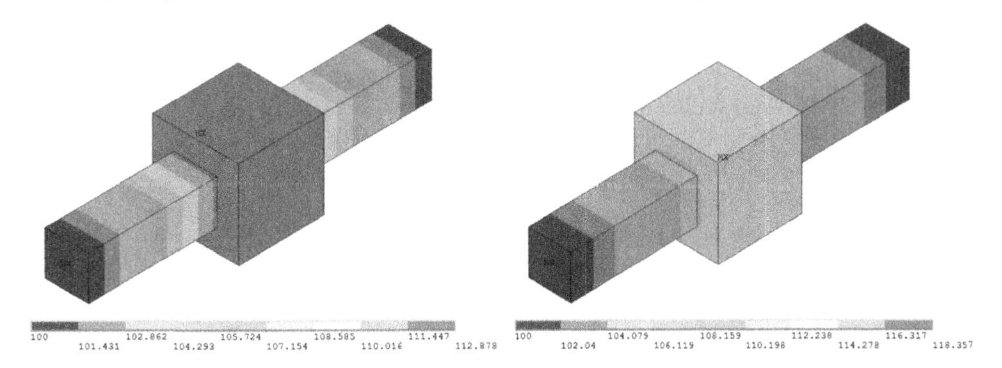

Figure 9-2-5 Plot of Nodal Temperature with PowerGraphics (left) and Full Graphics (right) (Win32 Graphics).

9-6. Turn PowerGraphics on

- Enter the following command into the command prompt:
  `/GRAPHICS,POWER`

9-7. Plot the nodal heat flux (thermal flux vector sum) in the model

Note that the maximum heat flux in the legend is 1367.46 W/m^2 *(Figure 9-2-6, left).*

9-8. Turn PowerGraphics off (turn Full Graphics on)

- Enter the following command into the command prompt:
  `/GRAPHICS,OFF`

9-9. Plot the nodal heat flux (thermal flux vector sum) in the model

Note that the maximum heat flux in the legend is now 3262.93 W/m^2 *(Figure 9-2-6, right). This value is significantly higher than the one obtained with PowerGraphics because the maximum heat flux occurs in the center of the bar.*

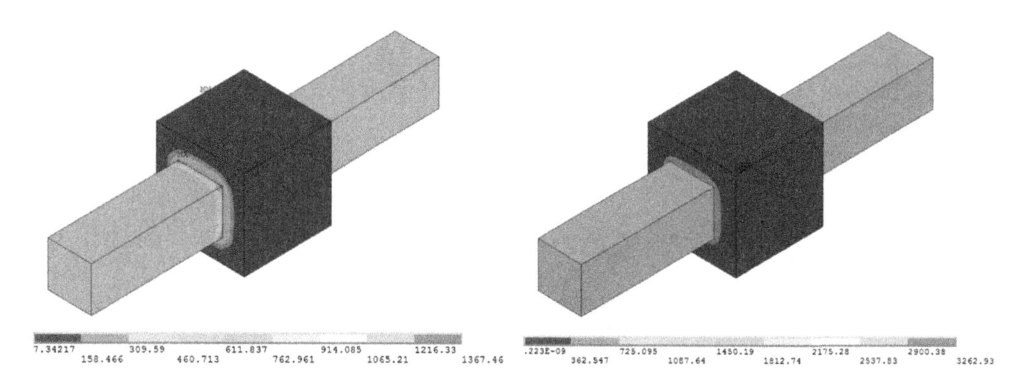

Figure 9-2-6 Plot of Nodal Heat Flux with PowerGraphics (left) and Full Graphics (right) (Win32 Graphics).

Step 10: Postprocess the Results to Explore the Impact of PowerGraphics vs. Full Graphics and the Surface Definition on a Half Model

10-1. Turn PowerGraphics on

10-2. Move the triad from the global origin to the lower left corner of the Graphics Window

- Utility Menu > PlotCtrls > Window Controls > Window Options . . .
- In the second section that says "[/TRIAD] Location of triad," choose "At bottom left" from the drop down menu
- Click OK
- Close any error messages that appear

By default, the triad is located at the global origin (0,0,0). In this model, the generation and therefore the maximum temperatures and heat fluxes are also centered at (0,0,0). Therefore, the triad must be moved so it does not interfere with the visualization of that data.

10-3. Select all nodes between $x = -0.2$ and $x = 0$ (the left half of the model) and then select all elements attached to those nodes

- Enter the following commands into the command prompt:

```
NSEL,S,LOC,X,-0.2,0
ESLN,S,1
```

10-4. Plot the nodal temperature distribution in the left half of the model

Note that the maximum temperature in the legend is now 118.357°C (Figure 9-2-7, left). This is the same as the Full Graphics plot of the full model (Figure 9-2-5, right) and confirms that the surfaces for PowerGraphics displays are defined by the active selection set.

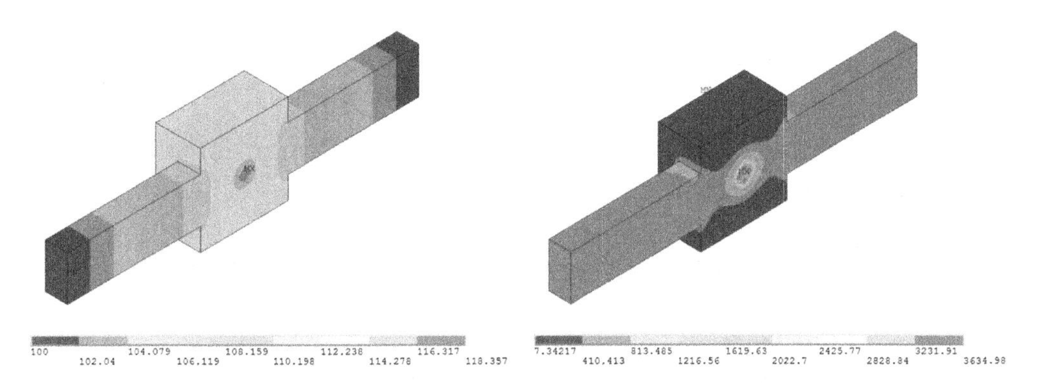

Figure 9-2-7 Plot of Nodal Temperature (left) and Nodal Heat Flux (right) with PowerGraphics (Win32 Graphics).

10-5. Plot the nodal heat flux (thermal flux vector sum) in the left half of the model

Note that the maximum heat flux in the legend is now 3634.98 W/m^2(Figure 9-2-7, right). This is approximately 11% higher than the maximum heat flux in the full model with Full Graphics (Figure 9-2-6, right). If you plot the nodal heat flux again using Full Graphics, you will get the same result. This indicates that results averaging is affected by the selection set.

Step 11: Postprocess the Results to Explore the Impact of PowerGraphics vs. Full Graphics on Results Listing vs. Plotting

11-1. Select everything

11-2. List the nodal temperature distribution in the model

- Utility Menu > List > Results > Nodal Solution ...

Note that the maximum temperature shown at the bottom of the file is 118.36°C. This is the same maximum value as the Full Graphics and half model contour plots. This confirms that PowerGraphics lists the DOF results for the entire selected set and not just for the surface.

11-3. Turn Full Graphics on

11-4. List the nodal temperature distribution in the model

Note that the maximum temperature shown at the bottom of the file is still 118.36°C.

11-5. Turn PowerGraphics on

11-6. List the nodal heat flux (thermal flux vector sum) in the model

If you scroll to the bottom of the window, you will see that the maximum heat flux in the last column (TFSUM) is 1089.5 W/m^2. This is lower than the value from the PowerGraphics contour plot legend (1367.46 W/m^2) and confirms that PowerGraphics uses different methods to calculate the averaged nodal results for listing and plotting operations.

11-7. Turn Full Graphics on

11-8. List the nodal heat flux (thermal flux vector sum) in the model

Note that the maximum heat flux in the last column (TFSUM) is now 3262.9 W/m^2. This is approximately the same value as the plotted solution for Full Graphics because Full Graphics uses the same averaging methods for both listing and plotting.

Step 12: Postprocess the Results to Explore the Impact of the Selection Set on Results Averaging at a Geometric Discontinuity

12-1. Select everything

12-2. Specify the range of values to use in the following contour plots

- Utility Menu > PlotCtrls > Style > Contours > Uniform Contours . . .
- For "Contour intervals" click the "User specified" radio button
- For "VMIN Min contour value" enter 520
- For "VMAX Max contour value" enter 690
- Click OK

These contour values were chosen by creating the following plots with automatic contours and then choosing the reduced contour range by inspection.

12-3. Select all nodes between *z* = −0.8 and 0.2 and all elements attached to those nodes

- Enter the following commands into the command prompt:

```
NSEL,S,LOC,Z,-0.8,0.2
ESLN,S,1
```

12-4. Plot elements to confirm the selection

12-5. Plot the nodal thermal flux showing the deformed shape with the undeformed model

12-6. Change to the front view

12-7. Reselect all nodes between *x* = −0.1 and 0.1, *y* = 0.1 and 0.1, and then all elements attached to those nodes

- Enter the following commands into the command prompt:

```
NSEL,R,LOC,X,-0.1,0.1
NSEL,R,LOC,Y,-0.1,0.1
ESLN,S,1
```

12-8. Plot the nodal thermal flux showing the deformed shape with the undeformed model

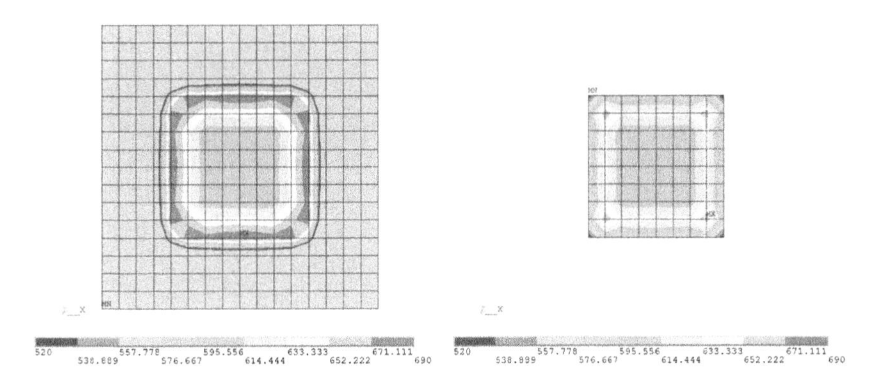

Figure 9-2-8 Plot of Nodal Thermal Flux Vector Sum at z = 0.2 (left) and Plot of Nodal Thermal Flux Vector Sum at x = -0.1 to 0.1, y = -0.1 to 0.1, and z = 0.2 with Constant Contour Colors (right) (Win32 Graphics).

Despite the fact that the plots in Figure 9-2-8 show the same results for the same set of nodes and elements, there are substantial differences, especially near the boundary between the 40x40 and 20x20 cross sections. This is because nodal results averaging is affected by the selection set.

The differences due to the nodal averaging are visible in this exercise because the mesh is very coarse. If the element edge length of the mesh is reduced from 0.025 to 0.01 m, the differences are no longer perceptible.

12-9. Specify the range of values to use in the following contour plots

- Utility Menu > PlotCtrls > Style > Contours > Uniform Contours . . .
- For "Contour intervals" click the "User specified" radio button
- For "VMIN Min contour value" enter 340
- For "VMAX Max contour value" enter 740
- Click OK

12-10. Select all nodes between *z* = −0.8 and 0.2 and all elements attached to those nodes

- Enter the following commands into the command prompt:

```
NSEL,S,LOC,Z,-0.8,0.2
ESLN,S,1
```

12-11. Plot the element solution thermal flux vector sum showing the deformed shape with the undeformed model

12-12. Reselect all nodes between *x* = −0.1 and 0.1, *y* = −0.1 and 0.1, and then all elements attached to those nodes

- Enter the following commands into the command prompt:

```
NSEL,R,LOC,X,-0.1,0.1
NSEL,R,LOC,Y, 0.1,0.1
ESLN,S,1
```

12-13. Plot the element solution thermal flux vector sum showing the deformed shape with the undeformed model

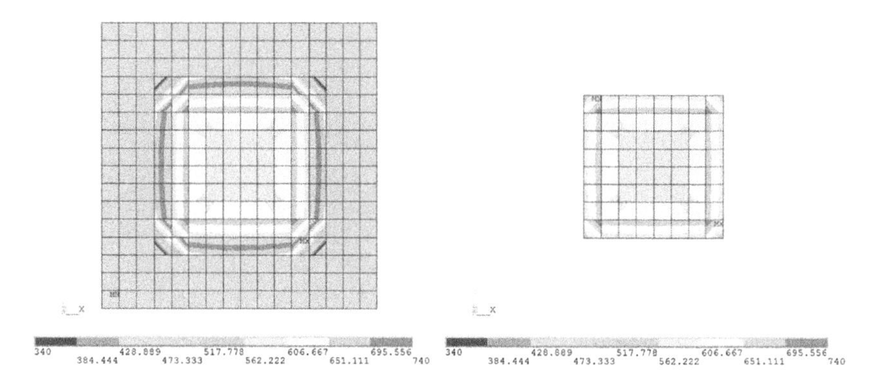

Figure 9-2-9 Plot of Element Solution Thermal Flux Vector Sum at z = 0.2 (left) and at x = -0.1 to 0.1, y = -0.1 to 0.1, and z = 0.2 with Constant Contour Colors (right) (Win32 Graphics).

Figure 9-2-9 shows that the element solution is not affected by the selection set. This is because the element solution is not averaged.

Step 13: Postprocess the Results to Explore the Impact of PowerGraphics vs. Full Graphics at a Geometric Discontinuity

13-1. Select everything

13-2. Ensure that PowerGraphics is turned on

13-3. Update the manual contour colors

- Utility Menu > PlotCtrls > Style > Contours > Uniform Contours . . .
- For "Contour intervals" click the "User specified" radio button
- For "VMIN Min contour value" enter 0
- For "VMAX Max contour value" enter 1370
- Click OK

13-4. Plot the nodal thermal flux showing the undeformed model

13-5. Change to the isometric view

13-6. Zoom in to view the detail at the interface between the 20x20 cm and the 40x 40 section of the geometry

13-7. Turn PowerGraphics off

13-8. Plot the nodal thermal flux showing the undeformed model

The plot with PowerGraphics (Figure 9-2-10, left) indicates a very high heat flux at the four corners where the smaller cross section meets the larger cross section. There is a thermal constriction in two dimensions at these corners, therefore a higher heat flux should be expected in these regions. In addition, the heat flux tapers quickly and evenly from the corners into the smaller part body (from red, to orange, yellow, and then green) as expected.

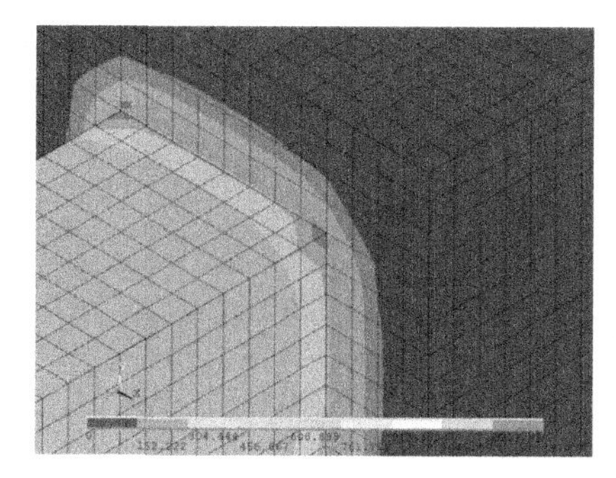 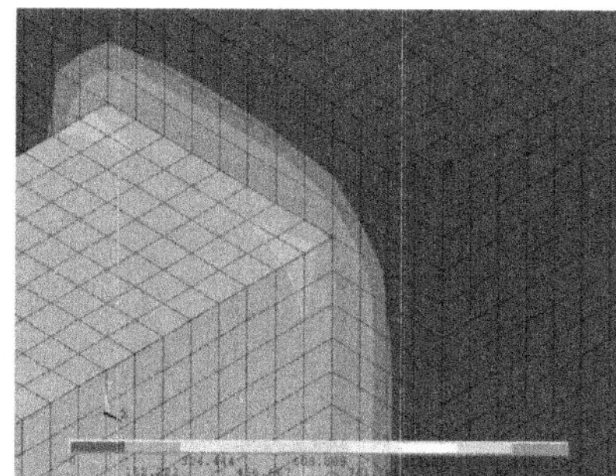

Figure 9-2-10 Plot of Nodal Thermal Flux Vector Sum with PowerGraphics (left) and Full Graphics (right) (Win32 Graphics).

The plot with Full Graphics (Figure 9-2-10, right) shows an even heat flux across all parts of the interface with no indication of increased heat flux at the corners. Instead, hot spots appear one element away from the interface along the corner lines (shown in orange). This heat is shown being averaged back into the surrounding body by the yellow triangles. These hot spots seem to be misplaced, and their behavior is not consistent with our engineering expectations. Therefore, we should treat these results with skepticism.

This example partially explains the documentation's statement that PowerGraphics provides more realistic results at discontinuities.

Step 14: Postprocess Each Side of a Discontinuity Separately

As noted in chapter 9, the Mechanical APDL Basic Analysis Guide recommends postprocessing each side of a discontinuity separately. In the last section of this exercise, we will follow that recommendation and explore the behavior of the model on both sides of the interface.

Because this section investigates the surfaces on each side of the interface, you should get the same results using Full Graphics and PowerGraphics.

14-1. Select all nodes between $z = -0.8$ and 0.2 and all elements attached to those nodes

- Enter the following commands into the command prompt:

```
NSEL,S,LOC,Z,-0.8,0.2
ESLN,S,1
```

14-2. Plot elements

14-3. Zoom to fit

14-4. Plot the nodal heat flux

14-5. Change to the front view

Figure 9-2-11 (left) shows the nodal heat flux for the 40x40 cm side of the left-hand interface. You can see that there is little or no heat flux on the part of the surface that does not contact the 20x20 section. There is a higher heat flux near the edges of the interface and a lower heat flux in the center of the bar.

14-6. Change to the isometric view

14-7. Select all nodes between $z = 0.2$ and 0.8 and all elements attached to those nodes

- Enter the following commands into the command prompt:

```
NSEL,S,LOC,Z,0.2,0.8
ESLN,S,1
```

14-8. Plot elements

14-9. Zoom to fit

14-10. Plot the nodal heat flux

14-11. Change to the back view

Figure 9-2-11 (right) shows that there are much higher heat fluxes on this side of the interface, with the highest heat fluxes around the edge of the 20x20 section. Nowhere do the values on the 20x20 plot perfectly match those from the 40x40 plot. This supports the documentation's recommendation to postprocess both sides of a discontinuity separately.

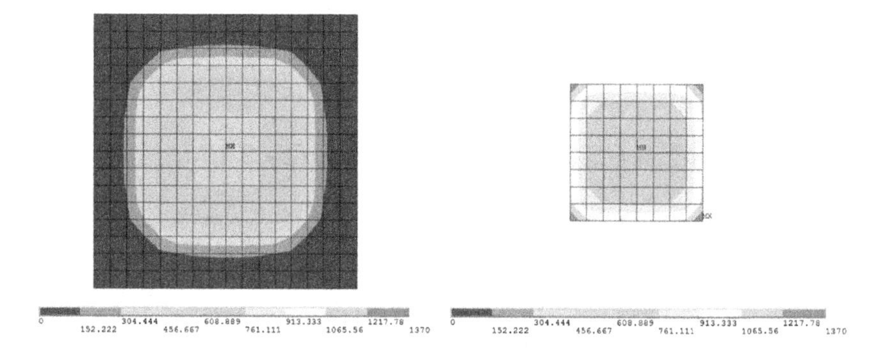

Figure 9-2-11 Plot of Nodal Thermal Flux Vector Sum on the 40x40 (left) and 20x20 (right) Sides of a Geometric Discontinuity (Element Edge Length of 0.025) (Win32 Graphics).

If you reduce the element edge length to 0.005 and rerun the model, you will see that the differences between the two halves of the interface decrease to the point where they are barely noticeable (Figure 9-2-12). However, there are still heat flux concentrations in the four corners of the 20x20 side of the interface. Further refinement of the mesh will reduce the size and impact of these concentrations, but they cannot be removed entirely.

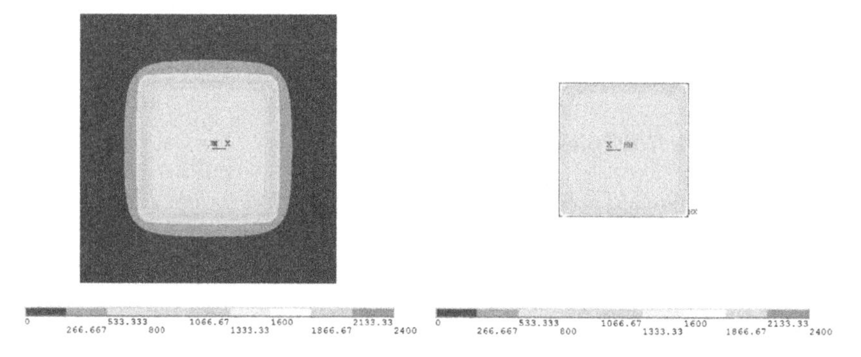

Figure 9-2-12 Plot of Nodal Thermal Flux Vector Sum on the 40x40 (left) and 20x20 (right) Sides of a Geometric Discontinuity (Element Edge Length of 0.005) (Win32 Graphics).

Step 15: Compare and Verify the Results

The mesh in this exercise is far too coarse to provide good results. This model and its results cannot be used for engineering decision-making.

Close the Program

- Utility Menu > File > Exit . . .
- Choose "Quit - No save!"
- Click OK

Sample Input File

```
/PREP7                          ! Enter the Preprocessor
K,1,-0.2,-0.2                   ! Create a keypoint at (-0.2,-0.2)
K,2,-0.2,-0.1                   ! Create a keypoint at (-0.2,-0.1)
K,3,-0.2,-0.025                 ! Create a keypoint at (-0.2,-0.025)
K,4,-0.2,0.025                  ! Create a keypoint at (-0.2,0.025)
K,5,-0.2,0.1                    ! Create a keypoint at (-0.2,0.1)
K,6,-0.2,0.2                    ! Create a keypoint at (-0.2,0.2)
L,1,2                           ! Create a line between keypoints 1 and 2
L,2,3                           ! Create a line between keypoints 2 and 3
L,3,4                           ! Create a line between keypoints 3 and 4
L,4,5                           ! Create a line between keypoints 4 and 5
L,5,6                           ! Create a line between keypoints 5 and 5
K,7,-0.1,0.2                    ! Create a keypoint at (-0.1,0.2)
K,8,-0.025,0.2                  ! Create a keypoint at (-0.025,0.2)
K,9,0.025,0.2                   ! Create a keypoint at (0.025,0.2)
K,10,0.1,0.2                    ! Create a keypoint at (0.1,0.2)
K,11,0.2,0.2                    ! Create a keypoint at (0.2,0.2)
L,6,7                           ! Create a line between keypoints 6 and 7
L,7,8                           ! Create a line between keypoints 7 and 8
L,8,9                           ! Create a line between keypoints 8 and 9
L,9,10                          ! Create a line between keypoints 9 and 10
L,10,11                         ! Create a line between keypoints 10 and 11
ADRAG,1,2,3,4,5,,6,7,8,9,10     ! Drag lines 1-5 along lines 6-10
NUMMRG,ALL                      ! Merge all entities
NUMCMP,LINE                     ! Compress entity numbering
```

```
! Create volumes for the model geometry
VEXT,ALL,,,0,0,0.025          ! Extrude all areas by 0.025 in z direction
ASEL,S,LOC,Z,0                ! Select areas at z=0
VEXT,ALL,,,0,0,-0.025         ! Extrude selected areas by -0.025 in z dir.
ASEL,S,LOC,Z,0.025            ! Select areas at z=0.025
VEXT,ALL,,,0,0,0.175          ! Extrude selected areas by 0.175 in z dir.
ASEL,S,LOC,Z,-0.025           ! Select areas at z=-0.025
VEXT,ALL,,,0,0,-0.175         ! Extrude selected areas by -0.175 in z dir.
ASEL,S,LOC,X,-0.1,0.1         ! Select areas between x=-0.1 and 0.1
ASEL,R,LOC,Y,-0.1,0.1         ! Reselect areas between y=-0.1 and 0.1
ASEL,R,LOC,Z,0.2              ! Reselect areas at z=0.2
VEXT,ALL,,,0,0,0.6            ! Extrude selected areas by 0.6 in z dir.
ASEL,S,LOC,X,-0.1,0.1         ! Select areas between x=-0.1 and 0.1
ASEL,R,LOC,Y,-0.1,0.1         ! Reselect areas between y=-0.1 and 0.1
ASEL,R,LOC,Z,-0.2             ! Reselect areas at z=-0.2
VEXT,ALL,,,0,0,-0.6           ! Extrude selected areas by -0.6 in z dir.

ALLSEL,ALL                    ! Select everything
NUMMRG,ALL                    ! Merge all entities in the model (again)
ET,1,70                       ! Use SOLID70 elements
MP,KXX,1,40                   ! Define thermal conductivity
ESIZE,0.025                   ! Use an element edge length of 0.025 m
VMESH,ALL                     ! Mesh the model

/SOL                          ! Enter the Solution processor
VSEL,S,LOC,X,-0.025,0.025     ! Select volumes between x=-0.025 and 0.025
VSEL,R,LOC,Y,-0.025,0.025     ! Reselect volumes between y=-0.025 and 0.025
VSEL,R,LOC,Z,-0.025,0.025     ! Reselect volumes between z=-0.025 and 0.025
BFV,ALL,HGEN,500000           ! Apply heat generation of 500,000 to volumes
NSEL,S,LOC,Z,0.8              ! Select all nodes at z=0.8
D,ALL,TEMP,100                ! Set the temperature of those nodes to 100
NSEL,S,LOC,Z,-0.8             ! Select all nodes at z=-0.8
D,ALL,TEMP,100                ! Set the temperature of those nodes to 100
ALLSEL,ALL                    ! Select everything
SOLVE                         ! Solve the model

/POST1                        ! Enter the General Postprocessor

! Postprocessing commands for step 9
/GRAPHICS,POWER               ! Turn PowerGraphics on
PLNSOL, TEMP,, 0              ! Plot the nodal temperature in the model
/GRAPHICS,FULL                ! Turn on Full Graphics
PLNSOL, TEMP,, 0              ! Plot the nodal temperature in the model
/GRAPHICS,POWER               ! Turn PowerGraphics on
PLNSOL, TF,SUM, 0             ! Plot the nodal heat flux vector sum
/GRAPHICS,FULL                ! Turn on Full Graphics
PLNSOL, TF,SUM, 0             ! Plot the nodal heat flux vector sum

! Postprocessing commands for step 10
/GRAPHICS,POWER               ! Turn PowerGraphics on
/TRIAD,LBOT                   ! Move the triad to the lower bottom corner
NSEL,S,LOC,x,-0.2,0           ! Select nodes between x=-0.2 and 0
ESLN,S,1                      ! Select elements attached to those nodes
PLNSOL, TEMP,, 0              ! Plot the nodal temperature in the model
PLNSOL, TF,SUM, 0             ! Plot the nodal heat flux vector sum
```

```
! Postprocessing commands for step 11
ALLSEL,ALL                  ! Select everything
PRNSOL,TEMP                 ! List the nodal temperature
/GRAPHICS,FULL              ! Turn on Full Graphics
PRNSOL,TEMP                 ! List the nodal temperature
/GRAPHICS,POWER             ! Turn PowerGraphics on
PRNSOL,TF,COMP              ! List the nodal heat flux vector sum
/GRAPHICS,FULL              ! Turn on Full Graphics
PRNSOL,TF,COMP              ! List the nodal heat flux vector sum

! Postprocessing commands for step 12
/CONT,1,9,520,,690          ! Set the values for the contour colors
ALLSEL,ALL                  ! Select everything
NSEL,S,LOC,Z,-0.8,0.2       ! Select nodes between z=-0.8 and 0.2
ESLN,S,1                    ! Select elements attached to those nodes
/VIEW,1,,,1                 ! Change to front view
PLNSOL, TF,SUM, 1           ! Plot the nodal heat flux vector sum
NSEL,R,LOC,X,-0.1,0.1       ! Reselect nodes between x=-0.1 and 0.1
NSEL,R,LOC,Y,-0.1,0.1       ! Reselect nodes between x=-0.1 and 0.1
ESLN,S,1                    ! Select elements attached to those nodes
PLNSOL, TF,SUM, 1           ! Plot the nodal heat flux vector sum

/CONT,1,9,340,,740          ! Set the values for the contour colors
ALLSEL,ALL                  ! Select everything
NSEL,S,LOC,Z,-0.8,0.2       ! Select nodes between z=-0.8 and 0.2
ESLN,S,1                    ! Select elements attached to those nodes
PLESOL, TF,SUM, 1           ! Plot the element solution heat flux vector sum
NSEL,R,LOC,X,-0.1,0.1       ! Reselect nodes between x=-0.1 and 0.1
NSEL,R,LOC,Y,-0.1,0.1       ! Reselect nodes between x=-0.1 and 0.1
ESLN,S,1                    ! Select elements attached to those nodes
PLNSOL, TF,SUM, 1           ! Plot the nodal heat flux vector sum

! Postprocessing commands for step 13
ALLSEL,ALL                  ! Select everything
/CONT,1,9,0,,1370           ! Set the values for the contour colors
/VIEW,1,1,1,1               ! Change to isometric view
/GRAPHICS,POWER             ! Turn PowerGraphics on
PLNSOL, TF,SUM, 1           ! Plot the nodal heat flux vector sum
/GRAPHICS,OFF               ! Turn on Full Graphics
PLNSOL, TF,SUM, 1           ! Plot the nodal heat flux vector sum

! Postprocessing commands for step 14
ALLSEL,ALL                  ! Select everything
NSEL,S,LOC,Z,-0.8,0.2       ! Select nodes between z=-0.8 and 0.2
ESLN,S,1                    ! Select elements attached to those nodes
/VIEW,1,,,1                 ! Change to the front view
PLESOL, TF,SUM, 1           ! Plot the element solution heat flux vector sum
NSEL,S,LOC,Z,0.2,0.8        ! Select nodes between z=0.2 and 0.8
ESLN,S,1                    ! Select elements attached to those nodes
/VIEW,1,,,-1                ! Change to the back view
PLNSOL, TF,SUM, 1           ! Plot the nodal heat flux vector sum

SAVE                        ! Save the database
!/EXIT                      ! Exit ANSYS
```

Exercise 9-3

Postprocessing a Cylindrical Structural Shell Using PowerGraphics, Results Coordinate Systems, and Load Case Combinations

Overview

In this exercise, you will focus on postprocessing a steady-state structural model of a cylindrical shell. The cylinder is made of 6061-T6 aluminum with a Young's modulus of 73.1 GPa and a Poisson's ratio of 0.33. It has a radius of 0.25 m, a width of 0.1 m, and a thickness of 0.025 m (Figure 9-3-1). The cylinder is clamped at the base. In the first loadstep, a downward force of −50 N will be applied to each of the nine nodes at (0,0.25) for a total force of −450 N. In the second loadstep, a sideways force of −50 N will be applied to each of the nine nodes at (0.25,0) for a total force of −450 N. In the third loadstep, a pressure of 1 MPa will be applied to all inner surfaces of the cylinder. The final loadstep will combine the first three loading conditions. The goal of this analysis is to understand the deformations and stresses in the cylinder for the various applied loads.

The postprocessing in this exercise will explore the effect of PowerGraphics vs. Full Graphics on 3D shell elements. It will also explore the impact of using the global Cartesian vs. global cylindrical results coordinate system. We will also define load cases for the first three solutions and operate on these results to gain a deeper understanding of the behavior of this model. Finally, we will compare the results of the combined loading conditions as-calculated using the load case operations and as-solved in the fourth loadstep.

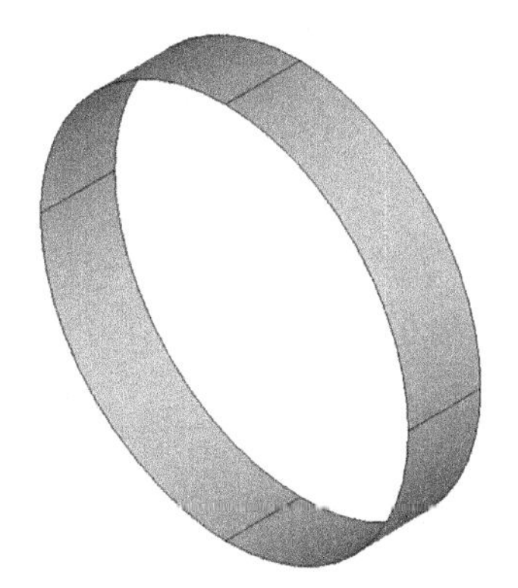

Figure 9-3-1 Schematic of the Cylinder.

ANSYS Mechanical APDL for Finite Element Analysis.
DOI: http://dx.doi.org/10.1016/B978-0-12-812981-4.00028-9

Model Attributes

Material Properties for 6061-T6 Aluminum

- Young's modulus—7.310e10 Pa
- Poisson's ratio—0.33

Loads

- Load case 1: Downward force (FY) of −50 N on nodes at (0,0.25) (−450 N total)
- Load case 2: Inward force (FX) of −50 N on nodes at (0.25,0) (−450 N total)
- Load case 3: Internal pressure of 1e6 Pa on all inner surfaces
- Load case 4: Combined loads from 1, 2, and 3

Constraints

- No displacement in x and y and no rotation about z for nodes at (0, − 0.25)
- No displacement in z for node at (0, − 0.25,0.05)

Postprocessing

- Load case 1
- Load case 1 with sign reversal
- Load case 2
- Load case 2 * 10
- Load case 3
- Load case 1 + Load case 2 + Load case 3
- Load case 4

File Management

Create a new folder in your "Intro-to-ANSYS" folder for Exercise9-3

Change the Working Directory and the Jobname

Start ANSYS

Step 1: Define Geometry

1-1. Change to the isometric view

1-2. Enter the Preprocessor

1-3. Create a keypoint at (0,0,0)

1-4. Create a keypoint at (0,0,0.1)

1-5. Create a line between Keypoints 1 and 2

1-6. Create a full circle with a radius of 0.25 centered at (0,0)

- Enter the following command into the command prompt:
```
CIRCLE,1,0.25
```

*It is far easer to issue the **CIRCLE** command directly than to use the GUI.*

1-7. Extrude the circle along line 1

- Main Menu > Preprocessor > Modeling > Operate > Extrude > Lines > Along Lines

Or

- Enter the following command into the command prompt:
```
ADRAG,2,3,4,5,,,1
```

The completed solid model geometry is shown in Figure 9-3-2 (left).

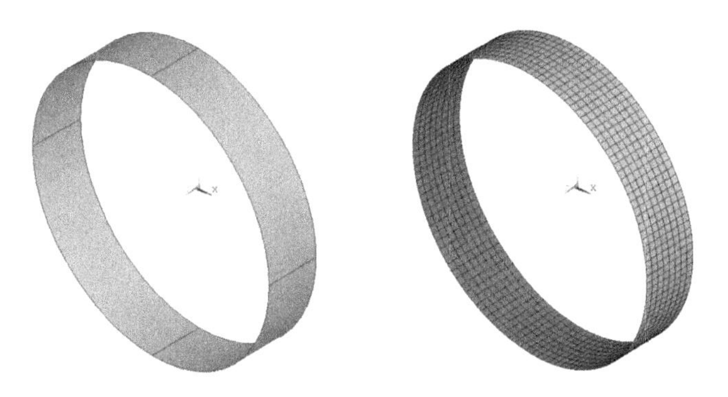

Figure 9-3-2 Completed Geometry (left) and Finite Element Mesh (right).

Step 2: Define Material Properties

2-1. Define the Young's modulus and the Poisson's ratio for this model

Usually we define the element types before the material properties. However, in this model, the material properties must be defined before the section properties can be defined.

Step 3: Define Element Types

3-1. Define SHELL181 as the element type #1 for this model

3-2. Define the key options for this element

- Preprocessor > Element Type > Add/Edit/Delete
- Click Options . . .
- Change Storage of layer data K8 to "All layers + Middle"
- Click OK

3-3. Define the section for this model

- Preprocessor > Sections > Shell > Lay-up > Add / Edit
- Enter "ShellThk" in the Name box (Figure 9-3-3)
- Enter "0.025" for the Shell thickness in the Thickness box
- Click OK

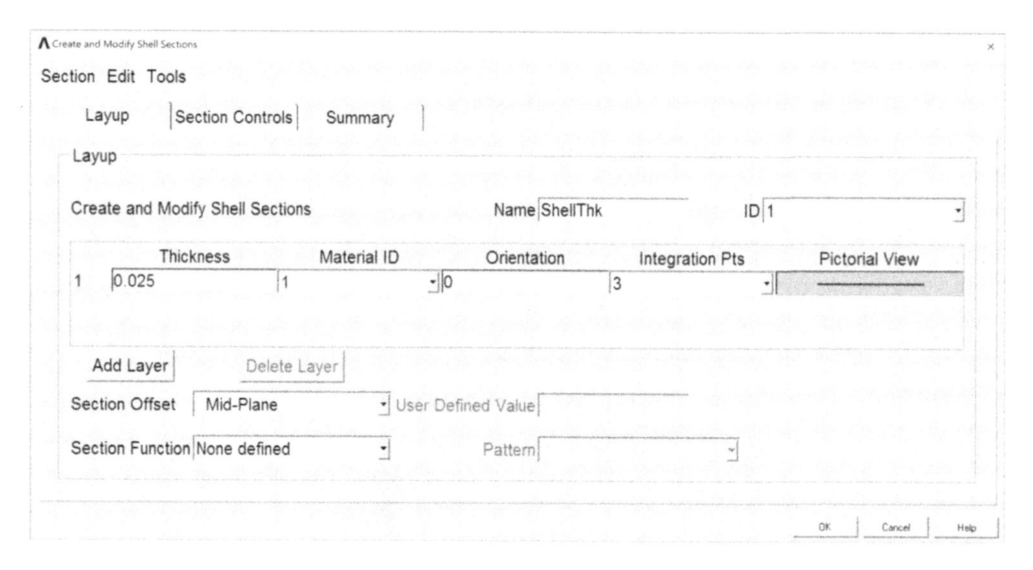

Figure 9-3-3 Create and Modify Shell Sections Dialog Box.

Step 4: Mesh

4-1. Mesh all areas in the model using an element edge length (ESIZE) of 0.0125

This should give you eight elements (nine nodes) across the width of the beam (Figure 9-3-2, right).

If you view the mesh from the isometric view, you should see that the outside of the cylinder has a turquoise color and the inside of the cylinder is purple. The turquoise color represents the top surfaces of the elements. The purple represents the bottom surfaces of the elements. If the colors are consistent for the entire model, then all elements will have the same orientation and we can apply boundary conditions and postprocess with confidence.

You will only be able to see the colors in the element plot if PowerGraphics is turned on.

Step 5: Apply Constraint Boundary Conditions

5-1. Select the nodes at the base of the cylinder

- Select the nodes at $x=0$
- Reselect the nodes at $y=-0.25$

5-2. Plot the nodes to confirm that the correct nodes have been selected

You should see a line containing nine nodes.

5-3. Set the UX, UY, and ROTZ DOFs for the selected nodes to zero

- Solution > Define Loads > Apply > Structural > Displacement > On Nodes

5-4. Reselect the node at $z=0.05$

5-5. Set the UZ DOF for the selected node to zero

Step 6: Apply Load Boundary Conditions for Loadstep 1 and Solve

6-1. Select the nodes at the top of the cylinder

- Select the nodes at $x = 0$
- Reselect the nodes at $y = 0.25$

6-2. Apply a download load (FY) of -50 N on the selected nodes

- Solution > Define Loads > Apply > Structural > Force/Moment > On Nodes

6-3. Set the Solution Options

The default solution options can be used for this analysis.

6-4. Select everything

6-5. Fit the image

6-6. Plot the nodes to confirm the applied boundary conditions

6-7. Solve

Step 7: Apply Load Boundary Conditions for Loadstep 2 and Solve

7-1. Delete all applied loads

- Enter the following command into the command prompt:
  `FDELE,ALL,ALL`

7-2. Select the nodes on the right side of the cylinder

- Select the nodes at $x = 0.25$
- Reselect the nodes at $y = 0$

7-3. Apply an inward load (FX) of -50 N on the selected nodes

7-4. Select everything

7-5. Solve

Step 8: Apply Load Boundary Conditions for Loadstep 3 and Solve

8-1. Delete all applied loads

8-2. Change pressure symbols to arrows

- Utility Menu > PlotCtrls > Symbols . . .
- In the second section, for [/PSF] Surface Load Symbols" choose "Pressures"
- In the second section, for "Show pres and convect as" choose "Arrows" from the drop down menu
- Click OK

8-3. Apply a pressure of 1e6 to all internal surfaces in the model

- Main Menu > Solution > Define Loads > Apply > Structural > Pressure > On Areas

Or

- Enter the following command into the command prompt:

```
SFA,ALL,1,PRES,1e6
```

8-4. Select everything

8-5. Solve

Step 9: Apply Load Boundary Conditions for Loadstep 4 and Solve

9-1. Add the loads from Steps 6 and 7, select everything, and solve

- Enter the following commands into the command prompt:

```
NSEL,S,LOC,X,0
NSEL,R,LOC,Y,0.25
F,ALL,FY,-50
NSEL,S,LOC,X,0.25
NSEL,R,LOC,Y,0
F,ALL,FX,-50
ALLSEL,ALL
SOLVE
```

Step 10: Postprocess the First Set of Results to Explore the Impact of PowerGraphics vs. Full Graphics

10-1. Read in the first set of results

- Main Menu > General Postproc > Read Results > By Pick

Or

- Enter the following commands into the command prompt:

```
/POST1
SET,1
```

10-2. Change to the front view

10-3. Plot the deformed shape

- Main Menu > General Postproc > Plot Results > Deformed Shape
- Choose "Def + undeformed"
- Click OK

This shows that the cylinder becomes shorter, wider, and less round as expected (Figure 9-3-4, left).

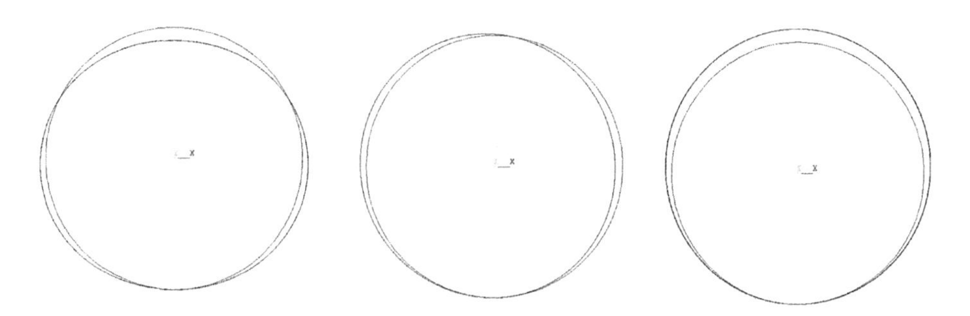

Figure 9-3-4 Deformed Shapes from Loadstep 1 (left), Loadstep 2 (center), and Loadstep 3 (right) (Front View) (Element Shape Display Off).

10-4. Return to the isometric view

10-5. Plot the nodal solution for the y-component of stress

The resulting stress plot is shown in Figure 9-3-5 (left). Note that the right side of the cylinder is red and has a positive stress, whereas the left side of the cylinder is blue and has a negative stress.

10-6. Rotate the view

This shows that the outside of the cylinder is red (i.e., experiences a positive stress) and the inside of the cylinder is blue (i.e., experiences a negative stress). This is consistent with expectations.

10-7. Return to the isometric view

10-8. Turn PowerGraphics off (turn Full Graphics on)

10-9. Plot the nodal solution for the y-component of stress

The resulting stress plot is shown in Figure 9-3-5 (right). You should now see that the plot is perfectly symmetric about all three axes. PowerGraphics plots the results for each surface on that surface. Thus, the outer surface of the cylinder shows the results for the top surface of the shell elements, and the inner surface of the cylinder shows the results for the bottom surface of the shell elements. This gives you a much more representative and realistic results plot by default. In contrast, Full Graphics shows you the results for the default surface (in this case, the top surface) on both surfaces of the cylinder.

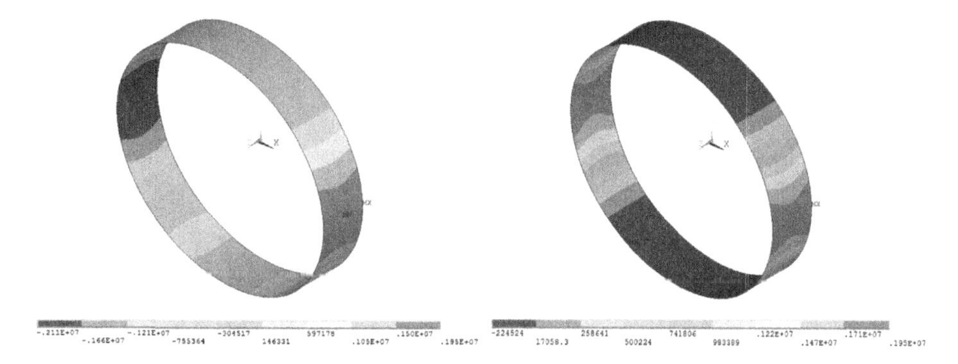

Figure 9-3-5 Nodal Stress in the Y Direction with PowerGraphics (left) and Full Graphics (right) (Win32 Graphics).

10-10. Turns on element shapes

- Utility Menu > PlotCtrls > Style > Size and Shape ...
- For "[/ESHAPE] Display of element," check the box to turn element shapes On
- Click OK

Note that there are no changes to the plot. Element shape display is not compatible with Full Graphics.

10-11. Turn PowerGraphics back on

We will use PowerGraphics for the remainder of this exercise.

10-12. Plot the nodal solution for the y-component of stress

10-13. Change to the front view

If you look carefully, you will see that there are some color variations at the right and left sides of the front face of the cylinder. This is because PowerGraphics is making its best effort to display the results for that "surface" of the model.

Step 11: Postprocess the First Set of Results to Explore the Impact of the Results Display Coordinate System

11-1. Plot the x-component of displacement

The resulting stress plot is shown in Figure 9-3-6 (left). Note that the right side of the model is red (indicating a positive displacement) and the left side of the model is blue (indicating a negative displacement). Rotating the model will not affect the color contour display. Changing from PowerGraphics to Full Graphics will also have no effect on this plot (beyond, of course, turning off the element shape display).

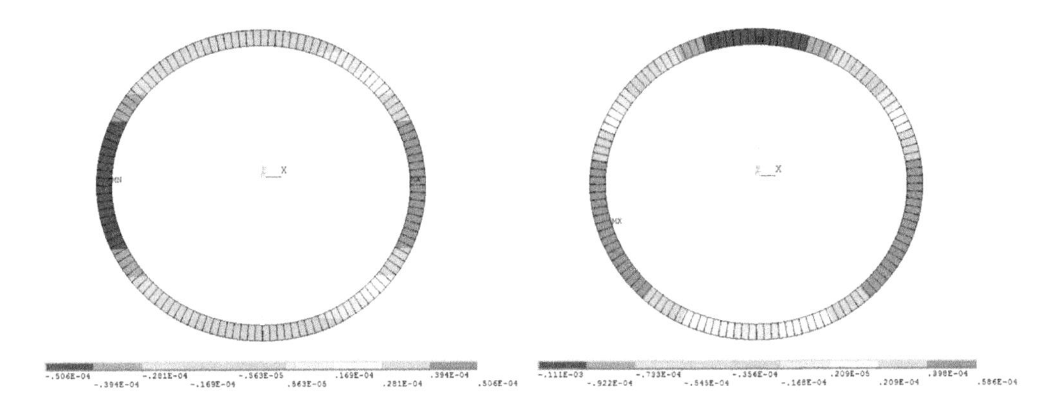

Figure 9-3-6 Nodal Deformation in the X Direction in a Global Cartesian Results Coordinate System (left) and in a Global Cylindrical Results Coordinate System (right) (Front View) (PowerGraphics) (Win32 Graphics).

11-2. Change the results display coordinate system to global cylindrical

- Main Menu > General Postproc > Options for Outp
- For "[RSYS] Results coord system," choose "Global cylindric" from the drop down menu

- Click OK

Note that in the third section of this dialog box, the option "[SHELL] Shell results are from" is set to "Top layer." Thus, Full Graphics will show the results for the top layer of the shell elements on all surfaces of the model. You can change this to "Middle layer" or "Bottom layer" if desired.

- Click OK

11-3. Plot the nodal x-component (now radial) displacement

The resulting stress plot is shown in Figure 9-3-6 (right). Note that the right and left sides of the plot are now both red. This is because the radial displacements on both sides of the cylinder are positive and equal in magnitude. As a result, you can now see more detail at the top and bottom of the plot.

Step 12: Find the Maximum Nodal Displacement Vector Sum and Equivalent Stresses for the First Three Solutions

The deformed shape, the displacement vector sum, the principle stresses, and the equivalent stress are all independent of the results coordinate system. Thus, these operations can be done either in RSYS 0 (global Cartesian) or in RSYS 1 (global cylindrical).

The model is currently set to RSYS 1. Because this is a more representative results coordinate system, we will continue to use RSYS 1 for the remainder of the exercise.

12-1. Plot the nodal displacement vector sum for the first solution (front view)

The resulting displacement plot is shown in Figure 9-3-7 (left). Note that the solution is symmetric and the maximum displacement in this model is 0.111e-3 m.

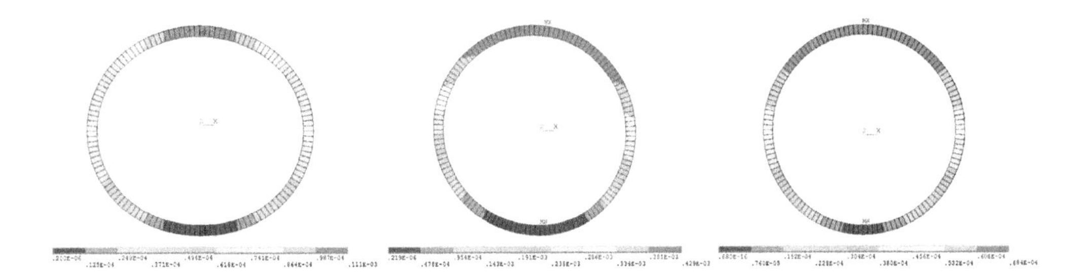

Figure 9-3-7 Nodal Displacement Vector Sum: First Solution (left), Second Solution (center), Third Solution (right) (front view) (RSYS 2) (PowerGraphics) (Win32 Graphics).

12-2. Plot the nodal equivalent (von Mises) stress for the first solution (Isometric View)

The resulting stress plot is shown in Figure 9-3-8 (left). Note that the solution is symmetric and the maximum equivalent stress for this model is 3.45 MPa. This is well below the yield stress (275 MPa).

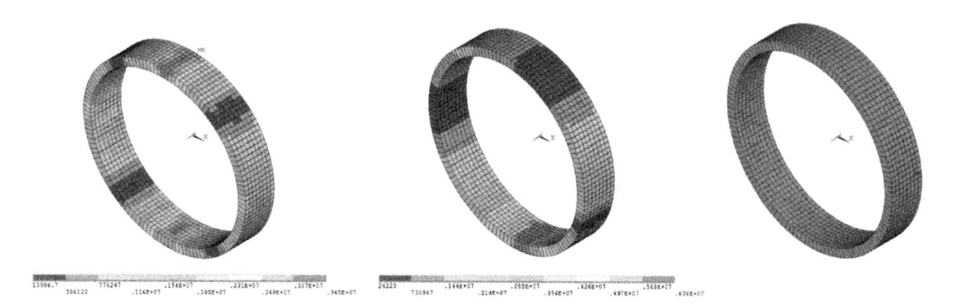

Figure 9-3-8 Equivalent Stress: First Solution (left), Second Solution (center), Third Solution (right) (Isometric View) (RSYS 2) (PowerGraphics) (Win32 Graphics).

12-3. Read in the second set of results

12-4. Plot the deformed shape for the second solution (front view)

The deformed shape for loadstep 2 is shown in Figure 9-3-4 (center). This shows that the entire cylinder has shifted to the left and a little bit up due to the applied load.

12-5. Plot the nodal displacement vector sum for the second solution (front view)

The resulting displacement plot is shown in Figure 9-3-7 (center). Note that the solution is asymmetric and the maximum displacement in this model is 0.429e-3 m.

12-6. Plot the nodal equivalent (von Mises) stress for the second solution (isometric view)

The resulting stress plot is shown in Figure 9-3-8 (center). Note that the solution is asymmetric and the maximum equivalent stress for this model is 6.38 MPa.

12-7. Read in the third set of results

12-8. Plot the deformed shape for the third solution (front view)

The deformed shape for loadstep 3 is shown in Figure 9-3-4 (right). This shows that the entire cylinder has expanded due to the internal pressure.

12-9. Plot the nodal displacement vector sum for the third solution (front view)

The resulting displacement plot is shown in Figure 9-3-7 (right). Note that the deformation is relatively uniform and the maximum displacement in this model is 0.684e-4 m.

12-10. Plot the nodal equivalent (von Mises) stress for the third solution (isometric view)

The resulting stress plot is shown in Figure 9-3-8 (right). Note that the equivalent stress for the entire model is 10 MPa.

Step 13: Use Load Case Operations to Change the Sign of the First Solution

13-1. Define a load case for the first solution

- Main Menu > General Postproc > Load Case > Create Load Case
- Ensure that the "Results file" radio button is selected
- Click OK
- For "LCNO Ref. no. for load case" enter 1
- For "LSTEP,SBSTEP Load step + substep nos." enter 1 in the first text box
- Leave the second text box empty
- Click OK

13-2. Define load cases for the second and third solutions

- Enter the following commands into the command prompt:

```
LCDEF,2,2
LCDEF,3,3
```

13-3. Zero the results database

- Main Menu > General Postproc > Load Case > Zero Load Case

13-4. Change the sign of the first load case

- Main Menu > General Postproc > Load Case > Calc Options > Scale Factor
- For "LCNO Ref. no. for load case" enter 1
- For "FACT Scale factor" enter −1
- Click OK

13-5. Read the first (now scaled) load case into the database

- Main Menu > General Postproc > Load Case > Read Load Case
- For "LCNA Ref. no. of load case" enter 1
- Click OK

13-6. Plot the deformed shape (Def + undeformed) (front view)

The cylinder should now be taller and thinner than before. This is because the effective applied load has changed direction and the cylinder is now being pulled up from the top rather than being pushed down.

13-7. Reset the sign for the first load case

- Main Menu > General Postproc > Load Case > Calc Options > Scale Factor
- For "LCNO Ref. no. for load case" enter 1
- For "FACT Scale factor" enter 1
- Click OK

Step 14: Use Load Case Operations to Change the Magnitude of the Second Solution

14-1. Zero the results database

- Main Menu > General Postproc > Load Case > Zero Load Case

14-2. Scale the results from the second load case by a factor of 10

- Main Menu > General Postproc > Load Case > Calc Options > Scale Factor

14-3. Read the second load case into the database

- Main Menu > General Postproc > Load Case > Read Load Case

14-4. Plot the nodal displacement vector sum (front view)

Note that the maximum displacement in this model is now 4.288e-3 m (10x greater than previously).

14-5. Plot the nodal equivalent (von Mises) stress (isometric view)

Note that the maximum equivalent stress in this model is now 63.8 MPa (10x greater than previously).

14-6. Reset the magnitude for the second load case

- Main Menu > General Postproc > Load Case > Calc Options > Scale Factor
- For "LCNO Ref. no. for load case" enter 2
- For "FACT Scale factor" enter 1
- Click OK

Step 15: Use Load Case Operations to Sum Solutions

15-1. Zero the results database

15-2. Read the first load case into the database

- Main Menu > General Postproc > Load Case > Read Load Case

15-3. Add the second load case to the database

- Main Menu > General Postproc > Load Case > Add
- For "LCASE1 1st Load case" enter 2
- Click Apply

15-4. Add the third load case to the database

- Main Menu > General Postproc > Load Case > Add
- For "LCASE1 1st Load case" enter 3
- Click OK

15-5. Plot the deformed shape (front view)

This plot has all of the expected characteristics of the combined load cases. There is a slight deformation to the left due to the applied load in load case 2. However, the internal pressure from load case 3 is mostly balancing the applied downward force from load case 1.

15-6. Plot the nodal displacement vector sum (front view)

Note that the maximum displacement for this combination of load cases is 0.430e-3 m. This is approximately the same as for the second solution and confirms that the vertical displacements from the first and third load cases are effectively canceling each other out.

15-7. Plot the nodal equivalent (von Mises) stress (isometric view)

Note that the equivalent stress for this combination of load cases model is 17.4 MPa. This is substantially higher than any of the solutions from the first three load cases and indicates that the stresses are combining rather than canceling.

Step 16: Compare the Results of the Load Case Summation with the Fourth Solution

16-1. Read in the fourth set of results

- Enter the following command into the command prompt:
  `SET,4`

The SET command automatically zeroes the results portion of the database before reading in the new results.

16-2. Plot the nodal displacement vector sum (front view)

The maximum displacement for the fourth load step is 0.430e-3 m. This is the same as for the load case combination.

16-3. Plot the nodal equivalent (von Mises) stress (isometric view)

The equivalent stress for the fourth load step is 17.4 MPa. This is also the same as for the load case combination. Therefore, we can use load case combinations and operations for this model with confidence.

Step 17: Compare and Verify the Results

For a cylindrical shell, the change in the vertical diameter (D_V) and the horizontal diameter (D_H) in response to a radial load are given by:

$$D_V = -0.1488\ WR^3/EI \qquad D_H = 0.1366WR^3/EI,$$

where W is the load (450 N), R is the radius of the cylinder (0.25 m), E is the modulus of elasticity (7.31e10 Pa), and I is the area moment of inertia of the ring cross section about the principal axis perpendicular to the plane of the ring. For a thin cylinder cross section loaded perpendicular to the longitudinal axis, the moment of inertial can be calculated as $I = bh^3/12$, where b is the width of the cylinder (0.1) and h is the thickness of the cylinder (0.025).

For load case 1, this results in a change in vertical diameter $D_V = -1.1e\text{-}4$ and a change in horizontal diameter D_H of 1.01e-4. For load case 1, the maximum displacement in y is −0.111e-3 and occurs at the top of the cylinder. The maximum displacement in *x* is 0.506e-4 at the right side of the model and the minimum displacement in *x* is −0.506e-4 at the left side of the model. Combining these results yields a change in horizontal diameter of 1.01e-4. Therefore, the model is in perfect agreement with the theory and can be used for engineering design and analysis.

Close the Program

- Utility Menu > File > Exit . . .
- Choose "Quit - No save!"
- Click OK

Sample Input File

```
/PREP7                    ! Enter the Preprocessor

K,1,0,0,0                 ! KP for center of the circle
K,2,0,0,0.1               ! KP for the drag path
L,1,2                     ! Create line 1 from KP 1 and 2
CIRCLE,1,0.25             ! Create a circle centered at KP1 w/rad 0.25
                          ! This creates KP 3-6, Lines 2-5
ADRAG,2,3,4,5,,,1         ! Extrude lines 2-5 along line 1
MP,EX,1,7.31e10           ! Define Young's modulus for material #1
MP,PRXY,1,0.3             ! Define Poisson's ratio for material #1
ET,1,181                  ! Define element #1 to be a SHELL181
KEYOPT,1,8,2              ! Output data for top, bottom, and middle
SECT,1,SHELL,,ShellThk    ! Create a shell section named ShellThk
SECDATA, 0.025,1,0.0,3    ! Define a shell thickness of 0.025
SECOFFSET,MID             ! Offset shell thickness from element midplane
ESIZE,0.0125              ! Use an element edge length of 0.0125
AMESH,ALL                 ! Mesh all areas
FINISH                    ! Finish and Exit the Preprocessor

/SOLU                     ! Enter the Solution processor
NSEL,S,LOC,X,0            ! Select nodes at x=0
NSEL,R,LOC,Y,-0.25        ! Reselect nodes at y=-0.25
NPLOT                     ! Plot nodes
D,ALL,UX,0                ! Constrain selected nodes in x
D,ALL,UY,0                ! Constrain selected nodes in y
D,ALL,ROTZ,0              ! Constrain selected nodes in rotation about z
NSEL,R,LOC,Z,0.05         ! Reselect nodes at z=;0.05
D,ALL,UZ,0                ! Constrain selected node in z

! Loadstep 1
NSEL,S,LOC,X,0            ! Select nodes at x=0
NSEL,R,LOC,Y,0.25         ! Reselect nodes at y=0.25
F,ALL,FY,-50              ! Apply a load of -50 in the y direction
ALLSEL                    ! Select everything
SOLVE                     ! Solve the model

! Loadstep 2
FDELE,ALL,ALL             ! Delete all applied forces
NSEL,S,LOC,X,0.25         ! Select nodes at x=0.25
NSEL,R,LOC,Y,0            ! Reselect nodes at y=0
F,ALL,FX,-50              ! Apply a load of -50 in the x direction
ALLSEL                    ! Select everything
SOLVE                     ! Solve the model

! Loadstep 3
FDELE,ALL,ALL             ! Delete all applied forces
/PSF,PRES,NORM,2,0,1      ! Change pressure symbols to arrows
SFA,ALL,1,PRES,1e6        ! Apply a pressure of 1e6 to all areas
ALLSEL                    ! Select everything
SOLVE                     ! Solve the model
```

```
! Loadstep 4
NSEL,S,LOC,X,0              ! Select nodes at x=0
NSEL,R,LOC,Y,0.25           ! Reselect nodes at y=0.25
F,ALL,FY,-50                ! Apply a load of -50 in the y direction
NSEL,S,LOC,X,0.25           ! Select nodes at x=;0.25
NSEL,R,LOC,Y,0              ! Reselect nodes at y=0
F,ALL,FX,-50                ! Apply a load of -50 in the x direction
ALLSEL                      ! Select everything
SOLVE                       ! Solve the model

! Postprocessing to explore PowerGraphics vs. Full Graphics
/POST1                      ! Enter the General Postprocessor
SET,1                       ! Read in the first set of results
/VIEW,1,,,1                 ! Change to front view
PLDISP,1                    ! Plot the deformed shape
/VIEW,1,1,1,1               ! Change to isometric view
PLNSOL,S,Y                  ! Plot the nodal y-component of stress
/GRAPHICS,FULL              ! Change to Full Graphics
PLNSOL,S,Y                  ! Plot the nodal y-component of stress
/ESHAPE,1                   ! Turn Element Shape Display on
/GRAPHICS,POWER             ! Turn PowerGraphics on
PLNSOL,S,Y                  ! Plot the nodal y-component of stress
/VIEW,1,,,1                 ! Change to front view

! Postprocessing to explore Results Display Coordinate Systems
PLNSOL,U,X                  ! Plot the nodal x-component of displacement
RSYS,1                      ! Use global cylindrical results coor. system
PLNSOL,U,X                  ! Plot the nodal x-component of displacement

! Postprocess the first three sets of results
PLNSOL,U,SUM                ! Plot the nodal displacement vector sum
PLNSOL, S,EQV               ! Plot the nodal equivalent stress
SET,2                       ! Read in the second set of results
PLNSOL,U,SUM                ! Plot the nodal displacement vector sum
PLNSOL, S,EQV               ! Plot the nodal equivalent stress
SET,3                       ! Read in the third set of results
PLNSOL,U,SUM                ! Plot the nodal displacement vector sum
PLNSOL, S,EQV               ! Plot the nodal equivalent stress

! Define the load cases
LCDEF,1,1                   ! Define load case 1
LCDEF,2,2                   ! Define load case 2
LCDEF,3,3                   ! Define load case 3

! Reverse the sign of the 1st results set
LCZERO                      ! Zero the results database
LCFACT,1,-1                 ! Scale the results for LC 1 by -1
LCASE,1                     ! Read the first load set into the database
/VIEW,1,,,1                 ! Change to front view
PLDISP,1                    ! Plot the deformed shape
LCFACT,1,1,                 ! Reset the results scaling factor
```

```
! Scale the 2nd results set by 10
LCZERO                      ! Zero the results database
LCFACT,2,10                 ! Scale the results from LC 2by 10
LCASE,2                     ! Read the second load case into the database
PLNSOL,U,SUM                ! Plot the nodal displacement vector sum
PLNSOL, S,EQV               ! Plot the nodal equivalent stress
LCFACT,2,1,                 ! Reset the results scaling factor

! Combine the first three results sets
LCZERO                      ! Zero the results database
LCASE,1                     ! Read the first load set into the database
LCOPER,ADD,2                ! Add the second load case to the database
LCOPER,ADD,3                ! Add the third load case to the database
/VIEW,1,,,1                 ! Change to front view
PLDISP,1                    ! Plot the deformed shape
/VIEW,1,1,1,1               ! Change to isometric view
PLNSOL,U,SUM                ! Plot the nodal displacement vector sum
PLNSOL, S,EQV               ! Plot the nodal equivalent stress

! Postprocess the 4th results set
SET,4
PLNSOL,U,SUM                ! Plot the nodal displacement vector sum
PLNSOL, S,EQV               ! Plot the nodal equivalent stress
FINISH                      ! Finish and Exit the Postprocessor

SAVE                        ! Save the database
!/EXIT                      ! Exit ANSYS
```

Chapter

10

Input Files

Suggested Reading Assignments:
Mechanical APDL Operations Guide: Chapter 7

CHAPTER OUTLINE

In this chapter, you will learn how to create, debug, and document input files. This requires a solid understanding of ANSYS commands and the ANSYS program structure. Therefore, we strongly recommend rereading chapter 2 before continuing with this chapter.

10.1. Approaches to Writing Input Files

As you have seen, an input file is simply a collection of commands written one per line in a plain text file. There are three basic approaches to creating an input file. In this book, they are referred to as the Direct Method, the Sequential Method, and the Concurrent Method.

10.1.1. The Direct Method for Preparing an Input File

The Direct Method is a computer programming approach. It involves opening a plain text editor and entering commands (or APDL code) until all instructions for building, solving, and postprocessing the model have been included. The commands can then be submitted to the program in either interactive or batch mode for processing. The Direct Method does not require an ANSYS license to prepare an input file, but it does require a license to run one.

In the early days of finite element modeling, all input files were batch files and all batch files were prepared this way. Today, the Direct Method is most useful for complex parametric models that are built using the ANSYS Parameter Design Language and that involve a lot of scripting. The authors use it most for research and development activities.

ANSYS Mechanical APDL for Finite Element Analysis.
DOI: http://dx.doi.org/10.1016/B978-0-12-812981-4.00010-1

The Direct Method requires you to have substantial experience with ANSYS and a strong understanding of how the program works. It also requires you to be comfortable reading and writing ANSYS commands and with using the documentation to fully understand command syntax and usage. For these reasons, the Direct Method is not recommended for new analysts who are creating new input or batch files. However, it is still a very powerful technique for new analysts who are modifying existing input or batch files. You will use the Direct Method to create a batch file from an existing input file in exercise 10-3.

10.1.2. The Sequential Method for Preparing an Input File

The Sequential Method is a GUI-based approach to creating input files. It involves building, solving, and postprocessing a model in the GUI, and then exporting and editing the log file to create the associated input or batch file.

In many ways, the Sequential Method is easier than the Direct Method. Because the Sequential Method always builds the model in interactive mode, you always have access to the help and feedback provided by the GUI. The Sequential Method does not require you to know the command names or syntax associated with the operations that you wish to perform: the GUI will automatically identify and issue the commands for you. In addition, it does not require you to choose which command to use when several commands can perform the same operation: the GUI menu paths will either guide you through the choices or the GUI will choose for you. Finally, the GUI performs many operations (such as entering and exiting the various processors) automatically. Therefore, you cannot 'forget' to perform these tasks even if you were unaware that you needed to. For these reasons, the Sequential Method is strongly recommended for new users.

However, the GUI does not always issue commands or perform tasks in the most intuitive way. Therefore, the Sequential Method requires you to understand how the GUI works and to use the documentation to look up various commands issued by the GUI. In addition, the log file captures all of the commands that were issued through the GUI—not just the useful ones. This includes commands that generated warnings and errors, commands for operations that were later reversed, and commands for solutions that were later overwritten. Therefore, the Sequential Method also requires you to remove invalid and unnecessary commands and to repair or streamline the commands that remain.

While the Sequential Method is substantially easier than the Direct Method, it also has more steps and therefore requires more work. The authors used the Sequential Method to create the input files for every exercise in this book. You will use the Sequential Method to create an input file in exercise 10-1.

10.1.3. The Concurrent Method for Preparing an Input File

The Concurrent Method is a hybrid approach. Simple commands and scripting information are entered directly into a plain text file. Less familiar and more complicated operations are performed in interactive mode, and then the associated commands are copied from the log file to the input file and edited if necessary. As you switch between these two approaches, you must manually update the model in interactive mode with the new information from the input file. The best way to do this is to clear the database, start a new analysis, and read in all known good commands from the input file (or copy and paste the known good commands into the command prompt). Once the GUI model has been synced with the input file, you can issue new commands in the GUI and continue to copy them into the log file. In this way, you create and debug the model and the input file at the same time (i.e. concurrently).

The Concurrent Method is a good option for both beginning and advanced users. The authors use it most for the design, analysis, and optimization of engineering components and systems (i.e. for engineering practice). You will use this approach to modify an existing input file in exercise 10-2.

10.2. Tools for Writing and Debugging Input Files

To write, run, and debug input files, you will need, at a minimum, an ANSYS license, a plain text editor, and the Mechanical APDL Command Dictionary. For complex analyses, we also recommend keeping a pen and some paper nearby for sketching. Finally, we suggest using other software program(s) to help visualize modeling options and to keep track of the state of the model as it is constructed.

10.2.1. Plain Text Editor

As noted in chapter 2, ANSYS has specific requirements about how commands and input files are formatted. Writing and storing commands in word processing software (e.g. Microsoft® Word) or saving input files in rich text formats (e.g., .rtf, .docx, .wpd, etc.) can add, remove, or embed information, formatting, and special characters. This information is not always visible and it can interfere with ANSYS's ability to recognize and execute your commands. Therefore, it is important that commands are only written and stored in plain text files (.txt).

There are many plain text editors that can be used to write and store input files. The authors use Microsoft Notepad because it is simple, robust, and included with the Windows® operating system. Some of our colleagues prefer the developer-friendly Notepad++. Other good options include Emacs, Vi, and Vim. You may choose any plain text editor that you are comfortable with. However, we suggest confirming that your program of choice is compatible with ANSYS (e.g. by testing it with a simple set of commands) before using it for a more important project.

10.2.2. Mechanical APDL Command Dictionary

The ANSYS Command Dictionary contains a full listing of all documented ANSYS commands. It is one of the most important and commonly used parts of the ANSYS documentation and is essential for writing and editing input files. It can be found at the end of the ANSYS Mechanical APDL Command Reference (Figure 10.1).

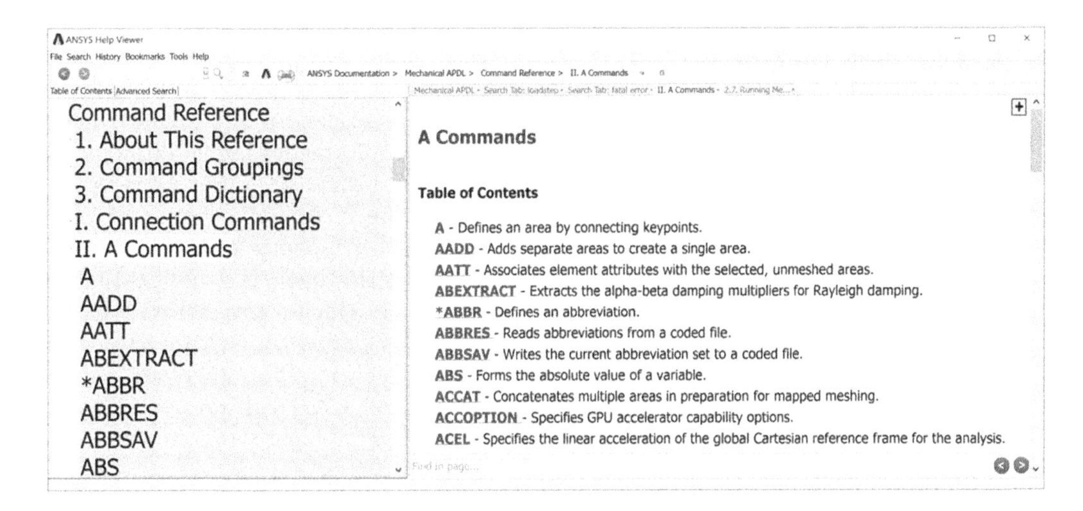

Figure 10.1 ANSYS Command Dictionary.

Commands in the ANSYS Command Dictionary are listed alphabetically starting with the **A** command (which defines an area by connecting keypoints) and ending with the **/ZOOM** command (which zooms a region in the graphics display window). In general, commands that begin with special characters (e.g. / or *) are alphabetized using the first letter after the special character. However, connection commands that start with a tilde (~) are grouped separately.

Each entry in the command dictionary includes the full command syntax, a description of the command's function, a list of compatible ANSYS products, detailed information about each of the command's arguments, notes that provide additional information about the command and its usage, and one or more menu paths that describe how to issue the command from the GUI. The entry for the **RECTNG** command is shown as an example in Figure 10.2.

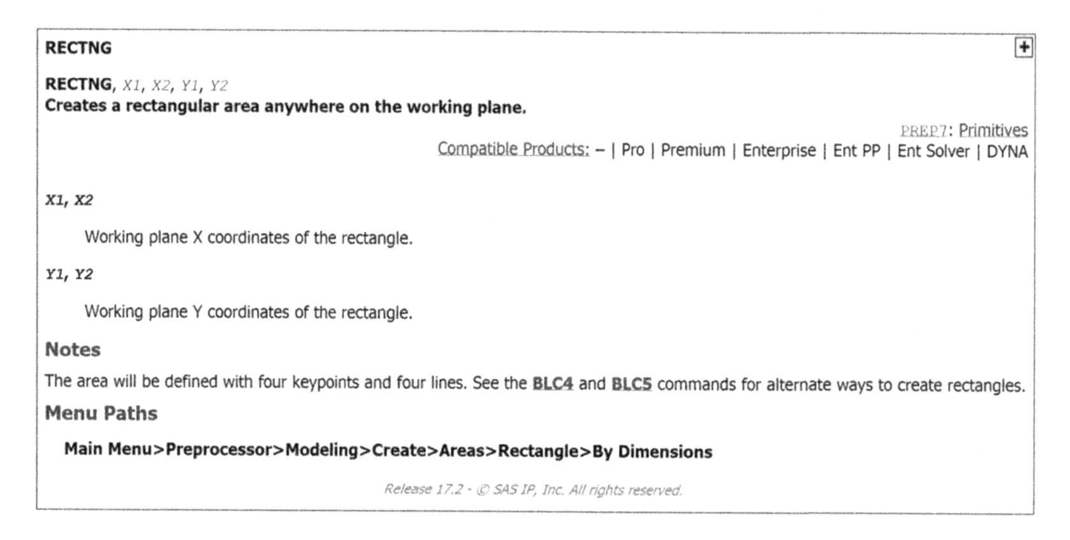

RECTNG

RECTNG, *X1*, *X2*, *Y1*, *Y2*
Creates a rectangular area anywhere on the working plane.

PREP7: Primitives
Compatible Products: – | Pro | Premium | Enterprise | Ent PP | Ent Solver | DYNA

X1, X2

Working plane X coordinates of the rectangle.

Y1, Y2

Working plane Y coordinates of the rectangle.

Notes

The area will be defined with four keypoints and four lines. See the **BLC4** and **BLC5** commands for alternate ways to create rectangles.

Menu Paths

Main Menu>Preprocessor>Modeling>Create>Areas>Rectangle>By Dimensions

Release 17.2 - © SAS IP, Inc. All rights reserved.

Figure 10.2 ANSYS Command Dictionary Entry for the RECTNG Command.

Some ANSYS commands have been undocumented and are no longer listed in the ANSYS Command Dictionary. In addition, some documented ANSYS commands have undocumented fields. As with all undocumented features, undocumented commands and fields are unsupported and should not be used.

10.2.3. Sketching Aids

As you have seen, ANSYS provides many tools for listing and plotting model information. However, sometimes it is helpful to visualize that information in a different way and to be able to view different types of information at the same time. For example, it is very convenient to create a line plot with keypoint numbering in ANSYS (Figure 10.3, left). It is also convenient to plot the nodes associated with those keypoints and to list the Cartesian coordinates associated with those keypoints and nodes. However, it is not possible to create a line plot with node numbers (Figure 10.3, center) or to label a line plot with the Cartesian coordinates associated with those keypoints and nodes (Figure 10.3, right). Those 'plots' must be created manually on paper or in another program. In general, the authors use a combination of blank paper, graph paper, presentation software, and CAD software to create sketches to support design and analysis in ANSYS. The images in Figure 10.3 were created using Microsoft PowerPoint.

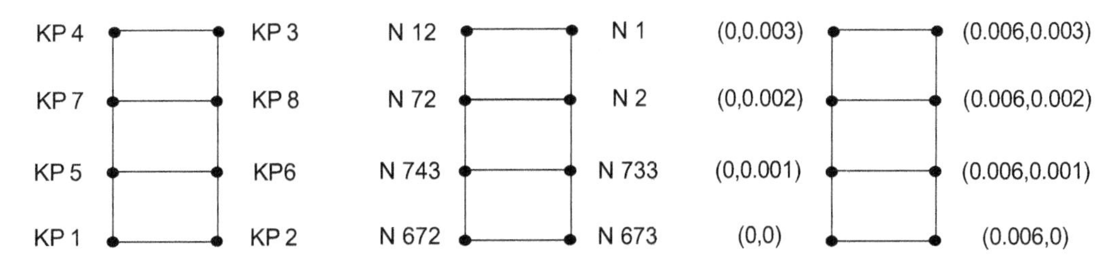

Figure 10.3 'Sketches' Showing Keypoint Numbering (left), Node Numbering (center), and Cartesian Coordinates (right) for Select Geometry from Exercise 10-2.

10.3. Accessing the Log File

Creating an input file often requires opening one or more log files. There are three ways to open a log file and access the commands inside: you can open the session log file from your working directory using a plain text editor, you can list the session log file from the Utility Menu, and you can write the database log file from the Utility Menu and then open it with a plain text editor from the directory in which it was saved.

10.3.1. Open the Session Log File From the Working Directory

To open the session log file from your working directory, open the folder that you are using as your working directory, look for a plain text file with the same name as the jobname, and open the file with a plain text editor. Depending on your operating system's file extension settings, the log file may be named jobname.log, jobname.txt, or jobname with no extension.

The session log file can be opened, copied, or saved under a different name at any time, but it cannot be modified or deleted while the program is using it (i.e. while ANSYS is running a job in that working directory and with that jobname). Accessing the session log file from your working directory is especially recommended for the Sequential Method.

10.3.2. List the Session Log File From Within ANSYS

When running ANSYS in interactive mode, you can list the session log file via the GUI path: **Utility Menu > File > List > Log File...**. The oldest commands are located at the top of the file. The most recent commands are at the bottom. Listing the log file is the most convenient option when using the Concurrent Method.

10.3.3. Export and Open the Database Log File

Finally, you can export the commands issued during the current session by writing the database command log file. This is done by following the GUI path: **Utility Menu > File > Write DB log file** ... or by issuing the **LGWRITE** command. This creates a plain text file with a .lgw extension with the name that you specify in the directory of your choice. The main advantage of writing the database log file is the fact that it allows you to omit nonessential commands such as listing, plotting, and graphics commands ("Write essential commands only") or to comment out the nonessential commands using exclamation points ("Write non-essential cmds as comments"). While new users may find these options useful, removing 'nonessential' commands reduces the autonomy of the analyst—who is ultimately responsible for deciding what is essential and what is not. In addition, removing commands that have been commented out increases the amount of work required to edit and format the final input file. Therefore, the authors rarely export the database log file and do not recommend that you do so.

10.4. Common Features of GUI-Generated Log Files

While it can be very helpful to have the GUI identify and issue commands for you, it often issues them in a nonintuitive way. This section describes some of the common features of log files and how the GUI behavior affects the process of editing a log file to create an input file.

10.4.1. The Header Block

All log files begin with a header block of commands. Log files are written with the expectation that they will be used as batch input files. Therefore, the header block begins with a **/BATCH** command. Next, the header block contains a release and time stamp. Finally, a series of graphics commands are issued to set up the program for interactive mode. A sample header is shown below. The comments have been added by the authors.

Example 10.1

```
/BATCH                          ! Enable batch processing
/COM,ANSYS RELEASE Release 17.2   BUILD 17.2    UP20160718     14:59:50
/input,menust,tmp,''            ! Load user-specific information if defined
/GRA,POWER                      ! Turn Power Graphics on
/GST,ON                         ! Turn graphical tracking on
/PLO,INFO,3                     ! Use multi-legend mode
/GRO,CURL,ON                    ! Place graph labels near the curve
/CPLANE,1                       ! Use the working plane as the cutting plane
/REPLOT,RESIZE                  ! Resize on replot
WPSTYLE,,,,,,,,0                ! Include the triad with working plane grid
```

Most of the header commands are optional (or "nonessential") because they represent the program defaults. However, the **/BATCH** command must be the first command issued in a batch run. Since this command is needed in batch mode and is ignored in interactive mode, it is good practice to leave the **/BATCH** command in an edited log file and to include one at the start of each new input file.

10.4.2. Release and Time Stamps

Each time you clear the database and start a new analysis (i.e. issue the **/CLEAR** command), a new time stamp is added to the log file. A sample time stamp is shown in Example 10.2.

Example 10.2

```
/COM,ANSYS RELEASE Release 17.2  BUILD 17.2  UP20160718  15:05:23
```

Release and time stamps can be useful for recovering your work and for tracking changes to a model. However, they are not needed for input files and can always be removed when editing a log file to create an input file.

10.4.3. LGWRITE Commands

If you write the database log file, the last command in both the session and database log files will be the **LGWRITE** command. A sample **LGWRITE** command is shown in Example 10.3.

Example 10.3

```
! LGWRITE,'logfile','lgw','C:\DESKTOP\INTRO-~1\CHAPTE~1\',COMMENT
```

The **LGWRITE** command is always commented out (i.e. preceded by an exclamation point) in both log files. This command is rarely necessary and should be deleted when editing a log file to create an input file.

10.4.4. Material Property Blocks

As noted in chapter 5, the GUI writes all linear material properties to the log file using the **MPTEMP** and **MPDATA** commands regardless of temperature dependence. Thus, it is common for two extraneous **MPTEMP** commands to appear at the beginning of a linear material model definition.

Example 10.4

```
MPTEMP,,,,,,,,          ! Zero the temperature table
MPTEMP,1,0              ! Create a new temperature table (T1 = 0)
MPDATA,EX,1,,7.31e10    ! Define Young's modulus for Material 1 at T1
MPDATA,PRXY,1,,0.33     ! Define Poisson's ratio for Material 1 at T1
```

Example 10.4 shows the commands issued by the GUI to define the Young's modulus and the Poisson's ratio of aluminum. The first **MPTEMP** command zeros out the temperature table for temperature-dependent material properties. The second **MPTEMP** reestablishes that temperature table and sets the first temperature equal to zero. For analyses that are temperature independent, these two commands have no effect on the analysis. You can replace this group of four commands with their **MP** counterparts as shown in Example 10.5.

Example 10.5

```
MP,EX,1,7.31e10         ! Define the Young's modulus for Material 1
MP,PRXY,1,0.33          ! Define the Poisson's ratio for Material 1
```

It is not necessary to streamline the material property command blocks. However, it is good practice to do so because it makes the model easier to understand at a glance and avoids confusion about the assumptions in the model.

10.4.5. Picker Blocks

If graphical picking was used in an interactive run, a Picker block will be present in the log file. As noted in chapter 7, the Picker identifies and operates on the picked items and then returns all entities to the selection status that they had before the picking operation began. A Picker block always begins with an **FLST** command. The **FLST** command is followed by one or more **FITEM** commands. A simple Picker block that does not use components ends with the command that operates on the picked items. A complex Picker block that uses components ends with a **CMDELE** command. Because Picker blocks rely on entity numbers for selection, they are not a robust selection method. When editing a log file to create an input file, you should always replace Picker blocks with selection commands that are independent of entity number.

For more detailed information on Picker commands and replacing Picker blocks with selection commands, see chapter 7 (section 7.6) of this book.

10.4.6. Nonessential Commands

It is common for log files from interactive sessions to contain numerous nonessential commands, such as plotting, listing, and graphics commands. For example, rotating the model in the GUI appends commands blocks such as the one shown in Example 10.6 to the log file.

Example 10.6

```
/ANG,1
/REP,FAST
/USER, 1
/VIEW, 1, 0.600652002689, 0.588870979613, 0.540784745565
/ANG, 1, -1.46102410334
/REPLO
/VIEW, 1, 0.800279316822, 0.581473850511, 0.146428058236
/ANG, 1, 18.1343262281
/REPLO
```

In general, you can remove:

- Entity plotting commands (**KPLOT**, **LPLOT**, **APLOT**, **VPLOT**, **NPLOT**, and **EPLOT**)
- Entity listing commands (**KLIST**, **LLIST**, **ALIST**, **VLIST**, **NLIST**, and **ELIST**)
- Other listing commands (**CMLIST**, **ETLIST**, **MPLIST**, **DLIST**, **FLIST**, etc.)
- Graphics Views commands (**/ANGLE**, **/AUTO**, **/REPLOT**, **/VIEW**, **/USER**, **/ZOOM**, etc.)
- Graphics Style commands (**/EDGE**, **/ESHAPE**, etc.)

unless these commands are used to tailor the display of the model or its results for images that will be used in reports, presentations, and other applications.

Input files are excellent ways to save detailed instructions for creating images from future analyses. However, it can be difficult to distinguish listing, plotting, and graphics commands that were used to build the model from similar commands that were used to tailor its output. Therefore, the Concurrent Method is strongly recommended when preparing input files with detailed or custom graphics output.

10.4.7. Extra Commands

When at least one option is changed in a dialog box, the GUI usually issues commands to change all of the options in the dialog box to reflect their updated states. For example, turning on keypoint numbering in the Symbols dialog box appends the commands shown in Example 10.7 to the log file:

Example 10.7

```
/PNUM,KP,1              ! Turn keypoint numbering on
/PNUM,LINE,0            ! Leave line numbering off
/PNUM,AREA,0            ! Leave area numbering off
/PNUM,VOLU,0            ! Leave volume numbering off
/PNUM,NODE,0            ! Leave node numbering off
/PNUM,TABN,0            ! Leave tabular boundary condition numbering off
/PNUM,SVAL,0            ! Leave stress (contour) values off
/NUMBER,0               ! Keep using colors and numbers
!*
/PNUM,ELEM,0            ! Leave element numbering off
/REPLOT                 ! Replot the image
!*
```

In this case, only the first command (`/PNUM,KP,1`) is wanted or needed. The others do not change the state of the model and therefore did not need to be issued.

If the Symbols dialog box is reopened and area numbering is turned on, the commands in Example 10.8 will be added to the log file. As before, the entire block of Symbols commands is supplied. In this case, only the third command (`/PNUM,AREA,1`) is wanted or needed. The others do not change the state of the model and therefore did not need to be reissued.

Example 10.8

```
/PNUM,KP,1              ! Turn keypoint numbering on (again)
/PNUM,LINE,0            ! Leave line numbering off
/PNUM,AREA,1            ! Turn area numbering on
/PNUM,VOLU,0            ! Leave volume numbering off
/PNUM,NODE,0            ! Leave node numbering off
/PNUM,TABN,0            ! Leave tabular boundary condition numbering off
/PNUM,SVAL,0            ! Leave stress (contour) numbering off
/NUMBER,0               ! Keep using colors and numbers
!*
/PNUM,ELEM,0            ! Leave element numbering off
/REPLOT                 ! Replot the image
!*
```

Extra commands should always be deleted when editing an input file.

10.4.8. Repeated Commands

Some actions that appear to be a single operation in the GUI are actually executed by issuing a single command multiple times. For example, when you move the Smart Size slider in the Mesh Tool from a Smart Size of 6 to a Smart Size of 4, three separate smart size commands are issued:

```
SMRT,6
SMRT,5
SMRT,4
```

Each repeated command supersedes the previous command. Therefore, only the final command is needed. The other repeated commands can and should be deleted when editing a log file to create an input file.

10.4.9. Reversed Commands

When experimenting with the best way to create a model with the GUI, it is common to create and then delete geometry, material models, boundary conditions, etc. It is also common to create and clear meshes repeatedly until an acceptable mesh is identified. If an operation is done and then undone without impacting the model, both the original command and the one that reversed it can and should be deleted from a draft input file.

It is important to note that there are many instances where pairs of commands that do something and then 'undo' it have a significant impact on the model. For example, concatenating solid model entities, meshing the model, and then deleting the concatenations has a major impact on the final mesh. Similarly, applying loads, solving the model, and then deleting those loads has a major impact on the model results. None of these commands can or should be deleted. Reversed commands should only be removed when their presence results in a zero-sum game.

10.4.10. Commands That Generate Warnings or Errors

Commands that generate warnings and errors must be removed from a batch file before it is run. If they are not removed and the file is run in batch mode, these commands will terminate the batch run and render the batch file useless. If these commands are not removed and the file is run in interactive mode, warnings will create pop-up dialog boxes that will have to be dismissed each time the input file is run. Similarly, any errors contained in the file will have to be corrected each time the input file is run. In all cases, commands that generate warnings and errors should be removed from input and batch files to prevent future problems.

10.4.11. Extra Spaces

Some commands that are issued by the GUI have extra spaces in the syntax. For example, a command in the log file may be listed as:

```
LSTR,     5,     6
```

instead of:

```
LSTR,5,6
```

As noted in chapter 2, extra spaces do not impact the interpretation or execution of ANSYS commands and therefore do not have to be removed. However, removing all extra spaces (or removing all but one extra space) can make your input files more compact and easier to read.

10.4.12. Extra Syntax

Finally, the GUI often issues the full syntax for commands even if some (or all) of the arguments are not needed. This usually takes the form of extra commas. For example, if you create a keypoint at (0,0,0) by leaving all entries in the Create Keypoints in Active Coordinate System dialog box blank, the GUI issues the command as:

```
K,,,,
```

The first blank argument assigns the keypoint number to the lowest available entity number by default. The following three arguments set the keypoint coordinates to zero by default. Therefore, all trailing commas can be discarded if desired and the command can be rewritten as:

```
K
```

Similarly, the GUI issues the **NUMMRG** command as:

```
NUMMRG,ALL, , , ,LOW
```

The “LOW” in this case specifies that the lowest number for the merged items will be kept and higher numbers will be discarded. Since this is the default, all trailing commas may be discarded and the command can be rewritten as:

```
NUMMRG,ALL
```

It is not necessary to remove any or all trailing commas when editing a log file to create an input file. In addition, you should never remove trailing commas without looking up the command in the ANSYS Command Dictionary to ensure that removing them will have no effect on the model. However, removing the extra syntax makes the final input file simpler and easier to read. Therefore, we recommend that more experienced users remove them for consistency and clarity.

10.5. Guidelines for the Sequential Method

The previous section outlines the major changes that need to be made to convert a log file into a clean input file. However, the extent of those changes depends on the nature of the log file. If the creation of the model was relatively straightforward, then there will be relatively few nonessential, extra, repeated, and reversed commands. In this case, the log file itself will be relatively clean and editing it will be the most efficient option. If the modeling process involved a large number of listing, plotting, and graphics commands, then it is sometimes easier to extract the useful commands from the log file by copying and pasting them into a new empty file rather than cleaning the old file. Finally, if the model went through multiple iterations and was created over the course of days or weeks, it is sometimes easiest to create a new log file, re-create the model using as few steps as possible, and then edit the new log file, rather than trying to decipher the old log file.

10.5.1. Procedure for Editing a Log File

There are many ways to edit a log file. You will eventually develop your own strategy and procedures. Until you do, we suggest the following procedure:

1. Archive (save) a copy of the log file in a dedicated backup folder
2. Make a working copy of the log file. (Do not modify the original or the archived copy!)
3. Rename the working copy to indicate that it will be the input file
4. Remove the header block, leaving the /**BATCH** command
5. Copy or add a /**PREP7** command
6. Edit the preprocessing commands:
   - Delete all release and time stamps
   - Delete all **LGWRITE** commands
   - Delete all **SAVE** commands
   - Delete all other nonessential commands (e.g. listing, plotting, graphics, etc.)
   - Delete all extra commands
   - Delete all repeated commands
   - Delete all reversed commands
   - Delete any commands that are known to generate error messages
   - Streamline the material property commands
   - Replace the picker blocks with selection commands based on location or attribute
   - Group standard commands based on their function and overall role in the modeling process. For example, group all solid modeling commands together, group all meshing commands together, etc.

This makes the input file easier to read and reduces the chance of duplicating commands.

- Remove extra spaces
- Remove extra syntax
- Add additional commands if necessary
- Comment the preprocessing commands
- Save the input file

7. Copy or add a **/SOL** command
8. Edit the solution commands using the same procedure as the preprocessing commands
9. Copy or add a **/POST1** or **/POST26** command
10. Edit the postprocessing commands using the same procedure as the solution commands
11. Test and debug the new input file (see section 10.6 for more details).

10.5.2. Procedure for Extracting Commands From a Log File

There are also many ways to extract commands from a log file to create a new input file. We suggest the following procedure for new analysts who do not have, or are still developing, a preferred method:

1. Archive (save) a copy of the log file in a dedicated backup folder
2. Make a working copy of the log file. (Do not modify the original or the archived copy!)
3. Create a blank input file to hold the extracted commands
4. Copy or add a **/BATCH** command into the input file
5. Copy or add a **/PREP7** command
6. Add a few blank lines. This is where you will put your preprocessing commands.
7. Copy or add a **/SOL** command
8. Add a few blank lines. This is where you will put your solution commands.
9. Copy or add a **/POST1** or **/POST26** command

   This creates a template for the input file and ensures that you will not forget to add the commands to enter and exit the various processors.

10. Scroll through the preprocessing commands to locate where the "good" commands begin. If there are multiple **/CLEAR** commands, the "good" preprocessing commands will be located below the last **/CLEAR**. If there are multiple **SAVE, /CLEAR**, and **RESUME** commands, the "good" commands may be distributed in several locations.

    If you cannot identify where the "good" commands begin, then you may have to re-create the model and generate a new cleaner log file.

11. Identify the "good" or "useful" commands using the guidelines in the previous section
12. Copy all "good" or "useful" preprocessing commands into the new input file
13. Edit the copied preprocessing commands in the input file as described above
14. Scroll through the solution commands to locate the "good" commands
15. Identify the "good" or "useful" solution commands
16. Copy all "good" or "useful" solution commands into the new input file
17. Edit the copied solution commands in the input file as described above
18. Scroll through the postprocessing commands to locate the "good" commands
19. Identify the "good" or "useful" postprocessing commands

20. Copy all "good" or "useful" postprocessing commands into the new input file
21. Edit the copied postprocessing commands in the input file as described above
22. Test and debug the new input file

10.6. Debugging an Input File

It is common for even the most carefully written or edited input files to contain errors. Therefore, it is almost always necessary to debug your input files.

10.6.1. Types of Input File Errors

There are three types of input file errors that are commonly observed during debugging: commands that do not work, individual commands that do not work as expected, and groups of commands that do not work together as expected.

The first type of error is due to a failure to correctly issue the command. The command name or one of the arguments may be misspelled, a required argument may be missing, the arguments may have been supplied in the wrong format or in the wrong field, or the command may have been issued in the wrong processor. Usually, these problems can be identified by executing the commands one-by-one until a warning or error message is generated. Once you have identified the problematic command, you can look it up in the ANSYS Command Dictionary and go field-by-field through the syntax until the error has been identified and corrected.

The second type of error is due to choosing the wrong command, misunderstanding how a command functions, or failing to supply appropriate argument values for the command. For example, this type of error may occur if you use the **BLC4** command (create a rectangular area or block by corners) instead of the **BLC5** command (create a rectangular area or block by center and corners) or the **BLOCK** command (create a rectangular area or block by two corners). It may also occur if the x and y coordinates (width and height) for the blocks are switched when they are supplied as arguments to the command. Both of these errors would result in rectangular areas or blocks that are incorrectly orientated. These errors can usually be identified immediately based on the graphical feedback from the program. If they are not found quickly, they can lead to the third type of error.

The third type of error in an input file can stem either from an undetected error in an earlier part of the model or from an invalid assumption or incorrect understanding of the state of the model that prevents a later command from functioning normally. For example, incorrectly generating the solid model geometry can cause subsequent Boolean operations to fail. Similarly, if a selection command fails to select any entities in the model, later commands that operate on the selected set will also fail. These errors are much more difficult to detect than the first two types because the source of the problem is located before the command that triggers the warning or error messages. These types of errors generally must be identified by going line-by-line through the input file and verifying the state of the model after each operation with list and plot commands.

10.6.2. End-of-File Commands

The **/EOF** command triggers an end-of-file exit when it is encountered in an input file. This means that ANSYS will stop reading commands in the input file where the **/EOF** is encountered. If the current input file was called from another input file, ANSYS will then begin to read and execute commands from the parent file. If there is no parent file and the file was run in interactive mode, control of the program will be returned to the user. If there is no parent file and the file was run in batch mode, the run will terminate.

The /**EOF** command is very useful for debugging input files. End-of-file commands were historically used in pairs because ANSYS would sometimes miss the first /**EOF** command and continued to read the file. This is not an issue today, but many analysts (including the authors) still use a pair of /**EOF**s to ensure an end-of-file exit.

10.6.3. The Debugging Process

Before beginning to debug an input file, you should double check that your material models, material property values, and choice of elements are correct, and that your system of units is consistent.

1. Make a working copy of the input file (Do not modify the original!)
2. Place a pair of /**EOF** commands (one per line) after the geometry commands
3. Save and run the input file
4. Use listing, plotting, and entity numbering controls to verify that the geometry is correct
5. Delete the /**EOF** commands
6. Insert new /**EOF** commands below the next command block (element definitions, material properties, etc.)
7. Repeat steps 3 through 6 until the input file stops running smoothly
8. Go line-by-line from the last known working part of the code and compare the syntax present with the syntax outlined in the ANSYS Command Dictionary until the problem has been located
9. Correct the problem
10. Remove all /**EOF** commands
11. Save and rerun the file to confirm that it is working

Sometimes it can be very difficult to debug an input file alone. Don't hesitate to ask a friend, classmate, or coworker to look at your input file for you. They may be able to see something that you have missed.

10.7. Documenting Your Work

The final step when creating an input file is to document your work. This includes documenting your modeling assumptions and decisions, and commenting your input files.

10.7.1. Documenting Modeling Assumptions and Decisions

Creating a finite element model requires you to make countless assumptions and decisions. It also requires you to gather and use information from a variety of sources. It is important to document that information so you can remember what you did months, or even years, after you created an input file. It is also important to document that information for the benefit of your colleagues and coworkers who may have to use or modify the model at a later time. Finally, it is important to document that information to support external validation and certification activities in academia and industry. The requirements for formal model documentation will be specific to your company, industry, or field. However, you can and should develop your own strategy or methods for informally documenting your work.

The authors usually document their models by creating informal presentations with different subsections for different aspects of the model (geometry, materials, loads, etc.). For complex models, it is not uncommon for these files to reach 100 slides or more. However, this process creates almost no extra work because most of the slides used for the informal documentation are the same slides that were created as part of the sketching activities described in section 10.2.3.

You can achieve a similar effect by sketching in or attaching printouts to a designer's notebook, or by creating informal reports in word processing software. The goal is to create a well-structured process in which modeling options are identified, compared, chosen, implemented, and documented in an organized, conscious, and rational manner. How you do that is entirely up to you.

10.7.2. Commenting an Input File

Commenting an input file is the same as commenting any other type of computer code. After each command, add an exclamation point (!) and write a short description of the command's function. If there are groups of commands whose function is not obvious, add a few empty lines above those commands and include one or more comments to describe the functions of the collective.

It is very important that comments do not wrap on to the next line of an input file. If they do, each new line must also begin with an exclamation point. Otherwise, ANSYS will try to interpret the comment as a command and an error will be generated.

Input file comments do not have to be extensive or grammatically correct. However, they must be clear enough for a colleague to understand what you did and how it affects the model. If possible, the comments should also provide some information about why you made the choices that you did.

Exercise 10-1

Using the Sequential Method to Create an Input File for 1D Steady-State Conduction Through a Steel Clad Copper Pan

Overview

In this exercise, you will use the Sequential Method to prepare an input file. First, you will perform a static thermal analysis to determine the temperature distribution through a section of steel clad copper cookware. Next, you will edit the log file to create an input file for the model. Finally, you will run the input file to verify that it was edited correctly.

The core of the plate is made of 2 mm thick copper with a thermal conductivity of 372 W/mK. Both sides of the copper are clad with 1 mm thick stainless steel, which has a thermal conductivity of 17 W/mK. We will model 6 mm of the plate width. The bottom surface is subject to a constant heat flux of 1200 kW/m^2 from the stove. The top surface of the plate is being cooled by boiling water and therefore has a constant temperature of 100°C (Figure 10-1-1). The goal is to find the temperature distribution through the plate, the maximum temperature of the plate, and the temperature of the copper at the top and bottom interfaces.

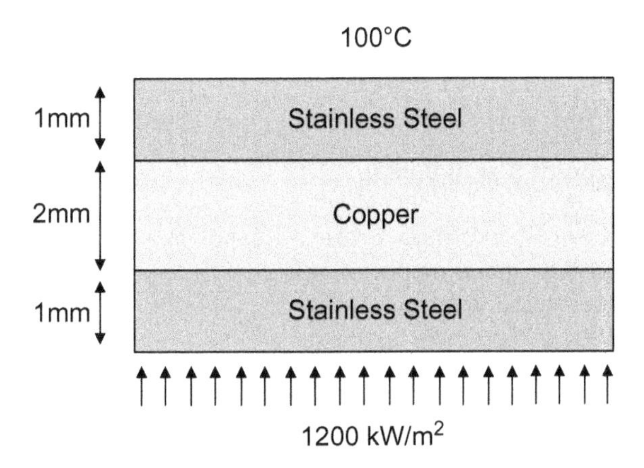

Figure 10-1-1 Schematic of the Cladded Plate.

ANSYS Mechanical APDL for Finite Element Analysis.
DOI: http://dx.doi.org/10.1016/B978-0-12-812981-4.00029-0

Model Attributes

Material Properties

- Stainless Steel—Thermal conductivity—17 W/mK
- Copper—Thermal conductivity—372 W/mK

Loads

- Heat flux of 1200 kW/m^2 on the bottom of the plate

Constraints

- Surface temperature T = 100°C on the top of the plate

File Management

Create a new folder in your "Intro-to-ANSYS" folder for Exercise 10-1

Change the Working Directory and the Jobname

Start ANSYS

In this exercise, more detailed instructions have been provided to ensure that your log file is as close as possible to the example log file shown in this exercise.

Step 1: Define Geometry

1-1. Create a rectangle 0.006 m wide and 0.004 m high with the lower left corner at (0,0)

- Preprocessor > Modeling > Create > Areas > Rectangle > By 2 Corners
- Enter 0 for WP X
- Enter 0 for WP Y
- Enter 0.006 for the Width
- Enter 0.004 for the Height
- Click OK

1-2. Create Keypoint 5 at (0,0.001)

- Preprocessor > Modeling > Create > Keypoints > In Active CS
- Leave NPT Keypoint number blank
- Enter 0 in the left text box
- Enter 0.001 in the center text box
- Click Apply

1-3. Create Keypoint 6 at (0.006,0.001)

1-4. Create Keypoint 7 at (0,0.003)

1-5. Create Keypoint 8 at (0.006,0.003)

1-6. Turn keypoint numbers on

- Utility Menu > PlotCtrls > Numbering . . .

1-7. Plot keypoints

- Utility Menu > Plot > Keypoints

1-8. Create a line between Keypoints 5 and 6

- Preprocessor > Modeling > Create > Lines > Lines > Straight Line
- Click Keypoint 5 and then click Keypoint 6
- Click OK

1-9. Create a line between Keypoints 7 and 8

1-10. Turn line numbers on

- Utility Menu > PlotCtrls > Numbering . . .

1-11. Plot lines

- Utility Menu > Plot > Lines

1-12. Divide the area by Line 5

- Preprocessor > Modeling > Operate > Booleans > Divide > Area by Line
- Click on the area
- Click OK
- Click on Line 5
- Click OK

1-13. Divide the top area by Line 6

- Preprocessor > Modeling > Operate > Booleans > Divide > Area by Line
- Click on the top area
- Click OK
- Click on Line 6
- Click OK

1-14. Turn area numbers on

- Utility Menu > PlotCtrls > Numbering . . .

1-15. Plot areas to confirm that the operations were successful

- Utility Menu > Plot > Areas

1-16. Save your geometry

- ANSYS Toolbar > SAVE_DB

Step 2: Define Element Types

2-1. Define PLANE55 as element type #1 for this model

- Preprocessor > Element Type > Add/Edit/Delete

Step 3: Define Material Properties

3-1. Define the material properties for the stainless steel (k = 17 W/mK)

- Preprocessor > Material Props > Material Models
- Thermal > Conductivity > Isotropic

3-2. Define the material properties for the copper (k = 372 Wm/K)

- Preprocessor > Material Props > Material Models
- Material > New Model . . .
- Thermal > Conductivity > Isotropic

Step 4: Mesh

4-1. Assign stainless steel as the material for the top and bottom areas

- Preprocessor > Meshing > Mesh Attributes > Picked Areas
- Click on the top and bottom areas
- Click OK
- Ensure that the MAT Material number is set to "1"
- Click OK

4-2. Assign copper as the material for the middle area

- Preprocessor > Meshing > Mesh Attributes > Picked Areas
- Click on the middle area
- Click OK
- Change the MAT Material number to "2"
- Click OK

4-3. Map mesh the model with an element edge length of 0.0001 m

- Preprocessor > Meshing > MeshTool
- In the third section, click the "Set" button for the Global Size Controls
- For SIZE Element edge length, enter 0.0001
- Click OK
- In the fourth section, check the "Mapped" radio button
- Click the "Mesh" button
- Click Pick All

Step 5: Apply Constraint Boundary Conditions

5-1. Apply a constant temperature of 100°C to the top line

- Solution > Define Loads > Apply > Thermal > Temperature > On Lines
- Pick the top line (Line 3)
- Click OK
- For LAB2 DOFs to be constrained, choose TEMP
- For VALUE Load TEMP value, enter 100
- Click OK

Step 6: Apply Load Boundary Conditions

6-1. Apply a constant heat flux of 1200 kW/m^2 to the bottom line

- Solution > Define Loads > Apply > Thermal > Heat Flux > On Lines
- Pick the bottom line (Line 1)
- Click OK
- For VALI Heat Flux, enter 1200e3
- Click OK

6-2. Save your progress

Step 7: Set the Solution Options

The default solution options can be used for this analysis.

Step 8: Solve

8-1. Select everything

- Utility Menu > Select > Everything

8-2. Solve the model

- Solution > Solve > Current LS

8-3. Save your results

Step 9: Postprocess the Results

9-1. Plot the nodal temperature distribution through the plate

- General Postprocessor > Plot Results > Contour Plot > Nodal Solu.

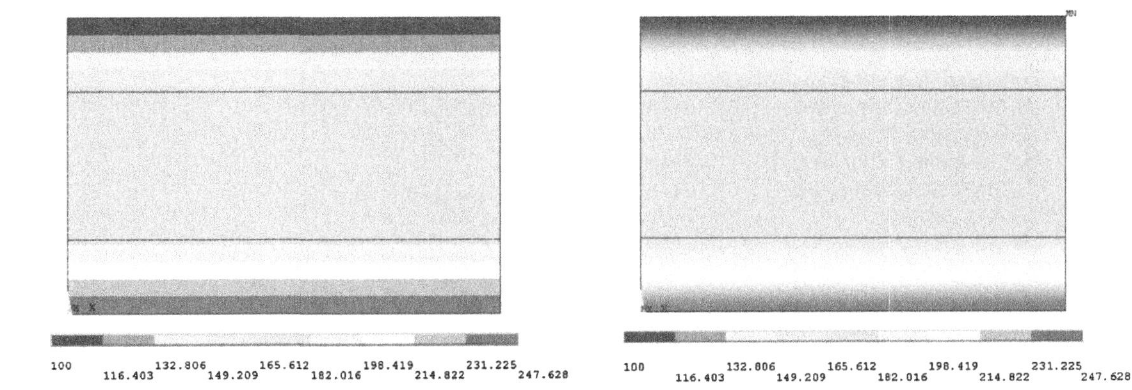

Figure 10-1-2 Nodal Temperature Contour Plot: Win32 Graphics (left), 3D Graphics (right).

The nodal temperature plot in Figure 10-1-2 shows that the maximum temperature in the pan is 247.6°C.

9-2. Select the elements associated with the copper

- Utility Menu > Select > Entities . . .
- Choose Elements, By Attributes, by Material num
- In the Min,Max,Inc text box, enter "2"
- Click OK

9-3. Select the nodes attached to those elements

- Utility Menu > Select > Entities . . .
- Choose Nodes, Attached to, Elements
- Click OK

9-4. Plot the nodal temperature distribution through the copper layer

- General Postprocessor > Plot Results > Contour Plot > Nodal Solu

This shows that the bottom of the copper is at approximately 177.04°C, and the top of the copper is at approximately 170.59°C.

Step 10: Compare and Verify the Results

The temperature drop across the various layers is related to the heat flux by Fourier's law: q = k (ΔT/L). For the copper layer, the expression becomes 1200e3 = (372)(ΔT)/(0.002), which yields a temperature drop of 6.45°C. Our model also predicts a drop of 6.45°C. Therefore, the model is in perfect agreement with the theory.

Step 11: File Management

11-1. Open the ANSYS Command Dictionary

- Utility Menu > Help > Help Topics
- Mechanical APDL > Command Reference (Figure 10-1-3).

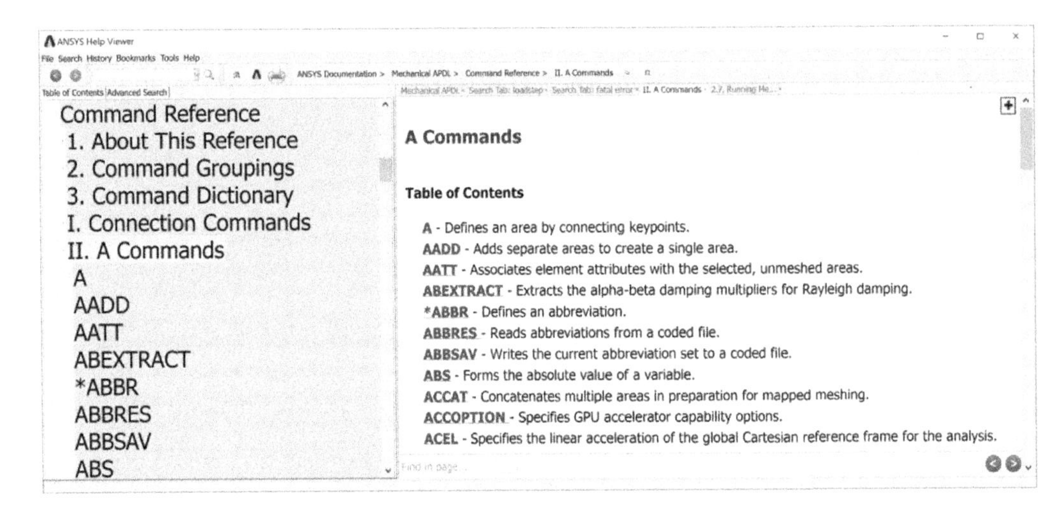

Figure 10-1-3 ANSYS Command Dictionary.

This will prepare you to look up commands and their syntax.

If the entry is too small to read, right click and follow the path Zoom > Zoom In to increase the display size.

11-2. Save a backup copy of the Session Log File

- Open the "Exercise 10-1" folder in the "Intro to ANSYS" directory (or the folder that you are using as your working directory)
- Identify the log file

The file may be named "Exercise10-1" (without an extension), "Exercise10-1.log," or "Exercise10-1.txt." If you view the folder Details, it should be the only file Type listed as a Text Document (Figure 10-1-4).

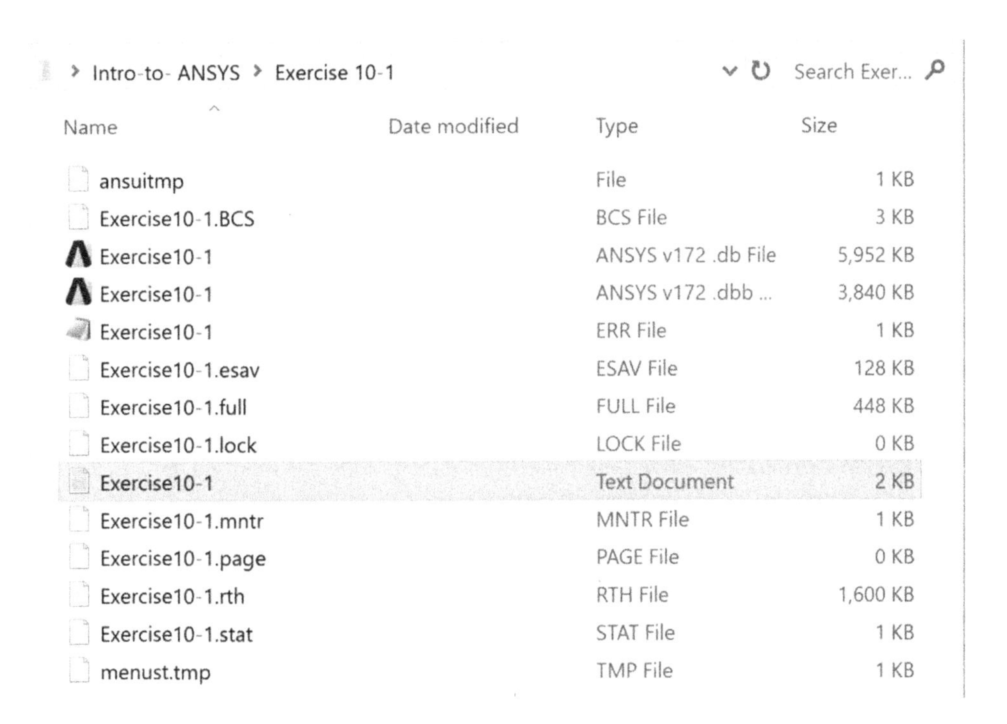

Figure 10-1-4 Working Directory with the Log File Highlighted.

- Create a folder named "Backup Files" in your working directory (Figure 10-1-5)
- Create a copy of the log file and place it in the Backup Files directory

This will allow you to restart the exercise from this point if necessary.

11-3. Save the Session Log File as Input10-1.txt

- Save the copy of the log file as Input10-1.txt in your working directory

This will be your working input file for the remainder of the exercise.

› Intro-to-ANSYS › Exercise 10-1 › Search Exer...

Name	Date modified	Type	Size
Backup Files		File folder	
ansuitmp		File	1 KB
Exercise10-1.BCS		BCS File	3 KB
Exercise10-1		ANSYS v172 .db File	5,952 KB
Exercise10-1		ANSYS v172 .dbb ...	3,840 KB
Exercise10-1		ERR File	1 KB
Exercise10-1.esav		ESAV File	128 KB
Exercise10-1.full		FULL File	448 KB
Exercise10-1.lock		LOCK File	0 KB
Exercise10-1		Text Document	2 KB
Exercise10-1.mntr		MNTR File	1 KB
Exercise10-1.page		PAGE File	0 KB
Exercise10-1.rth		RTH File	1,600 KB
Exercise10-1.stat		STAT File	1 KB
Input10-1		Text Document	2 KB
menust.tmp		TMP File	1 KB

Figure 10-1-5 Working Directory Showing the Backup Directory with the Working Input File Highlighted.

11-4. Open the new Input10-1 file

Use Notepad or another plain text editor. Do not use Word or another word processing program. Word processing programs usually include hidden information and formatting that can generate errors in ANSYS (Figure 10-1-6).

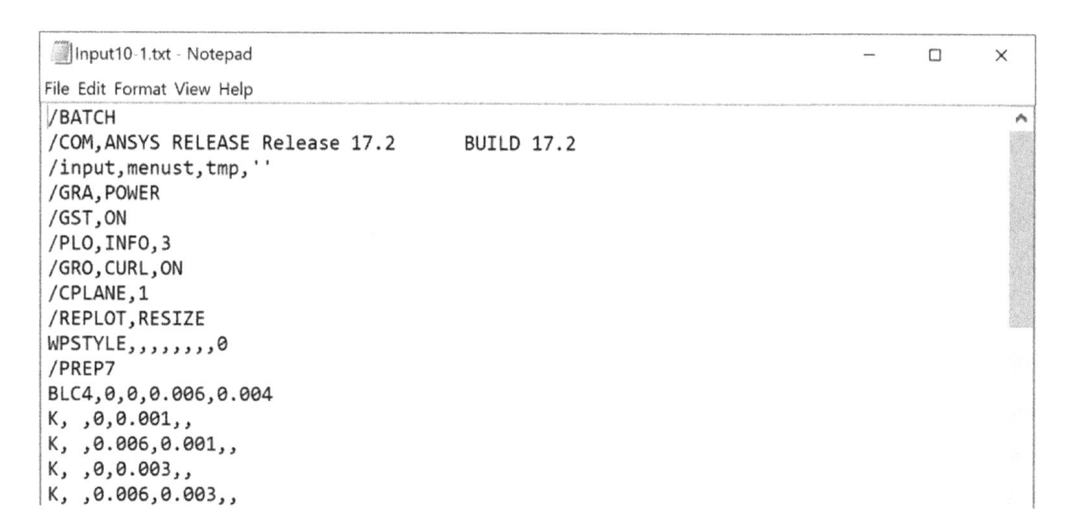

Input10-1.txt - Notepad

File Edit Format View Help

```
/BATCH
/COM,ANSYS RELEASE Release 17.2      BUILD 17.2
/input,menust,tmp,''
/GRA,POWER
/GST,ON
/PLO,INFO,3
/GRO,CURL,ON
/CPLANE,1
/REPLOT,RESIZE
WPSTYLE,,,,,,,,0
/PREP7
BLC4,0,0,0.006,0.004
K, ,0,0.001,,
K, ,0.006,0.001,,
K, ,0,0.003,,
K, ,0.006,0.003,,
```

Figure 10-1-6 Working Input File Opened with Microsoft Notepad.

The text in your working file should be similar to the following text. Your log file may have additional commands if you performed extra operations during the input process. You may work from the log file below if you prefer.

```
/BATCH

/COM,ANSYS RELEASE
Release 17.2
BUILD 17.2
UP20160718
14:59:50

/input,menust,tmp,''
/GRA,POWER
/GST,ON
/PLO,INFO,3
/GRO,CURL,ON
/CPLANE,1
/REPLOT,RESIZE
WPSTYLE,,,,,,,,0
/PREP7
BLC4,0,0,0.006,0.004
K,,0,0.001,,
K,,0.006,0.001,,
K,,0,0.003,,
K,,0.006,0.003,,
/PNUM,KP,1
/PNUM,LINE,0
/PNUM,AREA,0
/PNUM,VOLU,0
/PNUM,NODE,0
/PNUM,TABN,0
/PNUM,SVAL,0
/NUMBER,0
!*
/PNUM,ELEM,0
/REPLOT
!*
KPLOT
LSTR,   5,   6
LSTR,   7,   8
/PNUM,KP,1
/PNUM,LINE,1
/PNUM,AREA,0
/PNUM,VOLU,0
/PNUM,NODE,0
/PNUM,TABN,0
/PNUM,SVAL,0
/NUMBER,0
!*
/PNUM,ELEM,0
/REPLOT
!*
LPLOT
ASBL,   1,   5
ASBL,   3,   6
/PNUM,KP,1
/PNUM,LINE,1
/PNUM,AREA,1
/PNUM,VOLU,0
/PNUM,NODE,0
/PNUM,TABN,0
/PNUM,SVAL,0
/NUMBER,0
!*
/PNUM,ELEM,0
/REPLOT
!*
APLOT
SAVE
!*
ET,1,PLANE55
!*
!*
MPTEMP,,,,,,,,
MPTEMP,1,0
MPDATA,KXX,1,,17
MPTEMP,,,,,,,,
MPTEMP,1,0
MPDATA,KXX,2,,372
FLST,5,2,5,ORDE,2
FITEM,5,1
FITEM,5,-2
CM,_Y,AREA
ASEL,,,,P51X
CM,_Y1,AREA
CMSEL,S,_Y
!*
CMSEL,S,_Y1
AATT,1,,1,0,
CMSEL,S,_Y
CMDELE,_Y
CMDELE,_Y1
!*
CM,_Y,AREA
ASEL,,,,   4
CM,_Y1,AREA
CMSEL,S,_Y
!*
CMSEL,S,_Y1
AATT,2,,1,0,
CMSEL,S,_Y
CMDELE,_Y
CMDELE,_Y1
!*
ESIZE,0.0001,0,
MSHAPE,0,2D
MSHKEY,1
!*
FLST,5,3,5,ORDE,3
FITEM,5,1
FITEM,5,-2
FITEM,5,4
CM,_Y,AREA
ASEL,,,,P51X
CM,_Y1,AREA
CHKMSH,'AREA'
CMSEL,S,_Y
!*
AMESH,_Y1
!*
CMDELE,_Y
CMDELE,_Y1
CMDELE,_Y2
!*
FINISH
/SOL
FLST,2,1,4,ORDE,1
FITEM,2,3
!*
/GO
DL,P51X,,TEMP,100,0
FLST,2,1,4,ORDE,1
FITEM,2,1
/GO
!*
SFL,P51X,HFLUX,1200e3
SAVE
ALLSEL,ALL
/STATUS,SOLU
SOLVE
SAVE
FINISH
/POST1
!*
/EFACET,1
PLNSOL, TEMP,, 0
ESEL,S,MAT,,2
NSLE,S
!*
/EFACET,1
PLNSOL, TEMP,, 0
```

Step 12: Remove Nonessential Commands from the File

12-1. Remove the header block

- Remove all commands between the **/BATCH** command and the **/PREP7** command

12-2. Remove all SAVE commands

12-3. Remove all spacing comments (!*) but leave the extra lines for now

Step 13: Edit the Input File Geometry

13-1. Remove all graphics commands such as the commands to turn on numbering and to plot keypoints, lines, and areas

- Remove all **/PNUM** commands
- Remove all **/NUMBER** commands
- Remove all **/REPLOT** commands
- Remove all **KPLOT**, **LPLOT**, and **APLOT** commands

13-2. Format the solid modeling commands

- Remove any extra spaces within the solid modeling commands
- Removing trailing commas after the solid modeling commands
- Ensure that all values that start with a decimal point have a leading zero (e.g. 0.001)
- Remove any extra lines between the solid modeling commands

13-3. Comment the solid modeling commands

- At the end of each line, tab over twice and add an exclamation point
- After each exclamation point, add a brief explanation of each command
- If you are unfamiliar with a command, look it up in the Command Dictionary

The first part of your input file should now look like this:

```
/BATCH                  ! <Comment>
/PREP7                  ! <Comment>
BLC4,0,0,0.006,0.004    ! <Comment>
K,,0,0.001              ! <Comment>
K,,0.006,0.001          ! <Comment>
K,,0,0.003              ! <Comment>
K,,0.006,0.003          ! <Comment>
LSTR,5,6                ! <Comment>
LSTR,7,8                ! <Comment>
ASBL,1,5                ! <Comment>
ASBL,3,6                ! <Comment>
```

Step 14: Edit the Element and Mesh Attributes for the Input File

14-1. Streamline the material property commands

- Delete all **MPTEMP** commands
- Replace the first **MPDATA** command with: `MP,KXX,1,17`
- Replace the second **MPDATA** command with: `MP,KXX,2,372`

*It is good practice to use the **MPTEMP** and **MPDATA** commands only for temperature-dependent material properties. This makes the model easier to understand at a glance and avoids confusion about the assumptions in the model.*

14-2. Repair the first Picker block associated with the first AATT command

*The following group of commands selects the top and bottom areas and uses the **AATT** command to assign material number 1 to these areas.*

```
FLST,5,2,5,ORDE,2
FITEM,5,1
FITEM,5,-2
CM,_Y,AREA
ASEL,,,,P51X
CM,_Y1,AREA
CMSEL,S,_Y

CMSEL,S,_Y1
AATT,  1,,  1,  0,
CMSEL,S,_Y
CMDELE,_Y
CMDELE,_Y1
```

*At this point, you must make two choices. You must decide whether to use the **ASEL** and **AATT** commands to assign the material models to the areas in the model and then mesh the model as a whole (as was done in the exercise) or to use the **MAT** command and then mesh the areas individually.*

*Either way, you must replace the Picker blocks with **ASEL** commands. Therefore, you must also choose how to select those areas. The simplest option is to select the areas by entity number. However, the most robust option is to select the areas by location.*

For consistency, we will replicate the operations performed in the GUI. For robustness, we will select by location.

- Add a few blank lines above the first Picker block to create some space to work
- Open the **ASEL** entry in the ANSYS Command Dictionary to remind yourself of the command syntax

*There are two good ways to find the **ASEL** entry. The first option is to expand the menus on the left side of the Help Table of Contents following the path: Mechanical APDL > Command Reference > II. A Commands > ASEL. Clicking on the ASEL hyperlink in the left side of the window will open the documentation for that command. The second option is to use the search functionality. Type "ASEL" in the text box in the upper left corner of the Help Viewer and then either click on the magnifying glass or press enter (Figure 10-1-7). The **ASEL** entry should be the first search result (Figure 10-1-8).*

Figure 10-1-7 Searching the ANSYS Documentation.

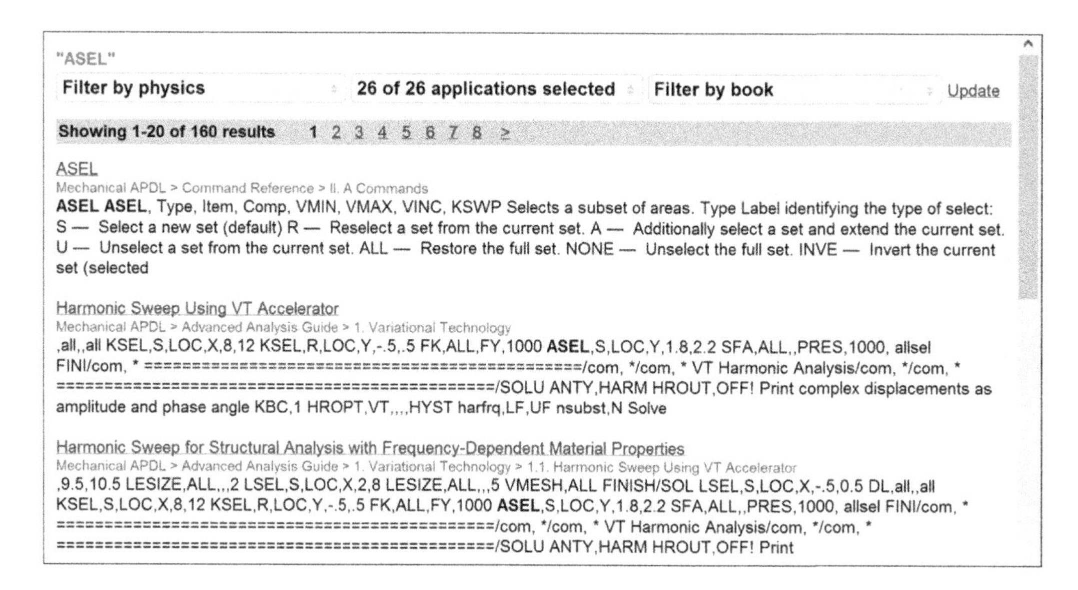

Figure 10-1-8 Search Results for the **ASEL** Command.

- Add the command `ASEL,S,LOC,Y,0,0.001`

Use the documentation to confirm that this creates a new set (S) by location (LOC) using the y coordinate (Y) in the range of 0 to 0.001 (i.e. selects the bottom area).

- Add the command `ASEL,A,LOC,Y,0.003,0.004`

Use the documentation to confirm that this adds to the existing set (A) of areas by location (LOC), using the y coordinate (Y) in the range of 0.003 to 0.004 (i.e. selects the top area).

- Open the **AATT** entry in the ANSYS Command Dictionary

The documentation shows that the AATT command has no arguments associated with the entity specification. Therefore, no repair of this command is needed.

- Delete all entries in the first Picker block except for the **AATT** command

14-3. Repair the second Picker block associated with the second AATT command

*Again, we must choose how to select the areas for the second **AATT** command. Now, we have three choices. We can select by entity number, by area, or by inverting the active set. In this case, we will select the area by inverting the active set.*

- Add the command `ASEL,INVE` after the first **AATT** command
- Delete all entries in the second Picker block except for the **AATT** command
- Remove any extra blank lines and blank spaces associated with these commands

14-4. Streamline the mesh control commands

- Open the **ESIZE** entry in the Mechanical APDL Command Dictionary

The NDIV entry notes that this field is "Not used if SIZE is input." Therefore the information in the third field is not used and can be deleted. In this input file, we will delete it for clarity. This makes it easier to look at the input file and quickly see that the element edge length (SIZE) was specified instead of the number of element divisions (NDIV). Otherwise, someone reading the input file would have to look at the values of both fields to determine which type of element sizing was specified.

- Delete the third field of the **ESIZE** command. It should now read: `ESIZE,0.0001`
- Open the **MSHAPE** entry in the Mechanical APDL Command Dictionary

*If you scroll down to the Command Default section, you will see that if "You do not specify an element shape but you do specify the type of meshing to be used [MSHKEY], then ANSYS uses the default shape of the element to mesh the model." In the exercise above, we specified mapped meshing; therefore the **MSHKEY** command is needed and the **MSHAPE** command is not. In this input file, we will delete it for clarity. Otherwise, someone reading the input file would have to look at the combination of the **MSHAPE** and **MSHKEY** commands to determine the meshing behavior of the system.*

- Remove the **MSHAPE** command

14-5. Repair the meshing Picker block

- Add an `ALLSEL` command below the **MSHKEY** command

*Because we selected areas by location for the **AATT** commands, the active set is no longer the full set of entities in the model. We need to select the full set so all areas in the model are meshed.*

- Open the **AMESH** entry in the Mechanical APDL Command Dictionary
- Verify that the Picker commands can be replaced with an "ALL" for the NA1 argument in the **AMESH** command
- Add the command `AMESH,ALL` below the **ALLSEL** command
- Delete all commands in the meshing Picker block including the **CMDELE** commands

14-6. Add an empty line below the FINISH command to separate the preprocessing commands from the solution commands

14-7. Format the element and mesh attribute commands

14-8. Remove any extra spaces within the element and mesh attribute commands

- Removing trailing commas after the element and mesh attribute commands
- Ensure that all values that start with a decimal point have a leading zero (e.g. 0.0001)
- Remove any extra lines between the element and mesh attribute commands

14-9. Comment the element and mesh attribute commands

- At the end of each line, tab over twice and add an exclamation point
- After each exclamation point, add a brief explanation of each command
- If you are unfamiliar with a command, look it up in the Command Dictionary

The second part of your input file should now look like this:

```
ET,1,PLANE55                ! <Comment>
MP,KXX,1,17                 ! <Comment>
MP,KXX,2,372                ! <Comment>
ASEL,S,LOC,Y,0,0.001        ! <Comment>
ASEL,A,LOC,Y,0.003,0.004    ! <Comment>
AATT,1,,1,0                 ! <Comment>
ASEL,INVE                   ! <Comment>
AATT,2,,1,0                 ! <Comment>
ESIZE,0.0001                ! <Comment>
MSHKEY,1                    ! <Comment>
ALLSEL                      ! <Comment>
AMESH,ALL                   ! <Comment>
FINISH                      ! <Comment>
```

Step 15: Edit the Solution Section of the Input File

15-1. Repair the Picker block associated with the applied constraints

*This Picker block selects a line and then applies a temperature constraint to that line using the **DL** command. We need to replace the Picker commands with an **LSEL** command and then update the **DL** command to act on the selected line.*

- Add the command `LSEL,S,LOC,Y,0.004` after the /**SOL** command
- In the **DL** command, replace the "P51X" argument with "ALL"
- Delete all remaining commands associated with that Picker block

15-2. Repair the Picker block associated with the applied loads

- Add the command `LSEL,S,LOC,Y,0` after the **DL** command
- In the **SFL** command, replace the "P51X" argument with "ALL"
- Delete all remaining commands associated with that Picker block

15-3. Remove the /STATUS command

15-4. Remove any SAVE commands that might still be present

15-5. Add an empty line below the FINISH command to separate the solution commands from the postprocessing commands

15-6. Format the solution commands

- Remove any extra spaces within the solution commands
- Removing trailing commas after the solution commands
- Ensure that all values that start with a decimal point have a leading zero (e.g. 0.001)
- Remove any extra lines between the solution commands

15-7. Comment the solution commands

- At the end of each line, tab over twice and add an exclamation point
- After each exclamation point, add a brief explanation of each command
- If you are unfamiliar with a command, look it up in the Command Dictionary

The third part of your input file should now look like this:

```
/SOL                    ! <Comment>
LSEL,S,LOC,Y,0.004      ! <Comment>
DL,ALL,,TEMP,100,0      ! <Comment>
LSEL,S,LOC,Y,0          ! <Comment>
SFL,ALL,HFLUX,1200e3,   ! <Comment>
ALLSEL,ALL              ! <Comment>
SOLVE                   ! <Comment>
FINISH                  ! <Comment>
```

Step 16: Edit the Postprocessing Section of the Input File

16-1. Remove unnecessary graphics commands

- Open the **/EFACET** entry in the Mechanical APDL Command Dictionary
- Confirm that it is not necessary for the input file to run
- Remove the /**EFACET** commands

16-2. Add a /WAIT command

- Add the command `/WAIT,5` after each PLNSOL command

*When you run you final input file, the plots will appear and then be replaced almost instantaneously. This will make it difficult to ensure that the plots were generated correctly. Instead, we will add a **/WAIT** command to delay the execution of the next command by the specified number of seconds.*

16-3. Add a FINISH command at the end of the file to exit the Postprocessor

16-4. Add a blank line after the FINISH command

16-5. Add a SAVE command at the end of the file to save the database and results

16-6. Add a /EXIT command that has been commented out (`!/EXIT`)

This command stops the run and returns control to the system. It is useful for more complicated analyses.

16-7. Format the postprocessing commands

- Remove any extra spaces within the postprocessing commands
- Remove any extra lines between the postprocessing commands

16-8. Comment the postprocessing commands

- At the end of each line, tab over twice and add an exclamation point
- After each exclamation point, add a brief explanation of each command
- If you are unfamiliar with a command, look it up in the Command Dictionary

The third part of your input file should now look like this.

```
/POST1            ! <Comment>
PLNSOL,TEMP,,0    ! <Comment>
/WAIT,5           ! <Comment>
ESEL,S,MAT,,2     ! <Comment>
NSLE,S            ! <Comment>
PLNSOL,TEMP,,0    ! <Comment>
/WAIT,5           ! <Comment>
FINISH            ! <Comment>

SAVE              ! <Comment>
!/EXIT            ! <Comment>
```

Step 17: Save and Run the Input File

17-1. Save your input file

17-2. Clear the database

- Utility Menu > File > Clear & Start New ...
- Click OK
- Click Yes

17-3. Read the input file into ANSYS in interactive mode

- Utility Menu > File > Read Input from ...
- Choose Input10-1.txt from your working directory
- Click OK

The program should now run unassisted until it reaches the end of the input file. The last image left on the screen should be of the temperature distribution in the copper because that was the last plot command issued in the file.

Step 18: Debug the Input File If Necessary

If your input file contains one or more errors, the program will display one or more Warning (Figure 10-1-9) or Error (Figure 10-1-10) dialog boxes. Some of these dialog boxes allow you to suspend processing of the input file. Others only allow you to acknowledge the problem.

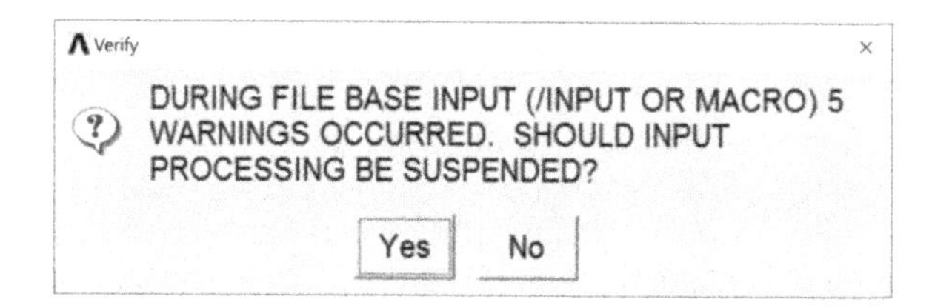

Figure 10-1-9 Example Verify Warning Dialog Box.

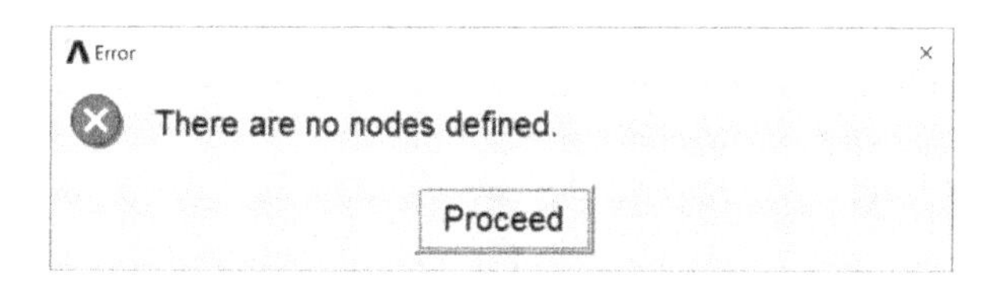

Figure 10-1-10 Example Error Dialog Box.

If this occurs:

- Suspend processing as soon as possible and
- Review the information about the error(s) in the Output Window and/or in the error file.

The error file is a plain text file and can be opened with any plain text editor. The error file for Exercise10-1 will be named Exercise10-1.err or Exercise10-1 without an extension. If there is no file with a .err extension in the working directory, list the directory contents in Details mode. Then, in the Type column, look for the file labeled "ERR File" (*Figure 10-1-11*).

› Intro-to- ANSYS › Exercise 10-1 Search Exer...

Name	Date modified	Type	Size
Backup Files		File folder	
ansuitmp		File	1 KB
Exercise10-1.BCS		BCS File	2 KB
Exercise10-1		ANSYS v172 .db File	5,952 KB
Exercise10-1		ANSYS v172 .dbb ...	5,952 KB
Exercise10-1		ERR File	1 KB
Exercise10-1.esav		ESAV File	128 KB
Exercise10-1.full		FULL File	448 KB
Exercise10-1.lock		LOCK File	0 KB
Exercise10-1		Text Document	2 KB
Exercise10-1.mntr		MNTR File	1 KB
Exercise10-1.page		PAGE File	0 KB
Exercise10-1.rth		RTH File	1,600 KB
Exercise10-1.stat		STAT File	1 KB
Input10-1		Text Document	1 KB
menust.tmp		TMP File	1 KB

Figure 10-1-11 Working Directory Shown with the Error File Highlighted.

- Correct the errors if they are easily identified

For example, if you are missing a ***/PREP7*** *command at the beginning of the file, you will see the error "K is not a recognized BEGIN command, abbreviation, or macro. This command will be ignored."*

- For more difficult errors, go line-by-line and compare your input file to the one at the end of this document to identify the differences
- Correct any discrepancies that you find
- Save and rerun the new version of the input file
- Continue this process until no more warnings or errors occur

Close the Program

- Utility Menu > File > Exit ...
- Choose “Quit - No save!”
- Click OK

Sample Input File

The input file that you created should be similar to the one below. The comments from the sample input file have been removed to encourage you to use the ANSYS Command Reference in this exercise.

```
/BATCH
/PREP7
BLC4,0,0,0.006,0.004
K,,0,0.001
K,,0.006,0.001
K,,0,0.003
K,,0.006,0.003
LSTR,5,6
LSTR,7,8
ASBL,1,5
ASBL,3,6

ET,1,PLANE55
MP,KXX,1,17
MP,KXX,2,372
ASEL,S,LOC,Y,0,0.001
ASEL,A,LOC,Y,0.003,0.004
AATT,1,,1,0
ASEL,INVE
AATT,2,,1,0
ESIZE,0.0001
MSHKEY,1
ALLSEL
AMESH,ALL
FINISH

/SOL
LSEL,S,LOC,Y,0.004
DL,ALL,,TEMP,100,0
LSEL,S,LOC,Y,0
SFL,ALL,HFLUX,1200e3,
ALLSEL,ALL
SOLVE
FINISH
```

```
/POST1
PLNSOL,TEMP,, 0
/WAIT,5
ESEL,S,MAT,,2
NSLE,S
PLNSOL,TEMP,,0
/WAIT,5
FINISH

SAVE
!/EXIT
```

Exercise 10-2

Using the Concurrent Method to Modify an Input File for Steady-State Conduction Through a Cladded Plate

Overview

In this exercise, you will use the Concurrent Method to modify the input file from exercise 10-1 to explore how various assumptions affect the behavior of the system.

Assume that the heat has been turned up and the pan is no longer centered on the stove. The heat flux on the bottom of the plate now varies linearly from 1500 to 1200 kW/m^2. You will modify the input file to reflect this change. You will then solve the model for three thicknesses of the copper layer—2 mm, 1 mm, and 0.5 mm—to determine if a thinner, and therefore lower cost, pan can be designed (Figure 10-2-1).

All other aspects of the problem will remain the same.

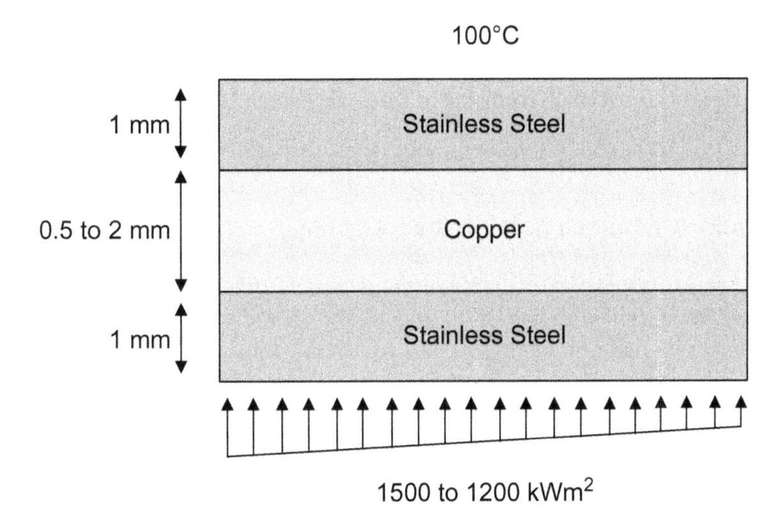

Figure 10-2-1 Schematic of the Cladded Plate.

ANSYS Mechanical APDL for Finite Element Analysis.
DOI: http://dx.doi.org/10.1016/B978-0-12-812981-4.00030-7

Model Attributes

Geometry for the Plate

- Stainless steel layers—1 mm
- Copper layer—0.5, 1, or 2 mm

Material Properties

- Stainless steel—Thermal conductivity—17 W/mK
- Copper—Thermal conductivity—372 W/mK

Loads

- Linearly varying heat flux of 1500 to 1200 kW/m^2 on the plate bottom

Constraints

- Surface temperature T = 100°C on the top of the plate

File Management

Create a new folder in your "Intro-to-ANSYS" folder for Exercise10-2

Change the Working Directory to the Exercise10-2 folder

Change the Jobname to Exercise10-2

Start ANSYS

Open the Input10-1.txt input file in your Exercise10-1 folder

Save the file as Input10-2-2mm.txt in your Exercise10-2 folder

Step 1: Prepare the Input File for Modification

1-1. Identify the branch point in the input file

The branch point is where the new analysis begins to differ from the old analysis. For the 2 mm thick copper layer, the analysis is the same until the load boundary conditions are applied. Thus, all commands up to and including the ***DL*** *command are still valid and can be reused.*

1-2. Add End of File commands to stop the input file from being read after the branch point

- Add a `/EOF` command after the second **LSEL** command
- Add another `/EOF` command after the first

1-3. Delete the old SFL command after the /EOF commands

1-4. Save your input file

Your Solution commands should now look like this:

```
/SOL
LSEL,S,LOC,Y,0.004
DL,ALL,,TEMP,100,0
LSEL,S,LOC,Y,0
/EOF
/EOF
ALLSEL,ALL
SOLVE
FINISH
```

1-5. Run the input file to load the model

- Utility Menu > File > Read Input from ...
- Choose Input10-2-2 mm.txt from your working directory
- Click OK

Step 2: Update the Load Boundary Conditions

2-1. Apply a linearly varying heat flux from 1500 to 1200 kW/m^2 to the bottom line

- Solution > Define Loads > Apply > Thermal > Heat Flux > On Lines
- Pick All
- For VALI Heat flux enter 1500e3
- For VALJ Heat flux enter 1200e3
- Click OK

2-2. Transfer this command to your input file

- Utility Menu > File > List > Log File ...
- Scroll to the bottom of the file
- Copy the new **SFL** command

While a Picker block was used to perform this operation, none of the Picker commands are needed for the new input file. Therefore, we will not copy them over.

- Paste this command above the /**EOF** commands in your input file

If you have trouble copying the commands, left click and drag the mouse to select the desired commands. To copy, use the keyboard shortcut "ctrl + c" (press the "ctrl" key and the "c" key at the same time). To paste, use the keyboard shortcut "ctrl + v."

You can also open the log file from the working directory and copy and paste normally.

2-3. Repair the SFL command

- In the **SFL** command, replace the "P51X" argument with "ALL"

Your Solution commands should now look like this:

```
/SOL
LSEL,S,LOC,Y,0.004
DL,ALL,,TEMP,100,0
LSEL,S,LOC,Y,0
SFL,ALL,HFLUX,1500e3,1200e3
/EOF
/EOF
ALLSEL,ALL
SOLVE
FINISH
```

Step 3: Run the Exercise10-2-2mm.txt Input File

3-1. Remove the /EOF commands

3-2. Comment the new commands

- At the end of each line, tab over twice and add an exclamation point
- After each exclamation point, add a brief explanation of each command
- If you are unfamiliar with a command, look it up in the Command Dictionary

3-3. Save your input file

3-4. Clear the database

- Utility Menu > File > Clear & Start New ...
- Click OK
- Click Yes

3-5. Run the input file to create the new model

- Utility Menu > File > Read Input from ...
- Choose Input10-2-2mm.txt from your working directory
- Click OK

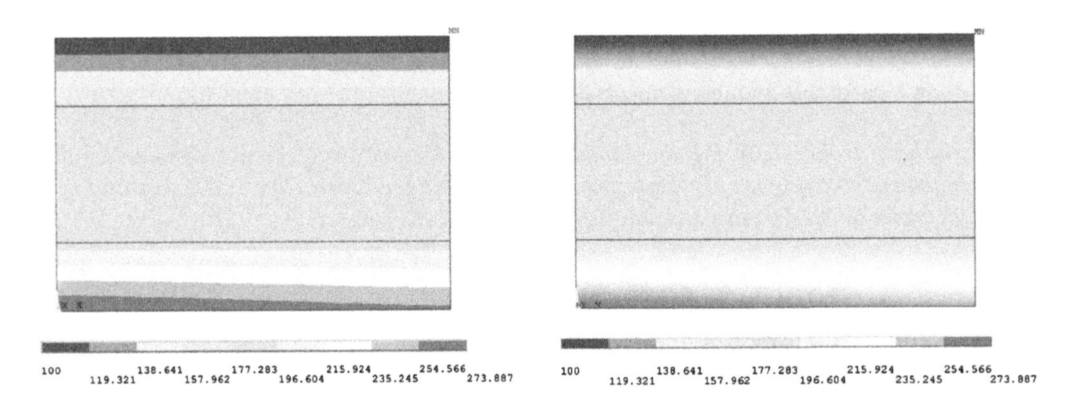

Figure 10-2-2 Nodal Temperature Contour Plot (Cu = 2 mm): Win32 Graphics (left), 3D Graphics (right).

In Figure 10-2-2, the asymmetry due to the linearly varying heat flux is apparent in the bottom steel layer but barely noticeable in the top steel layer. This is due to the influence of the copper. Note also that the maximum temperature in the model has increased from 247.6°C to 273.89°C due to the increased heat flux.

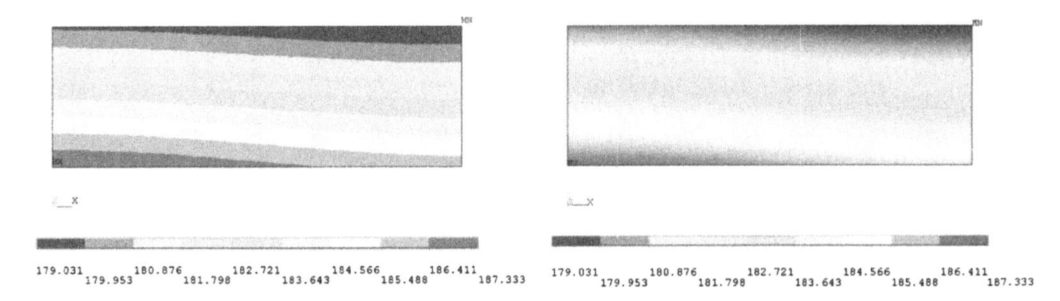

Figure 10-2-3 Nodal Temperature Contour Plot of the Copper Layer (Cu = 2 mm): Win32 Graphics (left), 3D Graphics (right)

The nodal temperature plot for the copper layer in Figure 10-2-3 confirms that 2 mm of copper is sufficient to evenly distribute the heat in the pan.

Step 4: Create a New Input File for a 1 mm Copper Layer

4-1. Save your input file as Input10-2-1mm.txt

4-2. Update the geometry for the thinner copper layer

Changing the thickness of the copper layer will require changing the height of the upper part of the copper layer and the y coordinates of the entire upper steel layer. The new coordinates for the keypoints and lines in the model are shown in Figure 10-2-4.

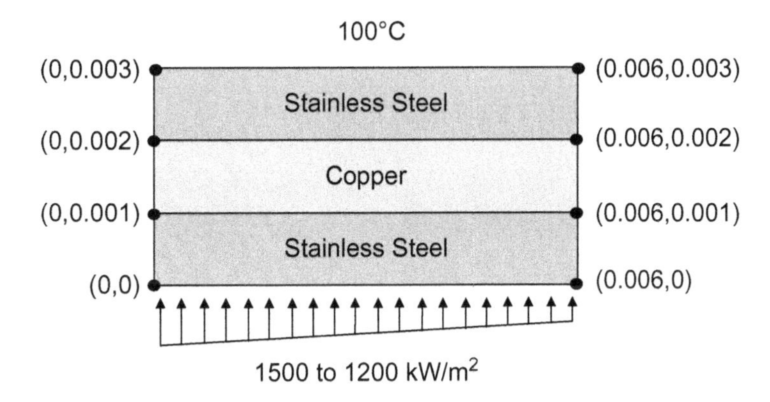

Figure 10-2-4 Updated Model Geometry (Cu = 1 mm).

- Third line: Change the height in the **BLC4** command from 0.004 to 0.003.
- Sixth and seventh lines: Change the heights of KPs 7 and 8 from 0.003 to 0.002.

```
BLC4,0,0,0.006,0.003
K,,0,0.001
K,,0.006,0.001
K,,0,0.002
K,,0.006,0.002
```

- Sixteenth line: Change the *y* coordinates of the second **ASEL** command from 0.003 and 0.004 to 0.002 and 0.003.

```
ASEL,S,LOC,Y,0,0.001
ASEL,A,LOC,Y,0.002,0.003
AATT,1,,1,0
```

4-3. Update the boundary conditions for the thinner copper layer

- Change the *y* coordinate for first **LSEL** below the /**SOL** command from 0.004 to 0.003.

```
/SOL
LSEL,S,LOC,Y,0.003
DL,ALL,,TEMP,100,0
```

4-4. Comment the new commands

Step 5: Run the Exercise10-2-1mm.txt Input File

5-1. Save your input file

5-2. Clear the database

- Utility Menu > File > Clear & Start New ...

5-3. Run the input file to create the new model

- Utility Menu > File > Read Input from ...

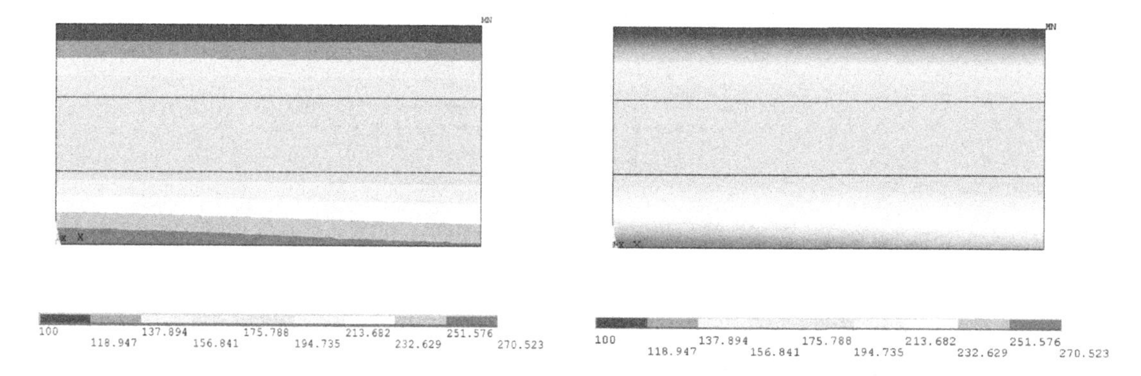

Figure 10-2-5 Nodal Temperature Contour Plot (Cu = 1 mm): Win32 Graphics (left), 3D Graphics (right)

Figure 10-2-5 shows that the thinner copper layer has lowered the thermal resistance of the pan and therefore lowered the maximum temperature from 273.89°C to 270.52°C. The temperature contours in the top layer of steel still appear to be uniform. However, there is now a difference of several degrees across the top of the copper layer (Figure 10-2-6). Thus, a thinner layer of copper can be used but not without some loss in performance.

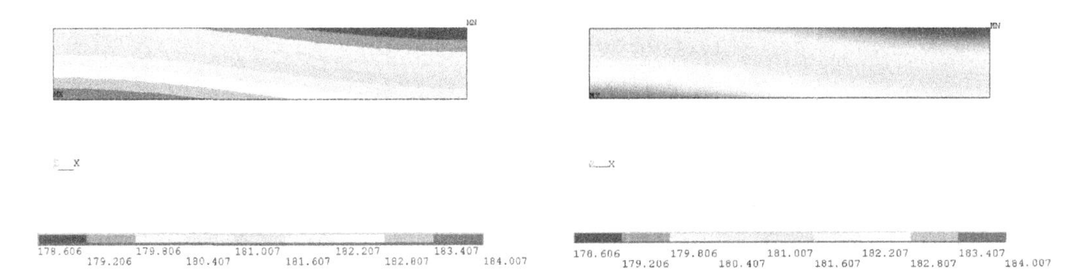

Figure 10-2-6 Nodal Temperature Contour Plot of the Copper Layer (Cu = 1 mm): Win32 Graphics (left), 3D Graphics (right)

Step 6: Create a New Input File for a 0.5 mm Copper Layer

6-1. Save your input file as Input10-2-05mm.txt

6-2. Update the geometry for the thinnest copper layer (Figure 10-2-7).

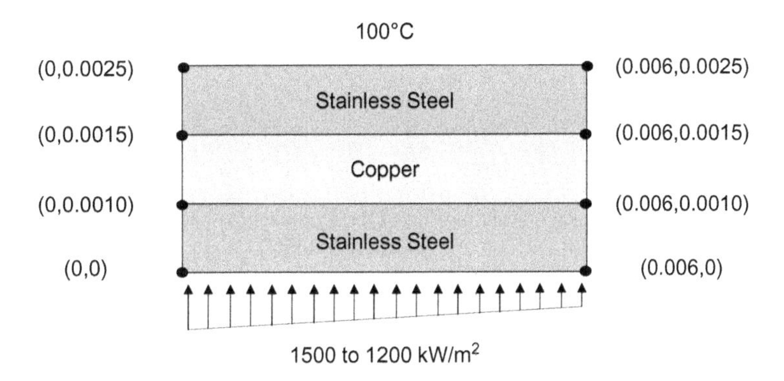

Figure 10-2-7 Updated Model Geometry (Cu = 0.5 mm).

- Third line: Change the height in the **BLC4** command from 0.003 to 0.0025.
- Sixth and seventh lines: Change the heights of KPs 7 and 8 from 0.002 to 0.0015.
- Sixteenth line: Change the *y* coordinates of the second **ASEL** command from 0.002 and 0.003 to 0.0015 and 0.0025.

6-3. Update the boundary conditions for the thinner copper layer

- Change the *y* coordinate for first **LSEL** below the /**SOL** command from 0.003 to 0.0025.

6-4. Comment the new commands

Step 7: Run the Exercise10-2-05mm.txt Input File

7-1. Save your input file

7-2. Clear the database

7-3. Run the input file to create the new model

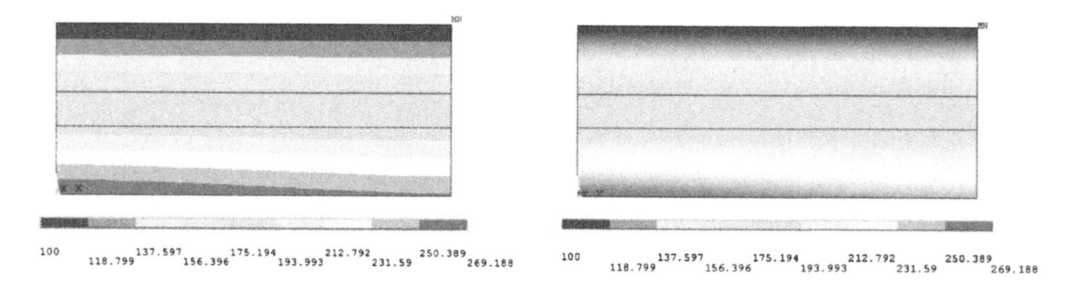

Figure 10-2-8 Nodal Temperature Contour Plot (Cu = 0.5 mm): Win32 Graphics (left), 3D Graphics (right).

Again, reducing the thickness of the copper layer has lowered the thermal resistance of the pan and therefore lowered the maximum temperature from 270.52°C to 269.19°C (Figure 10-2-8). However, the second temperature plot clearly shows that the copper layer is no longer able to evenly distribute the heat across the pan (Figure 10-2-9). Thus, the cost savings in this case may not be worth the reduction in performance.

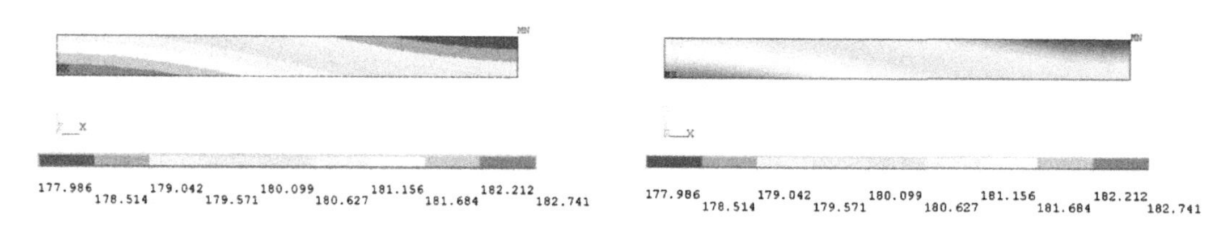

Figure 10-2-9 Nodal Temperature Contour Plot of the Copper Layer (Cu = 0.5 mm): Win32 Graphics (left), 3D Graphics (right).

Close the Program

- Utility Menu > File > Exit . . .
- Choose "Quit - No save!"
- Click OK

Sample Input File for Input10-2-2mm

The first input file that you created (Input10-2-2mm) should be similar to the one below. The comments from the sample input file have been removed to encourage you to use the ANSYS Command Reference in this exercise.

```
/BATCH
/PREP7
BLC4,0,0,0.006,0.004
K,,0,0.001
K,,0.006,0.001
K,,0,0.003
K,,0.006,0.003
LSTR,5,6
LSTR,7,8
ASBL,1,5
ASBL,3,6
ET,1,PLANE55
MP,KXX,1,17
MP,KXX,2,372
ASEL,S,LOC,Y,0,0.001
ASEL,A,LOC,Y,0.003,0.004
AATT,1,,1,0
ASEL,INVE
AATT,2,,1,0
ESIZE,0.0001
MSHKEY,1
ALLSEL
AMESH,ALL
FINISH

/SOL
LSEL,S,LOC,Y,0.004
DL,ALL,,TEMP,100,0
LSEL,S,LOC,Y,0
SFL,ALL,HFLUX,1500e3,1200e3
ALLSEL,ALL
SOLVE
FINISH

/POST1
PLNSOL, TEMP,, 0
/WAIT,5
ESEL,S,MAT,,2
NSLE,S
PLNSOL, TEMP,, 0
/WAIT,5
FINISH

SAVE
!/EXIT
```

Sample Input File for Input10-2-1mm

The second input file that you created (Input10-2-1mm) should be similar to the one below. The comments from the sample input file have been removed to encourage you to use the ANSYS Command Reference in this exercise.

```
/BATCH
/PREP7
BLC4,0,0,0.006,0.003
K,,0,0.001
K,,0.006,0.001
K,,0,0.002
K,,0.006,0.002
LSTR,5,6
LSTR,7,8
ASBL,1,5
ASBL,3,6
ET,1,PLANE55
MP,KXX,1,17
MP,KXX,2,372
ASEL,S,LOC,Y,0,0.001
ASEL,A,LOC,Y,0.002,0.003
AATT,1,,1,0
ASEL,INVE
AATT,2,,1,0
ESIZE,0.0001
MSHKEY,1
ALLSEL
AMESH,ALL
FINISH

/SOL
LSEL,S,LOC,Y,0.003
DL,ALL,,TEMP,100,0
LSEL,S,LOC,Y,0
SFL,ALL,HFLUX,1500e3,1200e3
ALLSEL,ALL
SOLVE
FINISH

/POST1
PLNSOL, TEMP,, 0
/WAIT,5
ESEL,S,MAT,,2
NSLE,S
PLNSOL, TEMP,, 0
/WAIT,5
FINISH

SAVE
!/EXIT
```

Sample Input File for Input10-2-05mm

The third input file that you created (Input10-2-05mm) should be similar to the one below. The comments from the sample input file have been removed to encourage you to use the ANSYS Command Reference in this exercise.

```
/BATCH
/PREP7
BLC4,0,0,0.006,0.0025
K,,0,0.001
K,,0.006,0.001
K,,0,0.0015
K,,0.006,0.0015
LSTR,5,6
LSTR,7,8
ASBL,1,5
ASBL,3,6
ET,1,PLANE55
MP,KXX,1,17
MP,KXX,2,372
ASEL,S,LOC,Y,0,0.001
ASEL,A,LOC,Y,0.0015,0.0025
AATT,1,,1,0
ASEL,INVE
AATT,2,,1,0
ESIZE,0.0001
MSHKEY,1
ALLSEL
AMESH,ALL
FINISH

/SOL
LSEL,S,LOC,Y,0.0025
DL,ALL,,TEMP,100,0
LSEL,S,LOC,Y,0
SFL,ALL,HFLUX,1500e3,1200e3
ALLSEL,ALL
SOLVE
FINISH

/POST1
PLNSOL,TEMP,, 0
/WAIT,5
ESEL,S,MAT,,2
NSLE,S
PLNSOL,TEMP,, 0
/WAIT,5
FINISH

SAVE
!/EXIT
```

Exercise 10-3

Using the Direct Method to Create a Batch File for Steady-State Conduction Through a Cladded Plate With Varying Surface Temperatures

Overview

In this exercise, you will use the Direct Method to modify the first input file from exercise 10-2 (Input10-2-2mm) to create a batch file with cooking surface temperatures of 100°C, 150°C, 200°C, 250°C, and 300°C (Figure 10-3-1). Finally, you will run the batch file in batch mode and examine the output.

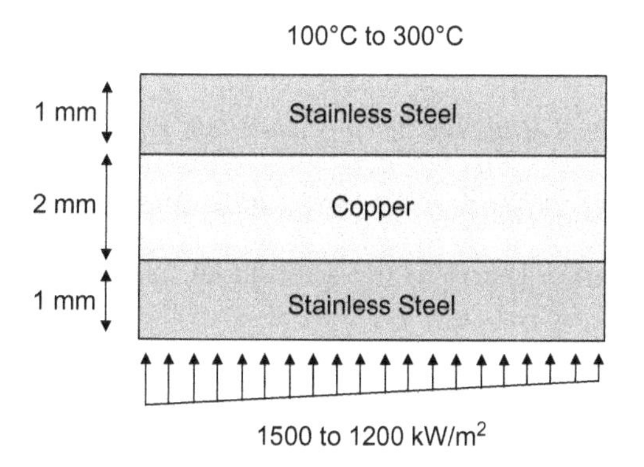

Figure 10-3-1 Schematic of the Cladded Plate.

Model Attributes

Geometry for the Plate

- Copper layer—2 mm

Loads

- Linearly varying heat flux of 1500−1200 kW/m^2 on the plate bottom

Constraints

- Surface temperatures of 100°C, 150°C, 200°C, 250°C, and 300°C on the top of the plate

ANSYS Mechanical APDL for Finite Element Analysis.
DOI: http://dx.doi.org/10.1016/B978-0-12-812981-4.00031-9

File Management

Create a new folder in your "Intro-to-ANSYS" folder for Exercise10-3

Create a draft of the Input10-3-2mm input file

- Copy the input file Input10-2-2mm.txt in the Exercise10-2 folder and paste it in the Exercise10-3 folder
- Create a second copy of this file in the Exercise10-3 folder and rename it Input10-3-2mm.txt

Open the Input10-3-2mm text file

Step 1: Update the Boundary Conditions

1-1. Identify the branch point in the input file

*The branch point is where the new analysis begins to differ from the old analysis. This analysis is the same until the second **FINISH** command (the one used to exit Solution).*

1-2. Place four blank lines between the SOLVE command and the FINISH command to create space to work

1-3. Copy the first LSEL command (just after the /SOL command) and paste it after the first SOLVE command, leaving one blank line between the SOLVE command and the LSEL command

1-4. Below the new LSEL command, add the command: `DLDELE,ALL,TEMP`

This command deletes all temperature constraints on the selected lines.

1-5. Copy the DL command and paste it after the DLDELE command

1-6. Change the temperature in the second DL command to 150

*The new **DL** command should replace the previous specification. Therefore, the **DLDELE** command added in Step 1-4 is not strictly necessary. However, it is good practice to remove the previous specification instead of relying on the program to replace it.*

1-7. Copy the ALLSEL and SOLVE commands after the SFL command and paste them after the second DL command

Your Solution commands should now look like this:

```
/SOL
LSEL,S,LOC,Y,0.004
DL,ALL,,TEMP,100,0
LSEL,S,LOC,Y,0
SFL,ALL,HFLUX,1500e3,1200e3
ALLSEL,ALL
SOLVE

LSEL,S,LOC,Y,0.004
DLDELE,ALL,TEMP
DL,ALL,,TEMP,150,0
ALLSEL,ALL
SOLVE

FINISH
```

1-8. Copy the second Solution command block and paste it below the second SOLVE 3 times, leaving one line blank in between each command block

1-9. Change the temperature in the third DL command to 200

1-10. Change the temperature in the fourth DL command to 250

1-11. Change the temperature in the fifth DL command to 300

1-12. Add a SAVE command after the last SOLVE command

Your Solution commands should now look like this:

```
/SOL
LSEL,S,LOC,Y,0.004
DL,ALL,,TEMP,100,0
LSEL,S,LOC,Y,0
SFL,ALL,HFLUX,1500e3,1200e3
ALLSEL,ALL
SOLVE

LSEL,S,LOC,Y,0.004
DLDELE,ALL,TEMP
DL,ALL,,TEMP,150,0
ALLSEL,ALL
SOLVE

LSEL,S,LOC,Y,0.004
DLDELE,ALL,TEMP
DL,ALL,,TEMP,200,0
ALLSEL,ALL
SOLVE

LSEL,S,LOC,Y,0.004
DLDELE,ALL,TEMP
DL,ALL,,TEMP,250,0
ALLSEL,ALL
SOLVE

LSEL,S,LOC,Y,0.004
DLDELE,ALL,TEMP
DL,ALL,,TEMP,300,0
ALLSEL,ALL
SOLVE

SAVE
FINISH
```

1-13. Save your batch file

Step 2: Test the Preprocessing and Solution Commands

2-1. Open the ANSYS Mechanical APDL Product Launcher

2-2. Change the Working Directory to the Exercise10-3 folder

2-3. Change the Jobname to Exercise10-3

2-4. Start ANSYS

2-5. Run the input file

- Utility Menu > File > Read Input from ...
- Choose Input10-3-2mm.txt from your working directory
- Click OK

The input file should run without any warnings or errors.

2-6. Open the error file to verify that there were no warnings or errors issued during the run (Figure 10-3-2)

› Intro-to- ANSYS › Exercise 10-3 Search Exer...

Name	Date modified	Type	Size
Exercise10-3.BCS		BCS File	2 KB
Exercise10-3		ANSYS v172 .db File	5,952 KB
Exercise10-3		ANSYS v172 .dbb ...	5,952 KB
Exercise10-3		ERR File	1 KB
Exercise10-3.esav		ESAV File	128 KB
Exercise10-3.full		FULL File	448 KB
Exercise10-3.lock		LOCK File	0 KB
Exercise10-3		Text Document	1 KB
Exercise10-3.mntr		MNTR File	2 KB
Exercise10-3.page		PAGE File	0 KB
Exercise10-3.rth		RTH File	4,800 KB
Exercise10-3.stat		STAT File	1 KB
Input10-2-2mm		Text Document	1 KB
Input10-3-2mm		Text Document	1 KB
menust.tmp		TMP File	1 KB

Figure 10-3-2 Working Directory with the Error File Highlighted.

The error file should only contain a time and release stamp (Figure 10-3-3).

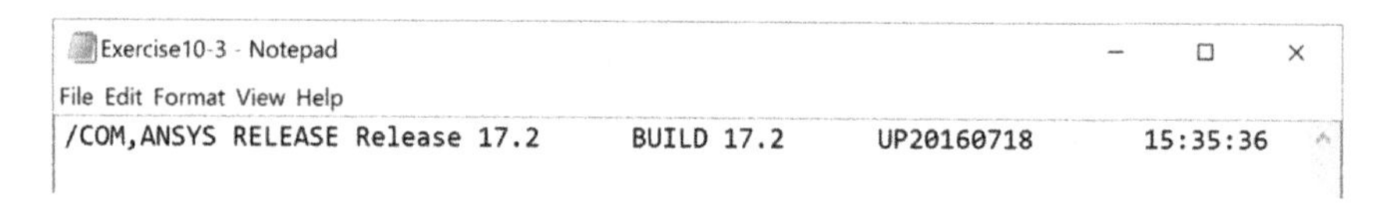

Figure 10-3-3 Contents of the Error File (No Warnings or Errors).

2-7. Close the error file

2-8. Close ANSYS

2-9. Reset the working directory

- Delete all files in your working directory except for the two input files (Input10-2-2mm.txt and Input10-3-2mm.txt) (Figure 10-3-4)

You will not be able to delete the files in your working directory until the interactive ANSYS session has been closed.

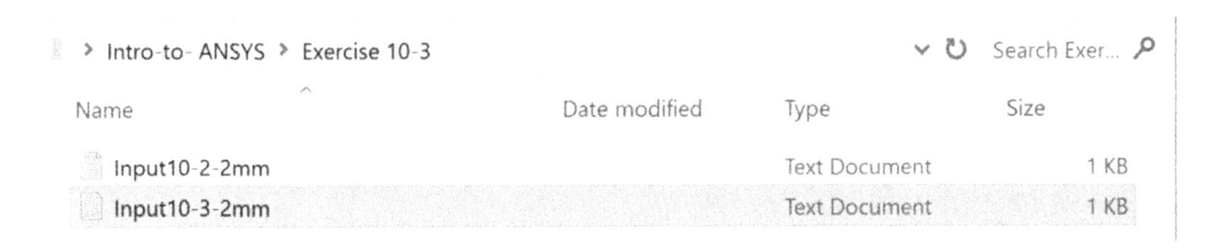

Figure 10-3-4 Reset Working Directory.

Step 3: Prepare the Existing Postprocessing Commands for Batch Mode

3-1. Delete the /WAIT commands

These commands will be executed in batch mode but will have no effect other than to increase the run time of the batch file.

3-2. Delete the final SAVE command

3-3. Add a blank line before the last FINISH command and remove the blank line after it to group it with the /EXIT command

3-4. Uncomment the /EXIT command

- Remove the exclamation point (!) in front of the **/EXIT** command at the end of the input file

3-5. Modify the /EXIT command to exit the program without saving

- Update the command to read `/EXIT,NOSAVE`

Your postprocessing commands should now look like this:

```
/POST1
PLNSOL, TEMP,, 0
ESEL,S,MAT,,2
NSLE,S
PLNSOL, TEMP,, 0

FINISH
/EXIT,NOSAVE
```

3-6. Save your batch file

Step 4: Modify the Postprocessing Commands for the First Plot From the First Load Step

*The postprocessing commands in the input file for exercise 10-2 were relatively simple. In this exercise, we will need four additional commands. The **/SHOW** command will redirect the graphical output for each plot to a separate file. The **SET** command will allow us to specify the results set on the results file to postprocess. The **/FILNAME** command will allow us to change the jobname and therefore to specify the file names for the graphical output. Finally, the **FILE** command will allow us to use a results file with a name that differs from the jobname.*

*However, the nature of these commands makes the structure of the postprocessing commands for exercise 10-3 more complicated. The **/SHOW** command directs plots to jobname00N.ext, where the file extension is determined by the first argument of the **/SHOW** command and where N starts at zero and increments by one for each plot in the working directory. For example, the first plot after issuing the command **/SHOW,JPEG** will be saved as jobname000.jpg. The second will be saved as jobname001.jpg. The third will be saved as jobname002.jpg, and so on.*

We could generate the graphics file names automatically and then go back and update the file names later. However, this is not a robust solution for batch files that will be used by other people or over long periods of time.

*The only way to save the plots with different file names is to change the jobname. However, the **/FILNAME** command is a Begin Level command. Each time you wish to change the jobname, you must exit the General Postprocessor, change the jobname, and then re-enter the Postprocessor.*

*Because we are changing the jobname, we must issue the **FILE** command to specify the results file to use each time we need to read in a new set of results (otherwise, we must change the jobname back to match the name of the results file).*

The combination of these restrictions leads to the more complicated postprocessing in this batch file.

4-1. Redirect graphical output to file

- Add the following command after the **FINISH** command to exit Solution and before the **/POST1** command, leaving one blank line above and below: `/SHOW,JPEG,REV`

*The first argument of the **/SHOW** command can specify many types of behavior. In this exercise, it is used to specify the graphics file type and extension. We have chosen to create JPEG files. However, you can also create TIFFs, PNGs, postscript files (PSCRs), and more.*

*There is also a <filename> option for the first argument of the **/SHOW** command. This writes the graphics output to a file with the specified name in an ANSYS-neutral format that is intended to be read by the ANSYS DISPLAY program. Processing the data though the DISPLAY program is more complicated than the procedure used in this exercise. For more information on the DISPLAY program, see Section 17.3.1 of the ANSYS Mechanical APDL Basic Analysis Guide.*

*The "REV" for the second argument of the **/SHOW** command specifies that the output should be in reverse video (i.e. the plots should have a white background).*

4-2. Read the results from the first load step into the database

- Add the following command after the **/POST1** command: `SET,1`

4-3. Change the jobname to reflect the file name for the first plot and then re-enter the General Postprocessor

- Add the following commands after the **SET** command:

```
FINISH
/FILNAME,TempAll2mm100C,0
/POST1
```

*The **FINISH** command exits the General Postprocessor and returns us to the Begin Level so the **/FILNAME** command can be issued. The second field of the **/FILNAME** command indicates that the first plot will show the nodal temperature distribution for the full geometry in the 2 mm model with a surface temperature of 100°C. The "0" in third field of the **/FILNAME** command instructs the program to use the existing error and log files. We then return to the General Postprocessor (POST1) to generate the plot.*

4-4. Add a blank line before the ESEL command

Your postprocessing commands should now look like the following. Comments for these commands have been added for clarity.

```
/SHOW,JPEG,REV                ! Direct graphical output to JPEG files

/POST1                        ! Enter the General Postprocessor
SET,1                         ! Load the results for load step #1
FINISH                        ! Exit the General Postprocessor
/FILNAME,TempAll2mm100C,0     ! Change the jobname
/POST1                        ! Enter the General Postprocessor
PLNSOL, TEMP,, 0              ! Plot the nodal temperature for the model

ESEL,S,MAT,,2                 ! Select elements associated with material #2
NSLE,S                        ! Select nodes attached to selected elements
PLNSOL, TEMP,, 0              ! Plot nodal temperature for selected nodes

FINISH                        ! Exit the General Postprocessor
/EXIT,NOSAVE                  ! Exit the program without saving
```

4-5. Save your batch file

Step 5: Modify the Postprocessing Commands for the Second Plot from the First Load Step

5-1. Change the jobname to reflect the file name for the second plot

- Add the following commands before the **ESEL** command:

```
FINISH
/FILNAME,TempCopper2mm100C,0
```

5-2. Re-enter the General Postprocessor

- Add the following command after the new **/FILNAME** command and before the **ESEL** command: `/POST1`

5-3. Remove the blank line between the command blocks for the first and second plots if it is still present

5-4. Save your batch file

Your postprocessing commands should now look like this:

```
/SHOW,JPEG,REV                      ! Direct graphical output to JPEG files

/POST1                              ! Enter the General Postprocessor
SET,1                               ! Load the results for load step #1
FINISH                              ! Exit the General Postprocessor
/FILNAME,TempAll2mm100C,0           ! Change the jobname (define the plot file name)
/POST1                              ! Enter the General Postprocessor
PLNSOL, TEMP,, 0                    ! Plot the nodal temperature for the model
FINISH                              ! Exit the General Postprocessor
/FILNAME,TempCopper2mm100C,0        ! Change the jobname (define the plot file name)
/POST1                              ! Enter the General Postprocessor
ESEL,S,MAT,,2                       ! Select elements associated with material #2
NSLE,S                              ! Select nodes attached to selected elements
PLNSOL, TEMP,, 0                    ! Plot nodal temperature for selected nodes

FINISH                              ! Exit the General Postprocessor
/EXIT,NOSAVE                        ! Exit the program without saving
```

Step 6: Prepare the Postprocessing Commands for the Second Load Step

6-1. Copy all postprocessing commands from the first postprocessing command block (from the /POST1 to the last PLNSOL)

This includes everything between the ***/SHOW*** *and the final* ***FINISH****.*

6-2. Paste these commands once, leaving one blank line in between the first and second command blocks

6-3. Remove the first /POST1 command in the second command block

There is no harm in leaving this command in the batch file. However, is not needed because the program is already operating in the General Postprocessor.

6-4. Reset the selection status of the model

- Add the following command to the beginning of the second block of commands: `ALLSEL,ALL`

6-5. Specify the results file to use

- Add the following command to the beginning of the second block of commands after the new **ALLSEL** command: `FILE,Exercise10-3-2mm,rth`

Your second block of postprocessing commands should now look like this:

```
ALLSEL,ALL                        ! Select everything in the model
FILE,Exercise10-3-2mm,rth         ! Specify the results file to use
SET,1                             ! Load the results for load step #1
FINISH                            ! Exit the General Postprocessor
/FILNAME,TempAll2mm100C,0         ! Change the jobname (define the plot file name)
/POST1                            ! Enter the General Postprocessor
PLNSOL, TEMP,, 0                  ! Plot the nodal temperature for the model
FINISH                            ! Exit the General Postprocessor
/FILNAME,TempCopper2mm100C,0      ! Change the jobname (define the plot file name)
/POST1                            ! Enter the General Postprocessor
ESEL,S,MAT,,2                     ! Select elements associated with material #2
NSLE,S                            ! Select nodes attached to selected elements
PLNSOL, TEMP,, 0                  ! Plot nodal temperature for selected nodes
```

Step 7: Postprocessing Commands for the Last Four Load Steps

7-1. Copy all of the postprocessing commands from the second postprocessing block

7-2. Paste these commands three times (once per additional load step), leaving one blank line between each command block

You should now have identical postprocessing command blocks for load steps 2–5.

7-3. Update the SET commands for each of the last four load steps

- In the second command block, update the SET command to say: SET,2
- In the third command block, update the SET command to say: SET,3
- In the fourth command block, update the SET command to say: SET,4
- In the fifth command block, update the SET command to say: SET,5

7-4. Update the first plot title for each of the last four load steps

- In the second command block, update the first jobname to read: TempAll2mm150
- In the third command block, update the first jobname to read: TempAll2mm200
- In the fourth command block, update the first jobname to read: TempAll2mm250
- In the fifth command block, update the first jobname to read: TempAll2mm300

7-5. Update the second plot title for each of the last four load steps

- In the second command block, update the second jobname to read: TempCopper2mm150
- In the third command block, update the second jobname to read: TempCopper2mm200
- In the fourth command block, update the second jobname to read: TempCopper2mm250
- In the fifth command block, update the second jobname to read: TempCopper2mm300

Your second postprocessing command block should now look like this:

```
ALLSEL,ALL                       ! Select everything in the model
FILE,Exercise10-3-2mm,rth        ! Specify the results file to use
SET,2                            ! Load the results for load step #2
FINISH                           ! Exit the General Postprocessor
/FILNAME,TempAll2mm150C,0        ! Change the jobname (define the plot file name)
/POST1                           ! Enter the General Postprocessor
PLNSOL, TEMP,, 0                 ! Plot the nodal temperature for the model
FINISH                           ! Exit the General Postprocessor
/FILNAME,TempCopper2mm150C,0     ! Change the jobname (define the plot file name)
/POST1                           ! Enter the General Postprocessor
ESEL,S,MAT,,2                    ! Select elements associated with material #2
NSLE,S                           ! Select nodes attached to selected elements
PLNSOL, TEMP,, 0                 ! Plot nodal temperature for selected nodes
```

7-6. Save your batch file

Step 8: Run Input10-3-2mm in Batch Mode

There should only be two files in your working directory at the moment: the input file from exercise 10-2 (Input10-2-2mm) and the new input file for exercise 10-3 (Input10-3-2mm).

8-1. Open the Mechanical APDL Product Launcher

8-2. In the "Simulation Environment" drop down menu, choose "ANSYS Batch" (Figure 10-3-5)

8-3. Change the Working Directory to the Exercise10-3 folder

8-4. Change the Jobname to Exercise10-3-2mm

8-5. Select the input file to use

- Click the Browse button
- Click on the drop down menu that says "ANSYS dat Files (*.dat) and choose "All files (*.*)
- Choose Input10-3-2mm
- Click Open

8-6. Set the Output file name to Exercise10-3-2mm-Output

8-7. Click Run to start the batch run

You should see a pop-up dialog box informing you that the "Requested batch run was submitted." Click OK to close this dialog box.

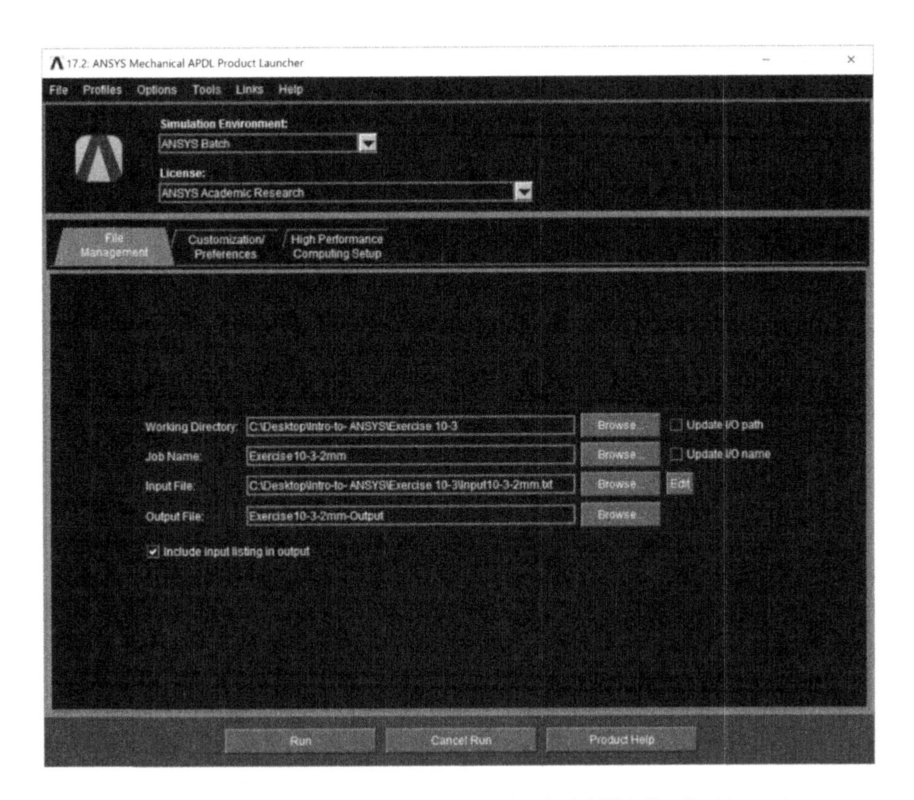

Figure 10-3-5 File Management in the Mechanical APDL Product Launcher.

8-8. Examine the files in your working directory

After the batch run is complete, you should see all of the normal files associated with an ANSYS run. For example, you should see the error file, the output file, and the results file. The batch file contained a ***SAVE*** *command; therefore, there is a database (.db) file in the working directory. In addition, there are 10 image (JPG) files—one per plot requested in the input file (Figure 10-3-6).*

Note that all of the image file names have a trailing "000." This is because the first graphics file name is always jobname000.ext.

> Intro-to- ANSYS > Exercise 10-3

Name	Date modi...	Type	Size
Exercise10-3-2mm.BCS		BCS File	2 KB
Exercise10-3-2mm		ANSYS v172 .db File	5,952 KB
Exercise10-3-2mm		ERR File	2 KB
Exercise10-3-2mm.esav		ESAV File	128 KB
Exercise10-3-2mm.full		FULL File	448 KB
Exercise10-3-2mm		Text Document	1 KB
Exercise10-3-2mm.mntr		MNTR File	2 KB
Exercise10-3-2mm.rth		RTH File	4,800 KB
Exercise10-3-2mm.stat		STAT File	1 KB
Exercise10-3-2mm-Output		File	60 KB
Input10-2-2mm		Text Document	1 KB
Input10-3-2mm		Text Document	3 KB
TempAll2mm100C000		JPG File	81 KB
TempAll2mm150C000		JPG File	81 KB
TempAll2mm200C000		JPG File	81 KB
TempAll2mm250C000		JPG File	81 KB
TempAll2mm300C000		JPG File	82 KB
TempCopper2mm100C000		JPG File	80 KB
TempCopper2mm150C000		JPG File	80 KB
TempCopper2mm200C000		JPG File	80 KB
TempCopper2mm250C000		JPG File	80 KB
TempCopper2mm300C000		JPG File	80 KB

Figure 10-3-6 Files in the Working Directory After Running the Batch File.

8-9. Open and compare the image files for the first and last load steps

In the first image file (TempAll2mm100C000.jpg) (Figure 10-3-7, left)., we can see that the plot of the nodal temperature for the entire model with the 100°C boundary condition is the same as in exercise 10-2. However, the legend is located to the right of the image instead of under it.

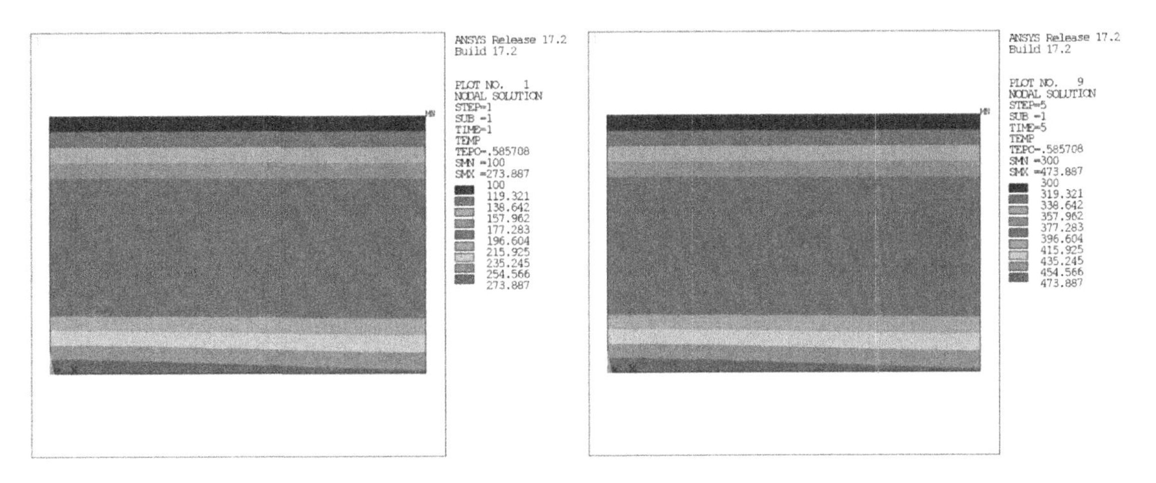

Figure 10-3-7 Nodal Temperature Plots of the Plate with Surface Temperatures of 100°C (left) and 300°C (right) (Win32 Graphics).

The plot for the full model with the 300°C boundary condition (TempAll2mm300C000.jpg) (Figure 10-3-7, right) has the same shape as the plot with the 100°C boundary condition. However, all values in the model have shifted by 200°C.

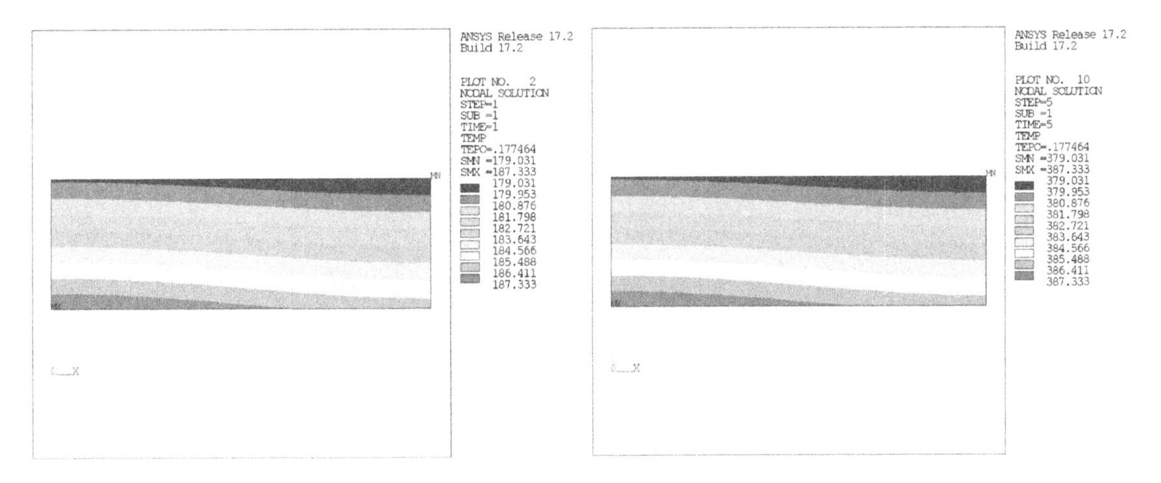

Figure 10-3-8 Nodal Temperature Plots of the Copper Layer with Surface Temperatures of 100°C (left) and 300°C (right) (Win32 Graphics).

The plot for the copper layer with the 100°C boundary condition (TempCopper2mm100C000.jpg) (Figure 10-3-8, left) is also the same as in exercise 10-2. The plot for the copper layer with the 300°C boundary condition (TempCopper2mm300C000.jpg) (Figure 10-3-8, right) again shows identical relative behavior and a shift in all values of 200°C.

Step 9: Compare and Verify the Results

The change in temperature across the cross section of the pan is related to the heat flux by Fourier's law: $q = k\ (\Delta T/L)$. Because the heat flux is the same for the five solutions that were calculated, we expected the change in temperature to be the same as well. This is confirmed by the results in Figures 10-3-7 and 10-3-8. Therefore, all five models are in perfect agreement with the theory.

Sample Input File With Manual Graphics File Names

```
/BATCH                              ! Enable the file to run in batch mode

/PREP7                              ! Enter the Preprocessor
BLC4,0,0,0.006,0.004                ! Create at 0.006x0.004 rectangle
K,,0,0.001                          ! Create keypoint at (0,0.001)
K,,0.006,0.001                      ! Create keypoint at (0.006,0.001)
K,,0,0.003                          ! Create keypoint at (0,0.003)
K,,0.006,0.003                      ! Create keypoint at (0.006,0.003)
LSTR,5,6                            ! Create line between keypoints
LSTR,7,8                            ! Create line between keypoints
ASBL,1,5                            ! Divide area 1 by line 5
ASBL,3,6                            ! Divide area 3 by line 6
ET,1,PLANE55                        ! Use Plane 55 elements
MP,KXX,1,17                         ! Set thermal conductivity of material #1
MP,KXX,2,372                        ! Set thermal conductivity of material #2
ASEL,S,LOC,Y,0,0.001                ! Select the bottom area by location
ASEL,A,LOC,Y,0.003,0.004            ! Select the top area by location
AATT,1,,1,0                         ! Use material 1 for these areas
ASEL,INVE                           ! Invert the selected set of areas
AATT,2,,1,0                         ! Use material 2 for this area
ESIZE,0.0001                        ! Use an element mesh size of 0.0001
MSHKEY,1                            ! Map mesh the model
ALLSEL                              ! Select everything
AMESH,ALL                           ! Mesh all areas
FINISH                              ! Exit the Preprocessor

/SOL                                ! Enter the Solution processor
LSEL,S,LOC,Y,0.004                  ! Select the line at y=0.004
DL,ALL,,TEMP,100,0                  ! Set temperature of selected line to 100
LSEL,S,LOC,Y,0                      ! Select the line at y=0
SFL,ALL,HFLUX,1500e3,1200e3         ! Apply a linearly varying heat flux to the line
ALLSEL,ALL                          ! Select everything
SOLVE                               ! Solve the model

LSEL,S,LOC,Y,0.004                  ! Select the line at y=0.004
DLDELE,ALL,TEMP                     ! Delete all temperature BCs on that line
DL,ALL,,TEMP,150,0                  ! Set temperature of selected line to 150
ALLSEL,ALL                          ! Select everything
SOLVE                               ! Solve the model

LSEL,S,LOC,Y,0.004                  ! Select the line at y=0.004
DLDELE,ALL,TEMP                     ! Delete all temperature BCs on that line
DL,ALL,,TEMP,200,0                  ! Set temperature of selected line to 200
ALLSEL,ALL                          ! Select everything
SOLVE                               ! Solve the model

LSEL,S,LOC,Y,0.004                  ! Select the line at y=0.004
DLDELE,ALL,TEMP                     ! Delete all temperature BCs on that line
DL,ALL,,TEMP,250,0                  ! Set temperature of selected line to 250
ALLSEL,ALL                          ! Select everything
SOLVE                               ! Solve the model
```

```
LSEL,S,LOC,Y,0.004              ! Select the line at y=0.004
DLDELE,ALL,TEMP                 ! Delete all temperature BCs on that line
DL,ALL,,TEMP,300,0              ! Set temperature of selected line to 300
ALLSEL,ALL                      ! Select everything
SOLVE                           ! Solve the model

SAVE                            ! Save the database
FINISH                          ! Exit the Solution processor

/SHOW,JPEG,REV                  ! Direct graphical output to JPEG files

/POST1                          ! Enter the General Postprocessor
SET,1                           ! Load the results for load step #1
FINISH                          ! Exit the General Postprocessor
/FILNAME,TempAll2mm100C,0       ! Change the jobname (define the plot file name)
/POST1                          ! Enter the General Postprocessor
PLNSOL, TEMP,, 0                ! Plot the nodal temperature for the model
FINISH                          ! Exit the General Postprocessor
/FILNAME,TempCopper2mm100C,0    ! Change the jobname (define the plot file name)
/POST1                          ! Enter the General Postprocessor
ESEL,S,MAT,,2                   ! Select elements associated with material #2
NSLE,S                          ! Select nodes attached to selected elements
PLNSOL, TEMP,, 0                ! Plot nodal temperature for selected nodes

ALLSEL,ALL                      ! Select everything in the model
FILE,Exercise10-3-2mm,rth       ! Specify the results file to use
SET,2                           ! Load the results for load step #2
FINISH                          ! Exit the General Postprocessor
/FILNAME,TempAll2mm150C,0       ! Change the jobname (define the plot file name)
/POST1                          ! Enter the General Postprocessor
PLNSOL, TEMP,, 0                ! Plot the nodal temperature for the model
FINISH                          ! Exit the General Postprocessor
/FILNAME,TempCopper2mm150C,0    ! Change the jobname (define the plot file name)
/POST1                          ! Enter the General Postprocessor
ESEL,S,MAT,,2                   ! Select elements associated with material #2
NSLE,S                          ! Select nodes attached to selected elements
PLNSOL, TEMP,, 0                ! Plot nodal temperature for selected nodes

ALLSEL,ALL                      ! Select everything in the model
FILE,Exercise10-3-2mm,rth       ! Specify the results file to use
SET,3                           ! Load the results for load step #3
FINISH                          ! Exit the General Postprocessor
/FILNAME,TempAll2mm200C,0       ! Change the jobname (define the plot file name)
/POST1                          ! Enter the General Postprocessor
PLNSOL, TEMP,, 0                ! Plot the nodal temperature for the model
FINISH                          ! Exit the General Postprocessor
/FILNAME,TempCopper2mm200C,0    ! Change the jobname (define the plot file name)
/POST1                          ! Enter the General Postprocessor
ESEL,S,MAT,,2                   ! Select elements associated with material #2
NSLE,S                          ! Select nodes attached to selected elements
PLNSOL, TEMP,, 0                ! Plot nodal temperature for selected nodes
ALLSEL,ALL                      ! Select everything in the model
FILE,Exercise10-3-2mm,rth       ! Specify the results file to use
SET,4                           ! Load the results for load step #4
```

```
FINISH                              ! Exit the General Postprocessor
/FILNAME,TempAll2mm250C,0           ! Change the jobname (define the plot file name)
/POST1                              ! Enter the General Postprocessor
PLNSOL, TEMP,, 0                    ! Plot the nodal temperature for the model
FINISH                              ! Exit the General Postprocessor
/FILNAME,TempCopper2mm250C,0 ! Change the jobname (define the plot file name)
/POST1                              ! Enter the General Postprocessor
ESEL,S,MAT,,2                       ! Select elements associated with material #2
NSLE,S                              ! Select nodes attached to selected elements
PLNSOL, TEMP,, 0                    ! Plot nodal temperature for selected nodes

ALLSEL,ALL                          ! Select everything in the model
FILE,Exercise10-3-2mm,rth           ! Specify the results file to use
SET,5                               ! Load the results for load step #5
FINISH                              ! Exit the General Postprocessor
/FILNAME,TempAll2mm300C,0           ! Change the jobname (define the plot file name)
/POST1                              ! Enter the General Postprocessor
PLNSOL, TEMP,, 0                    ! Plot the nodal temperature for the model
FINISH                              ! Exit the General Postprocessor
/FILNAME,TempCopper2mm300C,0 ! Change the jobname (define the plot file name)
/POST1                              ! Enter the General Postprocessor
ESEL,S,MAT,,2                       ! Select elements associated with material #2
NSLE,S                              ! Select nodes attached to selected elements
PLNSOL, TEMP,, 0                    ! Plot nodal temperature for selected nodes

FINISH                              ! Exit the General Postprocessor
/EXIT,NOSAVE                        ! Exit the program without saving
```

Sample Input File with Automatic Graphics File Names

Below is a sample input file that automatically generates the graphics file names. It is a much simpler input file. This will output plots named Exercise10-3-2mm000.jpg through Exercise10-3-2mm009.jpg. The even numbered files will be for the full model. The odd numbered files will be for the copper layer.

```
/BATCH                              ! Enable the file to run in batch mode

/PREP7                              ! Enter the Preprocessor
BLC4,0,0,0.006,0.004                ! Create at 0.006x0.004 rectangle
K,,0,0.001                          ! Create keypoint at (0,0.001)
K,,0.006,0.001                      ! Create keypoint at (0.006,0.001)
K,,0,0.003                          ! Create keypoint at (0,0.003)
K,,0.006,0.003                      ! Create keypoint at (0.006,0.003)
LSTR,5,6                            ! Create line between keypoints
LSTR,7,8                            ! Create line between keypoints
ASBL,1,5                            ! Divide Area 1 by Line 5
ASBL,3,6                            ! Divide Area 3 by Line 6
ET,1,PLANE55                        ! Use Plane 55 elements
MP,KXX,1,17                         ! Set thermal conductivity of material #1
MP,KXX,2,372                        ! Set thermal conductivity of material #2
ASEL,S,LOC,Y,0,0.001                ! Select the bottom area by location
ASEL,A,LOC,Y,0.003,0.004            ! Select the top area by location
```

```
AATT,1,,1,0                      ! Use material 1 for these areas
ASEL,INVE                        ! Invert the selected set of areas
AATT,2,,1,0                      ! Use material 2 for this area
ESIZE,0.0001                     ! Use an element mesh size of 0.0001
MSHKEY,1                         ! Map mesh the model
ALLSEL                           ! Select everything
AMESH,ALL                        ! Mesh all areas
FINISH                           ! Exit the Preprocessor

/SOL                             ! Enter the Solution processor
LSEL,S,LOC,Y,0.004               ! Select the line at y=0.004
DL,ALL,,TEMP,100,0               ! Set temperature of selected line to 100
LSEL,S,LOC,Y,0                   ! Select the line at y=;0
SFL,ALL,HFLUX,1500e3,1200e3      ! Apply a linearly varying heat flux to the line
ALLSEL,ALL                       ! Select everything
SOLVE                            ! Solve the model

LSEL,S,LOC,Y,0.004               ! Select the line at y=0.004
DLDELE,ALL,TEMP                  ! Delete all temperature BCs on that line
DL,ALL,,TEMP,150,0               ! Set temperature of selected line to 150
ALLSEL,ALL                       ! Select everything
SOLVE                            ! Solve the model

LSEL,S,LOC,Y,0.004               ! Select the line at y=0.004
DLDELE,ALL,TEMP                  ! Delete all temperature BCs on that line
DL,ALL,,TEMP,200,0               ! Set temperature of selected line to 200
ALLSEL,ALL                       ! Select everything
SOLVE                            ! Solve the model
LSEL,S,LOC,Y,0.004               ! Select the line at y=0.004
DLDELE,ALL,TEMP                  ! Delete all temperature BCs on that line
DL,ALL,,TEMP,250,0               ! Set temperature of selected line to 250
ALLSEL,ALL                       ! Select everything
SOLVE                            ! Solve the model

LSEL,S,LOC,Y,0.004               ! Select the line at y=0.004
DLDELE,ALL,TEMP                  ! Delete all temperature BCs on that line
DL,ALL,,TEMP,300,0               ! Set temperature of selected line to 300
ALLSEL,ALL                       ! Select everything
SOLVE                            ! Solve the model

SAVE                             ! Save the database
FINISH                           ! Exit the Solution processor

/SHOW,JPEG,REV                   ! Direct graphical output to JPEG files

/POST1                           ! Enter the General Postprocessor
SET,1                            ! Load the results for load step #1
PLNSOL, TEMP,, 0                 ! Plot nodal temperature for the model
ESEL,S,MAT,,2                    ! Select elements associated with material #2
NSLE,S                           ! Select nodes attached to selected elements
PLNSOL, TEMP,, 0                 ! Plot nodal temperature for selected nodes

ALLSEL,ALL                       ! Select everything in the model
SET,2                            ! Load the results for load step #2
```

```
PLNSOL, TEMP,, 0            ! Plot nodal temperature for the model
ESEL,S,MAT,,2               ! Select elements associated with material #2
NSLE,S                      ! Select nodes attached to selected elements
PLNSOL, TEMP,, 0            ! Plot nodal temperature for selected nodes

ALLSEL,ALL                  ! Select everything in the model
SET,3                       ! Load the results for load step #3
PLNSOL, TEMP,, 0            ! Plot nodal temperature for the model
ESEL,S,MAT,,2               ! Select elements associated with material #2
NSLE,S                      ! Select nodes attached to selected elements
PLNSOL, TEMP,, 0            ! Plot nodal temperature for selected nodes

ALLSEL,ALL                  ! Select everything in the model
SET,4                       ! Load the results for load step #4
PLNSOL, TEMP,, 0            ! Plot nodal temperature for the model
ESEL,S,MAT,,2               ! Select elements associated with material #2
NSLE,S                      ! Select nodes attached to selected elements
PLNSOL, TEMP,, 0            ! Plot nodal temperature for selected nodes

ALLSEL,ALL                  ! Select everything in the model
SET,5                       ! Load the results for load step #5
PLNSOL, TEMP,, 0            ! Plot nodal temperature for the model
ESEL,S,MAT,,2               ! Select elements associated with material #2
NSLE,S                      ! Select nodes attached to selected elements
PLNSOL, TEMP,, 0            ! Plot nodal temperature for selected nodes

FINISH                      ! Exit the General Postprocessor
/EXIT,NOSAVE                ! Exit the program without saving
```

Appendix: Chapter and Section Numbering for Selected ANSYS Mechanical APDL 17.2 Documentation

Throughout this book, we have made references to the ANSYS Mechanical APDL 17.2 documentation using the book name and the chapter and section numbers. The chapter and section numbers associated with this information can, and occasionally do, change with software revisions. However, the chapter and section titles associated with that information generally do not.

Below is a list of the references made to the ANSYS Mechanical APDL documentation by both chapter/section number and by title. This will help you to locate information in the documentation, even if the documentation numbering changes in the future.

Mechanical APDL Basic Analysis Guide

Chapter 1—Getting Started
- Section 1.1—Building the Model
  - Section 1.1.2—Defining Element Types
  - Section 1.1.3—Creating Cross Sections
  - Section 1.1.4—Defining Material Properties

Chapter 2—Loading
- Section 2.1—Understanding Loads
- Section 2.4—Stepped and Ramped Loads
- Section 2.5—Applying Loads
  - Section 2.5.3—Degree-of-Freedom Constraints
  - Section 2.5.6—Forces (Concentrated Loads)
  - Section 2.5.7—Surface Loads
  - Section 2.5.8—Applying Body Loads
  - Section 2.5.9—Applying Inertia Loads
  - Section 2.5.10—Applying Ocean Loads

Chapter 4—Solution
- Section 4.6—Solving Multiple Load Steps
  - Section 4.6.1—Using the Multiple SOLVE Method
- Section 4.8—Restarting an Analysis

Chapter 6—The General Postprocessor (POST1)
- Section 6.3—Additional POST1 Postprocessing
  - Section 6.3.3—Creating and Combining Load Cases

Chapter 7—The Time-History Postprocessor (POST26)

Chapter 8—Selecting and Components
- Section 8.2—Selecting for Meaningful Postprocessing

Chapter 11—PowerGraphics
Chapter 17—External Graphics
Section 17.3—Viewing and Translating Neutral Graphics Files
Section 17.3.1—Getting Started with the DISPLAY Program
Chapter 19—File Management and Files

Mechanical APDL Command Reference

Chapter 1—About This Reference
Chapter 2—Command Groupings
Section 2.6—SOLUTION Commands
Chapter 3—Command Dictionary

Mechanical APDL Coupled-Field Analysis Guide

Chapter 1—Coupled-Field Analyses
Section 1.2—System of Units

Mechanical APDL Element Reference

Chapter 2—Element Reference
Section 2.1—Overview of Element Characteristics
Section 2.2—Summary of Element Types
Section 2.3—Selecting Elements for Your Analysis
Section 2.4—Current-Technology Elements
Section 2.5—Degenerated Shape Elements
Chapter 3—General Element Features
Section 3.1—Element Input
Section 3.2—Solution Output
Section 3.3—Coordinate Systems
Chapter 4—Element Support for Material Models

Mechanical APDL Material Reference

Chapter 2—Material Model Element Support
Chapter 3—Linear Material Properties
Chapter 4—Nonlinear Material Properties
Chapter 7—Material Model Combinations
Chapter 9—GUI-Inaccessible Material Properties

Mechanical APDL Modeling and Meshing Guide

Chapter 1—Understanding Model Generation
Chapter 2—Planning Your Approach
Section 2.2—Choosing Between Linear and Higher Order Elements
Section 2.6—Determining the Appropriate Mesh Density
Chapter 3—Coordinate Systems
Section 3.3—Nodal Coordinate Systems
Section 3.4—Element Coordinate Systems
Section 3.5—The Results Coordinate System
Chapter 4—Using Working Planes

Chapter 5—Solid Modeling
Section 5.4—Sculpting Your Model with Boolean Operations
Chapter 6—Importing Solid Models from IGES Files
Chapter 7—Generating the Mesh
Chapter 8—Revising Your Model
Section 8.1—Refining a Mesh Locally
Section 8.4—Revising a Meshed Model: Clearing and Deleting
Chapter 9—Direct Generation

Mechanical APDL Operations Guide

Chapter 2—The Mechanical APDL Environment
Section 2.4—Program Files
Chapter 3—Running the Mechanical APDL Program
Section 3.4—Batch Mode
Section 3.4.1—Starting a Batch Job from the Command Line
Chapter 4—Using Interactive Mode
Section 4.3—Layout of the User Interface
Chapter 5—Graphical Picking
Chapter 6—Customizing Mechanical APDL and the GUI
Section 6.3—Customizing the GUI
Chapter 7—Using the Session and Command Logs

Mechanical APDL Theory Reference

Chapter 12—Element Tools
Section 12.1—Element Shape Testing
Chapter 14—Analysis Tools
Section 14.11—Newton-Raphson Procedure
Chapter 17—Postprocessing
Section 17.6—Error Approximation Technique

Index

Note: Page numbers followed by "*f*," "*t*," and "*b*" refer to figures, tables, and boxes, respectively.

M

N

O

P

Q

R

S

T

U